T0234859

Graduate Texts in Mathematics **84**

Springer
New York
Berlin
Heidelberg
Hong Kong
London
Milan
Paris
Tokyo

Graduate Texts in Mathematics

(continued after index)

Kenneth Ireland
Michael Rosen

A Classical Introduction to Modern Number Theory

Second Edition

Springer

Kenneth Ireland
(deceased)

Michael Rosen
Department of Mathematics
Brown University
Providence, RI 02912
USA

With 1 illustration.

Mathematics Subject Classification (2000): 11-01, 11-02

Library of Congress Cataloging-in-Publication Data
Ireland, Kenneth F.
   A classical introduction to modern number theory / Kenneth
Ireland, Michael Rosen.—2nd ed.
      p.   cm.—(Graduate texts in mathematics; 84)
      Includes bibliographical references and index.
      1. Number theory.   I. Rosen, Michael I.   II. Title.   III. Series.
QA241.I667   1990
512.7—dc20
                                                      90-9848
Printed on acid-free paper.

"A Classical Introduction to Modern Number Theory" is a revised and expanded version of "Elements of Number Theory" published in 1972 by Bogden and Quigley, Inc., Publishers.

(CH/SBA)

9

ISBN 978-1-4419-3094-1   ISBN 978-1-4757-2103-4 (eBook)
DOI 10.1007/978-1-4757-2103-4

Springer-Verlag is a part of *Springer Science+Business Media*

springeronline.com

Preface to the Second Edition

It is now 10 years since the first edition of this book appeared in 1980. The intervening decade has seen tremendous advances take place in mathematics generally, and in number theory in particular. It would seem desirable to treat some of these advances, and with the addition of two new chapters, we are able to cover some portion of this new material.

As examples of important new work that we have not included, we mention the following two results:

(1) The first case of Fermat's last theorem is true for infinitely many prime exponents p. This means that, for infinitely many primes p, the equation $x^p + y^p = z^p$ has no solutions in nonzero integers with $p \nmid xyz$. This was proved by L.M. Adelman and D.R. Heath-Brown and independently by E. Fouvry. An overview of the proof is given by Heath-Brown in the *Mathematical Intelligencer* (Vol. 7, No. 6, 1985).

(2) Let p_1, p_2, and p_3 be three distinct primes. Then at least one of them is a primitive root for infinitely many primes q. Recall that E. Artin conjectured that, if $a \in \mathbb{Z}$ is not 0, 1, -1, or a square, then there are infinitely many primes q such that a is a primitive root modulo q. The theorem we have stated was proved in a weaker form by R. Gupta and M.R. Murty, and then strengthened by the combined efforts of R. Gupta, M.R. Murty, V.K. Murty, and D.R. Heath-Brown. An exposition of this result, as well as an analogue on elliptic curves, is given by M.R. Murty in the *Mathematical Intelligencer* (Vol. 10, No. 4, 1988).

The new material that we have added falls principally within the framework of arithmetic geometry. In Chapter 19 we give a complete proof of L.J. Mordell's fundamental theorem, which asserts that the group of rational points on an elliptic curve, defined over the rational numbers, is finitely generated. In keeping with the spirit of the book, the proof (due in essence to A. Weil) is elementary. It makes no use of cohomology groups or any other advanced machinery. It does use finiteness of class number and a weak form of the Dirichlet unit theorem; both results are proved in the text.

The second new chapter, Chapter 20, is an overview of G. Faltings's proof of the Mordell conjecture and recent progress on the arithmetic of elliptic curves, especially the work of B. Gross, V.A. Kolyvagin, K. Rubin, and D. Zagier. Some of this work has surprising applications to other areas of number theory. We discuss one application to Fermat's last theorem, due to G. Frey, J.P. Serre, and K. Ribet. Another important application is the solution of an old problem due to K.F. Gauss about class numbers of imaginary quadratic number fields. This comes about by combining the work of B. Gross and D. Zagier with a result of D. Goldfeld. This chapter contains few proofs. Its main purpose is to give an informative survey in the hope that the reader will be inspired to learn the background necessary to a better understanding and appreciation of these important new developments.

The rest of the book is essentially unchanged. An attempt has been made to correct errors and misprints. In an effort to keep confusion to a minimum, we have not changed the bibliography at the end of the book. New references for the two new chapters, Chapters 19 and 20, will be found at the end of those chapters. We would like to thank Toru Nakahara and others for submitting a list of misprints from the first edition. Also, we thank Linda Guthrie for typing portions of the final chapters.

We have both been very pleased with the warm reception that the first edition of this book received. It is our hope that the new edition will continue to entice readers to delve deeper into the mysteries of this ancient, beautiful, and still vital subject.

February 1990   Kenneth Ireland
                                                         Michael Rosen

Addendum to Second Edition, Second Corrected Printing

The second printing of the second edition is unchanged except for corrections and the addition of a few clarifying comments. I would like to thank K. Conrad, M. Jastrzebski, F. Lemmermeyer and others who took the trouble to send us detailed lists of misprints.

November 1992   Michael Rosen

Notes for the Second Edition, Fifth Corrected Printing

In 1995 Andrew Wiles published a paper in the Annals of Mathematics which proved the Taniyama-Shimura-Weil conjecture is true for semi-stable elliptic curves over the rational numbers. Together with earlier results, principally the theorem of Ken Ribet mentioned on page 347, this proved Fermat's Last Theorem. The most famous conjecture in elementary number theory is finally a theorem!!!

April 1998   Michael Rosen

Preface

This book is a revised and greatly expanded version of our book *Elements of Number Theory* published in 1972. As with the first book the primary audience we envisage consists of upper level undergraduate mathematics majors and graduate students. We have assumed some familiarity with the material in a standard undergraduate course in abstract algebra. A large portion of Chapters 1–11 can be read even without such background with the aid of a small amount of supplementary reading. The later chapters assume some knowledge of Galois theory, and in Chapters 16 and 18 an acquaintance with the theory of complex variables is necessary.

Number theory is an ancient subject and its content is vast. Any introductory book must, of necessity, make a very limited selection from the fascinating array of possible topics. Our focus is on topics which point in the direction of algebraic number theory and arithmetic algebraic geometry. By a careful selection of subject matter we have found it possible to exposit some rather advanced material without requiring very much in the way of technical background. Most of this material is classical in the sense that is was discovered during the nineteenth century and earlier, but it is also modern because it is intimately related to important research going on at the present time.

In Chapters 1–5 we discuss prime numbers, unique factorization, arithmetic functions, congruences, and the law of quadratic reciprocity. Very little is demanded in the way of background. Nevertheless it is remarkable how a modicum of group and ring theory introduces unexpected order into the subject. For example, many scattered results turn out to be parts of the answer to a natural question: What is the structure of the group of units in the ring $\mathbb{Z}/n\mathbb{Z}$?

Reciprocity laws constitute a major theme in the later chapters. The law of quadratic reciprocity, beautiful in itself, is the first of a series of reciprocity laws which lead ultimately to the Artin reciprocity law, one of the major achievements of algebraic number theory. We travel along the road beyond quadratic reciprocity by formulating and proving the laws of cubic and biquadratic reciprocity. In preparation for this many of the techniques of algebraic number theory are introduced; algebraic numbers and algebraic integers, finite fields, splitting of primes, etc. Another important tool in this investigation (and in others!) is the theory of Gauss and Jacobi sums. This material is covered in Chapters 6–9. Later in the book we formulate and prove the more advanced partial generalization of these results, the Eisenstein reciprocity law.

A second major theme is that of diophantine equations, at first over finite fields and later over the rational numbers. The discussion of polynomial equations over finite fields is begun in Chapters 8 and 10 and culminates in Chapter 11 with an exposition of a portion of the paper "Number of solutions of equations over finite fields" by A. Weil. This paper, published in 1948, has been very influential in the recent development of both algebraic geometry and number theory. In Chapters 17 and 18 we consider diophantine equations over the rational numbers. Chapter 17 covers many standard topics from sums of squares to Fermat's Last Theorem. However, because of material developed earlier we are able to treat a number of these topics from a novel point of view. Chapter 18 is about the arithmetic of elliptic curves. It differs from the earlier chapters in that it is primarily an overview with many definitions and statements of results but few proofs. Nevertheless, by concentrating on some important special cases we hope to convey to the reader something of the beauty of the accomplishments in this area where much work is being done and many mysteries remain.

The third, and final, major theme is that of zeta functions. In Chapter 11 we discuss the congruence zeta function associated to varieties defined over finite fields. In Chapter 16 we discuss the Riemann zeta function and the Dirichlet L-functions. In Chapter 18 we discuss the zeta function associated to an algebraic curve defined over the rational numbers and Hecke L-functions. Zeta functions compress a large amount of arithmetic information into a single function and make possible the application of the powerful methods of analysis to number theory.

Throughout the book we place considerable emphasis on the history of our subject. In the notes at the end of each chapter we give a brief historical sketch and provide references to the literature. The bibliography is extensive containing many items both classical and modern. Our aim has been to provide the reader with a wealth of material for further study.

There are many exercises, some routine, some challenging. Some of the exercises supplement the text by providing a step by step guide through the proofs of important results. In the later chapters a number of exercises have been adapted from results which have appeared in the recent literature. We hope that working through the exercises will be a source of enjoyment as well as instruction.

In the writing of this book we have been helped immensely by the interest and assistance of many mathematical friends and acquaintances. We thank them all. In particular we would like to thank Henry Pohlmann who insisted we follow certain themes to their logical conclusion, David Goss for allowing us to incorporate some of his work into Chapter 16, and Oisin McGuiness for his invaluable assistance in the preparation of Chapter 18. We would like to thank Dale Cavanaugh, Janice Phillips, and especially Carol Ferreira, for their patience and expertise in typing large portions of the manuscript. Finally, the second author wishes to express his gratitude to the Vaughn Foundation Fund for financial support during his sabbatical year in Berkeley, California (1979/80).

July 25, 1981   Kenneth Ireland
                                                                     Michael Rosen

Contents

Chapter 1

Unique Factorization

The notion of prime number is fundamental in number theory. The first part of this chapter is devoted to proving that every integer can be written as a product of primes in an essentially unique way.

After that, we shall prove an analogous theorem in the ring of polynomials over a field.

On a more abstract plane, the general idea of unique factorization is treated for principal ideal domains.

Finally, returning from the abstract to the concrete, the general theory is applied to two special rings that will be important later in the book.

§1 Unique Factorization in $\mathbb{Z}$

As a first approximation, number theory may be defined as the study of the natural numbers 1, 2, 3, 4, ... L. Kronecker once remarked (speaking of mathematics generally) that God made the natural numbers and all the rest is the work of man. Although the natural numbers constitute, in some sense, the most elementary mathematical system, the study of their properties has provided generations of mathematicians with problems of unending fascination.

We say that a number a divides a number b if there is a number c such that $b = ac$. If a divides b, we use the notation $a|b$. For example, $2|8$, $3|15$, but $6 \nmid 21$. If we are given a number, it is tempting to factor it again and again until further factorization is impossible. For example, $180 = 18 \times 10 = 2 \times 9 \times 2 \times 5 = 2 \times 3 \times 3 \times 2 \times 5$. Numbers that cannot be factored further are called primes. To be more precise, we say that a number p is a prime if its only divisors are 1 and p. Prime numbers are very important because every number can be written as a product of primes. Moreover, primes are of great interest because there are many problems about them that are easy to state but very hard to prove. Indeed many old problems about primes are unsolved to this day.

The first prime numbers are 2, 3, 5, 7, 11, 13, 17, 19, 23, 29, 31, 37, 41, 43, ... One may ask if there are infinitely many prime numbers. The answer is yes. Euclid gave an elegant proof of this fact over 2000 years ago. We shall give his proof and several others in Chapter 2. One can ask other questions of this nature. Let $\pi(x)$ be the number of primes between 1 and x. What can be said about the function $\pi(x)$? Several mathematicians found by experiment that for large x the function $\pi(x)$ was approximately equal to $x/\ln(x)$. This assertion, known as the prime number theorem, was proved toward the end of the nineteenth century by J. Hadamard and independently by Ch.-J. de la Vallé Poussin. More precisely, they proved

$$\lim_{x \to \infty} \frac{\pi(x)}{x/\ln(x)} = 1.$$

Even from a small list of primes one can notice that they have a tendency to occur in pairs, for example, 3 and 5, 5 and 7, 11 and 13, 17 and 19. Do there exist infinitely many prime pairs? The answer is unknown.

Another famous unsolved problem is known as the Goldbach conjecture (C. H. Goldbach). Can every even number be written as the sum of two primes? Goldbach came to this conjecture experimentally. Nowadays electronic computers make it possible to experiment with very large numbers. No counterexample to Goldbach's conjecture has ever been found. Great progress toward a proof has been given by I. M. Vinogradov and L. Schnirel-mann. In 1937 Vinogradov was able to show that every sufficiently large odd number is the sum of three odd primes.

In this book we shall not study in depth the distribution of prime numbers or "additive" problems about them (such as the Goldbach conjecture). Rather our concern will be about the way primes enter into the multiplicative structure of numbers. The main theorem along these lines goes back essentially to Euclid. It is the theorem of unique factorization. This theorem is sometimes referred to as the fundamental theorem of arithmetic. It deserves the title. In one way or another almost all the results we shall discuss depend on it. The theorem states that every number can be factored into a product of primes in a unique way. What uniqueness means will be explained below.

As an illustration consider the number 180. We have seen that $180 = 2 \times 2 \times 3 \times 3 \times 5 = 2^2 \times 3^2 \times 5$. Uniqueness in this case means that the only primes dividing 180 are 2, 3, and 5 and that the exponents 2, 2, and 1 are uniquely determined by 180.

$\mathbb{Z}$ will denote the ring of integers, i.e., the set $0, \pm 1, \pm 2, \pm 3, \ldots$, together with the usual definition of sum and product. It will be more convenient to work with $\mathbb{Z}$ rather than restricting ourselves to the positive integers. The notion of divisibility carries over with no difficulty to $\mathbb{Z}$. If p is a positive prime, $-p$ will also be a prime. We shall not consider 1 or -1 as primes even though they fit the definition. This is simply a useful convention. Note that 1 and -1 divide everything and that they are the only integers with this property. They are called the units of $\mathbb{Z}$. Notice also that every nonzero integer divides zero. As is usual we shall exclude division by zero.

There are a number of simple properties of division that we shall simply list. The reader may wish to supply the proofs.

(1) $a|a$, $a \neq 0$.
(2) If $a|b$ and $b|a$, then $a = \pm b$.
(3) If $a|b$ and $b|c$, then $a|c$.
(4) If $a|b$ and $a|c$, then $a|b + c$.

Let $n \in \mathbb{Z}$ and let p be a prime. Then if n is not zero, there is a nonnegative integer a such that $p^a|n$ but $p^{a+1} \nmid n$. This is easy to see if both p and n are positive for then the powers of p get larger and larger and eventually exceed n. The other cases are easily reduced to this one. The number a is called the order of n at p and is denoted by $\text{ord}_p n$. Roughly speaking $\text{ord}_p n$ is the number of times p divides n. If $n = 0$, we set $\text{ord}_p 0 = \infty$. Notice that $\text{ord}_p n = 0$ if and only if (iff) $p \nmid n$.

Lemma 1. *Every nonzero integer can be written as a product of primes.*

PROOF. Assume that there is an integer that cannot be written as a product of primes. Let N be the smallest positive integer with this property. Since N cannot itself be prime we must have $N = mn$, where $1 < m, n < N$. However, since m and n are positive and smaller than N they must each be a product of primes. But then so is $N = mn$. This is a contradiction.

The proof can be given in a more positive way by using mathematical induction. It is enough to prove the result for all positive integers. 2 is a prime. Suppose that $2 < N$ and that we have proved the result for all numbers m such that $2 \leq m < N$. We wish to show that N is a product of primes. If N is a prime, there is nothing to do. If N is not a prime, then $N = mn$, where $2 \leq m, n < N$. By induction both m and n are products of primes and thus so is N. $\square$

By collecting terms we can write $n = p_1^{a_1} p_2^{a_2} \cdots p_m^{a_m}$, where the p_i are primes and the a_i are nonnegative integers. We shall use the following notation:

$$n = (-1)^{\varepsilon(n)} \prod_p p^{a(p)},$$

where $\varepsilon(n) = 0$ or 1 depending on whether n is positive or negative and where the product is over all positive primes. The exponents $a(p)$ are nonnegative integers and, of course, $a(p) = 0$ for all but finitely many primes. For example, if $n = 180$, we have $\varepsilon(n) = 0$, $a(2) = 2$, $a(3) = 2$, and $a(5) = 1$, and all other $a(p) = 0$.

We can now state the main theorem.

Theorem 1. *For every nonzero integer n there is a prime factorization*

$$n = (-1)^{\varepsilon(n)} \prod_p p^{a(p)},$$

with the exponents uniquely determined by n. In fact, we have $a(p) = \text{ord}_p n$.

The proof of this theorem is not as easy as it may seem. We shall postpone the proof until we have established a few preliminary results.

Lemma 2. *If $a, b \in \mathbb{Z}$ and $b > 0$, there exist $q, r \in \mathbb{Z}$ such that $a = qb + r$ with $0 \leq r < b$.*

PROOF. Consider the set of all integers of the form $a - xb$ with $x \in \mathbb{Z}$. This set includes positive elements. Let $r = a - qb$ be the least nonnegative element in this set. We claim that $0 \leq r < b$. If not, $r = a - qb \geq b$ and so $0 \leq a - (q + 1)b < r$, which contradicts the minimality of r. $\qquad\square$

Definition. If $a_1, a_2, \ldots, a_n \in \mathbb{Z}$, we define $(a_1, a_2, \ldots, a_n)$ to be the set of all integers of the form $a_1 x_1 + a_2 x_2 + \cdots + a_n x_n$ with $x_1, x_2, \ldots, x_n \in \mathbb{Z}$.

Let $A = (a_1, a_2, \ldots, a_n)$. Notice that the sum and difference of two elements in A are again in A. Also, if $a \in A$ and $r \in \mathbb{Z}$, then $ra \in A$. In ring-theoretic language, A is an *ideal* in the ring $\mathbb{Z}$.

Lemma 3. *If $a, b \in \mathbb{Z}$, then there is a $d \in \mathbb{Z}$ such that $(a, b) = (d)$.*

PROOF. We may assume that not both a and b are zero so that there are positive elements in (a, b). Let d be the smallest positive element in (a, b). Clearly $(d) \subseteq (a, b)$. We shall show that the reverse inclusion also holds.

Suppose that $c \in (a, b)$. By Lemma 2 there exist integers q and r such that $c = qd + r$ with $0 \leq r < d$. Since both c and d are in (a, b) it follows that $r = c - qd$ is also in (a, b). Since $0 \leq r < d$ we must have $r = 0$. Thus $c = qd \in (d)$. $\qquad\square$

Definition. Let $a, b \in \mathbb{Z}$. An integer d is called a *greatest common divisor* of a and b if d is a divisor of both a and b and if every other common divisor of a and b divides d.

Notice that if c is another greatest common divisor of a and b, then we must have $c \,|\, d$ and $d \,|\, c$ and so $c = \pm d$. Thus the greatest common divisor of two numbers, if it exists, is determined up to sign.

As an example, one may check that 14 is a greatest common divisor of 42 and 196. The following lemma will establish the existence of the greatest common divisor, but it will not give a method for computing it. In the Exercises we shall outline an efficient method of computation known as the Euclidean algorithm.

Lemma 4. *Let $a, b \in \mathbb{Z}$. If $(a, b) = (d)$ then d is a greatest common divisor of a and b.*

PROOF. Since $a \in (d)$ and $b \in (d)$ we see that d is a common divisor of a and b. Suppose that c is a common divisor. Then c divides every number of the form $ax + by$. In particular $c \,|\, d$. $\qquad\square$

Definition. We say that two integers a and b are *relatively prime* if the only common divisors are ± 1, the units.

It is fairly standard to use the notation (a, b) for the greatest common divisor of a and b. The way we have defined things, (a, b) is a set. However, since $(a, b) = (d)$ and d is a greatest common divisor (if we require d to be positive, we may use the article *the*) it will not be too confusing to use the symbol (a, b) for both meanings. With this convention we can say that a and b are relatively prime if $(a, b) = 1$.

Proposition 1.1.1. *Suppose that $a|bc$ and that $(a, b) = 1$. Then $a|c$.*

PROOF. Since $(a, b) = 1$ there exist integers r and s such that $ra + sb = 1$. Therefore, $rac + sbc = c$. Since a divides the left-hand side of this equation we have $a|c$. $\qquad\square$

This proposition is false if $(a, b) \neq 1$. For example, $6|24$ but $6\nmid 3$ and $6\nmid 8$.

Corollary 1. *If p is a prime and $p|bc$, then either $p|b$ or $p|c$.*

PROOF. The only divisors of p are ± 1 and $\pm p$. Thus $(p, b) = 1$ or p; i.e., either $p|b$ or p and b are relatively prime. If $p|b$, we are done. If not, $(p, b) = 1$ and so, by the proposition, $p|c$. $\qquad\square$

We can state the corollary in a slightly different form that is often useful: If p is a prime and $p\nmid b$ and $p\nmid c$, then $p\nmid bc$.

Corollary 2. *Suppose that p is a prime and that $a, b \in \mathbb{Z}$. Then $\mathrm{ord}_p ab = \mathrm{ord}_p a + \mathrm{ord}_p b$.*

PROOF. Let $\alpha = \mathrm{ord}_p a$ and $\beta = \mathrm{ord}_p b$. Then $a = p^\alpha c$ and $b = p^\beta d$, where $p \nmid c$ and $p \nmid d$. Then $ab = p^{\alpha+\beta}cd$ and by Corollary 1 $p \nmid cd$. Thus $\mathrm{ord}_p ab = \alpha + \beta = \mathrm{ord}_p a + \mathrm{ord}_p b$. $\qquad\square$

We are now in a position to prove the main theorem.
Apply the function ord_q to both sides of the equation

$$n = (-1)^{\varepsilon(n)} \prod_p p^{a(p)}$$

and use the property of ord_q given by Corollary 2. The result is

$$\mathrm{ord}_q n = \varepsilon(n)\, \mathrm{ord}_q(-1) + \sum_p a(p)\, \mathrm{ord}_q(p).$$

Now, from the definition of ord_q we have $\mathrm{ord}_q(-1) = 0$ and $\mathrm{ord}_q(p) = 0$ if $p \neq q$ and 1 if $p = q$. Thus the right-hand side collapses to the single term $a(q)$, i.e., $\mathrm{ord}_q n = a(q)$, which is what we wanted to prove.

It is to be emphasized that the key step in the proof is Corollary 1: namely, if $p|ab$, then $p|a$ or $p|b$. Whatever difficulty there is in the proof is centered about this fact.

§2 Unique Factorization in $k[x]$

The theorem of unique factorization can be formulated and proved in more general contexts than that of Section 1. In this section we shall consider the ring $k[x]$ of polynomials with coefficients in a field k. In Section 3 we shall consider principal ideal domains. It will turn out that the analysis of these situations will prove useful in the study of the integers.

If $f, g \in k[x]$, we say that f divides g if there is an $h \in k[x]$ such that $g = fh$.

If $\deg f$ denotes the degree of f, we have $\deg fg = \deg f + \deg g$. Also, remember that $\deg f = 0$ iff f is a nonzero constant. It follows that $f|g$ and $g|f$ iff $f = cg$, where c is a nonzero constant. It also follows that the only polynomials that divide all the others are the nonzero constants. These are the units of $k[x]$. A nonconstant polynomial p is said to be irreducible if $q|p$ implies that q is either a constant or a constant times p. Irreducible polynomials are the analog of prime numbers.

Lemma 1. *Every nonconstant polynomial is the product of irreducible polynomials.*

PROOF. The proof is by induction on the degree. It is easy to see that polynomials of degree 1 are irreducible. Assume that we have proved the result for all polynomials of degree less than n and that $\deg f = n$. If f is irreducible, we are done. Otherwise $f = gh$, where $1 \leq \deg g, \deg h < n$. By the induction assumption both g and h are products of irreducible polynomials. Thus so is $f = gh$.   □

It is convenient to define *monic polynomial*. A polynomial f is called monic if its leading coefficient is 1. For example, $x^2 + x - 3$ and $x^3 - x^2 + 3x + 17$ are monic but $2x^3 - 5$ and $3x^4 + 2x^2 - 1$ are not. Every polynomial (except zero) is a constant times a monic polynomial.

Let p be a monic irreducible polynomial. We define $\mathrm{ord}_p f$ to be the integer a defined by the property that $p^a|f$ but that $p^{a+1} \nmid f$. Such an integer must exist since the degree of the powers of p gets larger and larger. Notice that $\mathrm{ord}_p f = 0$ iff $p \nmid f$.

Theorem 2. *Let $f \in k[x]$. Then we can write*

$$f = c \prod_p p^{a(p)},$$

where the product is over all monic irreducible polynomials and c is a constant. The constant c and the exponents $a(p)$ are uniquely determined by f; in fact, $a(p) = \text{ord}_p\, f$.

The existence of such a product follows immediately from Lemma 1. As before, the uniqueness is more difficult and the proof will be postponed until we develop a few tools.

Lemma 2. *Let $f, g \in k[x]$. If $g \neq 0$, there exist polynomials $h, r \in k[x]$ such that $f = hg + r$, where either $r = 0$ or $r \neq 0$ and $\deg r < \deg g$.*

PROOF. If $g \mid f$, simply set $h = f/g$ and $r = 0$. If $g \nmid f$, let $r = f - hg$ be the polynomial of least degree among all polynomials of the form $f - lg$ with $l \in k[x]$. We claim that $\deg r < \deg g$. If not, let the leading term of r be ax^d and that of g be bx^m. Then $r - ab^{-1}x^{d-m}g = f - (h + ab^{-1}x^{d-m})g$ has smaller degree than r and is of the given form. This is a contradiction.   □

Definition. If $f_1, f_2, \ldots, f_n \in k[x]$, then $(f_1, f_2, \ldots, f_n)$ is the set of all polynomials of the form $f_1 h_1 + f_2 h_2 + \cdots + f_n h_n$, where $h_1, h_2, \ldots, h_n \in k[x]$.

In ring-theoretic language $(f_1, f_2, \ldots, f_n)$ is the ideal generated by $f_1, f_2, \ldots, f_n$.

Lemma 3. *Given $f, g \in k[x]$ there is a $d \in k[x]$ such that $(f, g) = (d)$.*

PROOF. In the set (f, g) let d be an element of least degree. We have $(d) \subseteq (f, g)$ and we want to prove the reverse inclusion. Let $c \in (f, g)$. If $d \nmid c$, then there exist polynomials h and r such that $c = hd + r$ with $\deg r < \deg d$. Since c and d are in (f, g) we have $r = c - hd \subseteq (f, g)$. Since r has smaller degree than d this is a contradiction. Therefore, $d \mid c$ and $c \in (d)$.   □

Definition. Let $f, g \in k[x]$. Then $d \in k[x]$ is said to be a *greatest common divisor* of f and g if d divides f and g and every common divisor of f and g divides d.

Notice that the greatest common divisor of two polynomials is determined up to multiplication by a constant. If we require it to be monic, it is uniquely determined and we may speak of *the* greatest common divisor.

Lemma 4. *Let $f, g \in k[x]$. By Lemma 3 there is a $d \in k[x]$ such that $(f, g) = (d)$. d is a greatest common divisor of f and g.*

PROOF. Since $f \in (d)$ and $g \in (d)$ we have $d \mid f$ and $d \mid g$. Suppose that $h \mid f$ and that $h \mid g$. Then h divides every polynomial of the form $fl + gm$ with $l, m \in k[x]$. In particular $h \mid d$, and we are done.   □

Definition. Two polynomials f and g are said to be *relatively prime* if the only common divisors of f and g are constants. In other words, $(f, g) = (1)$.

Proposition 1.2.1. *If f and g are relatively prime and $f \mid gh$, then $f \mid h$.*

PROOF. If f and g are relatively prime, we have $(f, g) = (1)$ so there are polynomials l and m such that $lf + mg = 1$. Thus $lfh + mgh = h$. Since f divides the left-hand side of this equation f must divide h. ☐

Corollary 1. *If p is an irreducible polynomial and $p \mid fg$, then $p \mid f$ or $p \mid g$.*

PROOF. Since p is irreducible $(p, f) = (p)$ or (1). In the first case $p \mid f$ and we are done. In the second case p and f are relatively prime and the result follows from the proposition. ☐

Corollary 2. *If p is a monic irreducible polynomial and $f, g \in k[x]$, we have* $\operatorname{ord}_p fg = \operatorname{ord}_p f + \operatorname{ord}_p g$.

PROOF. The proof is almost word for word the same as the proof to Corollary 2 to Proposition 1.1.1. ☐

The proof of Theorem 2 is now easy. Apply the function ord_q to both sides of the relation

$$f = c \prod_p p^{a(p)}.$$

We find that

$$\operatorname{ord}_q f = \operatorname{ord}_q c + \sum_p a(p)\, \operatorname{ord}_q p.$$

Now, since c is a constant $q \nmid c$ and $\operatorname{ord}_q c = 0$. Moreover, $\operatorname{ord}_q p = 0$ if $q \neq p$ and 1 if $q = p$. Thus the above relation yields $\operatorname{ord}_q f = a(q)$. This shows that the exponents are uniquely determined. It is clear that if the exponents are uniquely determined by f, then so is c. This completes the proof. ☐

§3 Unique Factorization in a Principal Ideal Domain

The reader will not have failed to notice the great similarity in the methods of proof in Sections 1 and 2. In this section we shall prove an abstract theorem that includes the previous results as special cases.

Throughout this section R will denote an integral domain.

Definition 1. R is said to be a *Euclidean domain* if there is a function λ from the nonzero elements of R to the set $\{0, 1, 2, 3, \ldots\}$ such that if $a, b \in R$, $b \neq 0$, there exists $c, d \in R$ with the property $a = cb + d$ and either $d = 0$ or $\lambda(d) < \lambda(b)$.

The rings $\mathbb{Z}$ and $k[x]$ are both Euclidean domains. In $\mathbb{Z}$ we can take ordinary absolute value as the function λ; in the ring $k[x]$ the function that assigns to every polynomial its degree will serve the purpose.

Proposition 1.3.1. *If R is a Euclidean domain and $I \subseteq R$ is an ideal, then there is an element $a \in R$ such that $I = Ra = \{ra \mid r \in R\}$.*

PROOF. Consider the set of nonnegative integers $\{\lambda(b) \mid b \in I, b \neq 0\}$. Since every set of nonnegative integers has a least element there is an $a \in I$, $a \neq 0$, such that $\lambda(a) \leq \lambda(b)$ for all $b \in I$, $b \neq 0$. We claim that $I = Ra$. Clearly, $Ra \subseteq I$. Suppose that $b \in I$; then we know that there are elements $c, d \in R$ such that $b = ca + d$, where either $d = 0$ or $\lambda(d) < \lambda(a)$. Since $d = b - ca \in I$ we cannot have $\lambda(d) < \lambda(a)$. Thus $d = 0$ and $b = ca \in Ra$. Therefore, $I \subseteq Ra$ and we are done. $\qquad\square$

For elements $a_1, \ldots, a_n \in R$, define $(a_1, a_2, \ldots, a_n) = Ra_1 + Ra_2 + \cdots + Ra_n = \{\sum_{i=1}^{n} r_i a_i \mid r_i \in R\}$. $(a_1, a_2, \ldots, a_n)$ is an ideal. If an ideal I is equal to $(a_1, \ldots, a_n)$ for some elements $a_i \in I$, we say that I is finitely generated. If $I = (a)$ for some $a \in I$, we say that I is a principal ideal.

Definition 2. R is said to be a *principal ideal domain* (PID) if every ideal of R is principal.

Proposition 1.3.1 asserts that every Euclidean domain is a PID. The converse of this statement is false, although it is somewhat hard to provide examples.

The remaining discussion in this section is about PID's. The notion of Euclidean domain is useful because in practice one can show that many rings are PID's by first establishing that they are Euclidean domains. We shall give two further examples in Section 4.

We introduce some more terminology. If $a, b \in R$, $b \neq 0$, we say that b divides a if $a = bc$ for some $c \in R$. Notation: $b \mid a$. An element $u \in R$ is called a unit if u divides 1. Two elements $a, b \in R$ are said to be associates if $a = bu$ for some unit u. An element $p \in R$ is said to be irreducible if $a \mid p$ implies that a is either a unit or an associate of p. A nonunit $p \in R$ is said to be prime if $p \neq 0$ and $p \mid ab$ implies that $p \mid a$ or $p \mid b$.

The distinction between irreducible element and prime element is new. In general these notions do not coincide. As we have seen they do coincide in $\mathbb{Z}$ and $k[x]$, and we shall prove shortly that they coincide in a PID.

Some of the notions we are discussing can be translated into the language of ideals. Thus $a \mid b$ iff $(b) \subseteq (a)$. $u \in R$ is a unit iff $(u) = R$. a and b are associate iff $(a) = (b)$. p is prime iff $ab \in (p)$ implies that either $a \in (p)$ or

$b \in (p)$. All these assertions are easy exercises. The notion of irreducible element can be formulated in terms of ideals, but we will not need it.

Definition. $d \in R$ is said to be a *greatest common divisor* (gcd) of two elements $a, b \in R$ if

(a) $d|a$ and $d|b$.
(b) $d'|a$ and $d'|b$ implies that $d'|d$.

It is easy to see that if both d and d' are gcd's of a and b, then d is associate to d'.

The gcd of two elements need not exist in a general ring. However,

Proposition 1.3.2. *Let R be a PID and $a, b \in R$. Then a and b have a greatest common divisor d and $(a, b) = (d)$.*

PROOF. Form the ideal (a, b). Since R is a PID there is an element d such that $(a, b) = (d)$. Since $(a) \subseteq (d)$ and $(b) \subseteq (d)$ we have $d|a$ and $d|b$. If $d'|a$ and $d'|b$, then $(a) \subseteq (d')$ and $(b) \subseteq (d')$. Thus $(d) = (a, b) \subseteq (d')$ and $d'|d$. We have proved that d is a gcd of a and b and that $(a, b) = (d)$. $\qquad\square$

Two elements a and b are said to be relatively prime if the only common divisors are units.

Corollary 1. *If R is a PID and $a, b \in R$ are relatively prime, then $(a, b) = R$.*

Corollary 2. *If R is a PID and $p \in R$ is irreducible, then p is prime.*

PROOF. Suppose that $p|ab$ and that $p \nmid a$. Since $p \nmid a$ it follows that the only common divisors are units. By Corollary 1 $(a, p) = R$. Thus $(ab, pb) = (b)$. Since $ab \in (p)$ and $pb \in (p)$ we have $(b) \subseteq (p)$. Thus $p|b$.

It is easy to see that a prime is irreducible. $\qquad\square$

From now on R will be a PID and we shall use the words *prime* and *irreducible* interchangeably.

We want to show that every nonzero element of R is a product of irreducible elements. The proof is in two steps. First one shows that if $a \in R$, $a \neq 0$, there is an irreducible dividing a. Then we show that a is a product of irreducibles.

Lemma 1. *Let $(a_1) \subseteq (a_2) \subseteq (a_3) \subseteq \cdots$ be an ascending chain of ideals. Then there is an integer k such that $(a_k) = (a_{k+l})$ for $l = 0, 1, 2, \ldots$. In other words, the chain breaks off in finitely many steps.*

PROOF. Let $I = \bigcup_{i=1}^{\infty}(a_i)$. It is easy to see that I is an ideal. Thus $I = (a)$ for some $a \in R$. But $a \in \bigcup_{i=1}^{\infty}(a_i)$ implies that $a \in (a_k)$ for some k, which shows that $I = (a) \subseteq (a_k)$. It follows that $I = (a_k) = (a_{k+1}) = \cdots$. $\qquad\square$

Proposition 1.3.3. *Every nonzero nonunit of R is a product of irreducibles.*

PROOF. Let $a \in R$, $a \neq 0$, a not a unit. We wish to show, to begin with, that a is divisible by an irreducible element. If a is irreducible, we are done. Otherwise $a = a_1 b_1$, where a_1 and b_1 are nonunits. If a_1 is irreducible, we are done. Otherwise $a_1 = a_2 b_2$, where a_2 and b_2 are nonunits. If a_2 is irreducible, we are done. Otherwise continue as before. Notice that $(a) \subset (a_1) \subset (a_2) \subset \cdots$. By Lemma 1 this chain cannot go on indefinitely. Thus for some k, a_k is irreducible.

We now show that a is a product of irreducibles. If a is irreducible, we are done. Otherwise let p_1 be an irreducible such that $p_1 | a$. Then $a = p_1 c_1$. If c_1 is a unit, we are done. Otherwise let p_2 be an irreducible such that $p_2 | c_1$. Then $a = p_1 p_2 c_2$. If c_2 is a unit, we are done. Otherwise continue as before. Notice that $(a) \subset (c_1) \subset (c_2) \subset \cdots$. This chain cannot go on indefinitely by Lemma 1. Thus for some k, $a = p_1 p_2 \cdots p_k c_k$, where c_k is a unit. Since $p_k c_k$ is irreducible, we are done.   □

We now want to define an ord function as we have done in Sections 1 and 2.

Lemma 2. *Let p be a prime and $a \neq 0$. Then there is an integer n such that $p^n | a$ but $p^{n+1} \nmid a$.*

PROOF. If the lemma were false, then for each integer $m > 0$ there would be an element b_m such that $a = p^m b_m$. Then $p b_{m+1} = b_m$ so that $(b_1) \subset (b_2) \subset (b_3) \subset \cdots$ would be an infinite ascending chain of ideals that does not break off. This contradicts Lemma 1.   □

The integer n, which is defined in Lemma 2, is uniquely determined by p and a. We set $n = \mathrm{ord}_p a$.

Lemma 3. *If $a, b \in R$ with $a, b \neq 0$, then $\mathrm{ord}_p ab = \mathrm{ord}_p a + \mathrm{ord}_p b$.*

PROOF. Let $\alpha = \mathrm{ord}_p a$ and $\beta = \mathrm{ord}_p b$. Then $a = p^\alpha c$ and $b = p^\beta d$ with $p \nmid c$ and $p \nmid d$. Thus $ab = p^{\alpha+\beta} cd$. Since p is prime $p \nmid cd$. Consequently, $\mathrm{ord}_p ab = \alpha + \beta = \mathrm{ord}_p a + \mathrm{ord}_p b$.   □

We are now in a position to formulate and prove the main theorem of this section.

Let S be a set of primes in R with the following two properties:

(a) Every prime in R is associate to a prime in S.
(b) No two primes in S are associate.

To obtain such a set choose one prime out of each class of associate primes. There is clearly a great deal of arbitrariness in this choice. In $\mathbb{Z}$ and $k[x]$ there were natural ways to make the choice. In $\mathbb{Z}$ we chose S to be the set of positive primes. In $k[x]$ we chose S to be the set of monic irreducible polynomials. In general there is no neat way to make the choice and this occasionally leads to complications (see Chapter 9).

Theorem 3. *Let R be a PID and S a set of primes with the properties given above. Then if $a \in R$, $a \neq 0$, we can write*

$$a = u \prod_p p^{e(p)}, \tag{1}$$

where u is a unit and the product is over all $p \in S$. The unit u and the exponents $e(p)$ are uniquely determined by a. In fact, $e(p) = \mathrm{ord}_p\, a$.

PROOF. The existence of such a decomposition follows immediately from Proposition 1.3.3.

To prove the uniqueness, let q be a prime in S and apply ord_q to both sides of Equation (1). Using Lemma 3 we get

$$\mathrm{ord}_q\, a = \mathrm{ord}_q\, u + \sum_p e(p)\, \mathrm{ord}_q\, p.$$

Now, from the definition of ord_q we see that $\mathrm{ord}_q\, u = 0$ and that $\mathrm{ord}_q\, p = 0$ if $q \neq p$ and 1 if $q = p$. Thus $\mathrm{ord}_q\, a = e(q)$. Since the exponents $e(q)$ are uniquely determined so is the unit u. This completes the proof. $\square$

§4 The Rings $\mathbb{Z}[i]$ and $\mathbb{Z}[\omega]$

As an application of the results in Section 3 we shall consider two examples that will be useful to us in later chapters.

Let $i = \sqrt{-1}$ and consider the set of complex numbers $\mathbb{Z}[i]$ defined by $\{a + bi \,|\, a, b \in \mathbb{Z}\}$. This set is clearly closed under addition and subtraction. Moreover, if $a + bi, c + di \in \mathbb{Z}[i]$, then $(a + bi)(c + di) = ac + adi + bci + bdi^2 = (ac - bd) + (ad + bc)i \in \mathbb{Z}[i]$. Thus $\mathbb{Z}[i]$ is closed under multiplication and is a ring. Since $\mathbb{Z}[i]$ is contained in the complex numbers it is an integral domain.

Proposition 1.4.1. $\mathbb{Z}[i]$ *is a Euclidean domain.*

PROOF. For $a + bi \in \mathbb{Q}[i]$ define $\lambda(a + bi) = a^2 + b^2$.

Let $\alpha = a + bi$ and $\gamma = c + di$ and suppose that $\gamma \neq 0$. $\alpha/\gamma = r + si$, where r and s are real numbers (they are, in fact, rational). Choose integers $m, n \in \mathbb{Z}$ such that $|r - m| \leq \frac{1}{2}$ and $|s - n| \leq \frac{1}{2}$. Set $\delta = m + ni$. Then $\delta \in \mathbb{Z}[i]$ and $\lambda((\alpha/\gamma) - \delta) = (r - m)^2 + (s - n)^2 \leq \frac{1}{4} + \frac{1}{4} = \frac{1}{2}$. Set $\rho = \alpha - \gamma\delta$. Then $\rho \in \mathbb{Z}[i]$ and either $\rho = 0$ or $\lambda(\rho) = \lambda(\gamma((\alpha/\gamma) - \delta)) = \lambda(\gamma)\lambda((\alpha/\gamma) - \delta) \leq \frac{1}{2}\lambda(\gamma) < \lambda(\gamma)$.

It follows that λ makes $\mathbb{Z}[i]$ into a Euclidean domain. $\square$

The ring $\mathbb{Z}[i]$ is called the ring of Gaussian integers after C. F. Gauss, who first studied its arithmetic properties in detail.

The numbers ± 1, $\pm i$ are the roots of $x^4 = 1$ over the complex numbers. Consider the equation $x^3 = 1$. Since $x^3 - 1 = (x - 1)(x^2 + x + 1)$ the roots of this equation are $1, (-1 \pm \sqrt{-3})/2$. Let $\omega = (-1 + \sqrt{-3})/2$. Then it is easy to check that $\omega^2 = (-1 - \sqrt{-3})/2$ and that $1 + \omega + \omega^2 = 0$.

Consider the set $\mathbb{Z}[\omega] = \{a + b\omega \mid a, b \in \mathbb{Z}\}$. $\mathbb{Z}[\omega]$ is closed under addition and subtraction. Moreover, $(a + b\omega)(c + d\omega) = ac + (ad + bc)\omega + bd\omega^2 = (ac - bd) + (ad + bc - bd)\omega$. Thus $\mathbb{Z}[\omega]$ is a ring. Again, since $\mathbb{Z}[\omega]$ is a subset of the complex numbers it is an integral domain.

We remark that $\mathbb{Z}[\omega]$ is closed under complex conjugation. In fact, since $\overline{\sqrt{-3}} = \overline{\sqrt{3}i} = -\sqrt{3}i = -\sqrt{-3}$ we see that $\bar{\omega} = \omega^2$. Thus if $\alpha = a + b\omega \in \mathbb{Z}[\omega]$, then $\bar{\alpha} = a + b\bar{\omega} = a + b\omega^2 = (a - b) - b\omega \in \mathbb{Z}[\omega]$.

Proposition 1.4.2. $\mathbb{Z}[\omega]$ *is a Euclidean domain.*

PROOF. For $\alpha = a + b\omega \in \mathbb{Z}[\omega]$ define $\lambda(\alpha) = a^2 - ab + b^2$. A simple calculation shows that $\lambda(\alpha) = \alpha\bar{\alpha}$.

Now, let $\alpha, \beta \in \mathbb{Z}[\omega]$ and suppose that $\beta \neq 0$. Then $\alpha/\beta = \alpha\bar{\beta}/\beta\bar{\beta} = r + s\omega$, where r and s are rational numbers. We have used the fact that $\beta\bar{\beta} = \lambda(\beta)$ is a positive integer and that $\alpha\bar{\beta} \in \mathbb{Z}[\omega]$ since α and $\bar{\beta} \in \mathbb{Z}[\omega]$.

Find integers m and n such that $|r - m| \leq \frac{1}{2}$ and $|s - n| \leq \frac{1}{2}$. Then put $\gamma = m + n\omega$. $\lambda((\alpha/\beta) - \gamma) = (r - m)^2 - (r - m)(s - n) + (s - n)^2 \leq \frac{1}{4} + \frac{1}{4} + \frac{1}{4} < 1$.

Let $\rho = \alpha - \gamma\beta$. Then either $\rho = 0$ or $\lambda(\rho) = \lambda(\beta((\alpha/\beta) - \gamma)) = \lambda(\beta)\lambda((\alpha/\beta) - \gamma) < \lambda(\beta)$. $\square$

From the analysis of Section 3 we know that the theorem of unique factorization is true in both $\mathbb{Z}[i]$ and $\mathbb{Z}[\omega]$. To go further with the analysis of these rings we would have to investigate the units and the prime elements. There are some results of this nature in the exercises.

NOTES

Rings for which the theorem of unique factorization into irreducibles holds are called unique factorization domains (UFD). The fact that $\mathbb{Z}$ is a UFD is already implicit in Euclid, but the first explicit and clear statement of the result seems to be in C. F. Gauss' masterpiece *Disquisitiones Arithmeticae* (available in English translation by A. A. Clark, Yale University Press, New Haven, Conn., 1966). Zermelo gave a clever proof by contradiction, which is reproduced in the excellent book of G. H. Hardy and Wright [40]. See also Davis and Shisha [120].

We have shown that every PID is a UFD. The converse is not true. For example, the ring of polynomials over a field in more than one variable is a

UFD but not a PID. P. Samuel has an excellent expository article on UFD's in [67]. A more elementary introduction may be found in the book of H. Rademacher and O. Toeplitz [65].

The reader may find it profitable to read the introductory material in several books on number theory. Chapter 3 of A. Frankel [32] and the introduction to H. Stark [73] are particularly good. There is also an early lecture by Hardy [39] that is highly recommended.

The ring $\mathbb{Z}[i]$ was introduced by Gauss in his second memoir on biquadratic reciprocity [34]. G. Eisenstein considered the ring $\mathbb{Z}[\omega]$ in connection with his work on cubic reciprocity. He mentions that to investigate the properties of this ring one need only consult Gauss' work on $\mathbb{Z}[i]$ and modify the proofs [28]. A thorough treatment of these two rings is given in Chapter 12 of Hardy and Wright [40]. In Chapter 14 they treat a generalization, namely, rings of integers in quadratic number fields. Stark's Chapter 8 deals with the same subject [73]. In 1966 Stark resolved a long-outstanding problem in the theory of numbers by showing that the ring of integers (see Chapter 6 of this book) in the field $\mathbb{Q}(\sqrt{d})$, with d negative, is a UFD when $d = -1, -2, -3, -7, -11, -19, -43, -67$, and -163 and for no other values of d.

The student who is familiar with a little algebra will notice that a "generic" non-UFD is given by the ring $k[x, y, z, w]$, with $xy = zw$, where k is a field. Another example of a non-UFD is $\mathbb{C}[x, y, z]$, with $x^2 + y^2 + z^2 = 1$, where $\mathbb{C}$ is the field of complex numbers. To see this notice that $(x + iy)(x - iy) = (1 - z)(1 + z)$.

EXERCISES

1. Let a and b be nonzero integers. We can find nonzero integers q and r such that $a = qb + r$, where $0 \leq r < b$. Prove that $(a, b) = (b, r)$.

2. (continuation) If $r \neq 0$, we can find q_1 and r_1 such that $b = q_1 r + r_1$ with $0 \leq r_1 < r$. Show that $(a, b) = (r, r_1)$. This process can be repeated. Show that it must end in finitely many steps. Show that the last nonzero remainder must equal (a, b). The process looks like

$$a = qb + r, \qquad 0 \leq r < b,$$
$$b = q_1 r + r_1, \qquad 0 \leq r_1 < r,$$
$$r = q_2 r_1 + r_2, \qquad 0 \leq r_2 < r_1,$$
$$\vdots$$
$$r_{k-1} = q_{k+1} r_k + r_{k+1}, \qquad 0 \leq r_{k+1} < r_k,$$
$$r_k = q_{k+2} r_{k+1}.$$

Then $r_{k+1} = (a, b)$. This process of finding (a, b) is known as the Euclidean algorithm.

3. Calculate $(187, 221)$, $(6188, 4709)$, and $(314, 159)$.

4. Let $d = (a, b)$. Show how one can use the Euclidean algorithm to find numbers m and n such that $am + bn = d$. (*Hint*: In Exercise 2 we have that $d = r_{k+1}$. Express r_{k+1} in terms of r_k and r_{k-1}. then in terms of r_{k-1} and r_{k-2}, etc.)

5. Find m and n for the pairs a and b given in Exercise 3.

6. Let $a, b, c \in \mathbb{Z}$. Show that the equation $ax + by = c$ has solutions in integers iff $(a, b) | c$.

7. Let $d = (a, b)$ and $a = da'$ and $b = db'$. Show that $(a', b') = 1$.

8. Let x_0 and y_0 be a solution to $ax + by = c$. Show that all solutions have the form $x = x_0 + t(b/d)$, $y = y_0 - t(a/d)$, where $d = (a, b)$ and $t \in \mathbb{Z}$.

9. Suppose that $u, v \in \mathbb{Z}$ and that $(u, v) = 1$. If $u | n$ and $v | n$, show that $uv | n$. Show that this is false if $(u, v) \neq 1$.

10. Suppose that $(u, v) = 1$. Show that $(u + v, u - v)$ is either 1 or 2.

11. Show that $(a, a + k) | k$.

12. Suppose that we take several copies of a regular polygon and try to fit them evenly about a common vertex. Prove that the only possibilities are six equilateral triangles, four squares, and three hexagons.

13. Let $n_1, n_2, \ldots, n_s \in \mathbb{Z}$. Define the greatest common divisor d of $n_1, n_2, \ldots, n_s$ and prove that there exist integers $m_1, m_2, \ldots, m_s$ such that $n_1 m_1 + n_2 m_2 + \cdots + n_s m_s = d$.

14. Discuss the solvability of $a_1 x_1 + a_2 x_2 + \cdots + a_r x_r = c$ in integers. (*Hint*: Use Exercise 13 to extend the reasoning behind Exercise 6.)

15. Prove that $a \in \mathbb{Z}$ is the square of another integer iff $\mathrm{ord}_p a$ is even for all primes p. Give a generalization.

16. If $(u, v) = 1$ and $uv = a^2$, show that both u and v are squares.

17. Prove that the square root of 2 is irrational, i.e., that there is no rational number $r = a/b$ such that $r^2 = 2$.

18. Prove that $\sqrt[n]{m}$ is irrational if m is not the nth power of an integer.

19. Define the least common multiple of two integers a and b to be an integer m such that $a | m$, $b | m$, and m divides every common multiple of a and b. Show that such an m exists. It is determined up to sign. We shall denote it by $[a, b]$.

20. Prove the following:
    (a) $\mathrm{ord}_p[a, b] = \max(\mathrm{ord}_p a, \mathrm{ord}_p b)$.
    (b) $(a, b)[a, b] = ab$.
    (c) $(a + b, [a, b]) = (a, b)$.

21. Prove that $\mathrm{ord}_p(a + b) \geq \min(\mathrm{ord}_p a, \mathrm{ord}_p b)$ with equality holding if $\mathrm{ord}_p a \neq \mathrm{ord}_p b$.

22. Almost all the previous exercises remain valid if instead of the ring $\mathbb{Z}$ we consider the ring $k[x]$. Indeed, in most we can consider any Euclidean domain. Convince yourself of this fact. For simplicity we shall continue to work in $\mathbb{Z}$.

23. Suppose that $a^2 + b^2 = c^2$ with $a, b, c \in \mathbb{Z}$. For example, $3^2 + 4^2 = 5^2$ and $5^2 + 12^2 = 13^2$. Assume that $(a, b) = (b, c) = (c, a) = 1$. Prove that there exist integers u and v such that $c - b = 2u^2$ and $c + b = 2v^2$ and $(u, v) = 1$ (there is no loss in generality in assuming that b and c are odd and that a is even). Consequently $a = 2uv$, $b = v^2 - u^2$, and $c = v^2 + u^2$. Conversely show that if u and v are given, then the three numbers a, b, and c given by these formulas satisfy $a^2 + b^2 = c^2$.

24. Prove the identities
    (a) $x^n - y^n = (x - y)(x^{n-1} + x^{n-2}y + \cdots + y^{n-1})$.
    (b) For n odd, $x^n + y^n = (x + y)(x^{n-1} - x^{n-2}y + x^{n-3}y^2 - \cdots + y^{n-1})$.

25. If $a^n - 1$ is a prime, show that $a = 2$ and that n is a prime. Primes of the form $2^p - 1$ are called Mersenne primes. For example, $2^3 - 1 = 7$ and $2^5 - 1 = 31$. It is not known if there are infinitely many Mersenne primes.

26. If $a^n + 1$ is a prime, show that a is even and that n is a power of 2. Primes of the form $2^{2^t} + 1$ are called Fermat primes. For example, $2^{2^1} + 1 = 5$ and $2^{2^2} + 1 = 17$. It is not known if there are infinitely many Fermat primes.

27. For all odd n show that $8 \mid n^2 - 1$. If $3 \nmid n$, show that $6 \mid n^2 - 1$.

28. For all n show that $30 \mid n^5 - n$ and that $42 \mid n^7 - n$.

29. Suppose that $a, b, c, d \subseteq \mathbb{Z}$ and that $(a, b) = (c, d) = 1$. If $(a/b) + (c/d) = $ an integer, show that $b = \pm d$.

30. Prove that $\frac{1}{2} + \frac{1}{3} + \cdots + \frac{1}{n}$ is not an integer.

31. Show that 2 is divisible by $(1 + i)^2$ in $\mathbb{Z}[i]$.

32. For $\alpha = a + bi \in \mathbb{Z}[i]$ we defined $\lambda(\alpha) = a^2 + b^2$. From the properties of λ deduce the identity $(a^2 + b^2)(c^2 + d^2) = (ac - bd)^2 + (ad + bc)^2$.

33. Show that $\alpha \in \mathbb{Z}[i]$ is a unit iff $\lambda(\alpha) = 1$. Deduce that $1, -1, i,$ and $-i$ are the only units in $\mathbb{Z}[i]$.

34. Show that 3 is divisible by $(1 - \omega)^2$ in $\mathbb{Z}[\omega]$.

35. For $\alpha = a + b\omega \in \mathbb{Z}[\omega]$ we defined $\lambda(\alpha) = a^2 - ab + b^2$. Show that α is a unit iff $\lambda(\alpha) = 1$. Deduce that $1, -1, \omega, -\omega, \omega^2,$ and $-\omega^2$ are the only units in $\mathbb{Z}[\omega]$.

36. Define $\mathbb{Z}[\sqrt{-2}]$ as the set of all complex numbers of the form $a + b\sqrt{-2}$, where $a, b \in \mathbb{Z}$, Show that $\mathbb{Z}[\sqrt{-2}]$ is a ring. Define $\lambda(\alpha) = a^2 + 2b^2$ for $\alpha = a + b\sqrt{-2}$. Use λ to show that $\mathbb{Z}[\sqrt{-2}]$ is a Euclidean domain.

37. Show that the only units in $\mathbb{Z}[\sqrt{-2}]$ are 1 and -1.

38. Suppose that $\pi \in \mathbb{Z}[i]$ and that $\lambda(\pi) = p$ is a prime in $\mathbb{Z}$. Show that π is a prime in $\mathbb{Z}[i]$. Show that the corresponding result holds in $\mathbb{Z}[\omega]$ and $\mathbb{Z}[\sqrt{-2}]$.

39. Show that in any integral domain a prime element is irreducible.

Chapter 2

Applications of Unique Factorization

The importance of the notion of prime number should be evident from the results of Chapter 1.

In this chapter we shall give several proofs of the fact that there are infinitely many primes in $\mathbb{Z}$. We shall also consider the analogous question for the ring $k[x]$.

The theorem of unique prime decomposition is sometimes referred to as the fundamental theorem of arithmetic. We shall begin to demonstrate its usefulness by using it to investigate the properties of some natural number-theoretic functions.

§1 Infinitely Many Primes in $\mathbb{Z}$

Theorem 1 (Euclid). *In the ring $\mathbb{Z}$ there are infinitely many prime numbers.*

PROOF. Let us consider positive primes. Label them in increasing order $p_1, p_2, p_3, \ldots$. Thus $p_1 = 2$, $p_2 = 3$, $p_3 = 5$, etc. Let $N = (p_1 p_2 \cdots p_n) + 1$. N is greater than 1 and not divisible by any p_i, $i = 1, 2, \ldots, n$. On the other hand, N is divisible by some prime, p, and p must be greater than p_n.

We have shown that given any positive prime there is another prime that is greater. It follows that the set of primes is infinite. ☐

The analogous theorem for $k[x]$ is that there are infinitely many monic, irreducible polynomials. If k is infinite, this is trivial since $x - a$ is monic and irreducible for all $a \in k$. This proof does not work if k is finite, but Euclid's proof may easily be adapted to this case. We leave this as an exercise.

Recall that in an integral domain two elements are called associate if they differ only by multiplication by a unit. We now know that in $\mathbb{Z}$ and $k[x]$ there are infinitely many nonassociate primes. It is instructive to consider a ring where all primes are associate, so that in essence there is only one prime.

Let $p \in \mathbb{Z}$ be a prime number and let $\mathbb{Z}_p$ be the set of all rational numbers a/b, where $p \nmid b$. One easily checks using the remark following Corollary 1 to Proposition 1.1.1 that $\mathbb{Z}_p$ is a ring. $a/b \in \mathbb{Z}_p$ is a unit if there is a $c/d \in \mathbb{Z}_p$ such that $a/b \cdot c/d = 1$. Then $ac = bd$, which implies $p \nmid a$ since $p \nmid b$ and $p \nmid d$. Conversely, any rational number a/b is a unit in $\mathbb{Z}_p$ if $p \nmid a$ and $p \nmid b$.

If $a/b \in \mathbb{Z}_p$, write $a = p^l a'$, where $p \nmid a'$. Then $a/b = p^l a'/b$. Thus every element of $\mathbb{Z}_p$ is a power of p times a unit. From this it is easy to see that the only primes in $\mathbb{Z}_p$ have the form pc/d, where c/d is a unit. Thus all the primes of $\mathbb{Z}_p$ are associate.

EXERCISE

If $a/b \in \mathbb{Z}_p$ is not a unit, prove that $a/b + 1$ is a unit. This phenomenon shows why Euclid's proof breaks down in general for integral domains.

§2 Some Arithmetic Functions

In the remainder of this chapter we shall give some applications of the unique factorization theorem.

An integer $a \in \mathbb{Z}$ is said to be square-free if it is not divisible by the square of any other integer greater than 1.

Proposition 2.2.1. *If $n \in \mathbb{Z}$, n can be written in the form $n = ab^2$, where $a, b \in \mathbb{Z}$ and a is square-free.*

PROOF. Let $n = p_1^{a_1} p_2^{a_2} \cdots p_l^{a_l}$. One can write $a_i = 2b_i + r_i$, where $r_i = 0$ or 1 depending on whether a_i is even or odd. Set $a = p_1^{r_1} p_2^{r_2} \cdots p_l^{r_l}$ and $b = p_1^{b_1} p_2^{b_2} \cdots p_l^{b_l}$. Then $n = ab^2$ and a is clearly square-free.   □

This lemma can be used to give another proof that there are infinitely many primes in $\mathbb{Z}$. Assume that there are not, and let $p_1, p_2, \ldots, p_l$ be a complete list of positive primes. Consider the set of positive integers less than or equal to N. If $n \leq N$, then $n = ab^2$, where a is square-free and thus equal to one of the 2^l numbers $p_1^{\varepsilon_1} p_2^{\varepsilon_2} \cdots p_l^{\varepsilon_l}$, where $\varepsilon_i = 0$ or 1, $i = 1, \ldots, l$. Notice that $b \leq \sqrt{N}$. There are at most $2^l \sqrt{N}$ numbers satisfying these conditions and so $N \leq 2^l \sqrt{N}$, or $\sqrt{N} \leq 2^l$, which is clearly false for N large enough. This contradiction proves the result.

It is possible to give a similar proof that there are infinitely many monic irreducibles in $k[x]$, where k is a finite field.

There are a number of naturally defined functions on the integers. For example, given a positive integer n let $v(n)$ be the number of positive divisors of n and $\sigma(n)$ the sum of the positive divisors of n. For example, $v(3) = 2$, $v(6) = 4$, and $v(12) = 6$ and $\sigma(3) = 4$, $\sigma(6) = 12$, and $\sigma(12) = 28$. Using unique factorization it is possible to obtain rather simple formulas for these functions.

Proposition 2.2.2. *If n is a positive integer, let $n = p_1^{a_1} p_2^{a_2} \cdots p_l^{a_l}$ be its prime decomposition. Then*

(a) $v(n) = (a_1 + 1)(a_2 + 1) \cdots (a_l + 1)$.
(b) $\sigma(n) = ((p_1^{a_1+1} - 1)/(p_1 - 1))((p_2^{a_2+1} - 1)/(p_2 - 1)) \cdots$
    $((p_l^{a_l+1} - 1)/(p_l - 1))$.

PROOF. To prove part (a) notice that $m | n$ iff $m = p_1^{b_1} p_2^{b_2} \cdots p_l^{b_l}$ and $0 \le b_i \le a_i$ for $i = 1, 2, \ldots, l$. Thus the positive divisors of n are one-to-one correspondence with the n-tuples $(b_1, b_2, \ldots, b_l)$ with $0 \le b_i \le a_i$ for $i = 1, \ldots, l$, and there are exactly $(a_1 + 1)(a_2 + 1) \cdots (a_l + 1)$ such n-tuples.

To prove part (b) notice that $\sigma(n) = \sum p_1^{b_1} p_2^{b_2} \cdots p_l^{b_l}$, where the sum is over the above set of n-tuples. Thus, $\sigma(n) = (\sum_{b_1=0}^{a_1} p_1^{b_1})(\sum_{b_2=0}^{a_2} p_2^{b_2}) \cdots (\sum_{b_l=0}^{a_l} p_l^{b_l})$, from which the result follows by use of the summation formula for the geometric series. □

There is an interesting and unsolved problem connected with the function $\sigma(n)$. A number n is said to be perfect if $\sigma(n) = 2n$. For example, 6 and 28 are perfect. In general, if $2^{m+1} - 1$ is a prime, then $n = 2^m(2^{m+1} - 1)$ is perfect, as can be seen by applying part (b) of Proposition 2.2.2. This fact is already in Euclid. L. Euler showed that any even perfect number has this form. Thus the problem of even perfect numbers is reduced to that of finding primes of the form $2^{m+1} - 1$. Such primes are called Mersenne primes. The two outstanding problems involving perfect numbers are the following: Are there infinitely many perfect numbers? Are there any odd perfect numbers?

The multiplicative analog of this problem is trivial. An integer n is called multiplicatively perfect if the product of the positive divisors of n is n^2. Such a number cannot be a prime or a square of a prime. Thus there is a proper divisor d such that $d \ne n/d$. The product of the divisors 1, d, n/d, and n is already n^2. Thus n is multiplicatively perfect iff there are exactly two proper divisors. The only such numbers are cubes of primes or products of two distinct primes. For example, 27 and 10 are multiplicatively perfect.

We now introduce a very important arithmetic function, the Möbius μ function. For $n \in \mathbb{Z}^+$, $\mu(1) = 1$, $\mu(n) = 0$ if n is not square-free, and $\mu(p_1 p_2 \cdots p_l) = (-1)^l$, where the p_i are distinct positive primes.

Proposition 2.2.3. *If $n > 1$, $\sum_{d|n} \mu(d) = 0$.*

PROOF. If $n = p_1^{a_1} p_2^{a_2} \cdots p_l^{a_l}$, then $\sum_{d|n} \mu(d) = \sum_{(\varepsilon_1, \ldots, \varepsilon_l)} \mu(p_1^{\varepsilon_1} \cdots p_l^{\varepsilon_l})$, where the ε_i are zero or 1. Thus

$$\sum_{d|n} \mu(d) = 1 - l + \binom{l}{2} - \binom{l}{3} + \cdots + (-1)^l = (1 - 1)^l = 0. \qquad \square$$

The full significance of the Möbius μ function can be understood most clearly when its connection with Dirichlet multiplication is brought to light.

Let f and g be complex valued functions on $\mathbb{Z}^+$. The Dirichlet product of f and g is defined by the formula $f \circ g(n) = \sum f(d_1)g(d_2)$, where the sum is over all pairs (d_1, d_2) of positive integers such that $d_1 d_2 = n$. This product is associative, as one can see by checking that $f \circ (g \circ h)(n) = (f \circ g) \circ h(n) = \sum f(d_1)g(d_2)h(d_3)$, where the sum is over all 3-tuples (d_1, d_2, d_3) of positive integers such that $d_1 d_2 d_3 = n$.

Define the function $\mathbb{I}$ by $\mathbb{I}(1) = 1$ and $\mathbb{I}(n) = 0$ for $n > 1$. Then $f \circ \mathbb{I} = \mathbb{I} \circ f = f$. Define I by $I(n) = 1$ for all $n \in \mathbb{Z}^+$. Then $f \circ I(n) = I \circ f(n) = \sum_{d|n} f(d)$.

Lemma. $I \circ \mu = \mu \circ I = \mathbb{I}$.

PROOF. $\mu \circ I(1) = \mu(1)I(1) = 1$. If $n > 1$, $\mu \circ I(n) = \sum_{d|n} \mu(d) = 0$. The same proof works for $I \circ \mu$. $\qquad\square$

Theorem 2 (Möbius Inversion Theorem). *Let* $F(n) = \sum_{d|n} f(d)$. *Then* $f(n) = \sum_{d|n} \mu(d)F(n/d)$.

PROOF. $F = f \circ I$. Thus $F \circ \mu = (f \circ I) \circ \mu = f \circ (I \circ \mu) = f \circ \mathbb{I} = f$. This shows that $f(n) = F \circ \mu(n) = \sum_{d|n} \mu(d)F(n/d)$. $\qquad\square$

Remark. We have considered complex-valued functions on the positive integers. It is useful to notice that Theorem 2 is valid whenever the functions take their value in an abelian group. The proof goes through word for word.

If the group law in the abelian group is written multiplicatively, the theorem takes the following form: If $F(n) = \prod_{d|n} f(d)$, then $f(n) = \prod_{d|n} F(n/d)^{\mu(d)}$.

The Möbius inversion theorem has many applications. We shall use it to obtain a formula for yet another arithmetic function, the Euler ϕ function. For $n \in \mathbb{Z}^+$, $\phi(n)$ is defined to be the number of integers between 1 and n relatively prime to n. For example, $\phi(1) = 1$, $\phi(5) = 4$, $\phi(6) = 2$, and $\phi(9) = 6$. If p is a prime, it is clear that $\phi(p) = p - 1$.

Proposition 2.2.4. $\sum_{d|n} \phi(d) = n$.

PROOF. Consider the n rational numbers $1/n, 2/n, 3/n, \ldots, (n-1)/n, n/n$. Reduce each to lowest terms; i.e., express each number as a quotient of relatively prime integers. The denominators will all be divisors of n. If $d|n$, exactly $\phi(d)$ of our numbers will have d in the denominator after reducing to lowest terms. Thus $\sum_{d|n} \phi(d) = n$. $\qquad\square$

Proposition 2.2.5. *If* $n = p_1^{a_1} p_2^{a_2} \cdots p_l^{a_l}$, *then*

$$\phi(n) = n(1 - (1/p_1))(1 - (1/p_2)) \cdots (1 - (1/p_l)).$$

PROOF. Since $n = \sum_{d|n} \phi(d)$ the Möbius inversion theorem implies that $\phi(n) = \sum_{d|n} \mu(d)n/d = n - \sum_i n/p_i + \sum_{i<j} n/p_i p_j \cdots = n(1 - (1/p_1))(1 - (1/p_2)) \cdots (1 - (1/p_l))$. $\qquad\square$

Later we shall give a more insightful proof of this formula. We shall also use the Möbius function to determine the number of monic irreducible polynomials of fixed degree in $k[x]$, where k is a finite field.

§3 $\sum 1/p$ Diverges

We began this chapter by proving that there are infinitely many prime numbers in $\mathbb{Z}$. We shall conclude by proving a somewhat stronger statement. The proof will assume some elementary facts from the theory of infinite series.

Theorem 3. $\sum 1/p$ *diverges, where the sum is over all positive primes in* $\mathbb{Z}$.

PROOF. Let $p_1, p_2, \ldots, p_{l(n)}$ be all the primes less than n and define $\lambda(n) = \prod_{i=1}^{l(n)} (1 - 1/p_i)^{-1}$. Since $(1 - 1/p_i)^{-1} = \sum_{a_i = 0}^{\infty} 1/p_i^{a_i}$ we see that

$$\lambda(n) = \sum (p_1^{a_1} p_2^{a_2} \cdots p_l^{a_l})^{-1},$$

where the sum is over all l-tuples of nonnegative integers $(a_1, a_2, \ldots, a_l)$. In particular, we see that $1 + \frac{1}{2} + \frac{1}{3} + \cdots + 1/n < \lambda(n)$. Thus $\lambda(n) \to \infty$ as $n \to \infty$. This already gives a new proof that there are infinitely many primes.

Next, consider $\log \lambda(n)$. We have

$$\log \lambda(n) = -\sum_{i=1}^{l} \log(1 - p_i^{-1}) = \sum_{i=1}^{l} \sum_{m=1}^{\infty} (m p_i^m)^{-1}$$

$$= p_1^{-1} + p_2^{-1} + \cdots + p_l^{-1} + \sum_{i=1}^{l} \sum_{m=2}^{\infty} (m p_i^m)^{-1}.$$

Now, $\sum_{m=2}^{\infty} (m p_i^m)^{-1} < \sum_{m=2}^{\infty} p_i^{-m} = p_i^{-2}(1 - p_i^{-1})^{-1} \leq 2 p_i^{-2}$. Thus $\log \lambda(n) < p_1^{-1} + p_2^{-1} + \cdots + p_l^{-1} + 2(p_1^{-2} + p_2^{-2} + \cdots + p_l^{-2})$. It is well known that $\sum_{n=1}^{\infty} n^{-2}$ converges. It follows that $\sum_{i=1}^{l} p_i^{-2}$ converges. Thus if $\sum p^{-1}$ converged, there would be a constant M such that $\log \lambda(n) < M$, or $\lambda(n) < e^M$. This, however, is impossible since $\lambda(n) \to \infty$ as $n \to \infty$. Thus $\sum p^{-1}$ diverges. $\square$

It is instructive to try to construct an analog of Theorem 3 for the ring $k[x]$, where k is a finite field with q elements. The role of the positive primes p is taken by the monic irreducible polynomials $p(x)$. The "size" of a monic polynomial $f(x)$ is given by the quantity $q^{\deg f(x)}$.

This is reasonable because for a positive integer n, n is the number of nonnegative integers less than n, i.e., the number of elements in the set $\{0, 1, 2, \ldots, n - 1\}$. Analogously, $q^{\deg f(x)}$ is the number of polynomials of degree less than $\deg f(x)$. This is easy to see. Any such polynomial has the form $a_0 x^m + a_1 x^{m-1} + \cdots + a_m$, where $m = \deg f(x) - 1$ and $a_i \in k$. There are q choices for a_i and the choice for each index is independent of the others. Thus there are $q^{m+1} = q^{\deg f(x)}$ such polynomials.

Theorem 4. $\sum q^{-\deg p(x)}$ *diverges, where the sum is over all monic irreducibles* $p(x)$ *in* $k[x]$.

PROOF. We first show that $\sum q^{-\deg f(x)}$ diverges and that $\sum q^{-2\deg f(x)}$ converges, where both sums are over all monic polynomials $f(x)$ in $k[x]$. Both results follow from the fact that there are exactly q^n monic polynomials of degree n in $k[x]$. Consider $\sum_{\deg f(x) \le n} q^{-\deg f(x)}$. This sum is equal to $\sum_{m=0}^{n} q^m q^{-m}$ $= n + 1$. Thus $\sum q^{-\deg f(x)}$ diverges. Similarly, $\sum_{\deg f(x) \le n} q^{-2\deg f(x)} =$ $\sum_{m=0}^{n} q^m q^{-2m} < (1 - 1/q)^{-1}$. Thus $\sum q^{-2\deg f(x)}$ converges.

The rest of the proof is an exact imitation of the proof of Theorem 2. The reader should fill in the details.   □

§4 The Growth of $\pi(x)$

In the introduction to Chapter 1 we defined $\pi(x)$ as the number of primes p, $1 < p \le x$. The study of the behavior of $\pi(x)$ for large x involves analytic techniques. We will prove in this section several results that require a minimum of results from analysis. In fact only the simplest properties of the logarithmic function are used.

We begin with the following simple consequence of Euclid's argument (Theorem 1) which gives a weak lower bound for $\pi(x)$. Throughout $\log x$ denotes the natural logarithm of x.

Proposition 2.4.1. $\pi(x) \ge \log(\log x)$, $x \ge 2$.

PROOF. Let p_n denote the nth prime. Then since any prime dividing $p_1 p_2 \cdots p_n$ $+ 1$ is distinct from $p_1, \ldots, p_n$ it follows that $p_{n+1} \le p_1 \cdots p_n + 1$. Now $p_1 < 2^{(2^1)}$, $p_2 < 2^{(2^2)}$ and if $p_n < 2^{(2^n)}$ then $p_{n+1} \le 2^{(2^1)} \cdot 2^{(2^2)} \cdots 2^{(2^n)} + 1 =$ $2^{2^{n+1}-2} + 1 < 2^{(2^{n+1})}$. It follows that $\pi(2^{(2^n)}) \ge n$. For $x > e$ choose an integer n so that $e^{(e^{n-1})} < x \le e^{(e^n)}$. If $n > 3$ then $e^{n-1} > 2^n$ so that

$$\pi(x) \ge \pi(e^{(e^{n-1})}) \ge \pi(e^{2^n}) \ge \pi(2^{2^n}) \ge n \ge \log(\log x).$$

This proves the result for $x > e^e$. If $x \le e^e$ the inequality is obvious.   □

The method employed in the paragraph following Proposition 2.2.1 to show that $\pi(x) \to \infty$ can also be used to obtain the following improvement of the above proposition. If n is a positive integer let $\gamma(n)$ denote the set of primes dividing n.

Proposition 2.4.2. $\pi(x) \ge \log x / 2 \log 2$.

PROOF. For any set of primes S define $f_S(x)$ to be the number of integers n, $1 \le n \le x$, with $\gamma(n) \subset S$. Suppose that S is a finite set with t elements. Writing such an n in the form $n = m^2 s$ with s square free we see that $m \le \sqrt{x}$

while s has at most 2^t choices corresponding to the various subsets of S. Thus $f_S(x) \leq 2^t \sqrt{x}$. Put $\pi(x) = m$ so that $p_{m+1} > x$. If $S = \{p_1, \ldots, p_m\}$ then clearly $f_S(x) = x$ which implies that $x \leq 2^m \sqrt{x} = 2^{\pi(x)} \sqrt{x}$. The result follows immediately.   □

It is interesting to note that the above method can also be used to give another proof to Theorem 2. For if $\sum 1/p_n$ converged then there is an n such that $\sum_{j>n} 1/p_j < \frac{1}{2}$. If $S = \{p_1, \ldots, p_n\}$ then $x - f_S(x)$ is the number of positive integers $m \leq x$ with $\gamma(m) \not\subseteq S$. That is, there exists a prime $p_j, j > n$ such that $p_j | m$. For such a prime there are $[x/p_j]$ multiples of p_j not exceeding x. Thus

$$x - f_S(x) \leq \sum_{j>n} \left[\frac{x}{p_j}\right] \leq \sum_{j>n} \frac{x}{p_j} < \frac{x}{2},$$

so that $f_S(x) \geq x/2$. On the other hand, $f_S(x) \leq 2^n \sqrt{x}$. These inequalities imply $2^n \geq \sqrt{x}/2$ which is false for n fixed and large x.

A function closely related to $\pi(x)$ is defined by $\theta(x) = \sum_{p \leq x} \log p$, the sum being over all primes at most x. We will use $\theta(x)$ to bound $\pi(x)$ from above. Put $\theta(1) = 0$.

Proposition 2.4.3. $\theta(x) < (4 \log 2)x$.

PROOF. Consider the binomial coefficient

$$\binom{2n}{n} = \frac{(n+1) \cdots (2n)}{1 \cdot 2 \cdots n}.$$

Clearly this integer is divisible by all primes p, $n < p < 2n$. Furthermore, since

$$(1+1)^{2n} = \sum_{j=0}^{2n} \binom{2n}{j}, \qquad 2^{2n} > \binom{2n}{n}.$$

Hence

$$2^{2n} > \binom{2n}{n} > \prod_{\substack{p > n \\ p < 2n}} p$$

and therefore

$$2n \log 2 > \sum_{\substack{p > n \\ p < 2n}} \log p = \theta(2n) - \theta(n).$$

Summing this relation for $n = 1, 2, 4, 8, \ldots, 2^{m-1}$ gives

$$\theta(2^m) < (\log 2)(2^{m+1} - 2)$$
$$< (\log 2)2^{m+1}.$$

If $2^{m-1} < x \le 2^m$ we obtain

$$\theta(x) \le \theta(2^m) < (\log 2)2^{m+1} = (4 \log 2)2^{m-1}$$
$$< (4 \log 2)x. \qquad \square$$

Corollary 1. *There is a positive constant c_1 such that $\pi(x) < c_1 x/\log x$ for $x \ge 2$.*

PROOF.
$$\theta(x) \ge \sum_{\substack{p \le x \\ p > \sqrt{x}}} \log p$$

$$\ge (\log \sqrt{x})(\pi(x) - \pi(\sqrt{x}))$$

$$\ge (\log \sqrt{x})\pi(x) - \sqrt{x} \log \sqrt{x}.$$

Thus

$$\pi(x) \le \frac{2\theta(x)}{\log x} + \sqrt{x}$$

$$\le (8 \log 2)\frac{x}{\log x} + \sqrt{x}.$$

The result follows by noting that $\sqrt{x} < 2x/\log x$ for $x \ge 2$. $\qquad \square$

Corollary 2. $\pi(x)/x \to 0$ *as* $x \to \infty$.

To bound $\pi(x)$ from below we begin by examining further the binomial coefficient $\binom{2n}{n}$. First of all

$$\binom{2n}{n} = \left(\frac{n+1}{1}\right)\left(\frac{n+2}{2}\right) \cdots \left(\frac{n+n}{n}\right) \ge 2^n.$$

On the other hand by Exercise 6 at the end of this chapter we have

$$\text{ord}_p\binom{2n}{n} = \text{ord}_p\frac{(2n)!}{(n!)^2} = \sum_{j=1}^{t_p} \left(\left[\frac{2n}{p^j}\right] - 2\left[\frac{n}{p^j}\right]\right)$$

where t_p is the largest integer such that $p^{t_p} \le 2n$. Thus $t_p = [\log 2n/\log p]$. Now it is easy to see that $[2x] - 2[x]$ is always 1 or 0. It follows that

$$\text{ord}_p\binom{2n}{n} \le \frac{\log 2n}{\log p}.$$

Proposition 2.4.4. *There is a positive constant c_2 such that $\pi(x) > c_2(x/\log x)$.*

PROOF. By the above we have

$$2^n \le \binom{2n}{n} \le \prod_{p < 2n} p^{t_p}.$$

Thus

$$n \log 2 \leq \sum_{p < 2n} t_p \log p = \sum_{p < 2n} \left[\frac{\log 2n}{\log p} \right] \log p.$$

If $\log p > \frac{1}{2} \log 2n$, i.e., $p > \sqrt{2n}$, then $[\log 2n/\log p] = 1$. Thus

$$n \log 2 \leq \sum_{p \leq \sqrt{2n}} \left[\frac{\log 2n}{\log p} \right] \log p + \sum_{\substack{p > \sqrt{2n} \\ p < 2n}} \log p$$

$$\leq \sqrt{2n} \log 2n + \theta(2n).$$

Therefore $\theta(2n) > n \log 2 - \sqrt{2n} \log 2n$. But $\sqrt{2n} \log 2n/n$ approaches 0 as $n \to \infty$, so that $\theta(2n) > Tn$ for some $T > 0$ and all n sufficiently large. Writing, for large x, $2n \leq x < 2n + 1$ we have $\theta(x) \geq \theta(2n) > Tn > T(x - 1)/2 > Cx$ for a suitable constant C. Thus there is a constant $c_2 > 0$ such that $\theta(x) > c_2 x$ for all $x \geq 2$. To complete the proof we observe that

$$\theta(x) = \sum_{p \leq x} \log p \leq \pi(x) \log x.$$

Thus

$$\pi(x) \geq \frac{\theta(x)}{\log x} > c_2 \frac{x}{\log x}. \qquad \square$$

The preceding two propositions were first proven by Tchebychef in 1852. These results are subsumed under the famous prime number theorem which asserts that in fact $\pi(x)(\log x/x) \to 1$ as $x \to \infty$. It is not difficult to see that this is equivalent to $\theta(x)/x \to 1$ as $x \to \infty$. The prime number theorem was conjectured, in a slightly different form by Gauss at the age of 15 or 16. The proof of the conjecture was not achieved until 1896 when J. Hadamard and Ch. de la Vallé Poussin established the result independently. Their proofs utilize complex analytic properties of the Riemann zeta function. In 1948 Atle Selberg was able to prove the result without the use of complex analysis.

NOTES

There are a multitude of unsolved problems in the theory of prime numbers. For example, it is not known if there are infinitely many primes of the form $n^2 + 1$. On the other hand we will prove in Chapter 16 that the linear polynomial $an + b$ always represents an infinite number of primes when $(a, b) = 1$. This is the celebrated theorem of Dirichlet on primes in an arithmetic progression.

It is not known whether there exist infinitely many primes of the form $2^N + 1$, the so-called Fermat primes, or if there are infinitely many primes of the form $2^N - 1$, the Mersenne primes.

Another outstanding problem is to decide whether there are an infinite number of primes p such that $p + 2$ is also prime. It is known that the sum of the reciprocals of the set of such primes converges, a result due to Viggo Brun [52].

Good discussions of unsolved problems about primes may be found in W. Sierpinski [71] and Shanks [70]. Readers with a background in analysis should read the paper by P. Erdös [31] as well as those of Hardy [38] and [39].

The key idea behind the proof of Theorem 2 is due to L. Euler. A pleasant account of this for the beginner is found in Rademacher and Toeplitz [65].

Theorem 3 gives a proof in the spirit of Euler that $k[x]$ contains infinitely many irreducibles. This already suggests that many of the theorems in classical number theory have analogs in the ring $k[x]$. This is indeed the case. An interesting reference along these lines is L. Carlitz [10].

The theorem of Dirichlet mentioned above has been proved for $k[x]$, k a finite field, by H. Kornblum [50]. Kornblum had his promising career cut short after he enlisted as Kriegsfreiwilliger in 1914. The prime number theorem also has an analog in $k[x]$. This was proved by E. Artin in his doctoral thesis [2].

A good introduction to analytic number theory is Chandrasekharan [112]. In the last chapter of this very readable text a proof of the prime number theorem is given that uses complex analysis. Proofs that are free of complex analysis (but not of subtlety) have been given by A. Selberg [215] and P. Erdös [133]. For an interesting account of the history of this theorem see L. J. Goldstein [139]. Finally we recommend the remarkable tract Primzahlen by E. Trost [229]; this 95 page book contains, in addition to many elementary results concerning the distribution of primes, Selberg's proof of the prime number theorem as well as an "elementary" proof of Dirichlet's theorem mentioned above. See also D. J. Newman [198].

EXERCISES

1. Show that $k[x]$, with k a finite field, has infinitely many irreducible polynomials.

2. Let $p_1, p_2, \ldots, p_t \in \mathbb{Z}$ be primes and consider the set of all rational numbers $r = a/b$, $a, b \in \mathbb{Z}$, such that $\mathrm{ord}_{p_i} a \geq \mathrm{ord}_{p_i} b$ for $i = 1, 2, \ldots, t$. Show that this set is a ring and that up to taking associates $p_1, p_2, \ldots, p_t$ are the only primes.

3. Use the formula for $\phi(n)$ to give a proof that there are infinitely many primes. [Hint: If $p_1, p_2, \ldots, p_t$ were all the primes, then $\phi(n) = 1$, where $n = p_1 p_2 \cdots p_t$.]

4. If a is a nonzero integer, then for $n > m$ show that $(a^{2^n} + 1, a^{2^m} + 1) = 1$ or 2 depending on whether a is odd or even. (Hint: If p is an odd prime and $p | a^{2^m} + 1$, then $p | a^{2^n} - 1$ for $n > m$.)

5. Use the result of Exercise 4 to show that there are infinitely many primes. (This proof is due to G. Polya.)

6. For a rational number r let $[r]$ be the largest integer less than or equal to r, e.g., $[\frac{1}{2}] = 0$, $[2] = 2$, and $[3\frac{1}{3}] = 3$. Prove $\mathrm{ord}_p n! = [n/p] + [n/p^2] + [n/p^3] + \cdots$.

7. Deduce from Exercise 6 that $\mathrm{ord}_p n! \leq n/(p-1)$ and that $\sqrt[n]{n!} \leq \prod_{p|n!} p^{1/(p-1)}$.

8. Use Exercise 7 to show that there are infinitely many primes. [*Hint*: $(n!)^2 \geq n^n$.] (This proof is due to Eckford Cohen.)

9. A function on the integers is said to be multiplicative if $f(ab) = f(a)f(b)$ whenever $(a, b) = 1$. Show that a multiplicative function is completely determined by its value on prime powers.

10. If $f(n)$ is a multiplicative function, show that the function $g(n) = \sum_{d|n} f(d)$ is also multiplicative.

11. Show that $\phi(n) = n \sum_{d|n} \mu(d)/d$ by first proving that $\mu(d)/d$ is multiplicative and then using Exercises 9 and 10.

12. Find formulas for $\sum_{d|n} \mu(d)\phi(d)$, $\sum_{d|n} \mu(d)^2\phi(d)^2$, and $\sum_{d|n} \mu(d)/\phi(d)$.

13. Let $\sigma_k(n) = \sum_{d|n} d^k$. Show that $\sigma_k(n)$ is multiplicative and find a formula for it.

14. If $f(n)$ is multiplicative, show that $h(n) = \sum_{d|n} \mu(n/d)f(d)$ is also multiplicative.

15. Show that
    (a) $\sum_{d|n} \mu(n/d)v(d) = 1$ for all n.
    (b) $\sum_{d|n} \mu(n/d)\sigma(d) = n$ for all n.

16. Show that $v(n)$ is odd iff n is a square.

17. Show that $\sigma(n)$ is odd iff n is a square or twice a square.

18. Prove that $\phi(n)\phi(m) = \phi((n, m))\phi([n, m])$.

19. Prove that $\phi(mn)\phi((m, n)) = (m, n)\phi(m)\phi(n)$.

20. Prove that $\prod_{d|n} d = n^{v(n)/2}$.

21. Define $\wedge(n) = \log p$ if n is a power of p and zero otherwise. Prove that $\sum_{d|n} \mu(n/d) \log d = \wedge(n)$. [*Hint*: First calculate $\sum_{d|n} \wedge(d)$ and then apply the Möbius inversion formula.]

22. Show that the sum of all the integers t such that $1 \leq t \leq n$ and $(t, n) = 1$ is $\frac{1}{2}n\phi(n)$.

23. Let $f(x) \in \mathbb{Z}[x]$ and let $\psi(n)$ be the number of $f(j), j = 1, 2, \ldots, n$, such that $(f(j), n) = 1$. Show that $\psi(n)$ is multiplicative and that $\psi(p^t) = p^{t-1}\psi(p)$. Conclude that $\psi(n) = n \prod_{p|n} \psi(p)/p$.

24. Supply the details to the proof of Theorem 3.

25. Consider the function $\zeta(s) = \sum_{n=1}^{\infty} 1/n^s$. $\zeta(s)$ is called the Riemann zeta function. It converges for $s > 1$. Prove the formal identity (Euler's identity) $\zeta(s) = \prod_p (1 - (1/p^s))^{-1}$. If we let s assume complex values, it can be shown that $\zeta(s)$ has an analytic continuation to the whole complex plane. The famous Riemann hypothesis states that the only zeros of $\zeta(s)$ lying in the strip $0 \leq \mathrm{Re}\ s \leq 1$ lie on the line $\mathrm{Re}\ s = \frac{1}{2}$.

26. Verify the formal identities
    (a) $\zeta(s)^{-1} = \sum_{n=1}^{\infty} \mu(n)/n^s$.
    (b) $\zeta(s)^2 = \sum_{n=1}^{\infty} v(n)/n^s$.
    (c) $\zeta(s)\zeta(s - 1) = \sum_{n=1}^{\infty} \sigma(n)/n^s$.

27. Show that $\sum' 1/n$, the sum being over square free integers, diverges. Conclude that $\prod_{p<N} (1 + 1/p) \to \infty$ as $N \to \infty$. Since $e^x > 1 + x$, conclude that $\sum_{p<N} 1/p \to \infty$. (This proof is due to I. Niven.)

Chapter 3

Congruence

Gauss first introduced the notion of congruence in Dis-
quisitiones Arithmeticae (see Notes in Chapter 1). It is
an extremely simple idea. Nevertheless, its importance
and usefulness in number theory cannot be exaggerated.

This chapter is devoted to an exposition of the simplest
properties of congruence. In Chapter 4, we shall go into
the subject in more depth.

§1 Elementary Observations

It is a simple observation that the product of two odd numbers is odd, the
product of two even numbers is even, and the product of an odd and even
number is even. Also, notice that an odd plus an odd is even, an even plus an
even is even, and an even plus an odd is odd. This information is summarized
in Tables 1 and 2. Table 1 is like a multiplication table and Table 2 like an
addition table.

Table 1		
	e	o
e	e	e
o	e	o

Table 2		
	e	o
e	e	o
o	o	e

These observations are so elementary one might ask if anything interesting
can be deduced from them. The answer, surprisingly, is yes.

Many problems in number theory have the form; if f is a polynomial in
one or several variables with integer coefficients, does the equation $f = 0$
have integer solutions? Such questions were considered by the Greek
mathematician Diophantus and are called Diophantine problems in his
honor.

Consider the equation $x^2 - 117x + 31 = 0$. We claim that there is no
solution that is an integer. Let n be any integer. n is either even or odd. If n
is even, so is n^2 and $117n$. Thus $n^2 - 117n + 31$ is odd. If n is odd, then n^2

and $117n$ are both odd. Thus $n^2 - 117n + 31$ is odd in this case also. Since every integer is even or odd, this shows that $n^2 - 117n + 31$ is never zero.

In Chapter 2 we showed that there are infinitely many prime numbers. We shall now show that there are infinitely many prime numbers that leave a remainder of 3 when divided by 4. Examples of such primes are 3, 7, 19, and 59.

An integer divided by 4 leaves a remainder of 0, 1, 2, or 3. Thus odd numbers are either of the form $4k + 1$ or $4l + 3$. The product of two numbers of the form $4k + 1$ is again of that form: $(4k + 1)(4k' + 1) = 4(4kk' + k + k') + 1$. It follows that an integer of the form $4l + 3$ must be divisible by a prime of the form $4l + 3$.

Now, suppose that there were only finitely many positive primes of the form $4l + 3$. This list begins $3, 7, 11, 19, 23, \ldots$. Let $p_1 = 7, p_2 = 11, p_3 = 19$, etc. Suppose that p_m is the largest prime of this form and set $N = 4p_1 p_2 \cdots p_m + 3$. N is not divisible by any of the p_i. However, N is of the form $4l + 3$ and so must be divisible by a prime p of the form $4l + 3$. We have $p > p_m$, which is a contradiction.

There is clearly some common principle underlying both arguments. We explore this in Section 2.

§2 Congruence in ℤ

Definition. If $a, b, m \in \mathbb{Z}$ and $m \neq 0$, we say that a is *congruent to b modulo m* if m divides $b - a$. This relation is written $a \equiv b\ (m)$.

Proposition 3.2.1.

(a) $a \equiv a\ (m)$.
(b) $a \equiv b\ (m)$ *implies that* $b \equiv a\ (m)$.
(c) *If* $a \equiv b\ (m)$ *and* $b \equiv c\ (m)$, *then* $a \equiv c\ (m)$.

PROOF.

(a) $a - a = 0$ and $m \mid 0$.
(b) If $m \mid b - a$, then $m \mid a - b$.
(c) If $m \mid b - a$ and $m \mid c - b$, then $m \mid c - a = (c - b) + (b - a)$.   □

Proposition 3.2.1 shows that congruence modulo m is an equivalence relation on the set of integers. If $a \in \mathbb{Z}$, let $\bar{a}$ denote the set of integers congruent to a modulo m, $\bar{a} = \{n \in \mathbb{Z} \mid n \equiv a\ (m)\}$. In other words $\bar{a}$ is the set of integers of the form $a + km$.

If $m = 2$, then $\bar{0}$ is the set of even integers and $\bar{1}$ is the set of odd integers.

Definition. A set of the form $\bar{a}$ is called a *congruence class modulo m*.

Proposition 3.2.2.

(a) $\bar{a} = \bar{b}$ iff $a \equiv b \, (m)$.
(b) $\bar{a} \neq \bar{b}$ iff $\bar{a} \cap \bar{b}$ is empty.
(c) *There are precisely m distinct congruence classes modulo m.*

PROOF.

(a) If $\bar{b} = \bar{a}$, then $a \in \bar{a} = \bar{b}$. Thus $a \equiv b \, (m)$. Conversely, if $a \equiv b \, (m)$, then
    $a \in \bar{b}$. If $c \equiv a \, (m)$, then $c \equiv b \, (m)$, which shows $\bar{a} \subseteq \bar{b}$. Since $a \equiv b \, (m)$
    implies that $b \equiv a \, (m)$, we also have $\bar{b} \subseteq \bar{a}$. Therefore $\bar{a} = \bar{b}$.
(b) Clearly, if $\bar{a} \cap \bar{b}$ is empty, then $\bar{a} \neq \bar{b}$. We shall show that $\bar{a} \cap \bar{b}$ not empty
    implies that $\bar{a} = \bar{b}$. Let $c \in \bar{a} \cap \bar{b}$. Then $c \equiv a \, (m)$ and $c \equiv b \, (m)$. It
    follows that $a \equiv b \, (m)$ and so by part (a) we have $\bar{a} = \bar{b}$.
(c) We shall show that $\bar{0}, \bar{1}, \bar{2}, \ldots, \overline{m-1}$ are all distinct and are a complete
    set of congruence classes modulo m. Suppose that $0 \leq k < l < m$. $\bar{k} = \bar{l}$
    implies that $k \equiv l \, (m)$ or that m divides $l - k$. Since $0 < l - k < m$ this
    is a contradiction. Therefore $\bar{k} \neq \bar{l}$. Now let $a \in \mathbb{Z}$. We can find integers
    q and r such that $a = qm + r$, where $0 \leq r < m$. It follows that $a \equiv r \, (m)$
    and that $\bar{a} = \bar{r}$.   □

Definition. The set of congruence classes modulo m is denoted by $\mathbb{Z}/m\mathbb{Z}$.

If $\bar{a}_1, \bar{a}_2, \ldots, \bar{a}_m$ are a complete set of congruence classes modulo m, then
$\{a_1, a_2, \ldots, a_m\}$ is called a *complete set of residues modulo m.*

For example, $\{0, 1, 2, 3\}$, $\{4, 9, 14, -1\}$, and $\{0, 1, -2, -1\}$ are complete
sets of residues modulo 4.

The set $\mathbb{Z}/m\mathbb{Z}$ can be made into a ring by defining in a natural way addition
and multiplication. This is accomplished by means of the following proposition.

Proposition 3.2.3. *If $a \equiv c \, (m)$ and $b \equiv d \, (m)$, then $a + b \equiv c + d \, (m)$ and*
$ab \equiv cd \, (m)$.

PROOF. If $m | c - a$ and $m | d - b$, then $m | (c - a) + (d - b) = (c + d) -$
$(a + b)$. Thus $a + b \equiv c + d \, (m)$.

Notice that $cd - ab = c(d - b) + b(c - a)$. Thus $m | cd - ab$ and $ab \equiv$
$cd \, (m)$.   □

If $\bar{a}, \bar{b} \in \mathbb{Z}/m\mathbb{Z}$, we define $\bar{a} + \bar{b}$ to be $\overline{a + b}$ and $\bar{a}\bar{b}$ to be $\overline{ab}$.

This definition seems to depend on a and b. We have to show that they
depend only on the congruence classes defined by a and b. This is easy.
Assume that $\bar{c} = \bar{a}$ and that $\bar{d} = \bar{b}$. We must show that $\overline{a + b} = \overline{c + d}$ and
that $\overline{ab} = \overline{cd}$, but this follows immediately from Propositions 3.2.2 and 3.2.3.

With these definitions $\mathbb{Z}/m\mathbb{Z}$ becomes a ring. The verification of this fact is
left to the reader.

| Table 3 | | | | Table 4 | | |
| Addition | | | | Multiplication | | |

	0	1	2		0	1	2
0	0	1	2	0	0	0	0
1	1	2	0	1	0	1	2
2	2	0	1	2	0	2	1

Tables 3 and 4 give explicitly the addition and multiplication in $\mathbb{Z}/3\mathbb{Z}$. (Bars over the numbers are omitted.) The reader should construct similar tables for $m = 4, 5$, and 6.

In discussing arithmetic problems it is sometimes more convenient to work with the ring $\mathbb{Z}/m\mathbb{Z}$ than with the notion of congruence modulo m. On the other hand, it is sometimes more convenient the other way around. We shall switch back and forth between the two viewpoints as the situation demands.

We proved earlier that the polynomial $x^2 - 117x + 31$ has no integer roots. It is possible to generalize this result using some of the material we have developed.

If $a \equiv b \ (m)$, then $a^2 \equiv b^2 \ (m)$, $a^3 \equiv b^3 \ (m)$, and in general $a^n \equiv b^n \ (m)$. It follows that if $p(x) \in \mathbb{Z}[x]$, then $p(a) \equiv p(b) \ (m)$. All this is a consequence of Proposition 3.2.3.

Take $m = 2$. Then a is congruent to either 0 or 1 modulo 2 and we have $p(a) \equiv p(0) \ (2)$ or $p(a) \equiv p(1) \ (2)$.

If $p(x) = a_0 x^n + a_1 x^{n-1} + \cdots + a_{n-1}x + a_n$, then $p(0) = a_n$ and $p(1) = a_0 + a_1 + \cdots + a_n$. Our calculations yield the following result: If $p(x) \in \mathbb{Z}[x]$ and $p(0)$ and $p(1)$ are both odd, then $p(x)$ has no integer roots.

$x^2 - 117x + 31$ has constant term 31, and the sum of the coefficients is -85, both of which are odd. Other examples are $2x^2 + 2x + 1$ and $3x^3 + 2x^2 + x + 3$.

§3 The Congruence $ax \equiv b \ (m)$

The simplest congruence is $ax \equiv b \ (m)$. In this section we shall develop a criterion to test this congruence for solvability, and if it is solvable, give a formula for the number of solutions.

Before beginning we must give a definition of what we mean by the number of solutions to a congruence. Quite generally, let $f(x_1, \ldots, x_n)$ be a polynomial in n variables with integer coefficients and consider the congruence $f(x_1, \ldots, x_n) \equiv 0 \ (m)$. A solution is an n-tuple of integers $(a_1, \ldots, a_n)$ such that $f(a_1, a_2, \ldots, a_n) \equiv 0 \ (m)$. If $(b_1, \ldots, b_n)$ is another n-tuple such that

$b_i \equiv a_i$ (m) for $i = 1, \ldots, n$, then it is easy to see that $f(b_1, \ldots, b_n) \equiv 0$ (m). We do not want to consider these two solutions as being essentially different. Thus two solutions $(a_1, \ldots, a_n)$ and $(b_1, \ldots, b_n)$ are called equivalent if $a_i \equiv b_i$ for $i = 1, \ldots, n$. The number of solutions to $f(x_1, \ldots, x_n) \equiv 0$ (m) is defined to be the number of inequivalent solutions.

For example, 3, 8, and 13 are solutions to $6x \equiv 3$ (15). 18 is also a solution, but the solution $x = 18$ is equivalent to the solution $x = 3$.

It is useful to consider the matter from another point of view. The map from $\mathbb{Z}$ to $\mathbb{Z}/m\mathbb{Z}$ given by $a \to \bar{a}$ is a homomorphism. If $f(a_1, \ldots, a_n) \equiv 0$ (m), then $\bar{f}(\bar{a}_1, \ldots, \bar{a}_n) = \bar{0}$. Here $\bar{f}(x_1, \ldots, x_n) \in \mathbb{Z}/m\mathbb{Z}[x_1, \ldots, x_n]$ is the polynomial obtained from f by putting a bar over each coefficient of f. One can now see that equivalence classes of solutions to $f(x_1, \ldots, x_n) = 0$ are in one-to-one correspondence with solutions to $\bar{f}(x_1, \ldots, x_n) = \bar{0}$ in the ring $\mathbb{Z}/m\mathbb{Z}$. This interpretation of the number of solutions arises frequently.

We now return to the number of solutions of the congruence $ax \equiv b$ (m).

Let $d > 0$ be the greatest common divisor of a and m. Set $a' = a/d$ and $m' = m/d$. Then a' and m' are relatively prime.

Proposition 3.3.1. *The congruence $ax \equiv b$ (m) has solutions iff $d | b$. If $d | b$, then there are exactly d solutions. If x_0 is a solution, then the other solutions are given by $x_0 + m', x_0 + 2m', \ldots, x_0 + (d - 1)m'$.*

PROOF. If x_0 is a solution, then $ax_0 - b = my_0$ for some integer y_0. Thus $ax_0 - my_0 = b$. Since d divides $ax_0 - my_0$, we must have $d | b$.

Conversely, suppose that $d | b$. By Lemma 4 on page 4 there exist integers x_0' and y_0' such that $ax_0' - my_0' = d$. Let $c = b/d$ and multiply both sides of the equation by c. Then $a(x_0'c) - m(y_0'c) = b$. Let $x_0 = x_0'c$. Then $ax_0 \equiv b$ (m).

We have shown that $ax \equiv b$ (m) has a solution iff $d | b$.

Suppose that x_0 and x_1 are solutions. $ax_0 \equiv b$ (m) and $ax_1 \equiv b$ (m) imply that $a(x_1 - x_0) \equiv 0$ (m). Thus $m | a(x_1 - x_0)$ and $m' | a'(x_1 - x_0)$, which implies that $m' | x_1 - x_0$ or $x_1 = x_0 + km'$ for some integer k. One easily checks that any number of the form $x_0 + km'$ is a solution and that the solutions $x_0, x_0 + m', \ldots, x_0 + (d - 1)m'$ are inequivalent. Let $x_1 = x_0 + km'$ be another solution. There are integers r and s such that $k = rd + s$ and $0 \le s < d$. Thus $x_1 = x_0 + sm' + rm$ and x_1 is equivalent to $x_0 + sm'$. This completes the proof.   □

As an example, let us consider the congruence $6x \equiv 3$ (15) once more. We first solve $6x - 15y = 3$. Dividing by 3, we have $2x - 5y = 1$. $x = 3, y = 1$ is a solution. Thus $x_0 = 3$ is a solution to $6x \equiv 3$ (15). Now, $m = 15$ and $d = 3$ so that $m' = 5$. The three inequivalent solutions are 3, 8, and 13.

We have two important corollaries.

Corollary 1. *If a and m are relatively prime, then $ax \equiv b$ (m) has one and only one solution.*

PROOF. In this case $d = 1$ so clearly $d \mid b$, and there are $d = 1$ solutions.   □

Corollary 2. *If p is a prime and $a \not\equiv 0 \ (p)$, then $ax \equiv b \ (p)$ has one and only one solution.*

PROOF. Immediate from Corollary 1.   □

Corollaries 1 and 2 can be interpreted in terms of the ring $\mathbb{Z}/m\mathbb{Z}$. The congruence $ax \equiv b \ (m)$ is equivalent to the equation $\bar{a}x = \bar{b}$ over the ring $\mathbb{Z}/m\mathbb{Z}$.

What are the units of $\mathbb{Z}/m\mathbb{Z}$? $\bar{a} \in \mathbb{Z}/m\mathbb{Z}$ is a unit iff $\bar{a}x = \bar{1}$ is solvable. $ax \equiv 1 \ (m)$ is solvable iff $d \mid 1$, i.e., iff a and m are relatively prime. Thus $\bar{a}$ is a unit iff $(a, m) = 1$, and it follows easily that there are exactly $\phi(m)$ units in $\mathbb{Z}/m\mathbb{Z}$ [see page 20 for the definition of $\phi(m)$].

If p is a prime and $\bar{a} \neq \bar{0}$ is in $\mathbb{Z}/p\mathbb{Z}$, then $(a, p) = 1$. Thus every nonzero element of $\mathbb{Z}/p\mathbb{Z}$ is a unit, which shows that $\mathbb{Z}/p\mathbb{Z}$ is a field.

If m is not a prime, then $m = m_1 m_2$, where $0 < m_1, m_2 < m$. Thus $\overline{m_1} \neq \bar{0}$, $\overline{m_2} \neq \bar{0}$, but $\overline{m_1 m_2} = \overline{m_1} \overline{m_2} = \bar{m} = \bar{0}$. Therefore $\mathbb{Z}/m\mathbb{Z}$ is not a field.

Summarizing we have

Proposition 3.3.2. *An element $\bar{a}$ of $\mathbb{Z}/m\mathbb{Z}$ is a unit iff $(a, m) = 1$. There are exactly $\phi(m)$ units in $\mathbb{Z}/m\mathbb{Z}$. $\mathbb{Z}/m\mathbb{Z}$ is a field iff m is a prime.*

Corollary 1 (Euler's Theorem). *If $(a, m) = 1$, then $a^{\phi(m)} \equiv 1 \ (m)$.*

PROOF. The units in $\mathbb{Z}/m\mathbb{Z}$ form a group of order $\phi(m)$. If $(a, m) = 1$, $\bar{a}$ is a unit. Thus $\bar{a}^{\phi(m)} = \bar{1}$ or $a^{\phi(m)} \equiv 1 \ (m)$.   □

Corollary 2 (Fermat's Little Theorem). *If p is a prime and $p \nmid a$, then $a^{p-1} \equiv 1 \ (p)$.*

PROOF. If $p \nmid a$, then $(a, p) = 1$. Thus $a^{\phi(p)} \equiv 1 \ (p)$. The result follows, since for a prime p, $\phi(p) = p - 1$.   □

It is possible to generalize many of the results in this section to principal ideal domains.

The notions of congruence and residue class can be carried over to an arbitrary commutative ring. The first part of Proposition 3.3.1 is valid in a PID; i.e., $ax \equiv b \ (m)$ has a solution iff $d \mid b$ and the solution is unique iff a and m are relatively prime. The only difference is that the number of solutions need not be finite. In any case, using this result one proves in analogy to part of Proposition 3.3.2 that if R is a PID and $m \in R$ is not zero or a unit, then $R/(m)$ is a field iff m is a prime.

In particular, if k is a field, then $k[x]/(f(x))$ is a field iff $f(x)$ is irreducible.

§4 The Chinese Remainder Theorem

When the modulus m of a congruence is composite it is sometimes possible to reduce a congruence modulo m to a system of simpler congruences. The main theorem of this type is the so-called Chinese remainder theorem (Theorem 1), which we prove below. This theorem is valid for any PID (in fact, even more generally). However, we shall continue to work in $\mathbb{Z}$ and leave to the reader the relatively simple exercise of carrying over the proof for PID's.

Lemma 1. *If* $a_1, \ldots, a_t$ *are all relatively prime to* m, *then so is* $a_1 a_2 \cdots a_t$.

PROOF. $\bar{a}_i \in \mathbb{Z}/m\mathbb{Z}$ is a unit. Thus so is $\bar{a}_1 \bar{a}_2 \cdots \bar{a}_t = \overline{a_1 a_2 \cdots a_t}$. By Proposition 3.3.2, $a_1 a_2 \cdots a_t$ is relatively prime to m. □

Another proof goes as follows. If $a_1 a_2 \cdots a_t$ was not prime to m, there would be a prime p that divides them both. $p | a_1 a_2 \cdots a_t$ implies that $p | a_i$ for some i. It follows that $(a_i, m) \neq 1$, which contradicts the hypothesis.

Lemma 2. *Suppose that* $a_1, \ldots, a_t$ *all divide* n *and that* $(a_i, a_j) = 1$ *for* $i \neq j$. *Then* $a_1 a_2 \cdots a_t$ *divides* n.

PROOF. The proof is by induction on t. If $t = 1$, there is nothing to do. Suppose that $t > 1$ and that the lemma is true for $t - 1$. Then $a_1 a_2 \cdots a_{t-1}$ divides n. By Lemma 1, a_t is prime to $a_1 a_2 \cdots a_{t-1}$. Thus there are integers r and s such that $r a_t + s a_1 a_2 \cdots a_{t-1} = 1$. Multiply both sides by n. Inspection shows that the left-hand side is divisible by $a_1 a_2 \cdots a_t$ and the result follows. □

Theorem 1 (Chinese Remainder Theorem). *Suppose that* $m = m_1 m_2 \cdots m_t$ *and that* $(m_i, m_j) = 1$ *for* $i \neq j$. *Let* $b_1, b_2, \ldots, b_t$ *be integers and consider the system of congruences*:

$$x \equiv b_1 \ (m_1), x \equiv b_2 \ (m_2), \ldots, x \equiv b_t \ (m_t).$$

This system always has solutions and any two solutions differ by a multiple of m.

PROOF. Let $n_i = m/m_i$. By Lemma 1, $(m_i, n_i) = 1$. Thus there are integers r_i and s_i such that $r_i m_i + s_i n_i = 1$. Let $e_i = s_i n_i$. Then $e_i \equiv 1 \ (m_i)$ and $e_i \equiv 0 \ (m_j)$ for $j \neq i$.

Set $x_0 = \sum_{i=1}^{t} b_i e_i$. Then we have $x_0 \equiv b_i e_i \ (m_i)$ and consequently $x_0 \equiv b_i \ (m_i)$. x_0 is a solution.

Suppose that x_1 is another solution. Then $x_1 - x_0 \equiv 0 \ (m_i)$ for $i = 1, 2, \ldots, t$. In other words, $m_1, m_2, \ldots, m_t$ divide $x_1 - x_0$. By Lemma 2, m divides $x_1 - x_0$. □

We wish to interpret Theorem 1 from a ring-theoretic point of view. If $R_1, R_2, \ldots, R_n$ are rings, then $R_1 \oplus R_2 \oplus \cdots \oplus R_n = S$, the direct sum of the R_i, is defined to be the set of n-tuples $(r_1, r_2, \ldots, r_n)$ with $r_i \in R_i$. Addition and multiplication are defined by $(r_1, r_2, \ldots, r_n) + (r'_1, r'_2, \ldots, r'_n) = (r_1 + r'_1, \ldots, r_n + r'_n)$ and $(r_1, r_2, \ldots, r_n) \cdot (r'_1, r'_2, \ldots, r'_n) = (r_1 r'_1, r_2 r'_2, \ldots, r_n r'_n)$. The zero element is $(0, 0, \ldots, 0)$ and the identity is $(1, 1, \ldots, 1)$. $u \in S$ is a unit iff there is a $v \in S$ such that $uv = 1$. If $u = (u_1, \ldots, u_n)$ and $v = (v_1, \ldots, v_n)$, then $uv = 1$ implies that $u_i v_i = 1$ for $i = 1, \ldots, n$. Thus u_i is a unit for each i. Conversely, if u_i is a unit for each i, then $u = (u_1, u_2, \ldots, u_n)$ is a unit. For a ring R we denote the group of units by $U(R)$. $U(R_1) \times U(R_2) \times \cdots \times U(R_n)$ is the set of n-tuples $(u_1, u_2, \ldots, u_n)$, where $u_i \in R_i$. This is a group under component-wise multiplication. We have shown

Proposition 3.4.1. *If* $S = R_1 \oplus R_2 \oplus \cdots \oplus R_n$, *then* $U(S) = U(R_1) \times U(R_2) \times U(R_3) \times \cdots \times U(R_n)$.

Let $m_1, m_2, \ldots, m_t$ be pairwise relatively prime integers. ψ_i will denote the natural homomorphism from $\mathbb{Z}$ to $\mathbb{Z}/m_i\mathbb{Z}$. We construct a map ψ from $\mathbb{Z}$ to $\mathbb{Z}/m_1\mathbb{Z} \oplus \mathbb{Z}/m_2\mathbb{Z} \oplus \cdots \oplus \mathbb{Z}/m_t\mathbb{Z}$ as follows: $\psi(n) = (\psi_1(n), \psi_2(n), \ldots, \psi_t(n))$ for all $n \in \mathbb{Z}$. It is easy to check that ψ is a ring homomorphism. What are the kernel and image of ψ?

$(\bar{b}_1, \bar{b}_2, \ldots, \bar{b}_t) = \psi(n)$ iff $\psi_i(n) = \bar{b}_i$ for $i = 1, \ldots, t$; i.e., $n \equiv b_i \ (m_i)$ for $i = 1, \ldots, t$. The Chinese Remainder Theorem assures us that such an n always exists. Thus ψ is onto.

$\psi(n) = 0$ iff $n \equiv 0 \ (m_i)$, $i = 1, \ldots, t$, iff n is divisible by $m = m_1 m_2 \cdots m_t$. This is immediate from Lemma 2. Thus the kernel of ψ is the ideal $m\mathbb{Z}$.

We have shown

Theorem 1'. *The map* ψ *induces an isomorphism between* $\mathbb{Z}/m\mathbb{Z}$ *and* $\mathbb{Z}/m_1\mathbb{Z} \oplus \mathbb{Z}/m_2\mathbb{Z} \oplus \cdots \oplus \mathbb{Z}/m_t\mathbb{Z}$.

Corollary. $U(\mathbb{Z}/m\mathbb{Z}) \approx U(\mathbb{Z}/m_1\mathbb{Z}) \times U(\mathbb{Z}/m_2\mathbb{Z}) \times \cdots \times U(\mathbb{Z}/m_t\mathbb{Z})$.

PROOF. Immediate from Theorem 1' and Proposition 3.4.1. $\qquad\qquad\qquad \square$

Both sides of the isomorphism in the above corollary are finite groups. The order of the left-hand side is $\phi(m)$ and the order of the right-hand side is $\phi(m_1)\phi(m_2) \cdots \phi(m_t)$. Thus $\phi(m) = \phi(m_1)\phi(m_2) \cdots \phi(m_t)$.

Let $m = p_1^{a_1} p_2^{a_2} \cdots p_t^{a_t}$ be the prime decomposition of m. We have $\phi(m) = \phi(p_1^{a_1})\phi(p_2^{a_2}) \cdots \phi(p_t^{a_t})$. For a prime power, p^a, $\phi(p^a) = p^a - p^{a-1}$, because the numbers less than p^a and prime to p^a are prime to p. Since $p^a/p = p^{a-1}$ numbers less than p^a are divisible by p, $p^a - p^{a-1}$ numbers are prime to p. Notice that $p^a - p^{a-1} = p^a(1 - 1/p)$. It follows that $\phi(m) = m \prod (1 - 1/p)$. We proved this formula in Chapter 2 in a different manner.

Let us summarize. In treating a number of arithmetical questions, the notion of congruence is extremely useful. This notion led us to consider the ring $\mathbb{Z}/m\mathbb{Z}$ and its group of units $U(\mathbb{Z}/m\mathbb{Z})$. To go more deeply into the structure of these algebraic objects we write $m = p_1^{a_1} p_2^{a_2} \cdots p_t^{a_t}$ and are led, via the Chinese Remainder Theorem, to the following isomorphisms:

$$\mathbb{Z}/m\mathbb{Z} \approx \mathbb{Z}/p_1^{a_1}\mathbb{Z} \oplus \mathbb{Z}/p_2^{a_2}\mathbb{Z} \oplus \cdots \oplus \mathbb{Z}/p_t^{a_t}\mathbb{Z},$$

$$U(\mathbb{Z}/m\mathbb{Z}) \approx U(\mathbb{Z}/p_1^{a_1}\mathbb{Z}) \times U(\mathbb{Z}/p_2^{a_2}\mathbb{Z}) \times \cdots \times U(\mathbb{Z}(p_t^{a_t}\mathbb{Z}).$$

For prime powers it is possible to push the investigation much further. This is the subject of Chapter 4.

NOTES

It would be useful for the reader to consult other treatments of the basic material given here. See, for example, the very readable book of Davenport [22] and (again) Hardy and Wright [40]. See also Niven and Zuckerman [61], T. Nagell [60], E. Landau [52] and Vinogradov [77].

An interesting discussion of the various possible ways of arranging this material can be found in P. Samuel, "Sur l'organization d'un cours d'arithmetique," *L'Enseignment Math.*, **13**, (1967), 223–231. A more advanced discussion of congruences is given in the first chapter of Borevich and Shafarevich [9]; this book also shows how the theory of congruences is useful in determining whether equations can be solved in integers. We mention also the beautiful treatment by J. P. Serre [69].

Historically the notion of congruences was first introduced and used systematically in Gauss' *Disquisitiones Arithmeticae*. The notion of congruence is a wonderful example of the usefulness of employing the "right" notation.

As far as the Chinese Remainder Theorem is concerned we note that Hardy and Wright [40] note that R. Bachman [4] notes that Sun Tsu was aware of this result in the first century A.D. The theorem is capable of vast generalizations. Properly formulated it holds in any ring with identity. Surprisingly it is no more difficult to prove in general than in the special case we have given (see Proposition 12.3.1).

EXERCISES

1. Show that there are infinitely many primes congruent to -1 modulo 6.

2. Construct addition and multiplication tables for $\mathbb{Z}/5\mathbb{Z}$, $\mathbb{Z}/8\mathbb{Z}$, and $\mathbb{Z}/10\mathbb{Z}$.

3. Let abc be the decimal representation for an integer between 1 and 1000. Show that abc is divisible by 3 iff $a + b + c$ is divisible by 3. Show that the same result is true if we replace 3 by 9. Show that abc is divisible by 11 iff $a - b + c$ is divisible by 11. Generalize to any number written in decimal notation.

4. Show that the equation $3x^2 + 2 = y^2$ has no solution in integers.

5. Show that the equation $7x^3 + 2 = y^3$ has no solution in integers.

6. Let an integer $n > 0$ be given. A set of integers $a_1, a_2, \ldots, a_{\phi(n)}$ is called a reduced residue system modulo n if they are pairwise incongruent modulo n and $(a_i, n) = 1$ for all i. If $(a, n) = 1$, prove that $aa_1, aa_2, \ldots, aa_{\phi(n)}$ is again a reduced residue system modulo n.

7. Use Exercise 6 to give another proof of Euler's theorem, $a^{\phi(n)} \equiv 1 \ (n)$ for $(a, n) = 1$.

8. Let p be an odd prime. If $k \in \{1, 2, \ldots, p - 1\}$, show that there is a unique b_k in this set such that $kb_k \equiv 1 \ (p)$. Show that $k \neq b_k$ unless $k = 1$ or $k = p - 1$.

9. Use Exercise 7 to prove that $(p - 1)! \equiv -1 \ (p)$. This is known as Wilson's theorem.

10. If n is not a prime, show that $(n - 1)! \equiv 0 \ (n)$, except when $n = 4$.

11. Let $a_1, a_2, \ldots, a_{\phi(n)}$ be a reduced residue system modulo n and let N be the number of solutions to $x^2 \equiv 1 \ (n)$. Prove that $a_1 a_2 \cdots a_{\phi(n)} \equiv (-1)^{N/2} \ (n)$.

12. Let $\binom{p}{k} = p!/(k!(p - k)!)$ be a binomial coefficient, and suppose that p is a prime.

    If $1 \leq k \leq p - 1$, show that p divides $\binom{p}{k}$. Deduce $(a + 1)^p \equiv a^p + 1 \ (p)$.

13. Use Exercise 12 to give another proof of Fermat's theorem, $a^{p-1} \equiv 1 \ (p)$ if $p \nmid a$.

14. Let p and q be distinct odd primes such that $p - 1$ divides $q - 1$. If $(n, pq) = 1$, show that $n^{q-1} \equiv 1 \ (pq)$.

15. For any prime p show that the numerator of $1 + \frac{1}{2} + \frac{1}{3} + \cdots + 1/p - 1$ is divisible by p. (*Hint*: Make use of Exercises 8 and 9.)

16. Use the proof of the Chinese Remainder Theorem to solve the system $x \equiv 1 \ (7)$, $x \equiv 4 \ (9)$, $x \equiv 3 \ (5)$.

17. Let $f(x) \in Z[x]$ and $n = p_1^{a_1} p_2^{a_2} \cdots p_t^{a_t}$. Show that $f(x) \equiv 0 \ (n)$ has a solution iff $f(x) \equiv 0 \ (p_i^{a_i})$ has a solution for $i = 1, 2, \ldots, t$.

18. Let N be the number of solutions to $f(x) \equiv 0 \ (n)$ and N_i be the number of solutions to $f(x) \equiv 0 \ (p_i^{a_i})$. Prove that $N = N_1 N_2 \cdots N_t$.

19. If p is an odd prime, show that 1 and -1 are the only solutions to $x^2 \equiv 1 \ (p^a)$.

20. Show that $x^2 \equiv 1 \ (2^b)$ has one solution if $b = 1$, two solutions if $b = 2$, and four solutions if $b \geq 3$.

21. Use Exercises 18–20 to find the number of solutions to $x^2 \equiv 1 \ (n)$.

22. Formulate and prove the Chinese Remainder Theorem in a principal ideal domain.

23. Extend the notion of congruence to the ring $Z[i]$ and prove that $a + bi$ is always congruent to 0 or 1 modulo $1 + i$.

24. Extend the notion of congruence to the ring $Z[\omega]$ and prove that $a + b\omega$ is always congruent to either -1, 1, or 0 modulo $1 - \omega$.

25. Let $\lambda = 1 - \omega \in \mathbb{Z}[\omega]$. If $\alpha \in \mathbb{Z}[\omega]$ and $\alpha \equiv 1 \ (\lambda)$, prove that $\alpha^3 \equiv 1 \ (9)$. (*Hint*: Show first that $3 = -\omega^2 \lambda^2$.)

26. Use Exercise 25 to show that if $\xi, \eta, \zeta \in \mathbb{Z}[\omega]$ are not zero and $\xi^3 + \eta^3 + \zeta^3 = 0$, then λ divides at least one of the elements ξ, η, ζ.

The Structure of $U(\mathbb{Z}/n\mathbb{Z})$

Having introduced the notion of congruence and discussed some of its properties and applications we shall now go more deeply into the subject. The key result is the existence of primitive roots modulo a prime. This theorem was used by mathematicians before Gauss but he was the first to give a proof. In the terminology introduced in Chapter 3 the existence of primitive roots is equivalent to the fact that $U(\mathbb{Z}/p\mathbb{Z})$ is a cyclic group when p is a prime. Using this fact we shall find an explicit description of the group $U(\mathbb{Z}/n\mathbb{Z})$ for arbitrary n.

§1 Primitive Roots and the Group Structure of $U(\mathbb{Z}/n\mathbb{Z})$

If $n = p_1^{a_1} p_2^{a_2} \cdots p_l^{a_l}$, then, as was shown in Chapter 3, $U(\mathbb{Z}/n\mathbb{Z}) \approx U(\mathbb{Z}/p_1^{a_1}\mathbb{Z}) \times \cdots \times U(\mathbb{Z}/p_l^{a_l}\mathbb{Z})$. Thus to determine the structure of $U(\mathbb{Z}/n\mathbb{Z})$ it is sufficient to consider the case $U(\mathbb{Z}/p^a\mathbb{Z})$, where p is a prime. We begin by considering the simplest case, $U(\mathbb{Z}/p\mathbb{Z})$.

Since $\mathbb{Z}/p\mathbb{Z}$ is a field, it will be helpful to have available the following simple lemma about fields.

Lemma 1. *Let $f(x) \in k[x]$, k a field. Suppose that $\deg f(x) = n$. Then f has at most n distinct roots.*

PROOF. The proof goes by induction on n. For $n = 1$ the assertion is trivial. Assume that the lemma is true for polynomials of degree $n - 1$. If $f(x)$ has no roots in k, we are done. If α is a root, $f(x) = q(x)(x - \alpha) + r$, where r is a constant. Setting $x = \alpha$ we see that $r = 0$. Thus $f(x) = q(x)(x - \alpha)$ and $\deg q(x) = n - 1$. If $\beta \neq \alpha$ is another root of $f(x)$, then $0 = f(\beta) = (\beta - \alpha)q(\beta)$, which implies that $q(\beta) = 0$. Since by induction $q(x)$ has at most $n - 1$ distinct roots, $f(x)$ has at most n distinct roots. $\qquad\square$

Corollary. *Let $f(x)$, $g(x) \in k[x]$ and $\deg f(x) = \deg g(x) = n$. If $f(\alpha_i) = g(\alpha_i)$ for $n + 1$ distinct elements $\alpha_1, \alpha_2, \ldots, \alpha_n, \alpha_{n+1}$, then $f(x) = g(x)$.*

PROOF. Apply the lemma to the polynomial $f(x) - g(x)$. $\qquad\square$

Proposition 4.1.1. $x^{p-1} - 1 \equiv (x - 1)(x - 2) \cdots (x - p + 1) \, (p)$.

PROOF. If $\bar{a}$ denotes the residue class of an integer a in $\mathbb{Z}/p\mathbb{Z}$, an equivalent way of stating the proposition is $x^{p-1} - \bar{1} = (x - \bar{1})(x - \bar{2}) \cdots (x - \overline{(p-1)})$ in $\mathbb{Z}/p\mathbb{Z}[x]$. Let $f(x) = (x^{p-1} - \bar{1}) - (x - \bar{1})(x - \bar{2}) \cdots (x - \overline{(p-1)})$. $f(x)$ has degree less than $p - 1$ (the leading terms cancel) and has the $p - 1$ roots $\bar{1}, \bar{2}, \ldots, \overline{p-1}$ (Fermat's Little Theorem). Thus $f(x)$ is identically zero. $\square$

Corollary. $(p - 1)! \equiv -1 \, (p)$.

PROOF. Set $x = 0$ in Proposition 4.1.1. $\square$

This result is known as Wilson's theorem. It is not hard to prove that if $n > 4$ is not prime, then $(n - 1)! \equiv 0 \, (n)$ (see Exercise 10 of Chapter 3). Thus the congruence $(n - 1)! \equiv -1 \, (n)$ is characteristic for primes. We shall make use of Wilson's theorem later when discussing quadratic residues.

Proposition 4.1.2. *If $d \mid p - 1$, then $x^d \equiv 1 \, (p)$ has exactly d solutions.*

PROOF. Let $dd' = p - 1$. Then

$$\frac{x^{p-1} - 1}{x^d - 1} = \frac{(x^d)^{d'} - 1}{x^d - 1} = (x^d)^{d'-1} + (x^d)^{d'-2} + \cdots + x^d + 1 = g(x).$$

Therefore

$$x^{p-1} - 1 = (x^d - 1)g(x)$$

and

$$x^{p-1} - \bar{1} = (x^d - \bar{1})\bar{g}(x).$$

If $x^d - \bar{1}$ had less than d roots, then by Lemma 1 the right-hand side would have less than $p - 1$ roots. However, the left-hand side has the $p - 1$ roots $\bar{1}, \bar{2}, \ldots, \overline{p-1}$. Thus $x^d \equiv 1 \, (p)$ has exactly d roots as asserted. $\square$

Theorem 1. *$U(\mathbb{Z}/p\mathbb{Z})$ is a cyclic group.*

PROOF. For $d \mid p - 1$ let $\psi(d)$ be the number of elements in $U(\mathbb{Z}/p\mathbb{Z})$ of order d. By Proposition 4.1.2 we see that the elements of $U(\mathbb{Z}/p\mathbb{Z})$ satisfying $x^d \equiv \bar{1}$ form a group of order d. Thus $\sum_{c \mid d} \psi(c) = d$. Applying the Möbius inversion theorem we obtain $\psi(d) = \sum_{c \mid d} \mu(c)d/c$. The right-hand side of this equation is equal to $\phi(d)$, as was seen in the proof of Proposition 2.2.5. In particular, $\psi(p - 1) = \phi(p - 1)$, which is greater than 1 if $p > 2$. Since the case $p = 2$ is trivial, we have shown in all cases the existence of an element [in fact, $\phi(p - 1)$ elements] of order $p - 1$. $\square$

Theorem 1 is of fundamental importance. It was first proved by Gauss. After giving some new terminology we shall outline two more proofs.

Definition. An integer a is called a *primitive root* mod p if $\bar{a}$ generates the group $U(\mathbb{Z}/p\mathbb{Z})$. Equivalently, a is a primitive root mod p if $p - 1$ is the smallest positive integer such that $a^{p-1} \equiv 1\ (p)$.

As an example, 2 is a primitive root mod 5, since the least positive residues of $2, 2^2, 2^3$, and 2^4 are 2, 4, 3, and 1. Thus $4 = 5 - 1$ is the smallest positive integer such that $2^n \equiv 1\ (5)$.

For $p = 7$, 2 is not a primitive root since $2^3 \equiv 1\ (7)$, but 3 is since $3, 3^2, 3^3, 3^4, 3^5$, and 3^6 are congruent to 3, 2, 6, 4, 5, and 1 mod 7.

Although Theorem 1 shows the existence of primitive roots for a given prime, there is no simple way of finding one. For small primes trial and error is probably as good a method as any.

A celebrated conjecture of E. Artin states that if $a > 1$ is not a square, then there are infinitely many primes for which a is a primitive root. Some progress has been made in recent years, but the conjecture still seems far from resolution. See [35].

Because of its importance, we outline two more proofs of Theorem 1. The reader is invited to fill in the details.

Let $p - 1 = q_1^{e_1} q_2^{e_2} \cdots q_t^{e_t}$ be the prime decomposition of $p - 1$. Consider the congruences

(1) $x^{q_i^{e_i-1}} \equiv 1\ (p)$.
(2) $x^{q_i^{e_i}} \equiv 1\ (p)$.

Every solution to congruence 1 is a solution of congruence 2. Moreover, congruence 2 has more solutions than congruence 1. Let g_i be a solution to congruence 2 that is not a solution to congruence 1 and set $g = g_1 g_2 \cdots g_t$. $\bar{g}_i$ generates a subgroup of $U(\mathbb{Z}/p\mathbb{Z})$ of order $q_i^{e_i}$. It follows that $\bar{g}$ generates a subgroup of $U(\mathbb{Z}/p\mathbb{Z})$ of order $q_1^{e_1} q_2^{e_2} \cdots q_t^{e_t} = p - 1$. Thus g is a primitive root and $U(\mathbb{Z}/p\mathbb{Z})$ is cyclic.

Finally, on group-theoretic grounds we can see that $\psi(d) \le \phi(d)$ for $d \mid p - 1$. Both $\sum_{d \mid p-1} \psi(d)$ and $\sum_{d \mid p-1} \phi(d)$ are equal to $p - 1$. It follows that $\psi(d) = \phi(d)$ for all $d \mid p - 1$. In particular, $\psi(p - 1) = \phi(p - 1)$. For $p > 2$, $\phi(p - 1) > 1$, implying that $\psi(p - 1) > 1$. The result follows.

The notion of primitive root can be generalized somewhat.

Definition. Let $a, n \in \mathbb{Z}$. a is said to be a *primitive root* mod n if the residue class of a mod n generates $U(\mathbb{Z}/n\mathbb{Z})$. It is equivalent to require that a and n be relatively prime and that $\phi(n)$ be the smallest positive integer such that $a^{\phi(n)} \equiv 1\ (n)$.

In general, it is not true that $U(\mathbb{Z}/n\mathbb{Z})$ is cyclic. For example, the elements of $U(\mathbb{Z}/8\mathbb{Z})$ are $\bar{1}, \bar{3}, \bar{5}, \bar{7}$, and $\bar{1}^2 = \bar{1}, \bar{3}^2 = \bar{1}, \bar{5}^2 = \bar{1}, \bar{7}^2 = \bar{1}$. Thus there is no element of order $4 = \phi(8)$. It follows that not every integer possesses primitive roots. We shall shortly determine those integers that do.

Lemma 2. *If p is a prime and $1 \leq k < p$, then the binomial coefficient $\binom{p}{k}$ is divisible by p.*

PROOF. We give two proofs.

(a) By definition

$$\binom{p}{k} = \frac{p!}{k!\,(p-k)!} \quad \text{so that } p! = k!\,(p-k)!\,\binom{p}{k}.$$

Now, p divides $p!$, but p does not divide $k!\,(p-k)!$ since this expression is a product of integers less than, and thus relatively prime to p. Thus p divides $\binom{p}{k}$.

(b) By Fermat's Little Theorem $a^{p-1} \equiv 1\ (p)$ if $p \nmid a$. It follows that $a^p \equiv a\ (p)$ for all a. In particular, $(1+a)^p \equiv 1 + a \equiv 1 + a^p\ (p)$ for all a. Thus $(1+x)^p - 1 - x^p \equiv 0\ (p)$ has p solutions. Since the polynomial has degree less than p it follows from the corollary to Lemma 1 that $(\bar{1} + x)^p - \bar{1} - x^p$ is identically zero in $\mathbb{Z}/p\mathbb{Z}[x]$

$$(1+x)^p - 1 - x^p = \sum_{k=1}^{p-1} \binom{p}{k} x^k.$$

Thus $\overline{\binom{p}{k}} = \bar{0}$ for $1 \leq k \leq p-1$, implying that $p|\binom{p}{k}$. The only interest in this proof is that we do not assume any information on $\binom{p}{k}$. $\square$

Lemma 3. *If $l \geq 1$ and $a \equiv b\ (p^l)$, then $a^p \equiv b^p\ (p^{l+1})$.*

PROOF. We may write $a = b + cp^l$, $c \in \mathbb{Z}$. Thus $a^p = b^p + \binom{p}{1}b^{p-1}cp^l + A$, where A is an integer divisible by p^{l+2}. The second term is clearly divisible by p^{l+1}. Thus $a^p \equiv b^p\ (p^{l+1})$. $\square$

Corollary 1. *If $l \geq 2$ and $p \neq 2$, then $(1 + ap)^{p^{l-2}} \equiv 1 + ap^{l-1}\ (p^l)$ for all $a \in \mathbb{Z}$.*

PROOF. The proof is by induction on l. For $l = 2$ the assertion is trivial. Suppose that it is true for some $l \geq 2$. We show that it is then true for $l + 1$. Applying Lemma 3 we obtain

$$(1 + ap)^{p^{l-1}} \equiv (1 + ap^{l-1})^p\ (p^{l+1}).$$

By the binomial theorem

$$(1 + ap^{l-1})^p = 1 + \binom{p}{1}ap^{l-1} + B,$$

where B is a sum of $p - 2$ terms. Using Lemma 2 it is easy to see that all these terms are divisible by $p^{1+2(l-1)}$ except perhaps for the last term, $a^p p^{p(l-1)}$. Since $l \geq 2$, $1 + 2(l-1) \geq l + 1$, and since also $p \geq 3$, $p(l-1) \geq l + 1$. Thus $p^{l+1}|B$ and $(1 + ap)^{p^{l-1}} \equiv 1 + ap^l\ (p^{l+1})$, which is as required. $\square$

Before starting a second corollary we need a definition.

Definition. Let $a, n \in \mathbb{Z}$ and $(a, n) = 1$. We say a has *order* e mod n if e is the smallest positive integer such that $a^e \equiv 1\ (n)$. This is equivalent to saying that $\bar{a}$ has order e in the group $U(\mathbb{Z}/n\mathbb{Z})$.

Corollary 2. *If $p \neq 2$ and $p \nmid a$, then p^{l-1} is the order of $1 + ap$ mod p^l.*

PROOF. By Corollary 1, $(1 + ap)^{p^{l-1}} \equiv 1 + ap^l\ (p^{l+1})$, implying that $(1 + ap)^{p^{l-1}} \equiv 1\ (p^l)$ and thus that $1 + ap$ has order dividing p^{l-1}. $(1 + ap)^{p^{l-2}} \equiv 1 + ap^{l-1}\ (p^l)$ shows that p^{l-2} is not the order of $1 + ap$ (it is here we use the hypothesis $p \nmid a$). The result follows. $\square$

We are now in a position to extend Theorem 1. It turns out that we shall have to treat the prime 2 separately from the odd primes. The necessity of treating 2 differently from the other primes occurs repeatedly in number theory.

Theorem 2. *If p is an odd prime and $l \in \mathbb{Z}^+$, then $U(\mathbb{Z}/p^l\mathbb{Z})$ is cyclic; i.e., there exist primitive roots mod p^l.*

PROOF. By Theorem 1 there exist primitive roots mod p. If $g \in \mathbb{Z}$ is a primitive root mod p, then so is $g + p$. If $g^{p-1} \equiv 1\ (p^2)$, then $(g + p)^{p-1} \equiv g^{p-1} + (p - 1)g^{p-2}p \equiv 1 + (p - 1)g^{p-2}p\ (p^2)$. Since p^2 does not divide $(p - 1) \times g^{p-2}p$ we may assume from the beginning that g is a primitive root mod p and that $g^{p-1} \not\equiv 1\ (p^2)$.

We claim that such a g is already a primitive root mod p^l. To prove this it is sufficient to prove that if $g^n \equiv 1\ (p^l)$, then $\phi(p^l) = p^{l-1}\ (p - 1)|n$.

$g^{p-1} = 1 + ap$, where $p \nmid a$. By Corollary 2 to Lemma 3, p^{l-1} is the order of $1 + ap$ mod p^l. Since $(1 + ap)^n \equiv 1\ (p^l)$ we have $p^{l-1}|n$.

Let $n = p^{l-1}n'$. Then $g^n = (g^{p^{l-1}})^{n'} \equiv g^{n'}\ (p)$, and therefore $g^{n'} \equiv 1\ (p)$. Since g is a primitive root mod p, $p - 1|n'$. We have proved that $p^{l-1}(p - 1)|n$, as required. $\square$

Theorem 2′. *2^l has primitive roots for $l = 1$ or 2 but not for $l \geq 3$. If $l \geq 3$, then $\{(-1)^a 5^b | a = 0, 1 \text{ and } 0 \leq b < 2^{l-2}\}$ constitutes a reduced residue system mod 2^l. It follows that for $l \geq 3$, $U(\mathbb{Z}/2\mathbb{Z})$ is the direct product of two cyclic groups, one of order 2, the other of order 2^{l-2}.*

PROOF. 1 is a primitive root mod 2, and 3 is a primitive root mod 4. From now on let us assume that $l \geq 3$.

We claim that (1) $5^{2^{l-3}} \equiv 1 + 2^{l-1}\ (2^l)$. This is true for $l = 3$. Assume that it is true for $l \geq 3$ and we shall prove it is true for $l + 1$. First notice that $(1 + 2^{l-1})^2 = 1 + 2^l + 2^{2l-2}$ and that $2l - 2 \geq l + 1$ for $l \geq 3$. Applying Lemma 3 to congruence (1), we get (2) $5^{2^{l-2}} \equiv 1 + 2^l\ (2^{l+1})$. Our claim is now established by induction.

From (2) we see that $5^{2^{l-2}} \equiv 1 \ (2^l)$, whereas from (1) we see that $5^{2^{l-3}} \not\equiv 1 \ (2^l)$. Thus 2^{l-2} is the order of 5 mod 2^l.

Consider the set $\{(-1)^a 5^b | a = 1, 2 \text{ and } 0 \le b < 2^{l-2}\}$. We claim that these 2^{l-1} numbers are incongruent mod 2^l. Since $\phi(2^l) = 2^{l-1}$ this will show that our set is in fact a reduced residue system mod 2^l.

If $(-1)^a 5^b \equiv (-1)^{a'} 5^{b'} \ (2^l)$, $l \ge 3$, then $(-1)^a \equiv (-1)^{a'} \ (4)$, implying that $a \equiv a' \ (2)$. Thus $a = a'$. Going further, $a = a'$ implies that $5^b \equiv 5^{b'} (2^l)$ or that $5^{b-b'} \equiv 1 \ (2^l)$. Therefore, $b \equiv b' \ (2^{l-2})$, which yields $b = b'$.

Finally, notice that $(-1)^a 5^b$ raised to the 2^{l-2} power is congruent to 1 mod 2^l. Thus 2^l has no primitive roots if $l \ge 3$.   □

Consider the situation mod 8. 1, 3, 5, and 7 constitute a reduced residue system. We have $5^0 \equiv 1$, $5^1 \equiv 5$, $-5^0 \equiv 7$, and $-5^1 \equiv 3$. Table 1 represents the situation mod 16. The second row contains the least positive residues of the powers of 5, and the third row those of the negative powers of 5.

Table 1

	5^0	5^1	5^2	5^3
+	1	5	9	13
−	15	11	7	3

Theorems 2 and 2′ permit us to give a fairly complete description of the group $U(\mathbb{Z}/n\mathbb{Z})$ for arbitrary n.

Theorem 3. *Let $n = 2^a p_1^{a_1} p_2^{a_2} \cdots p_l^{a_l}$ be the prime decomposition of n. Then*

$$U(\mathbb{Z}/n\mathbb{Z}) \approx U(\mathbb{Z}/2^a\mathbb{Z}) \times U(\mathbb{Z}/p_1^{a_1}\mathbb{Z}) \times \cdots \times U(\mathbb{Z}/p_l^{a_l}\mathbb{Z}).$$

$U(\mathbb{Z}/p_i^{a_i}\mathbb{Z})$ is a cyclic group of order $p_i^{a_i-1}(p_i - 1)$. $U(\mathbb{Z}/2^a\mathbb{Z})$ is cyclic of order 1 and 2 for $a = 1$ and 2, respectively. If $a \ge 3$, then it is the product of two cyclic groups, one of order 2, the other of order 2^{a-2}.

PROOF. Theorems 2, 2′, and Theorem 1′ of Chapter 3.   □

We conclude this section by giving an answer to the question of which integers possess primitive roots.

Proposition 4.1.3. *n possesses primitive roots iff n is of the form 2, 4, p^a, or $2p^a$, where p is an odd prime.*

PROOF. By Theorem 2′ we can assume that $n \ne 2^l, l \ge 3$. If n is not of the given form, it is easy to see that n can be written as a product $m_1 m_2$, where $(m_1, m_2) = 1$ and $m_1, m_2 > 2$. We then have that $\phi(m_1)$ and $\phi(m_2)$ are both even and that $U(\mathbb{Z}/n\mathbb{Z}) \approx U(\mathbb{Z}/m_1\mathbb{Z}) \times U(\mathbb{Z}/m_2\mathbb{Z})$. Both $U(\mathbb{Z}/m_1\mathbb{Z})$ and $U(\mathbb{Z}/m_2\mathbb{Z})$ have elements of order 2, but this shows that $U(\mathbb{Z}/n\mathbb{Z})$ is not cyclic since a cyclic group contains at most one element of order 2. Thus n does not possess primitive roots.

We already know that 2, 4, and p^a possess primitive roots. Since $U(\mathbb{Z}/2p^a\mathbb{Z})$ $\approx U(\mathbb{Z}/2\mathbb{Z}) \times U(\mathbb{Z}/p^a\mathbb{Z}) \approx U(\mathbb{Z}/p^a\mathbb{Z})$ it follows that $U(\mathbb{Z}/2p^a\mathbb{Z})$ is cyclic; i.e., $2p^a$ possesses primitive roots. □

§2 nth Power Residues

Definition. If $m, n \in \mathbb{Z}^+$, $a \in \mathbb{Z}$, and $(a, m) = 1$, then we say that a is an nth *power residue* mod m if $x^n \equiv a\ (m)$ is solvable.

Proposition 4.2.1. *If $m \in \mathbb{Z}^+$ possesses primitive roots and $(a, m) = 1$, then a is an nth power residue mod m iff $a^{\phi(m)/d} \equiv 1\ (m)$, where $d = (n, \phi(m))$.*

PROOF. Let g be a primitive root mod m and $a = g^b$, $x = g^y$. Then the congruence $x^n \equiv a\ (m)$ is equivalent to $g^{ny} \equiv g^b\ (m)$, which in turn is equivalent to $ny \equiv b\ (\phi\ (m))$. The latter congruence is solvable iff $d|b$. Moreover, it is useful to notice that if there is one solution, there are exactly d solutions.

If $d|b$, then $a^{\phi(m)/d} \equiv g^{b\phi(m)/d} \equiv 1\ (m)$. Conversely, if $a^{\phi(m)/d} \equiv 1\ (m)$, then $g^{b\phi(m)/d} \equiv 1\ (m)$, which implies that $\phi(m)$ divides $b\phi(m)/d$ or $d|b$. This proves the result. □

The proof yields the following additional information. If $x^n \equiv a\ (m)$ is solvable, there are exactly $(n, \phi(m))$ solutions.

Now suppose that $m = 2^e p_1^{e_1} \cdots p_l^{e_l}$. Then $x^n \equiv a\ (m)$ is solvable iff the system of congruences

$$x^n \equiv a\ (2^e),\ x^n \equiv a\ (p_1^{e_1}), \ldots, x^n \equiv a\ (p_l^{e_l})$$

is solvable. Since odd prime powers possess primitive roots we may apply Proposition 4.2.1 to the last l congruences. We are reduced to a consideration of the congruence $x^n \equiv a\ (2^e)$. Since 2 and 4 possess primitive roots we may further assume that $e \geq 3$.

Proposition 4.2.2. *Suppose that a is odd, $e \geq 3$, and consider the congruence $x^n \equiv a\ (2^e)$. If n is odd, a solution always exists and it is unique.*

If n is even, a solution exists iff $a \equiv 1\ (4)$, $a^{2^{e-2}/d} \equiv 1\ (2^e)$, where $d = (n, 2^{e-2})$. When a solution exists there are exactly $2d$ solutions.

PROOF. We leave the proof as an exercise. One begins by writing $a \equiv (-1)^s 5^t$ (2^e) and $x \equiv (-1)^y 5^z\ (2^e)$. □

Propositions 4.2.1 and 4.2.2 give a fairly satisfactory answer to the question; When is an integer a an nth power residue mod m? It is possible to go a bit further in some cases.

Proposition 4.2.3. *If p is an odd prime, $p \nmid a$, and $p \nmid n$, then if $x^n \equiv a\ (p)$ is solvable, so is $x^n \equiv a\ (p^e)$ for all $e \geq 1$. All these congruences have the same number of solutions.*

PROOF. If $n = 1$, the assertion is trivial, so we may assume $n \geq 2$. Suppose that $x^n \equiv a\ (p^e)$ is solvable. Let x_0 be a solution and set $x_1 = x_0 + bp^e$. A short computation shows $x_1^n \equiv x_0^n + nbp^e x_0^{n-1}\ (p^{e+1})$. We wish to solve $x_1^n \equiv a\ (p^{e+1})$. This is equivalent to finding an integer b such that $nx_0^{n-1}b \equiv ((a - x_0^n)/p^e)\ (p)$. Notice that $(a - x_0^n)/p^e$ is an integer and that $p \nmid nx_0^{n-1}$. Thus this congruence is uniquely solvable for b, and with this value of b, $x_1^n \equiv a\ (p^{e+1})$.

If $x^n \equiv a\ (p)$ has no solutions, then $x^n \equiv a\ (p^e)$ has no solutions. On the other hand, if $x^n \equiv a\ (p)$ has a solution, so do all the congruences $x^n \equiv a\ (p^e)$, as we have just seen. By the remark following Proposition 4.2.1 the number of solutions to $x^n \equiv a\ (p^e)$ is $(n, \phi\ (p^e))$ provided one solution exists. If $p \nmid n$, it is easy to see that $(n, \phi\ (p)) = (n, \phi\ (p^e))$ for all $e \geq 1$. This concludes the proof. $\qquad\qquad\square$

As usual the result for the powers of 2 is more complicated.

Proposition 4.2.4. *Let 2^l be the highest power of 2 dividing n. Suppose that a is odd and that $x^n \equiv a\ (2^{2l+1})$ is solvable. Then $x^n \equiv a\ (2^e)$ is solvable for all $e \geq 2l + 1$ (and consequently for all $e \geq 1$). Moreover, all these congruences have the same number of solutions.*

PROOF. We leave the proof as an exercise. One begins by assuming that $x^n \equiv a\ (2^m)$, $m \geq 2l + 1$, has a solution x_0. Let $x_1 = x_0 + b2^{m-l}$. One shows, by an appropriate choice of b, that $x_1^n \equiv a\ (2^{m+1})$. $\qquad\qquad\square$

Notice that $x^2 \equiv 5\ (2^2)$ is solvable (for example, $x = 1$) but that $x^2 \equiv 5\ (2^3)$ is not. On the other hand, one can prove easily from the proposition that if $a \equiv 1\ (8)$, then $x^2 \equiv a\ (2^e)$ is solvable for all e and conversely.

NOTES

Lemma 1 and its important consequence, Proposition 4.1.1, are due to J. Lagrange (1768).

Fermat's theorem [that $a^{p-1} \equiv 1\ (p)$ if $p \nmid a$] was first proved by Euler. Wilson's theorem was stated by E. Waring and proved by Lagrange.

The important result on the existence of primitive roots modulo a prime was asserted by Euler and, as we have mentioned, was first proved by Gauss. The proofs of this result can be modified to prove the more general assertion that a finite subgroup of the multiplicative group of a field is cyclic, i.e., is generated by one element.

There are a number of interesting conjectures related to primitive roots. The celebrated conjecture of E. Artin asserts that given an integer a that is not a square, and not -1, there are infinitely many primes for which a is a primitive root. In the case $a = 10$ this goes back to Gauss and amounts to asserting the existence of infinitely many primes p such that the period of the decimal expansion of $1/p$ has length $p - 1$. (See Chapter 4 of Rademacher [64] for an introduction to the theory of decimal expansions.) For an excellent survey article devoted to the Artin conjecture and related questions, see Goldstein [35].

Lehmer [54] discovered the following curious result. The first prime of the form $326n^2 + 3$ for which 326 is not a primitive root must be bigger than 10 million. He mentions other results of the same nature. It would be interesting to see what is responsible for this strange behavior.

Given a prime p, what can be said about the size of the smallest positive integer that is a primitive root mod p? This problem has given rise to a lot of research. One contribution, due to L. K. Hua, is that the number in question is less than $2^{m+1}p^{1/2}$, where m is the number of distinct primes dividing $p - 1$. For a discussion of this problem and a good bibliography, see Erdös [31]. For other interesting results and problems see [76] and [12].

There exist many investigations into the existence of sequences of consecutive integers each of which is a kth power modulo p. Consider primes of the form $kt + 1$. A basic result due to A. Brauer asserts that if m is a given positive integer, then for all primes p sufficiently large there are m consecutive integers $r, r + 1, \ldots, r + m - 1$ all of which are kth powers modulo p. The question of finding the least such r for given p and m is a problem of current interest. For this, and a discussion of other open questions in this area, see the article by Mills [59].

Given a prime p, what can be said about the size of the smallest positive integer that is a nonsquare modulo p? An interesting conjecture is the following: For a given n the integer in question is smaller than $\sqrt[n]{p}$ for all sufficiently large p. For more discussion, see P. Erdös [31] and Chapter 3 of Chowla [18].

Finally, we mention that an analog of the Artin conjecture on primitive roots has actually been proved in the ring $k[x]$ by H. Bilharz [8]. Bilharz proved his theorem under the assumption that the Riemann hypothesis holds for the so-called congruence zeta function (see Chapter 11). This was actually proved several years later by A. Weil. In recent years C. Hooley was able to prove that Artin's orginal conjecture was correct under the assumption that the extended Riemann hypothesis holds in algebraic number fields [46]. For a discussion of the classical Riemann hypothesis and its consequences, see Chowla [18]. No one at present seems to have the slightest idea as to how to prove the Riemann hypothesis for number fields so that it seems clear that Hooley is not about to have the same good luck that Bilharz enjoyed.

EXERCISES

1. Show that 2 is a primitive root modulo 29.

2. Compute all primitive roots for $p = 11, 13, 17$, and 19.

3. Suppose that a is a primitive root modulo p^n, p an odd prime. Show that a is a primitive root modulo p.

4. Consider a prime p of the form $4t + 1$. Show that a is a primitive root modulo p iff $-a$ is a primitive root modulo p.

5. Consider a prime p of the form $4t + 3$. Show that a is a primitive root modulo p iff $-a$ has order $(p - 1)/2$.

6. If $p = 2^n + 1$ is a Fermat prime, show that 3 is a primitive root modulo p.

7. Suppose that p is a prime of the form $8t + 3$ and that $q = (p - 1)/2$ is also a prime. Show that 2 is a primitive root modulo p.

8. Let p be an odd prime. Show that a is a primitive root module p iff $a^{(p-1)/q} \not\equiv 1\ (p)$ for all prime divisors q of $p - 1$.

9. Show that the product of all the primitive roots modulo p is congruent to $(-1)^{\phi(p-1)}$ modulo p.

10. Show that the sum of all the primitive roots modulo p is congruent to $\mu(p - 1)$ modulo p.

11. Prove that $1^k + 2^k + \cdots + (p - 1)^k \equiv 0\ (p)$ if $p - 1 \nmid k$ and $-1\ (p)$ if $p - 1 | k$.

12. Use the existence of a primitive root to give another proof of Wilson's theorem $(p - 1)! \equiv -1\ (p)$.

13. Let G be a finite cyclic group and $g \in G$ a generator. Show that all the other generators are of the form g^k, where $(k, n) = 1$, n being the order of G.

14. Let A be a finite abelian group and $a, b \in A$ elements of order m and n, respectively. If $(m, n) = 1$, prove that ab has order mn.

15. Let K be a field and $G \subseteq K^*$ a finite subgroup of the multiplicative group of K. Extend the arguments used in the proof of Theorem 1 to show that G is cyclic.

16. Calculate the solutions to $x^3 \equiv 1\ (19)$ and $x^4 \equiv 1\ (17)$.

17. Use the fact that 2 is a primitive root modulo 29 to find the seven solutions to $x^7 \equiv 1\ (29)$.

18. Solve the congruence $1 + x + x^2 + \cdots + x^6 \equiv 0\ (29)$.

19. Determine the numbers a such that $x^3 \equiv a\ (p)$ is solvable for $p = 7, 11$, and 13.

20. Let p be a prime and d a divisor of $p - 1$. Show that the dth powers form a subgroup of $U(\mathbb{Z}/p\mathbb{Z})$ of order $(p - 1)/d$. Calculate this subgroup for $p = 11, d = 5$; $p = 17$, $d = 4$; $p = 19, d = 6$.

21. If g is a primitive root modulo p and $d | p - 1$, show that $g^{(p-1)/d}$ has order d. Show also that a is a dth power iff $a \equiv g^{kd}\ (p)$ for some k. Do Exercises 16–20 making use of these observations.

22. If a has order 3 modulo p, show that $1 + a$ has order 6.

23. Show that $x^2 \equiv -1 \ (p)$ has a solution iff $p \equiv 1 \ (4)$ and that $x^4 \equiv -1 \ (p)$ has a solution iff $p \equiv 1 \ (8)$.

24. Show that $ax^m + by^n \equiv c \ (p)$ has the same number of solutions as $ax^{m'} + by^{n'} \equiv c(p)$, where $m' = (m, p - 1)$ and $n' = (n, p - 1)$.

25. Prove Propositions 4.2.2 and 4.2.4.

Chapter 5

Quadratic Reciprocity

If p is a prime, the discussion of the congruence $x^2 \equiv a\ (p)$ is fairly easy. It is solvable iff $a^{(p-1)/2} \equiv 1\ (p)$. With this fact in hand a complete analysis is a simple matter. However, if the question is turned around, the problem is much more difficult. Suppose that a is an integer. For which primes p is the congruence $x^2 \equiv a\ (p)$ solvable? The answer is provided by the law of quadratic reciprocity. This law was formulated by Euler and A. M. Legendre but Gauss was the first to provide a complete proof. Gauss was extremely proud of this result. He called it the Theorema Aureum, *the golden theorem.*

§1 Quadratic Residues

If $(a, m) = 1$, a is called a quadratic residue mod m if the congruence $x^2 \equiv a\ (m)$ has a solution. Otherwise a is called a quadratic nonresidue mod m.

For example, 2 is a quadratic residue mod 7, but 3 is not. In fact, $1^2, 2^2$, $3^2, 4^2, 5^2$, and 6^2 are congruent to 1, 4, 2, 2, 4, and 1, respectively. Thus 1, 2, and 4 are quadratic residues, and 3, 5, and 6 are not.

Given any fixed positive integer m it is possible to determine the quadratic residues by simply listing the positive integers less than and prime to m, squaring them, and reducing mod m. This is what we have just done for $m = 7$.

The following proposition gives a less tedious way of deciding when a given integer is a quadratic residue mod m.

Proposition 5.1.1. *Let $m = 2^e p_1^{e_1} \cdots p_l^{e_l}$ be the prime decomposition of m, and suppose that $(a, m) = 1$. Then $x^2 \equiv a\ (m)$ is solvable iff the following conditions are satisfied:*

(a) *If $e = 2$, then $a \equiv 1\ (4)$.*
   If $e \geq 3$, then $a \equiv 1\ (8)$.
(b) *For each i we have $a^{(p_i-1)/2} \equiv 1\ (p_i)$.*

PROOF. By the Chinese Remainder Theorem the congruence $x^2 \equiv a\ (m)$ is equivalent to the system $x^2 \equiv a\ (2^e)$, $x^2 \equiv a\ (p_1^{e_1}), \ldots, x^2 \equiv a\ (p_l^{a_l})$.

Consider $x^2 \equiv a \, (2^e)$. 1 is the only quadratic residue mod 4, and 1 is the only quadratic residue mod 8. Thus we have solvability iff $a \equiv 1 \, (4)$ if $e = 2$ and $a \equiv 1 \, (8)$ if $e = 3$. A direct application of Proposition 4.2.4 shows that $x^2 \equiv a \, (8)$ is solvable iff $x^2 \equiv a \, (2^e)$ is solvable for all $e \geq 3$.

Now consider $x^2 \equiv a \, (p_i^{e_i})$. Since $(2, p_i) = 1$ it follows from Proposition 4.2.3 that this congruence is solvable iff $x^2 \equiv a \, (p_i)$ is solvable. To this congruence apply Proposition 4.2.1 with $n = 2$, $m = p$, and $d = (n, \phi \, (m)) = (2, p - 1) = 2$. We obtain that $x^2 \equiv a \, (p_i)$ is solvable iff $a^{(p_i - 1)/2} \equiv 1 \, (p_i)$. $\square$

This result reduces questions about quadratic residues to the corresponding questions for prime moduli. In what follows p will denote an odd prime.

Definition. The symbol (a/p) will have the value 1 if a is a quadratic residue mod p, -1 if a is a quadratic nonresidue mod p, and zero if $p | a$. (a/p) is called the *Legendre symbol*.

The Legendre symbol is an extremely convenient device for discussing quadratic residues. We shall list some of its properties.

Proposition 5.1.2.

(a) $a^{(p-1)/2} \equiv (a/p) \, (p)$.
(b) $(ab/p) = (a/p)(b/p)$.
(c) *If $a \equiv b \, (p)$, then $(a/p) = (b/p)$.*

PROOF. If p divides a or b, all three assertions are trivial. Assume that $p \nmid a$ and that $p \nmid b$.

We know that $a^{p-1} \equiv 1 \, (p)$; thus $(a^{(p-1)/2} + 1)(a^{(p-1)/2} - 1) = a^{p-1} - 1 \equiv 0 \, (p)$. It follows that $a^{(p-1)/2} \equiv \pm 1 \, (p)$. By Proposition 5.1.1, $a^{(p-1)/2} \equiv 1 \, (p)$ iff a is a quadratic residue mod p. This proves part (a).

To prove part (b) we apply part (a). $(ab)^{(p-1)/2} \equiv (ab/p) \, (p)$ and $(ab)^{(p-1)/2} = a^{(p-1)/2} b^{(p-1)/2} \equiv (a/p)(b/p) \, (p)$. Thus $(ab/p) \equiv (a/p)(b/p) \, (p)$, which implies that $(ab/p) = (a/p)(b/p)$.

Part (c) is obvious from the definition. $\square$

Corollary 1. *There are as many residues as nonresidues* mod p.[*]

PROOF. $a^{(p-1)/2} \equiv 1 \, (p)$ has $(p - 1)/2$ solutions. Thus there are $(p - 1)/2$ residues and $p - 1 - ((p - 1)/2) = (p - 1)/2$ nonresidues. $\square$

Corollary 2. *The product of two residues is a residue, the product of two nonresidues is a residue, and the product of a residue and a nonresidue is a nonresidue.*

PROOF. This all follows easily from part (b). $\square$

---

[*] In the remainder of this chapter "residues" and "nonresidues" refer to quadratic residues and quadratic nonresidues.

Corollary 3. $(-1)^{(p-1)/2} = (-1/p)$.

PROOF. Substitute $a = -1$ in part (a).   ☐

Corollary 3 is particularly interesting. Every odd integer has the form $4k + 1$ or $4k + 3$. Using this one can restate Corollary 3 as follows: $x^2 \equiv -1$ (p) has a solution iff p is of the form $4k + 1$. Thus -1 is a residue of the primes 5, 13, 17, 29, ... and a nonresidue of the primes 3, 7, 11, 19, ... The reader should check some of these assertions numerically.

One is led by this result to ask a more general question. If a is an integer, for which primes p is a a quadratic residue mod p? The answer to this question is provided by the law of quadratic reciprocity to whose statement and proof we shall soon devote a great deal of attention.

Corollary 3 enables us to prove that there are infinitely many primes of the form $4k + 1$. Suppose that $p_1, p_2, \ldots, p_m$ are a finite set of such primes and consider $(2p_1 p_2 \cdots p_m)^2 + 1$. Suppose that p divides this integer. -1 will then be a quadratic residue mod p and thus p will be of the form $4k + 1$. p is not among the p_i since $(2p_1 p_2 \cdots p_m)^2 + 1$ leaves a remainder of 1 when divided by p_i. We have shown that every finite set of primes of the form $4k + 1$ excludes some primes of that form. Thus the set of such primes is infinite.

To return to the theory of quadratic residues, we are now going to introduce another characterization of the symbol (a/p) due to Gauss.

Consider $S = \{-(p - 1)/2, -(p - 3)/2, \ldots, -1, 1, 2, \ldots, (p - 1)/2\}$. This is called the set of least residues mod p. If $p \nmid a$, let μ be the number of negative least residues of the integers $a, 2a, 3a, \ldots, ((p - 1)/2)a$. For example, let $p = 7$ and $a = 4$. Then $(p - 1)/2 = 3$, and $1 \cdot 4$, $2 \cdot 4$, and $3 \cdot 4$ are congruent to -3, 1, and -2, respectively. Thus in this case $\mu = 2$.

Lemma (Gauss' Lemma). $(a/p) = (-1)^\mu$.

PROOF. Let $\pm m_l$ be the least residue of la, where m_l is positive. As l ranges between 1 and $(p - 1)/2$, μ is clearly the number of minus signs that occur in this way. We claim that $m_l \neq m_k$ if $l \neq k$ and $1 \leq l, k \leq (p - 1)/2$. For, if $m_l = m_k$, then $la \equiv \pm ka$ (p), and since $p \nmid a$ this implies that $l \pm k \equiv 0$ (p). The latter congruence is impossible since $l \neq k$ and $|l \pm k| \leq |l| + |k| \leq p - 1$. It follows that the sets $\{1, 2, \ldots, (p - 1)/2\}$ and $\{m_1, m_2, \ldots, m_{(p-1)/2}\}$ coincide. Multiply the congruences $1 \cdot a \equiv \pm m_1$ (p), $2 \cdot a \equiv \pm m_2$ (p), ..., $((p - 1)/2)a \equiv \pm m_{(p-1)/2}$ (p). We obtain

$$\left(\frac{p - 1}{2}\right)! \, a^{(p-1)/2} \equiv (-1)^\mu \left(\frac{p - 1}{2}\right)! \, (p).$$

This yields $a^{(p-1)/2} \equiv (-1)^\mu$ (p). By Proposition 5.1.2, $a^{(p-1)/2} \equiv (a/p)$ (p). The result follows.   ☐

Gauss's lemma is an extremely powerful tool. We shall base our first proof of the quadratic reciprocity law on it. Before getting to that, however, we can use it immediately to get a characterization of those primes for which 2 is a quadratic residue.

Proposition 5.1.3. *2 is a quadratic residue of primes of the form $8k + 1$ and $8k + 7$. 2 is a quadratic nonresidue of primes of the form $8k + 3$ and $8k + 5$. This information is summarized in the formula*

$$\left(\frac{2}{p}\right) = (-1)^{(p^2-1)/8}.$$

PROOF. We leave to the reader the task of showing that the formula is equivalent to the first two assertions.

Let p be an odd prime (as usual) and notice that the number μ is equal to the number of elements of the set $2 \cdot 1, 2 \cdot 2, \ldots, 2 \cdot (p-1)/2$ that exceed $(p-1)/2$. Let m be determined by the two conditions $2m \le (p-1)/2$ and $2(m+1) > (p-1)/2$. Then $\mu = ((p-1)/2) - m$.

If $p = 8k + 1$, then $(p-1)/2 = 4k$ and $m = 2k$. Thus $\mu = 4k - 2k = 2k$ is even and $(2/p) = 1$.

If $p = 8k + 7$, then $(p-1)/2 = 4k + 3$, $m = 2k + 1$, and $\mu = 4k + 3 - (2k+1) = 2k + 2$ is even. Thus $(2/p) = 1$ in this case as well.

If $p = 8k + 3$, then $(p-1)/2 = 4k + 1$, $m = 2k$, and $\mu = 4k + 1 - 2k = 2k + 1$ is odd. Thus $(2/p) = -1$.

Finally, if $p = 8k + 5$, then $(p-1)/2 = 4k + 2$, $m = 2k + 1$, and $\mu = 4k + 2 - (2k+1) = 2k + 1$ is odd. Thus $(2/p) = -1$ and we are done. $\square$

As an example, consider $p = 7$ and $p = 17$. These primes are congruent to 7 and 1, respectively, mod 8, and indeed $3^2 \equiv 2$ (7) and $6^2 \equiv 2$ (17). On the other hand, $p = 19$ and $p = 5$ are congruent to 3 and 5, respectively, and it is easily checked numerically that 2 is a quadratic nonresidue for both primes.

One can use Proposition 5.1.3 to prove that there are infinitely many primes of the form $8k + 7$. Let $p_1, \ldots, p_m$ be a finite collection of such primes, and consider $(4p_1 p_2 \cdots p_m)^2 - 2$. The odd prime divisors of this number have the form $8k + 1$ or $8k + 7$, since for such prime divisors 2 is a quadratic residue. Not all the odd prime divisors can have the form $8k + 1$ (prove it). Let p be a prime divisor of the form $8k + 7$. Then p is not in the set $\{p_1, p_2, \ldots, p_n\}$ and we are done.

§2 Law of Quadratic Reciprocity

Theorem 1 (Law of Quadratic Reciprocity). *Let p and q be odd primes. Then*

(a) $(-1/p) = (-1)^{(p-1)/2}$.
(b) $(2/p) = (-1)^{(p^2-1)/8}$.
(c) $(p/q)(q/p) = (-1)^{((p-1)/2)((q-1)/2)}$.

We are going to postpone the proof until Section 3. In Chapter 6 we shall prove the theorem once again from a different standpoint, and also indicate something of its history. It is among the deepest and most beautiful results of elementary number theory and the beginning of a line of reciprocity theorems that culminate in the very general Artin reciprocity law, perhaps the most impressive theorem in all number theory. It would take us far outside the compass of this book to even state the Artin reciprocity law, but in Chapter 9 we shall state and prove the laws of cubic and biquadratic reciprocity.

Parts (a) and (b) of Theorem 1 have already been proven and some of their consequences discussed. Let us turn our attention to part (c).

If either p or q are of the form $4k + 1$, then $((p - 1)/2)((q - 1)/2) \equiv 0 \ (2)$. If both p and q are of the form $4k + 3$, then $((p - 1)/2)((q - 1)/2) \equiv 1 \ (2)$. This permits us to restate part (c) as follows:

(1) If either p or q is of the form $4k + 1$, then p is a quadratic residue mod q iff q is a quadratic residue mod p.
(2) If both p and q are of the form $4k + 3$, then p is a quadratic residue mod q iff q is a quadratic nonresidue mod p.

As a first application of quadratic reciprocity we show how, in conjunction with Proposition 5.1.2, it can be used in numerical computations of the Legendre symbol. A single example should suffice to illustrate the method.

We propose to calculate $(79/101)$. Since $101 \equiv 1 \ (4)$ we have $(79/101) = (101/79) = (22/79)$. The last step follows from $101 \equiv 22 \ (79)$. Further, $(22/79) = (2/79)(11/79)$. Now $79 \equiv 7 \ (8)$. Thus $(2/79) = 1$. Since both 11 and 79 are congruent to 3 mod 4 we have $(11/79) = -(79/11) = -(2/11)$. Finally $11 \equiv 3 \ (8)$ implies that $(2/11) = -1$. Therefore $(79/101) = 1$; i.e., 79 is a quadratic residue mod 101. Indeed, $33^2 \equiv 79 \ (101)$.

The next application is perhaps more significant. We noticed earlier that -1 is a quadratic residue of primes of the form $4k + 1$ and that 2 is a quadratic residue of primes that are either of the form $8k + 1$ or $8k + 7$. If a is an arbitrary integer, for what primes p is a a quadratic residue mod p? We are now in a position to give an answer. To begin with, we consider the case where $a = q$, an odd prime.

Theorem 2. *Let q be an odd prime.*

(a) *If $q \equiv 1 \ (4)$, then q is a quadratic residue mod p iff $p \equiv r \ (q)$, where r is a quadratic residue mod q.*
(b) *If $q \equiv 3 \ (4)$, then q is a quadratic residue mod p iff $p \equiv \pm b^2 \ (4q)$, where b is an odd integer prime to q.*

PROOF. If $q \equiv 1 \ (4)$, then by Theorem 1 we have $(q/p) = (p/q)$. Part (a) is thus clear.

If $q \equiv 3 \ (4)$, Theorem 1 yields $(q/p) = (-1)^{(p-1)/2}(p/q)$. Assume first that $p \equiv \pm b^2 \ (4q)$, where b is odd. If we take the plus sign, we get $p \equiv b^2 \equiv 1 \ (4)$ and $p \equiv b^2 \ (q)$. Thus $(-1)^{(p-1)/2} = 1$ and $(p/q) = 1$, giving $(q/p) = 1$. If we take the minus sign, then $p \equiv -b^2 \equiv -1 \equiv 3 \, (4)$ and $p \equiv -b^2 \, (q)$. The first congruence shows that $(-1)^{(p-1)/2} = -1$. The second shows that $(p/q) = (-b^2/q) = (-1/q)(b/q)^2 = (-1/q) = -1$ since $q \equiv 3 \, (4)$. Once again we have $(q/p) = 1$.

To go the other way, assume that $(q/p) = 1$. We have two cases to deal with:

(1) $(-1)^{(p-1)/2} = -1$ and $(p/q) = -1$.
(2) $(-1)^{(p-1)/2} = 1$ and $(p/q) = 1$.

In case 2 we have $p \equiv b^2 \, (q)$ and $p \equiv 1 \, (4)$. b can be assumed to be odd since if it is even we can use $b' = b + q$ instead. If b is odd, then $b^2 \equiv 1 \, (4)$ and $p \equiv b^2 \, (4)$ and thus $p \equiv b^2 \, (4q)$, as required.

In case 1 we have $p \equiv 3 \, (4)$ and $p \equiv -b^2 \, (q)$. The last congruence follows since $q \equiv 3 \, (4)$ implies that every nonresidue is the negative of a residue (prove it). Again, we may assume that b is odd. In that case $-b^2 \equiv 3 \, (4)$ so $p \equiv -b^2 \, (4)$ and $p \equiv -b^2 \, (4q)$. This concludes the proof. $\qquad\square$

Take $q = 3$ as a first illustration. By part (b) of Theorem 2 we must find the residues mod 12 of the squares of odd integers prime to 3. 1^2, 5^2, 7^2, and 11^2 are all congruent to 1. Thus 3 is a quadratic residue of primes p congruent to $\pm 1 \, (12)$ and a quadratic nonresidue of primes congruent to $\pm 5 \, (12)$.

Next consider $q = 5$. Since $5 \equiv 1 \, (4)$ we are in the simpler part (a) of Theorem 2. 1 and 4 are the residues mod 5, and 2 and 3 the nonresidues. Thus 5 is a residue of primes congruent to 1 or 4 mod 5 and a nonresidue of primes congruent to 2 or 3 mod 5.

"Numbers congruent to b mod m" and "numbers of the form $mk + b$" are shorthand expressions describing the set $\{b, b \pm m, b \pm 2m, \ldots\}$. This set is an arithmetic progression with initial term b and difference m. In our investigations so far we have seen that the answer to the question for which primes p is a a quadratic residue has been for those primes p that occur in a certain fixed, finite number of arithmetic progressions. This situation is entirely general. Instead of stating this result as a theorem (the statement would be very complicated) we shall work out a few numerical examples.

For $a = -3$, $(-3/p) = (-1/p)(3/p)$. Thus -3 is a quadratic residue mod p if either $(-1/p) = 1$ and $(3/p) = 1$ or $(-1/p) = -1$ and $(3/p) = -1$.

By our previous results the first case obtains when $p \equiv 1 \, (4)$ and $p \equiv \pm 1 \, (12)$. If $p \equiv -1 \, (12)$, then $p \equiv -1 \, (4)$. The only primes that satisfy both congruences are $\equiv 1 \, (12)$.

In the second case $p \equiv 3 \, (4)$ and $p \equiv \pm 5 \, (12)$. If $p \equiv 5 \, (12)$, then $p \equiv 1 \, (4)$. Thus the only primes that satisfy both these congruences are $\equiv -5 \, (12)$.

Summarizing, -3 is a quadratic residue mod p iff p is congruent to 1 or -5 mod 12.

Now consider $a = 6$. Since $(6/p) = (2/p)(3/p)$ we again have two cases: $(2/p) = 1$ and $(3/p) = 1$ or $(2/p) = -1$ and $(3/p) = -1$. The first case holds if $p \equiv 1, 7 \, (8)$ and $p \equiv 1, 11 \, (12)$. The only two pairs of congruences that are compatible are $p \equiv 1$ (8) and $p \equiv 1$ (12), and $p \equiv 7$ (8) and $p = 11$ (12). By standard techniques (see Chapter 3) the primes satisfying these congruences are congruent to 1 or 23 mod 24.

In the second case we have to consider $p \equiv 3, 5$ (8) and $p \equiv 5, 7$ (12). Separating these into four pairs of congruences we see that the only solutions are congruent to 5 and 19 mod 24.

Summarizing, 6 is a quadratic residue mod p iff $p \equiv 1, 5, 19, 23$ (24).

As a numerical check we see for the primes 73, 5, 19, and 23 that $15^2 \equiv 6$ (73), $1^2 \equiv 6$ (5), $5^2 \equiv 6$ (19), and $11^2 \equiv 6$ (23).

As a final application of the quadratic reciprocity law we investigate the question; if a is a quadratic residue mod all primes p not dividing a, what can be said about a? If a is a square, it is a residue for all primes not dividing a. It turns out that the converse of this statement is true as well. In fact, we shall soon prove an even stronger result. First, however, it is necessary to define and investigate briefly a new symbol.

Definition. Let b be an odd, positive integer and a any integer. Let $b = p_1 p_2 \cdots p_m$, where the p_i are (not necessarily distinct) primes. The symbol (a/b) defined by

$$\left(\frac{a}{b}\right) = \left(\frac{a}{p_1}\right)\left(\frac{a}{p_2}\right) \cdots \left(\frac{a}{p_m}\right)$$

is called the *Jacobi symbol.*

The Jacobi symbol has properties that are remarkably similar to the Legendre symbol, which it generalizes. A word of caution is useful. (a/b) may equal 1 without a being a quadratic residue mod b. For example, $(2/15) = (2/3)(2/5) = (-1)(-1) = 1$, but 2 is not a quadratic residue mod 15. It is true, however, that if $(a/b) = -1$, then a is a quadratic nonresidue mod b.

Proposition 5.2.1.

(a) $(a_1/b) = (a_2/b)$ if $a_1 \equiv a_2$ (b).
(b) $(a_1 a_2/b) = (a_1/b)(a_2/b)$.
(c) $(a/b_1 b_2) = (a/b_1)(a/b_2)$.

PROOF. Parts (a) and (b) are immediate from the corresponding properties of the Legendre symbol. Part (c) is obvious from the definition.   □

Lemma. *Let r and s be odd integers. Then*

(a) $(rs - 1)/2 \equiv ((r - 1)/2) + ((s - 1)/2)$ (2).
(b) $(r^2 s^2 - 1)/8 \equiv ((r^2 - 1)/8) + ((s^2 - 1)/8)$ (2).

PROOF. Since $(r - 1)(s - 1) \equiv 0$ (4) we have $rs - 1 \equiv (r - 1) + (s - 1)$ (4). Part (a) follows by dividing by 2.

$r^2 - 1$ and $s^2 - 1$ are both divisible by 4. Thus $(r^2 - 1)(s^2 - 1) \equiv 0 \ (16)$ and $r^2 s^2 - 1 \equiv (r^2 - 1) + (s^2 - 1) \ (16)$. Part (b) follows upon dividing by 8.   □

Corollary. *Let $r_1, r_2, \ldots, r_m$ be odd integers. Then*

(a) $\sum_{i=1}^{m} (r_i - 1)/2 \equiv (r_1 r_2 \cdots r_m - 1)/2 \ (2)$.
(b) $\sum_{i=1}^{m} (r_i^2 - 1)/8 \equiv (r_1^2 r_2^2 \cdots r_m^2 - 1)/8 \ (2)$.

PROOF. The proof is a simple induction on m, using the lemma.   □

Proposition 5.2.2.

(a) $(-1/b) = (-1)^{(b-1)/2}$.
(b) $(2/b) = (-1)^{(b^2-1)/8}$.
(c) *If a is odd and positive as well as b, then*

$$\left(\frac{a}{b}\right)\left(\frac{b}{a}\right) = (-1)^{((a-1)/2)((b-1)/2)}.$$

PROOF.

$$(-1/b) = (-1/p_1)(-1/p_2) \cdots (-1/p_m) = (-1)^{(p_1-1)/2} \cdots (-1)^{(p_m-1)/2}$$
$$= (-1)^{\sum (p_i-1)/2}.$$

By the lemma $\sum (p_i - 1)/2 \equiv (p_1 p_2 \cdots p_m - 1)/2 \equiv (b-1)/2 \ (2)$. This proves part (a).

Part (b) is proved in exactly the same way.

Now if $a = q_1 q_2 \cdots q_l$, then

$$\left(\frac{a}{b}\right)\left(\frac{b}{a}\right) = \prod_i \prod_j \left(\frac{q_i}{p_j}\right)\left(\frac{p_j}{q_i}\right) = (-1)^{\sum_i \sum_j ((q_i-1)/2)((p_j-1)/2)}.$$

The product and sum range over $1 \leq i \leq l$ and $1 \leq j \leq m$. Again using the lemma we have

$$\sum_i \sum_j \left(\frac{(p_j-1)}{2}\right)\left(\frac{(q_i-1)}{2}\right) \equiv \frac{(a-1)}{2} \sum_i \frac{(p_j-1)}{2}$$
$$\equiv \left(\frac{(a-1)}{2}\right)\left(\frac{(b-1)}{2}\right) (2).$$

This proves part (c).   □

The Jacobi symbol has many uses. For one thing, it is a convenient aid for computing the Legendre symbol. We now use it to prove the following theorem.

Theorem 3. *Let a be a nonsquare integer. Then there are infinitely many primes p for which a is a quadratic nonresidue.*

PROOF. It is easily seen that we may assume that a is square-free. Let $a = 2^e q_1 q_2$ $\cdots q_n$, where the q_i are distinct odd primes and $e = 0$ or 1. The case $a = 2$ has to be dealt with separately. We shall assume to begin with that $n \geq 1$, i.e., that a is divisible by an odd prime.

Let $l_1, l_2, \ldots, l_k$ be a finite set of odd primes not including any q_i. Let s be any nonresidue mod q_n, and find a simultaneous solution to the congruences

$$x \equiv 1 \ (l_i), \qquad i = 1, \ldots, k,$$

$$x \equiv 1 \ (8),$$

$$x \equiv 1 \ (q_i), \qquad i = 1, 2, \ldots, n - 1.$$

$$x \equiv s \ (q_n),$$

Call the solution b. b is odd. Suppose that $b = p_1 p_2 \cdots p_m$ is its prime decomposition. Since $b \equiv 1 \ (8)$ we have $(2/b) = 1$ and $(q_i/b) = (b/q_i)$ by Proposition 5.2.2. Thus $(a/b) = (2/b)^e(q_1/b) \cdots (q_{n-1}/b)(q_n/b) = (b/q_1) \cdots (b/q_{n-1})(b/q_n) = (1/q_1) \cdots (1/q_{n-1})(s/q_n) = -1$.

On the other hand, by the definition of (a/b), we have $(a/b) = (a/p_1)(a/p_2)$ $\cdots (a/p_m)$. It follows that $(a/p_i) = -1$ for some i.

Notice that l_j does not divide b. Thus $p_i \notin \{l_1, l_2, \ldots, l_k\}$.

To summarize, if a is a nonsquare, divisible by an odd prime, we have found a prime p, outside a given finite set of primes $\{2, l_1, l_2, \ldots, l_k\}$, such that $(a/p) = -1$. This proves Theorem 3 in this case.

It remains to consider the case $a = 2$. Let $l_1, \ldots, l_k$ be a finite set of primes, excluding 3, for which $(2/l_i) = -1$. Let $b = 8l_1 l_2 \cdots l_k + 3$. b is not divisible by 3 or any l_i. Since $b \equiv 3 \ (8)$ we have $(2/b) = (-1)^{(b^2-1)/8} = -1$. Suppose that $b = p_1 p_2 \cdots p_m$ is the prime decomposition of b. Then, as before, we see that $(2/p_i) = -1$ for some i. $p_i \notin \{3, l_1, l_2, \ldots, l_k\}$. This proves Theorem 3 for $a = 2$. $\qquad \square$

§3 A Proof of the Law of Quadratic Reciprocity

Gauss found eight separate proofs for the law of quadratic reciprocity. There are over a hundred now in existence. Of course, they are not all essentially different. Many just differ in small details from others. We shall present an ingenious proof due to Eisenstein. For a somewhat more elementary and standard proof, see [61].

A complex number ζ is called an nth root of unity if $\zeta^n = 1$ for some integer $n > 0$. If n is the least integer with this property, then ζ is called a primitive nth root of unity.

The nth roots of unity are $1, e^{2\pi i/n}, e^{(2\pi i/n)2}, \ldots, e^{(2\pi i/n)(n-1)}$. Among these the primitive nth roots of unity are $e^{(2\pi i/n)k}$, where $(k, n) = 1$.

If ζ is an nth root of unity and $m \equiv l \ (n)$, then $\zeta^m = \zeta^l$. If ζ is a primitive nth root of unity and $\zeta^m = \zeta^l$, then $m \equiv l \ (n)$.

These elementary properties are easy to prove.

Consider the function $f(z) = e^{2\pi i z} - e^{-2\pi i z} = 2i \sin 2\pi z$. This function satisfies $f(z + 1) = f(z)$ and $f(-z) = -f(z)$. Also, its only real zeros are the half integers. In other words, if r is a real number and $2r \notin \mathbb{Z}$, then $f(r) \neq 0$.

We wish to prove an important identity involving $f(z)$, but first we need an algebraic lemma.

Lemma. *If $n > 0$ is odd, we have*

$$x^n - y^n = \prod_{k=0}^{n-1} (\zeta^k x - \zeta^{-k} y), \quad \text{where } \zeta = e^{2\pi i/n}.$$

PROOF. $1, \zeta, \zeta^2, \ldots, \zeta^{n-1}$ are all roots of the polynomial $z^n - 1$. Since there are n of them and they are all distinct we have $z^n - 1 = \prod_{k=0}^{n-1}(z - \zeta^k)$. Let $z = x/y$ and multiply both sides by y^n. We get $x^n - y^n = \prod_{k=0}^{n-1}(x - \zeta^k y)$.

Since n is odd as k runs over a complete system of residues mod n, so does $-2k$. Thus

$$x^n - y^n = \prod_{k=0}^{n-1}(x - \zeta^{-2k}y)$$

$$= \zeta^{-(1+2+\cdots+n-1)} \prod_{k=0}^{n-1}(\zeta^k x - \zeta^{-k}y)$$

$$= \prod_{k=0}^{n-1}(\zeta^k x - \zeta^{-k}y).$$

In the last step we have used the fact that $1 + 2 + 3 + \cdots + (n - 1) = n((n - 1)/2)$ is divisible by n. □

Proposition 5.3.1. *If n is a positive odd integer and $f(z) = e^{2\pi i z} - e^{-2\pi i z}$, then*

$$\frac{f(nz)}{f(z)} = \prod_{k=1}^{(n-1)/2} f\left(z + \frac{k}{n}\right) f\left(z - \frac{k}{n}\right).$$

PROOF. In the lemma, substitute $x = e^{2\pi i z}$ and $y = e^{-2\pi i z}$. We see that

$$f(nz) = \prod_{k=0}^{n-1} f\left(z + \frac{k}{n}\right).$$

Notice that $f(z + k/n) = f(z + k/n - 1) = f(z - (n - k)/n)$. As k goes from $(n + 1)/2$ to $n - 1$, $n - k$ goes from $(n - 1)/2$ to 1. Thus

$$\frac{f(nz)}{f(z)} = \prod_{k=1}^{(n-1)/2} f\left(z + \frac{k}{n}\right) \prod_{k=(n+1)/2}^{n-1} f\left(z + \frac{k}{n}\right)$$

$$= \prod_{k=1}^{(n-1)/2} f\left(z + \frac{k}{n}\right) \prod_{k=(n+1)/2}^{n-1} f\left(z - \frac{n-k}{n}\right)$$

$$= \prod_{k=1}^{(n-1)/2} f\left(z + \frac{k}{n}\right) f\left(z - \frac{k}{n}\right).$$ □

Proposition 5.3.2. *If p is an odd prime, $a \in \mathbb{Z}$, and $p \nmid a$, then*

$$\prod_{l=1}^{(p-1)/2} f\left(\frac{la}{p}\right) = \left(\frac{a}{p}\right) \prod_{l=1}^{(p-1)/2} f\left(\frac{l}{p}\right).$$

PROOF. As in the lemma of Section 1, $la \equiv \pm m_l \ (p)$, where $1 \le m_l \le (p-1)/2$. Thus la/p and $\pm m_l/p$ differ by an integer. This implies that $f(la/p) = f(\pm m_l/p) = \pm f(m_l/p)$.

The result now follows by taking the product of both sides as l goes from 1 to $(p-1)/2$ and applying Gauss' lemma. ☐

We are now in a position to prove the law of quadratic reciprocity. Let p and q be odd primes. Then by Proposition 5.3.2

$$\prod_{l=1}^{(p-1)/2} f\left(\frac{lq}{p}\right) = \left(\frac{q}{p}\right) \prod_{l=1}^{(p-1)/2} f\left(\frac{l}{p}\right).$$

By Proposition 5.3.1

$$\frac{f(ql/p)}{f(l/p)} = \prod_{m=1}^{(q-1)/2} f\left(\frac{l}{p} + \frac{m}{q}\right) f\left(\frac{l}{p} - \frac{m}{q}\right).$$

Putting these two equations together we have

$$\left(\frac{q}{p}\right) = \prod_{m=1}^{(q-1)/2} \prod_{l=1}^{(p-1)/2} f\left(\frac{l}{p} + \frac{m}{q}\right) f\left(\frac{l}{p} - \frac{m}{q}\right).$$

In the same way we find

$$\left(\frac{p}{q}\right) = \prod_{m=1}^{(q-1)/2} \prod_{l=1}^{(p-1)/2} f\left(\frac{m}{q} + \frac{l}{p}\right) f\left(\frac{m}{q} - \frac{l}{p}\right).$$

Since $f(m/q - l/p) = -f(l/p - m/q)$ we see that

$$(-1)^{((p-1)/2)((q-1)/2)}\left(\frac{q}{p}\right) = \left(\frac{p}{q}\right)$$

and therefore that

$$\left(\frac{p}{q}\right)\left(\frac{q}{p}\right) = (-1)^{((p-1)/2)((q-1)/2)}.$$

The proof is complete. ☐

We conclude this chapter by giving an equivalent formulation of the law of quadratic reciprocity.

Proposition 5.3.3. *Let p and q be distinct odd primes and $a \ge 1$ an integer. Then the following assertions are equivalent:*
(a) $(p/q)(q/p) = (-1)^{((p-1)/2)((q-1)/2)}$.
(b) *If $p \equiv \pm q \ (4a)$, $p \nmid a$, then $(a/p) = (a/q)$.*

PROOF. In order to show (a) implies (b) it is enough, by multiplicativity, to show that (b) holds when a is prime. For $a = 2$ the result follows from Proposition 5.1.3. If a is an odd prime then by (a) $(a/p) = (-1)^{((p-1)/2)((a-1)/2)}(p/a)$. If $p \equiv q \ (4a)$ then $(p/a) = (q/a)$ so that

$$\left(\frac{a}{p}\right) = (-1)^{((p-1)/2)((a-1)/2)}\left(\frac{q}{a}\right) = (-1)^{((p-1)/2)((a-1)/2)}(-1)^{((q-1)/2)((a-1)/2)}\left(\frac{a}{q}\right)$$

$$= (-1)^{((a-1)/2)((p+q-2)/2)}\left(\frac{a}{q}\right).$$

But $p \equiv q \ (4a)$ implies $p + q - 2 \equiv 0 \ (4)$ and the result follows. If, on the other hand $p \equiv -q \ (4a)$, a similar calculation shows

$$\left(\frac{a}{p}\right) = (-1)^{((a-1)/2)((p+q)/2)}\left(\frac{a}{q}\right).$$

Since $p + q \equiv 0 \ (4)$ the result also holds in this case.

To show that (b) implies (a) suppose first of all that $p > q$ and $p \equiv q \ (4)$. The $p = q + 4a$, $a \geq 1$. Thus

$$\left(\frac{p}{q}\right) = \left(\frac{q+4a}{q}\right) = \left(\frac{a}{q}\right) = \left(\frac{a}{p}\right) = \left(\frac{4a}{p}\right) = \left(\frac{p-q}{p}\right) = \left(\frac{-q}{p}\right)$$

$$= (-1)^{(p-1)/2}\left(\frac{q}{p}\right).$$

If $p \equiv 1 \ (4)$ then $(p/q) = (q/p)$ which gives (a). If $p \equiv 3 \ (4)$ then $q \equiv 3 \ (4)$ and we obtain $(p/q) = -(q/p)$ which is part (a) in that case. Finally if $p \equiv -q \ (4)$ then, $p + q = 4a$ and

$$\left(\frac{p}{q}\right) = \left(\frac{-q+4a}{q}\right) = \left(\frac{a}{q}\right) = \left(\frac{a}{p}\right) = \left(\frac{4a}{p}\right) = \left(\frac{p+q}{p}\right) = \left(\frac{q}{p}\right).$$

Thus $(p/q) = (q/p)$ which is the assertion of part (a) since in this case at least one of p or q must be congruent to 1 modulo 4. The proof is complete. □

Note that by part (b) of the above proposition we see that if $(r, 4a) = 1$ the quadratic character of a is the same for all primes in the arithmetic progression $r + 4at$, $t \in \mathbb{Z}$. In Chapter 16 we will see that infinitely many such primes exist. Note also that the quadratic character of a prime of the form $r + 4at$ is the same as that for a prime of the form $-r + 4at$. It was in this form that Euler first discovered this most remarkable law.

NOTES

Kronecker has pointed out that the law of quadratic reciprocity follows immediately from a conjecture of Euler contained in the paper "Theoremata

circa divisores numerorum in hac forma $pa^2 \pm qb^2$ contentorum" (1744–
1746). It also appears explicitly in a later paper of Euler entitled "Observa-
tiones circa divisionem quadratorum per numeros primos." Using sufficient
conditions for the solvability of the equation $ax^2 + by^2 + cz^2 = 0$ (see
Proposition 17.3.2). Legendre (1785) was able to prove the result in special
cases. For example, the consideration of $x^2 + py^2 = qz^2$ where $p \equiv 1$ (4)
and $q \equiv 3$ (4) leads to the conclusion that if q is a square modulo p then p
is a square modulo q. The first complete proof of the theorem is due to Gauss
who recorded the date of the proof in his diary on April 8, 1796. During his
lifetime Gauss published six proofs of this remarkable law. The proof we
have given in this chapter is taken from Eisenstein's paper "Applications de
l'Algebre a l'Arithmetique transcendante." Kummer in an historical study
of the laws of reciprocity, refers to this proof as one of the most beautiful of
all the proofs ("... einen der schönsten Beweise dieses von den ausgezeich-
netsten Mathematikern viel bewiesenen Theorems ..."). Replacing the
trigonometric function by certain elliptic functions Eisenstein was able,
without much more difficulty, to prove the laws of cubic and biquadratic
reciprocity as well.

Throughout the nineteenth century various mathematicians including
Cauchy, Eisenstein, Dirichlet, Dedekind, and Kronecker gave new proofs
to the law of quadratic reciprocity. By 1921 there were, according to P.
Bachman, 56 known proofs. Even in recent times new proofs continue to
appear. See, for example, the papers by M. Gerstenhaber [128] and R. Swan
[75]. On the other hand, the first proof of Gauss has been reconsidered
recently by E. Brown [99].

The Jacobi symbol is one generalization of the Legendre symbol. For an
interesting generalization in another direction, see the paper of P. Cartier
[14].

Quadratic reciprocity can be formulated in rings other than $\mathbb{Z}$. Dirichlet
proved such a theorem for the ring of Gaussian integers $\mathbb{Z}[i]$. D. Hilbert was
able to prove that quadratic reciprocity held for any algebraic number field,
a result that was an important stepping stone to class field theory. In another
direction it can be shown that reciprocity holds for the ring $k[x]$, where k is a
finite field. See Artin [2] and Carlitz [10]. This result had already been stated
(though not proved) by Dedekind in 1857.

The generalization of Theorem 3 to higher powers was discovered first by
E. Trost in 1934.* Later it was stated as a conjecture by S. Chowla and subsequently proven by N. C. Ankeny and C. A. Rogers.† They proved that if
$x^n \equiv a$ (p) has a solution for all but a finite number of primes p, then either
$a = b^n$ or $n|8$ and $a = 2^{n/8}b^n$. When n is square-free and $(a, n) = 1$, the result
can be shown to follow from the Eisenstein reciprocity law as was done by
J. Kraft and M. Rosen [211]. Their proof will be given in Chapter 14. See

* Zur Theorie der Potenzreste. *Nieuw Arch. Wiskunde*, **18**, (1934), 15–61.

† A conjecture of Chowla. *Ann. Math.*, **53**, No. 3 (1951), 541–550.

also H. Flanders [134] where the result is generalized to the case of algebraic number fields and algebraic function fields of one variable over a finite field.

EXERCISES

1. Use Gauss' lemma to determine $(\frac{5}{7})$, $(\frac{3}{11})$, $(\frac{6}{13})$, and $(-1/p)$.

2. Show that the number of solutions to $x^2 \equiv a$ (p) is given by $1 + (a/p)$.

3. Suppose that $p \nmid a$. Show that the number of solutions to $ax^2 + bx + c \equiv 0$ (p) is given by $1 + ((b^2 - 4ac)/p)$.

4. Prove that $\sum_{a=1}^{p-1} (a/p) = 0$.

5. Prove that $\sum_{x=0}^{p-1}((ax + b)/p) = 0$ provided that $p \nmid a$.

6. Show that the number of solutions to $x^2 - y^2 \equiv a$ (p) is given by

$$\sum_{y=0}^{p-1} (1 + ((y^2 + a)/p)).$$

7. By calculating directly show that the number of solutions to $x^2 - y^2 \equiv a$ (p) is $p - 1$ if $p \nmid a$ and $2p - 1$ if $p \mid a$. (*Hint:* Use the change of variables $u = x + y$, $v = x - y$.)

8. Combining the results of Exercises 6 and 7 show that

$$\sum_{y=0}^{p-1} \left(\frac{y^2 + a}{p}\right) = \begin{cases} -1, & \text{if } p \nmid a, \\ p - 1, & \text{if } p \mid a. \end{cases}$$

9. Prove that $1^2 3^2 5^2 \cdots (p - 2)^2 \equiv (-1)^{(p+1)/2}$ (p) by using Wilson's theorem.

10. Let $r_1, r_2, \ldots, r_{(p-1)/2}$ be the quadratic residues between 1 and p. Show that their product is congruent to 1 (p) if $p \equiv 3$ (4) and congruent to -1 (p) if $p \equiv 1$ (4).

11. Suppose that $p \equiv 3$ (4) and that $q \equiv 2p + 1$ is also prime. Prove that $2^p - 1$ is not prime. (*Hint:* Use the quadratic character of 2 to show that $q | 2^p - 1$.) One must assume that $p > 3$.

12. Let $f(x) \in \mathbb{Z}[x]$. We say that a prime p divides $f(x)$ if there is an integer n such that $p | f(n)$. Describe the prime divisors of $x^2 + 1$ and $x^2 - 2$.

13. Show that any prime divisor of $x^4 - x^2 + 1$ is congruent to 1 modulo 12.

14. Use the fact that $U(\mathbb{Z}/p\mathbb{Z})$ is cyclic to give a direct proof that $(-3/p) = 1$ when $p \equiv 1$ (3). [*Hint:* There is a ρ in $U(\mathbb{Z}/p\mathbb{Z})$ of order 3. Show that $(2\rho + 1)^2 = -3$.]

15. If $p \equiv 1$ (5), show directly that $(5/p) = 1$ by the method of Exercise 14. [*Hint:* Let ρ be an element of $U(\mathbb{Z}/p\mathbb{Z})$ or order 5. Show that $(\rho + \rho^4)^2 + (\rho + \rho^4) - \bar{1} = \bar{0}$, etc.]

16. Using quadratic reciprocity find the primes for which 7 is a quadratic residue. Do the same for 15.

17. Supply the details to the proof of Proposition 5.2.1 and to the corollary to the lemma following it.

18. Let D be a square-free integer that is also odd and positive. Show that there is an integer b prime to D such that $(b/D) = -1$.

19. Let D be as in Exercise 18. Show that $\sum (a/D) = 0$, where the sum is over a reduced residue system modulo D (see Exercise 6 of Chapter 3). Conclude that exactly one half of the elements in $U(\mathbb{Z}/D\mathbb{Z})$ satisfy $(a/D) = 1$.

20. (continuation) Let $a_1, a_2, \ldots, a_{\phi(D)/2}$ be integers between 1 and D such that $(a_i, D) = 1$ and $(a_i/D) = 1$. Prove that D is a quadratic residue modulo a prime $p \nmid D$, $p \equiv 1 \ (4)$ iff $p \equiv a_i \ (D)$ for some i.

21. Apply the method of Exercises 19 and 20 to find those primes for which 21 is a quadratic residue. [*Answer*: Those $p \equiv 1, 4, 5, 16, 17,$ and 20 (21).]

22. Use the Jacobi symbol to determine $(113/997)$, $(215/761)$, $(514/1093)$, and $(401/757)$.

23. Suppose that $p \equiv 1 \ (4)$. Show that there exist integers s and t such that $pt = 1 + s^2$. Conclude that p is not a prime in $\mathbb{Z}[i]$. Remember that $\mathbb{Z}[i]$ has unique factorization.

24. If $p \equiv 1 \ (4)$, show that p is the sum of two squares; i.e., $p = a^2 + b^2$ with $a, b \in \mathbb{Z}$. (*Hint*: $p = \alpha\beta$ with α and β being nonunits in $\mathbb{Z}[i]$. Take the absolute value of both sides and square the result.) This important result was discovered by Fermat.

25. An integer is called a biquadratic residue modulo p if it is congruent to a fourth power. Using the identity $x^4 + 4 = ((x + 1)^2 + 1)((x - 1)^2 + 1)$ show that -4 is a biquadratic residue modulo p iff $p \equiv 1 \ (4)$.

26. This exercise and Exercises 27 and 28 give Dirichlet's beautiful proof that 2 is a biquadratic residue modulo p iff p can be written in the form $A^2 + 64B^2$, where $A, B \in \mathbb{Z}$. Suppose that $p \equiv 1 \ (4)$. Then $p = a^2 + b^2$ by Exercise 24. Take a to be odd. Prove the following statements:
    (a) $(a/p) = 1$.
    (b) $((a + b)/p) = (-1)^{((a+b)^2 - 1)/8}$.
    (c) $(a + b)^2 \equiv 2ab \ (p)$.
    (d) $(a + b)^{(p-1)/2} \equiv (2ab)^{(p-1)/4} \ (p)$.
    [*Hint*: $2p = (a + b)^2 + (a - b)^2$.]

27. Suppose that f is such that $b \equiv af \ (p)$. Show that $f^2 \equiv -1 \ (p)$ and that $2^{(p-1)/4} \equiv f^{ab/2} \ (p)$.

28. Show that $x^4 \equiv 2 \ (p)$ has a solution for $p \equiv 1 \ (4)$ iff p is of the form $A^2 + 64B^2$.

29. Let (RR) be the number of pairs $(n, n + 1)$ in the set $1, 2, 3, \ldots, p - 1$ such that n and $n + 1$ are both quadratic residues modulo p. Let (NR) be the number of pairs $(n, n + 1)$ in the set $1, 2, 3, \ldots, p - 1$ such that n is a quadratic nonresidue and $n + 1$ is a quadratic residue. Similarly, define (RN) and (NN). Determine the sums $(RR) + (RN), (NR) + (NN), (RR) + (NR),$ and $(RN) + (NN)$.

30. Show that $(RR) + (NN) - (RN) - (NR) = \sum_{n=1}^{p-1}(n(n + 1))/p$. Evaluate this sum and show that it is equal to -1. (*Hint*: The result of Exercise 8 is useful.)

31. Use the results of Exercises 29 and 30 to show that $(RR) = \frac{1}{4}(p - 4 - \varepsilon)$, where $\varepsilon = (-1)^{(p-1)/2}$.

32. If p is an odd prime show that $(2/p) = \prod_{j=1}^{(p-1)/2} 2 \cos(2\pi j/p)$. Use this result to give another proof to Proposition 5.1.3.

33. Use Proposition 5.3.2 to derive the quadratic character of -1.

34. If p is an odd prime distinct from 3 show that $(3/p) = \prod_{j=1}^{(p-1)/2} (3 - 4\sin^2(2\pi j/p))$.

35. Use the preceding exercise to show that 3 is a square modulo p iff p is congruent to 1 or -1 modulo 12.

36. Show that part (c) of Proposition 5.2.2 is true if a is negative and b is positive (both still odd).

37. Show that if a is negative then $p \equiv q$ (4a), $p \nmid a$ implies $(a/p) = (a/q)$.

38. Let p be an odd prime. Derive the quadratic character of 2 modulo p by verifying the following steps, involving the Jacobi symbol:

$$\left(\frac{2}{p}\right) = \left(\frac{8-p}{p}\right) = \left(\frac{p}{p-8}\right) = \left(\frac{8}{p-8}\right) = \left(\frac{2}{p-8}\right).$$

Generalize the argument to show that

$$\left(\frac{a}{p}\right) = \left(\frac{a}{p-4a}\right), \qquad a > 0, \, p \nmid a.$$

Chapter 6

Quadratic Gauss Sums

The method by which we proved the quadratic reciprocity in Chapter 5 is ingenious but is not easy to use in more general situations. We shall give a new proof in this chapter that is based on methods that can be used to prove higher reciprocity laws. In particular, we shall introduce the notion of a Gauss sum, which will play an important role in the latter part of this book.

Section 1 introduces algebraic numbers and algebraic integers. The proofs are somewhat technical. The reader may wish to simply skim this section on a first reading.

§1 Algebraic Numbers and Algebraic Integers

Definition. An *algebraic number* is a complex number α that is a root of a polynomial $a_0 x^n + a_1 x^{n-1} + a_2 x^{n-2} + \cdots + a_n = 0$, where $a_0, a_1, a_2, \ldots, a_n \in \mathbb{Q}$, and $a_0 \neq 0$.

An *algebraic integer* ω is a complex number that is a root of a polynomial $x^n + b_1 x^{n-1} + \cdots + b_n = 0$, where $b_1, b_2, \ldots, b_n \in \mathbb{Z}$.

Clearly every algebraic integer is an algebraic number. The converse is false, as we shall see.

Proposition 6.1.1. *A rational number $r \in \mathbb{Q}$ is an algebraic integer iff $r \in \mathbb{Z}$.*

PROOF. If $r \in \mathbb{Z}$, then r is a root of $x - r = 0$. Thus r is an algebraic integer.

Suppose that $r \in \mathbb{Q}$ and that r is an algebraic integer; i.e., r satisfies an equation $x^n + b_1 x^{n-1} + \cdots + b_n = 0$ with $b_1, \ldots, b_n \in \mathbb{Z}$. $r = c/d$, where $c, d \in \mathbb{Z}$ and we may assume that c and d are relatively prime. Substituting c/d into the equation and multiplying both sides by d^n yields

$$c^n + b_1 c^{n-1} d + \cdots + b_n d^n = 0.$$

It follows that d divides c^n and, since $(d, c) = 1$, that $d | c$. Again, since $(d, c) = 1$ it follows that $d = \pm 1$, and so $r = c/d$ is in $\mathbb{Z}$.

It follows, for example, that $\frac{2}{5}$ is not an algebraic integer. $\qquad\square$

The main results of this section are that the set of algebraic numbers forms a field and that the set of algebraic integers forms a ring. We need some preliminary work.

Definition. A subset $V \subset \mathbb{C}$ of the complex numbers is called a $\mathbb{Q}$ module if

(a) $\gamma_1, \gamma_2 \in V$ implies that $\gamma_1 \pm \gamma_2 \in V$.
(b) $\gamma \in V$ and $r \in \mathbb{Q}$ implies that $r\gamma \in V$.
(c) There exist elements $\gamma_1, \gamma_2, \ldots, \gamma_l \in V$ such that every $\gamma \in V$ has the form $\sum_{i=1}^{l} r_i \gamma_i$ with $r_i \in \mathbb{Q}$.

More briefly, $V \subset \mathbb{C}$ is a $\mathbb{Q}$ module if it is a finite dimensional vector space over $\mathbb{Q}$.

If $\gamma_1, \gamma_2, \ldots, \gamma_l \in \mathbb{C}$, the set of all expressions $\sum_{i=1}^{l} r_i \gamma_i$, $r_1, r_2, \ldots, r_l \in \mathbb{Q}$ is easily seen to be a $\mathbb{Q}$ module. We denote this $\mathbb{Q}$ module by $[\gamma_1, \gamma_2, \ldots, \gamma_l]$.

Proposition 6.1.2. *Let $V = [\gamma_1, \gamma_2, \ldots, \gamma_l]$, and suppose that $\alpha \in \mathbb{C}$ has the property that $\alpha\gamma \in V$ for all $\gamma \in V$. Then α is an algebraic number.*

PROOF. $\alpha\gamma_i \in V$ for $i = 1, 2, \ldots, l$. Thus $\alpha\gamma_i = \sum_{j=1}^{l} a_{ij}\gamma_j$, where $a_{ij} \in \mathbb{Q}$. It follows that $0 = \sum_{j=1}^{l} (a_{ij} - \delta_{ij}\alpha)\gamma_j$, where $\delta_{ij} = 0$ if $i \neq j$ and $\delta_{ij} = 1$ if $i = j$. By standard linear algebra we have that $\det(a_{ij} - \delta_{ij}\alpha) = 0$. Writing out the determinant we see that α satisfies a polynomial of degree l with rational coefficients. Thus α is an algebraic number. $\square$

Proposition 6.1.3. *The set of algebraic numbers forms a field.*

PROOF. Suppose that α_1 and α_2 are algebraic numbers. We shall show that $\alpha_1\alpha_2$ and $\alpha_1 + \alpha_2$ are algebraic numbers.

Suppose that $\alpha_1^n + r_1\alpha_1^{n-1} + r_2\alpha_1^{n-2} + \cdots + r_n = 0$ and that $\alpha_2^m + s_1\alpha_2^{m-1} + s_2\alpha_2^{m-2} + \cdots + s_m = 0$, where $r_i, s_j \in \mathbb{Q}$. Let V be the $\mathbb{Q}$ module obtained by forming all $\mathbb{Q}$ linear combinations of the elements $\alpha_1^i\alpha_2^j$, where $0 \leq i < n$ and $0 \leq j < m$. For $\gamma \in V$ we have $\alpha_1\gamma \in V$ and $\alpha_2\gamma \in V$ (prove it). Thus we also have $(\alpha_1 + \alpha_2)\gamma \in V$ and $(\alpha_1\alpha_2)\gamma \in V$. By Proposition 6.1.2 it follows that both $\alpha_1 + \alpha_2$ and $\alpha_1\alpha_2$ are algebraic numbers.

Finally, if α is an algebraic number, not zero, we must show that α^{-1} is an algebraic number. Suppose that $a_0\alpha^n + a_1\alpha^{n-1} + \cdots + a_n = 0$, where the $a_i \in \mathbb{Q}$. Then $a_n\alpha^{-n} + a_{n-1}\alpha^{-(n-1)} + \cdots + a_0 = 0$. The result follows. $\square$

To prove that the set of algebraic integers form a ring it is necessary only to alter the above proofs slightly.

Definition. A subset $W \subset \mathbb{C}$ is called a $\mathbb{Z}$ *module* if

(a) $\gamma_1, \gamma_2 \in W$ implies that $\gamma_1 \pm \gamma_2 \in W$.
(b) There exist elements $\gamma_1, \gamma_2, \ldots, \gamma_l \in W$ such that every $\gamma \in W$ is of the form $\sum_{i=1}^{l} b_i \gamma_i$ with $b_i \in \mathbb{Z}$.

Proposition 6.1.4. *Let W be a $\mathbb{Z}$ module and suppose that $\omega \in \mathbb{C}$ is such that $\omega\gamma \in W$ for all $\gamma \in W$. Then ω is an algebraic integer.*

PROOF. The proof proceeds exactly as in Proposition 6.1.2, except that now the $a_{ij} \in \mathbb{Z}$. The equation $\det(a_{ij} - \delta_{ij}\omega) = 0$ when written out shows that ω satisfies a monic equation of degree l with integer coefficients. Thus ω is an algebraic integer. $\square$

Proposition 6.1.5. *The set of algebraic integers forms a ring.*

PROOF. The proof follows from Proposition 6.1.4 in exactly the same way in which Proposition 6.1.3 follows from Proposition 6.1.2. We leave the details to the reader. $\square$

Let Ω denote the ring of algebraic integers. If ω_1, ω_2, $\gamma \in \Omega$, we say that $\omega_1 \equiv \omega_2 \ (\gamma)$ (ω_1 is congruent to ω_2 modulo γ) if $\omega_1 - \omega_2 = \gamma\alpha$ with $\alpha \in \Omega$. This notion of congruence satisfies all the formal properties of congruence in $\mathbb{Z}$.

If $a, b, c \in \mathbb{Z}, c \neq 0$, then $a \equiv b \ (c)$ is ambiguous since it denotes congruence in $\mathbb{Z}$ and in Ω. The ambiguity is only apparent, however. If $a - b = c\alpha$ with $\alpha \in \Omega$, then α is both a rational number and an algebraic integer. Thus α is an ordinary integer by Proposition 6.1.1.

The following proposition will be useful.

Proposition 6.1.6. *If ω_1, $\omega_2 \in \Omega$ and $p \in \mathbb{Z}$ is a prime, then*

$$(\omega_1 + \omega_2)^p \equiv \omega_1^p + \omega_2^p \ (p).$$

PROOF. $(\omega_1 + \omega_2)^p = \sum_{k=0}^{p} \binom{p}{k}\omega_1^k\omega_2^{p-k}$. By Lemma 2, Chapter 4, we have $p | \binom{p}{k}$ for $1 \leq k \leq p - 1$. The result follows from this and the fact that Ω is a ring. $\square$

A root of unity is a solution to an equation of the form $x^n - 1 = 0$. Thus roots of unity are algebraic integers, and so are $\mathbb{Z}$ linear combinations of roots of unity.

We conclude this section by presenting several important properties of algebraic numbers. If α is an algebraic number then clearly any nonzero polynomial $f(x)$ in $\mathbb{Q}[x]$ of smallest degree for which $f(\alpha) = 0$ must be irreducible.

Proposition 6.1.7. *If α is an algebraic number then α is the root of a unique monic irreducible $f(x)$ in $\mathbb{Q}[x]$. Furthermore if $g(x) \in \mathbb{Q}[x]$, $g(\alpha) = 0$ then $f(x)|g(x)$.*

PROOF. Let $f(x)$ be any monic irreducible with $f(\alpha) = 0$. We prove the second assertion first. If $f(x) \nmid g(x)$ then $(f(x), g(x)) = 1$. By Lemma 4, Section 2, Chapter 1 we may write $f(x)h(x) + g(x)t(x) = 1$ for polynomials $h(x), t(x) \in \mathbb{Q}[x]$. Putting $x = \alpha$ gives a contradiction. Uniqueness now follows immediately. $\square$

The polynomial defined in Proposition 6.1.7 depends therefore only upon α. It is called the minimal polynomial of α. If the degree of the minimal polynomial is n, then α is called an algebraic number of degree n. If $f(x)$ is irreducible of degree n, then, using the fundamental theorem of algebra and Exercise 16 we see that $f(x)$ is the minimal polynomial for each of its n roots. If α, β are roots of $f(x)$ then α and β are said to be conjugate.

The set of complex numbers $g(\alpha)/h(\alpha)$ where $g(x)$, $h(x) \in \mathbb{Q}[x]$, $h(\alpha) \neq 0$ forms a field denoted by $\mathbb{Q}(\alpha)$. Denote by $\mathbb{Q}[\alpha]$ the ring of polynomials in α with rational coefficients. Then one has the following important result.

Proposition 6.1.8. *If $\alpha \in \Omega$ then $\mathbb{Q}(\alpha) = \mathbb{Q}[\alpha]$.*

PROOF. Clearly $\mathbb{Q}[\alpha] \subset \mathbb{Q}(\alpha)$. If $h(\alpha) \in \mathbb{Q}[\alpha]$, $h(\alpha) \neq 0$, then by Proposition 6.1.7, $f(x) \nmid h(x)$, where $f(x)$ is the minimal polynomial of α. Thus $(f(x), h(x)) = 1$ so that by Lemma 4, Section 2, Chapter 1, $s(x)f(x) + t(x)h(x) = 1$ for elements $s(x)$, $t(x) \in \mathbb{Q}[x]$. Put $x = \alpha$ so that $t(\alpha)h(\alpha) = 1$. Thus $h(\alpha)^{-1} \in \mathbb{Q}[\alpha]$. If $\beta \in \mathbb{Q}(\alpha)$ then $\beta = g(\alpha)h(\alpha)^{-1}$ for $g(x)$, $h(x) \in \mathbb{Q}[x]$ and the above shows that $\beta \in \mathbb{Q}[\alpha]$. ☐

Corollary. *If α is an algebraic number of degree n then $[\mathbb{Q}(\alpha) : \mathbb{Q}] = n$.*

PROOF. By the proposition it is enough to show $[\mathbb{Q}[\alpha] : \mathbb{Q}] = n$. Since $f(\alpha) = 0$ it is easily seen that $1, \ldots, \alpha^{n-1}$ span $\mathbb{Q}[\alpha]$. If on the other hand $a_0 + a_1\alpha + \cdots + a_{n-1}\alpha^{n-1} = 0$, $a_i \in \mathbb{Q}$, then $g(\alpha) = 0$ for $g(x) = a_0 + a_1 x + \cdots + a_{n-1}x^{n-1}$. Then, by Proposition 6.1.7, $f(x) \mid g(x)$. But $\deg(g(x)) < \deg(f(x))$ which implies that $a_0 = a_1 = a_2 = \cdots = a_{n-1} = 0$. Therefore $1, \alpha, \ldots, \alpha^{n-1}$ are linearly independent over $\mathbb{Q}$. ☐

§2 The Quadratic Character of 2

Let $\zeta = e^{2\pi i/8}$. Then ζ is a primitive eighth root of unity. Thus $0 = \zeta^8 - 1 = (\zeta^4 - 1)(\zeta^4 + 1)$. Since $\zeta^4 \neq 1$ we have $\zeta^4 = -1$. Multiplying by ζ^{-2} and then adding ζ^{-2} to both sides yields $\zeta^2 + \zeta^{-2} = 0$. This equation is also easily derived from the observation that $\zeta^2 = e^{i(\pi/2)} = i$.

The quadratic character of 2 will now be derived from the relation

$$(\zeta + \zeta^{-1})^2 = \zeta^2 + 2 + \zeta^{-2} = 2.$$

Let $\tau = \zeta + \zeta^{-1}$ and notice that ζ and τ are algebraic integers. We may thus work with congruences in the ring of algebraic integers.

Let p be an odd prime in $\mathbb{Z}$ and notice that

$$\tau^{p-1} = (\tau^2)^{(p-1)/2} = 2^{(p-1)/2} \equiv (2/p) \ (p).$$

It follows that $\tau^p \equiv (2/p)\tau \ (p)$. By Proposition 6.1.6, $\tau^p = (\zeta + \zeta^{-1})^p \equiv \zeta^p + \zeta^{-p} \ (p)$.

Remembering that $\zeta^8 = 1$ we have $\zeta^p + \zeta^{-p} = \zeta + \zeta^{-1}$ for $p \equiv \pm 1$ (8) and $\zeta^p + \zeta^{-p} = \zeta^3 + \zeta^{-3}$ for $p \equiv \pm 3$ (8). The result in the latter case may be simplified by observing that $\zeta^4 = -1$ implies that $\zeta^3 = -\zeta^{-1}$. Thus $\zeta^p + \zeta^{-p} = -(\zeta + \zeta^{-1})$ if $p \equiv \pm 3$ (8). Summarizing,

$$\zeta^p + \zeta^{-p} = \begin{cases} \tau, & \text{if } p \equiv \pm 1 \ (8), \\ -\tau, & \text{if } p \equiv \pm 3 \ (8). \end{cases}$$

Substituting this result into the relation $\tau^p \equiv (2/p)\tau\ (p)$ yields

$$(-1)^{\varepsilon}\tau \equiv \left(\frac{2}{p}\right)\tau\ (p), \quad \text{where } \varepsilon \equiv \frac{p^2 - 1}{8}\ (2).$$

Multiply both sides of the congruence by τ. Then

$$(-1)^{\varepsilon}2 \equiv \left(\frac{2}{p}\right)2\ (p),$$

implying that

$$(-1)^{\varepsilon} \equiv \left(\frac{2}{p}\right)\ (p).$$

This last congruence implies that $(2/p) = (-1)^{\varepsilon}$, which is the desired result.

Euler (1707–1783), in an early paper, proved that 2 is a quadratic residue modulo primes $p \equiv 1$ (8). His method contains the key idea of the above proof.

Euler assumed that $U(\mathbb{Z}/p\mathbb{Z})$ is a cyclic group. Gauss was the first to give a rigorous proof of this fact (see Theorem 1, Chapter 4). Let λ be a generator of $U(\mathbb{Z}/p\mathbb{Z})$ and set $\gamma = \lambda^{(p-1)/8}$. Then γ has order 8, so that $\gamma^4 = -\bar{1}$ and $\gamma^2 + \gamma^{-2} = \bar{0}$. Therefore, $(\gamma + \gamma^{-1})^2 = \gamma^2 + \bar{2} + \gamma^{-2} = \bar{2}$. This shows that $\bar{2}$ is a square in $U(\mathbb{Z}/p\mathbb{Z})$, which is equivalent to 2 being a quadratic residue modulo p.

If $p \not\equiv 1$ (8), this proof cannot get started. However, the theory of finite fields enables us to carry through to a complete proof of quadratic reciprocity using Euler's idea. We shall develop the theory of finite fields in Chapter 7.

§3 Quadratic Gauss Sums

Given the relation $(\zeta + \zeta^{-1})^2 = 2$ of Section 2, one might ask if there is a similar relation when 2 is replaced by an odd prime p. The answer is yes, and, moreover, the full law of quadratic reciprocity follows from this new relation by using the method of Section 2.

Throughout this section ζ will denote $e^{2\pi i/p}$, a primitive pth root of unity.

Lemma 1. $\sum_{t=0}^{p-1} \zeta^{at}$ is equal to p if $a \equiv 0\ (p)$. Otherwise it is zero.

PROOF. If $a \equiv 0\ (p)$, then $\zeta^a = 1$, and so $\sum_{t=0}^{p-1} \zeta^{at} = p$. If $a \not\equiv 0\ (p)$, then $\zeta^a \neq 1$ and $\sum_{t=0}^{p-1} \zeta^{at} = (\zeta^{ap} - 1)/(\zeta^a - 1) = 0$. $\qquad\square$

Corollary. $p^{-1} \sum_{t=0}^{p-1} \zeta^{t(x-y)} = \delta(x, y)$, where $\delta(x, y) = 1$ if $x \equiv y\ (p)$ and $\delta(x, y) = 0$ if $x \not\equiv y\ (p)$.

PROOF. The proof is immediate from Lemma 1. $\qquad\square$

All summations for the remainder of this section will be over the range zero to $p - 1$. It will simplify notation to avoid writing out this fact each time.

Lemma 2. $\sum_t (t/p) = 0$, where (t/p) is the Legendre symbol.

PROOF. By definition $(0/p) = 0$. Of the remaining $p - 1$ terms in the summation, half are $+1$ and half are -1, since by Corollary 1 to Proposition 5.1.2, there are as many quadratic residues as quadratic nonresidues mod p. $\qquad\square$

We are now in a position to introduce the notion of Gauss sum.

Definition. $g_a = \sum_t (t/p)\zeta^{at}$ is called a *quadratic Gauss sum*.

Proposition 6.3.1. $g_a = (a/p)g_1$.

PROOF. If $a \equiv 0\ (p)$, then $\zeta^{at} = 1$ for all t, and $g_a = \sum (t/p) = 0$ by Lemma 2. This gives the result in the case that $a \equiv 0\ (p)$.

Now suppose that $a \not\equiv 0\ (p)$. Then

$$\left(\frac{a}{p}\right)g_a = \sum_t \left(\frac{at}{p}\right)\zeta^{at} = \sum_x \left(\frac{x}{p}\right)\zeta^x = g_1.$$

We have used the fact that at runs over a complete residue system mod p when t does and that (x/p) and ζ^x depend only on the residue class of x modulo p.

Since $(a/p)^2 = 1$ when $a \not\equiv 0\ (p)$ our result follows by multiplying the equation $(a/p)g_a = g_1$ on both sides by (a/p). $\qquad\square$

From now on we shall denote g_1 by g. It follows from Proposition 6.3.1 that $g_a^2 = g^2$ if $a \not\equiv 0\ (p)$. We shall now deduce this common value.

Proposition 6.3.2. $g^2 = (-1)^{(p-1)/2}p$.

PROOF. The idea of the proof is to evaluate the sum $\sum_a g_a g_{-a}$ in two ways.

If $a \not\equiv 0\ (p)$, then $g_a g_{-a} = (a/p)(-a/p)g^2 = (-1/p)g^2$. If follows that

$$\sum_a g_a g_{-a} = \left(\frac{-1}{p}\right)(p - 1)g^2.$$

Now, notice that

$$g_a g_{-a} = \sum_x \sum_y \left(\frac{x}{p}\right)\left(\frac{y}{p}\right)\zeta^{a(x-y)}.$$

Summing both sides over a and using the corollary to Lemma 1 yields

$$\sum_a g_a g_{-a} = \sum_x \sum_y \left(\frac{x}{p}\right)\left(\frac{y}{p}\right)\delta(x, y)p = (p - 1)p.$$

Putting these results together we obtain $(-1/p)(p - 1)g^2 = (p - 1)p$. Therefore, $g^2 = (-1/p)p$.   □

Let $p^* = (-1)^{(p-1)/2}p$. The equation $g^2 = p^*$ is the desired analog of the equation $\tau^2 = 2$. Let $q \neq p$ be another odd prime. We proceed to prove the law of quadratic reciprocity by working with congruences mod q in the ring of algebraic integers:

$$g^{q-1} = (g^2)^{(q-1)/2} = p^{*(q-1)/2} \equiv \left(\frac{p^*}{q}\right)(q).$$

Thus

$$g^q \equiv \left(\frac{p^*}{q}\right)g \ (q).$$

Using Proposition 6.1.6 we see that

$$g^q = \left(\sum \left(\frac{t}{p}\right)\zeta^t\right)^q \equiv \sum \left(\frac{t}{p}\right)^q \zeta^{qt} \equiv g_q \ (q).$$

It follows that $g^q \equiv g_q \equiv (q/p)g \ (q)$ and so

$$\left(\frac{q}{p}\right)g \equiv \left(\frac{p^*}{q}\right)g \ (q).$$

Multiply both sides by g, and use $g^2 = p^*$:

$$\left(\frac{q}{p}\right)p^* \equiv \left(\frac{p^*}{q}\right)p^* \ (q),$$

which implies that

$$\left(\frac{q}{p}\right) \equiv \left(\frac{p^*}{q}\right) \ (q)$$

and finally

$$\left(\frac{q}{p}\right) = \left(\frac{p^*}{q}\right).$$

To see that this result is what we want simply notice that

$$\left(\frac{p^*}{q}\right) = \left(\frac{-1}{q}\right)^{(p-1)/2}\left(\frac{p}{q}\right) = (-1)^{((q-1)/2)((p-1)/2)}\left(\frac{p}{q}\right).$$

The notion of quadratic Gauss sum that we have used can be considerably generalized. We shall present some of these generalizations after developing the theory of finite fields. Cubic Gauss sums will be used to prove the law of cubic reciprocity, and quartic Gauss sums will be used to prove biquadratic reciprocity.

§4 The Sign of the Quadratic Gauss Sum[*]

According to Proposition 6.3.2, the quadratic Gauss sum has value $\pm\sqrt{p}$ if $p \equiv 1 \ (4)$ and $\pm i\sqrt{p}$ if $p \equiv 3 \ (4)$. Thus the value of $g(\chi)$ is determined up to sign. The determination of the sign is a much more difficult problem. The conjecture that the plus sign holds in each case was made by Gauss and recorded in his diary in May 1801. It was not until four years later that he found a proof. On August 30, 1805 Gauss recorded in his diary that a proof the "very elegant theorem mentioned in 1801" had finally been achieved. He wrote to his friend W. Olbers on September 3, 1805 that seldom had a week passed for four years that he had not tried in vain to prove his conjecture. Finally according to Gauss "Wie der Blitz einschlägt, hat sich das Räthsel gelöst ..." (as lightning strikes was the puzzle solved).

Subsequently proofs were found by Dirichlet, Cauchy, Kronecker, Mertens, Schur, and others. In this section we present one of Kronecker's proofs.

As in the previous section $\zeta = e^{2\pi i/p}$. Then $1, \zeta, \ldots, \zeta^{p-1}$ are the roots of $x^p - 1$.

Proposition 6.4.1. *The polynomial* $1 + x + \cdots + x^{p-1}$ *is irreducible in* $\mathbb{Q}[x]$.

PROOF. By Exercise 4 at the end of this chapter ("Gauss' lemma") it is enough to show that $1 + x + \cdots + x^{p-1}$ has no nontrivial factorization in $\mathbb{Z}[x]$. Suppose, on the contrary, that $1 + x + x^2 + \cdots + x^{p-1} = f(x)g(x)$ where $f(x), g(x) \in \mathbb{Z}[x]$ and each has degree greater than one. Putting $x = 1$ gives $p = f(1)g(1)$. Therefore we may assume $g(1) = 1$. Using a bar to denote reduction modulo p we conclude that $\bar{g}(\bar{1}) \neq \bar{0}$. On the other hand since $p|\binom{p}{j}, j = 1, \ldots, p - 1$, we have $x^p - 1 \equiv (x - 1)^p \ (p)$ and division of both sides by $x - 1$ shows that $1 + x + \cdots + x^{p-1} \equiv (x - 1)^{p-1} \ (p)$. By Theorem 2, Chapter 1 and Proposition 3.3.2 it follows that $g(x) \equiv (x - 1)^s \ (p)$ for some positive integer s. However, this contradicts the fact that $\bar{g}(\bar{1}) \neq (\bar{0})$, and the proof is complete. $\qquad\square$

---

[*] In this section the Gauss sum g will be denoted by $g(\chi)$ with $\chi(t) = (t/p)$ by definition.

Combining the above proposition with Proposition 6.1.7 we see that if $g(\zeta) = 0$ for $g(x) \in \mathbb{Q}[x]$ then $1 + x + \cdots + x^{p-1} | g(x)$. This observation will be useful later.

Proposition 6.4.2. $\prod_{k=1}^{(p-1)/2} (\zeta^{2k-1} - \zeta^{-(2k-1)})^2 = (-1)^{(p-1)/2} p$.

PROOF. One has $x^p - 1 = (x - 1) \prod_{j=1}^{p-1} (x - \zeta^j)$. Divide by $x - 1$ and put $x = 1$ to obtain $p = \prod_r (1 - \zeta^r)$, where the product is over any complete set of representative of the nonzero cosets modulo p. The integers $\pm(4k - 2)$, $k = 1, 2, \ldots, (p - 1)/2$ are easily seen to be such a system of residues. Thus

$$p = \prod (1 - \zeta^{4k-2}) \prod (1 - \zeta^{-(4k-2)})$$
$$= \prod (\zeta^{-(2k-1)} - \zeta^{2k-1}) \prod (\zeta^{2k-1} - \zeta^{-(2k-1)})$$
$$= (-1)^{(p-1)/2} \prod (\zeta^{2k-1} - \zeta^{-(2k-1)})^2,$$

all the products being over $k = 1, 2, \ldots, (p - 1)/2$. $\qquad \square$

Proposition 6.4.3.

$$\prod_{k=1}^{(p-1)/2} (\zeta^{2k-1} - \zeta^{-(2k-1)}) = \begin{cases} \sqrt{p}, & \text{if } p \equiv 1 \ (4), \\ i\sqrt{p}, & \text{if } p \equiv 3 \ (4). \end{cases}$$

PROOF. By Proposition 6.4.2 we have only to compute the sign of the product. The product is

$$i^{(p-1)/2} \prod_{k=1}^{(p-1)/2} 2 \sin \frac{(4k - 2)\pi}{p}.$$

But $\sin((4k - 2)/p)\pi < 0$ if $(p + 2)/4 < k \leq (p - 1)/2$. It follows that the product has $(p - 1)/2 - [(p + 2)/4]$ negative terms and this is easily seen to be $(p - 1)/4$ or $(p - 3)/4$ according as $p \equiv 1 \ (4)$ or $p \equiv 3 \ (4)$, respectively. The result follows immediately. $\qquad \square$

By Proposition 6.3.2 and Proposition 6.4.2 we know that

$$g(\chi) = \varepsilon \prod_{k=1}^{(p-1)/2} (\zeta^{2k-1} - \zeta^{-(2k-1)}), \qquad (1)$$

where $\varepsilon = \pm 1$. The evaluation of the Gauss sum is completed by Proposition 6.4.3 if we can show that $\varepsilon = +1$. The following argument of Kronecker shows that this is the case. See also Exercise 22.

Proposition 6.4.4. $\varepsilon = +1$.

PROOF. Consider the polynomial

$$f(x) = \sum_{j=1}^{p-1} \chi(j)x^j - \varepsilon \prod_{k=1}^{(p-1)/2} (x^{2k-1} - x^{p-(2k-1)}). \qquad (2)$$

Then $f(\zeta) = 0$ by (1) and $f(1) = 0$ by Lemma 2. By the comment preceding Proposition 6.4.2 and the fact that $1 + x + \cdots + x^{p-1}$ and $x - 1$ are relatively prime we conclude that $x^p - 1 \mid f(x)$. Write $f(x) = (x^p - 1)h(x)$ and replace x by e^z to obtain

$$\sum_{j=1}^{p-1} \chi(j)e^{jz} - \varepsilon \prod_{k=1}^{(p-1)/2} (e^{(2k-1)z} - e^{z(p-(2k-1))}) = (e^{pz} - 1)h(e^z). \qquad (3)$$

The coefficient of $z^{(p-1)/2}$ on the left-hand side of (3) is easily seen to be

$$\frac{\sum_{j=1}^{p-1} \chi(j)j^{(p-1)/2}}{((p-1)/2)!} - \varepsilon \prod_{k=1}^{(p-1)/2} (4k - p - 2).$$

On the other hand by Exercise 21 the coefficient of $z^{(p-1)/2}$ on the right-hand side of (3) is pA/B where $p \nmid B$, A and B being integers. Equating coefficients, multiplying by $B((p-1)/2)!$ and reducing modulo p shows that

$$\sum_{j=1}^{p-1} \chi(j)j^{(p-1)/2} \equiv \varepsilon \left(\frac{p-1}{2}\right)! \prod_{k=1}^{(p-1)/2} (4k - 2) \ (p)$$

$$\equiv \varepsilon(2 \cdot 4 \cdot 6 \cdots (p-1)) \prod_{k=1}^{(p-1)/2} (2k - 1)$$

$$\equiv \varepsilon (p - 1)!$$

$$\equiv -\varepsilon (p)$$

using Wilson's theorem (corollary to Proposition 4.1.1).

By Proposition 5.1.2 $j^{(p-1)/2} \equiv \chi(j) \ (p)$ so one has

$$\sum_{j=1}^{p-1} \chi(j)^2 \equiv (p - 1) \equiv -\varepsilon (p)$$

and therefore

$$\varepsilon \equiv 1 \ (p).$$

Since $\varepsilon = \pm 1$ we conclude finally that $\varepsilon = 1$. This concludes the proof. □

The result may be stated as

Theorem 1. *The value of the quadratic Gauss sum $g(\chi)$ is given by*

$$g(\chi) = \begin{cases} \sqrt{p}, & \text{if } p \equiv 1 \ (4), \\ i\sqrt{p}, & \text{if } p \equiv 3 \ (4). \end{cases}$$

NOTES

In the famous eleventh supplement to L. Dirichlet's *Vorlesungen über Zahlentheorie* [127] (1893) R. Dedekind introduced the concept of an algebraic number (§164) as well as that of an algebraic integer (§173). However the use of certain algebraic integers such as Gauss sums to prove the law of quadratic reciprocity occurs much earlier with Eisenstein, Jacobi, and others. Among the various proofs of this theorem given by Gauss, the fourth (1811) and the sixth (1818) are of central importance. The fourth proof is a corollary to Gauss' remarkable calculation of the value of the classical Gauss sum. While, as we mentioned in Section 6 he proved this result in 1805, it was not until 1811 that he published the proof in his famous paper "Summierung gewisser Reihen von besonderer Art" [34]. In this paper he shows more generally that if n is any positive integer then $\sum_{t=0}^{n-1} \zeta^{t^2}$ has the value $\sqrt{n}$ or $i\sqrt{n}$ according as $n \equiv 1$ (4) or $n \equiv 3$ (4). Here $\zeta = e^{2\pi i/n}$. The argument is quite ingenious. The proof can be found in English in Nagell [60], pp. 174–180. It is not difficult to derive quadratic reciprocity from this result (see, for example, Dirichlet [125], pp. 253–256).

The sixth and last of Gauss' published proofs of the law of quadratic reciprocity was published in 1818 under the title "Neue Beweise und Erweiterungen des Fundamentalsatzes in der Lehre von den Quadratischen Resten" [34], pp. 496–510. He mentions in the introduction to this paper that for years he had searched for a method that would generalize to the cubic and biquadratic case and that finally his untiring efforts were crowned with success ("... die unermüdliche Arbeit wurde endlich von glücklichem Erfolge gekrönt."). The purpose of publishing this sixth proof, he states, was to bring to a close that part of the higher arithmetic dealing with quadratic residues and to say, in a sense, farewell ("... und so diesem Teile der höheren Arithmetik gewissermassen Lebewohl zu sagen.") In this proof Gauss considers the polynomial $f_k(x) = \sum_{t=0}^{p-1} \chi(t)x^{kt}$ and proves, without using roots of unity, that $1 + x + \cdots + x^{p-1}$ divides $f_1(x)^2 - (-1)^{(p-1)/2}p$ as well as $f_q(x) - (q/p)f_1(x)$. Reciprocity follows by noting that $f_q(x) \equiv f_1(x)^q$ (q). The proof we have given in Section 3 amounts to putting $x = \zeta_p$ in the above and working with congruences in the ring of algebraic integers. This observation was made (at least) by Cauchy, Eisenstein, and Jacobi (in alphabetical order) and represents the stepping stone to the study of the higher reciprocity laws via Gauss sums.

The beginning student will do well to study several of the classical introductions to the theory of algebraic numbers. Aside from Dirichlet and Dedekind mentioned above, we cite E. Landau [165] and E. Hecke [44]. In recent times there have appeared many texts of varying levels of difficulty. We mention here W. Adams and L. Goldstein [84], LeVeque [180], and H. Pollard and H. Diamond [63]. Hecke's book has just appeared in English (*Algebraic Number Theory*, Springer-Verlag, 1981).

EXERCISES

1. Show that $\sqrt{2} + \sqrt{3}$ is an algebraic integer.

2. Let α be an algebraic number. Show that there is an integer n such that $n\alpha$ is an algebraic integer.

3. If α and β are algebraic integers, prove that any solution to $x^2 + \alpha x + \beta = 0$ is an algebraic integer. Generalize this result.

4. A polynomial $f(x) \in \mathbb{Z}[x]$ is said to be primitive if the greatest common divisor of its coefficients is 1. Prove that the product of primitive polynomials is again primitive. [Hint: Let $f(x) = a_0 x^n + a_1 x^{n-1} + \cdots + a_n$ and $g(x) = b_0 x^m + b_1 x^{m-1} + \cdots + b_m$ be primitive. If p is a prime, let a_i and b_j be the coefficients with the smallest subscripts such that $p \nmid a_i$ and $p \nmid b_j$. Show that the coefficient of x^{i+j} in $f(x)g(x)$ is not divisible by p.] This is one of the many results known as Gauss' lemma.

5. Let α be an algebraic integer and $f(x) \in \mathbb{Q}[x]$ be the monic polynomial of least degree such that $f(\alpha) = 0$. Use Exercise 4 to show that $f(x) \in \mathbb{Z}[x]$.

6. Let $x^2 + mx + n \in \mathbb{Z}[x]$ be irreducible and α be a root. Show that $\mathbb{Q}[\alpha] = \{r + s\alpha \mid r, s \in \mathbb{Q}\}$ is a ring (in fact, it is a field). Let $m^2 - 4n = D_0^2 D$, where D is square-free. Show that $\mathbb{Q}[\alpha] = \mathbb{Q}[\sqrt{D}]$.

7. (continuation) If $D \equiv 2, 3 \ (4)$, show that all the algebraic integers in $\mathbb{Q}[\sqrt{D}]$ have the form $a + b\sqrt{D}$, where $a, b \in \mathbb{Z}$. If $D \equiv 1 \ (4)$, show that all the algebraic integers in $\mathbb{Q}[\sqrt{D}]$ have the form $a + b((-1 + \sqrt{D})/2)$, where $a, b \in \mathbb{Z}$. [Hint: Show that $r + s\sqrt{D}$ satisfies $x^2 - 2rx + (r^2 - Ds^2) = 0$. Thus by Exercise 5, $r + s\sqrt{D}$ is an algebraic integer iff $2r$ and $r^2 - Ds^2$ are in $\mathbb{Z}$].

8. Let $\omega = e^{2\pi i/3}$. ω satisfies $x^3 - 1 = 0$. Show that $(2\omega + 1)^2 = -3$ and use this to determine $(-3/p)$ by the method of Section 2.

9. Verify Proposition 6.3.2 explicitly for $p = 3$ and $p = 5$; i.e., write out the Gauss sum longhand and square.

10. What is $\sum_{a=1}^{p-1} g_a$?

11. By evaluating $\sum_t (1 + (t/p))\zeta^t$ in two ways prove that $g = \sum_t \zeta^{t^2}$.

12. Write $\psi_a(t) = \zeta^{at}$. Show that
    (a) $\overline{\psi_a(t)} = \psi_a(-t) = \psi_{-a}(t)$.
    (b) $(1/p) \sum_a \psi_a(t - s) = \delta(t, s)$.

13. Let f be a function from $\mathbb{Z}$ to the complex numbers. Suppose that p is a prime and that $f(n + p) = f(n)$ for all $n \in \mathbb{Z}$. Let $\hat{f}(a) = p^{-1} \sum_t f(t)\psi_{-a}(t)$. Prove that $f(t) = \sum_a \hat{f}(a)\psi_a(t)$. This result is directly analogous to a result in the theory of Fourier series.

14. In Exercise 13 take f to be the Legendre symbol and show that $\hat{f}(a) = p^{-1}g_{-a}$.

15. Show that $|\sum_{t=m}^{n} (t/p)| < \sqrt{p} \log p$. The inequality holds for the sum over any range. This remarkable inequality is associated with the names of Polya and Vinogradov. [Hint: Use the relation $(t/p)g = g_t$ and sum. The inequality $\sin x \geq (2/\pi)x$ for any acute angle x will be useful.]

16. Let α be an algebraic number with minimal polynomial $f(x)$. Show that $f(x)$ does not have repeated roots in $\mathbb{C}$.

17. Show that the minimal polynomial for $\sqrt[3]{2}$ is $x^3 - 2$.

18. Show that there exist algebraic numbers of arbitrarily high degree.

19. Find the conjugates of $\cos 2\pi/5$.

20. Let F be a subfield of $\mathbb{C}$ which is a finite dimensional vector space over $\mathbb{Q}$ of degree n. Show that every element of F is algebraic of degree at most n. [*Note*: That an element exists with degree exactly n is more difficult to prove (see Exercise 17, Chapter 12).]

21. Let $f(x) = \sum_{n=0}^{\infty} a_n x^n/n!$ and $g(x) = \sum_{n=0}^{\infty} b_n x^n/n!$ be power series with a_n and b_n integers. If p is a prime such that $p|a_i$ for $i = 0, \ldots, p-1$ show that each coefficient c_t of the product $f(x)g(x) = \sum_{n=0}^{\infty} c_n x^n$ for $t = 0, \ldots, p-1$ may be written in the form $p(A/B)$, $p \nmid B$.

22. Show that the relation $\varepsilon \equiv 1\ (p)$ in Proposition 6.4.4 can also be achieved by replacing x by $1 + t$ instead of e^z.

23. If $f(x) = x^n + a_1 x^{n-1} + \cdots + a_n$, $a_i \in \mathbb{Z}$ and p is a prime such that $p|a_i$, $i = 1, \ldots, n$, $p^2 \nmid a_n$ show that $f(x)$ is irreducible over $\mathbb{Q}$ (Eisenstein's irreducibility criterion).

Chapter 7

Finite Fields

We have already met with examples of finite fields, namely, the fields $\mathbb{Z}/p\mathbb{Z}$, where p is a prime number. In this chapter we shall prove that there are many more finite fields and shall investigate their properties. This theory is beautiful and interesting in itself and, moreover, is a very useful tool in number-theoretic investigations. As an illustration of the latter point, we shall supply yet another proof of the law of quadratic reciprocity. Other applications will come later.

One more comment. Up to now the great majority of our proofs have used very few results from abstract algebra. Although nowhere in this book will we use very sophisticated results from algebra, from now on we shall assume that the reader has some familiarity with the material in a standard undergraduate course in the subject.

§1 Basic Properties of Finite Fields

In this section we shall discuss properties of finite fields without worrying about questions of existence. The construction of finite fields will be taken up in Section 2.

Let F be a finite field with q elements. The multiplicative group F^* of F has $q - 1$ elements. Thus every element $\alpha \in F^*$ satisfies the equation $x^{q-1} = 1$ (in this context 1 stands for the multiplicative identity of F and not the integer 1), and every element in F satisfies $x^q = x$.

Proposition 7.1.1.

$$x^q - x = \prod_{a \in F}(x - \alpha).$$

PROOF. Both polynomials are to be considered as elements of $F[x]$.

Every element $\alpha \in F$ is a root of $x^q - x$. Since F has q elements and since the degree of $x^q - x$ is q, the result follows. $\qquad\square$

Corollary 1. *Let $F \subset K$, where K is a field. An element $\alpha \in K$ is in F iff $\alpha^q = \alpha$.*

PROOF. $\alpha^q = \alpha$ iff α is a root of $x^q - x$. By Proposition 7.1.1, the roots of $x^q - x$ are precisely the elements of F. $\qquad\square$

Corollary 2. *If $f(x)$ divides $x^q - x$, then $f(x)$ has d distinct roots, where d is the degree of $f(x)$.*

PROOF. Let $f(x)g(x) = x^q - x$. $g(x)$ has degree $q - d$. If $f(x)$ has fewer than d distinct roots, then by Lemma 1 of Chapter 4, $f(x)g(x)$ would have fewer than $d + (q - d) = q$ distinct roots, which is not the case.   □

Theorem 1. *The multiplicative group of a finite field is cyclic.*

PROOF. This theorem is a generalization of Theorem 1 in Chapter 4. The proof is almost identical.

If $d \mid q - 1$, then $x^d - 1$ divides $x^{q-1} - 1$ and it follows from Corollary 2 that $x^d - 1$ had d distinct roots. Thus the subgroup of F^* consisting of elements satisfying $x^d = 1$ has order d.

Let $\psi(d)$ be the number of elements in F^* of order d. Then $\sum_{c \mid d} \psi(c) = d$. By the Möbius inversion formula

$$\psi(d) = \sum_{c \mid d} \mu(c) \frac{d}{c} = \phi(d).$$

In particular, $\psi(q - 1) = \phi(q - 1) > 1$, unless we are in the trivial case $q = 2$. This concludes the proof.   □

The fact that F^* is cyclic when F is finite allows us to give the following partial generalization of Proposition 4.2.1.

Proposition 7.1.2. *Let $\alpha \in F^*$. Then $x^n = \alpha$ has solutions iff $\alpha^{(q-1)/d} = 1$, where $d = (n, q - 1)$. If there are solutions, then there are exactly d solutions.*

PROOF. Let γ be a generator of F^* and set $\alpha = \gamma^a$ and $x = \gamma^y$. Then $x^n = \alpha$ is equivalent to the congruence $ny \equiv a\,(q - 1)$. The result now follows by applying Proposition 3.3.1.   □

It is worthwhile to examine what happens in the extreme cases $n \mid q - 1$ and $(n, q - 1) = 1$.

If $n \mid q - 1$, then there are exactly $(q - 1)/n$ elements of F^* that are nth powers, and if α is an nth power, then $x^n = \alpha$ has n solutions.

If $(n, q - 1) = 1$, then every element is an nth power in a unique way; i.e., for $\alpha \in F^*$, $x^n = \alpha$ has one and only one solution.

We have investigated the structure of F^*. Now we turn our attention to the additive group of F.

Lemma 1. *Let F be a finite field. The integer multiples of the identity form a subfield of F isomorphic to $\mathbb{Z}/p\mathbb{Z}$ for some prime number p.*

PROOF. To avoid confusion, let us temporarily call e the identity of F^* instead of 1. Map $\mathbb{Z}$ to F by taking n to ne. This is easily seen to be a ring homomorphism. The image is a finite subring of F, and so in particular it is an integral domain. The kernel is a nonzero prime ideal. Therefore, the image is isomorphic to $\mathbb{Z}/p\mathbb{Z}$ for some prime p. $\square$

We shall identify $\mathbb{Z}/p\mathbb{Z}$ with its image in F and think of F as a finite dimensional vector space over $\mathbb{Z}/p\mathbb{Z}$. Let n denote that dimension and let $\omega_1, \omega_2, \ldots, \omega_n$ be a basis. Then every element $\omega \in F$ can be expressed uniquely in the form $a_1\omega_1 + a_2\omega_2 + \cdots + a_n\omega_n$, where $a_i \in \mathbb{Z}/p\mathbb{Z}$. It follows that F has p^n elements. We have proved

Proposition 7.1.3. *The number of elements in a finite field is a power of a prime.*

If e is the identity of the finite field F, let p be the smallest integer such that $pe = 0$. We have seen that p must be a prime number. It is called the characteristic of F. For $\alpha \in F$ we have $p\alpha = p(e\alpha) = (pe)\alpha = 0 \cdot \alpha = 0$. This observation leads to the following very useful proposition.

Proposition 7.1.4. *If F has characteristic p, then $(\alpha + \beta)^{p^d} = \alpha^{p^d} + \beta^{p^d}$ for all $\alpha, \beta \in F$ and all positive integers d.*

PROOF. The proof is by induction on d. For $d = 1$, we have

$$(\alpha + \beta)^p = \alpha^p + \sum_{k=1}^{p-1} \binom{p}{k}\alpha^{p-k}\beta^k + \beta^p = \alpha^p + \beta^p.$$

All the intermediate terms vanish because $p | \binom{p}{k}$ for $1 \le k \le p - 1$ by Lemma 2 of Chapter 4.

To pass from d to $d + 1$ just raise both sides of $(\alpha + \beta)^{p^d} = \alpha^{p^d} + \beta^{p^d}$ to the pth power. $\square$

Suppose that F is a finite field of dimension n over $\mathbb{Z}/p\mathbb{Z}$. We want to find out which fields E lie between $\mathbb{Z}/p\mathbb{Z}$ and F. If d is the dimension of E over $\mathbb{Z}/p\mathbb{Z}$, then it follows by general field theory that $d | n$. We shall give another proof below. It turns out that there is one and only one intermediate field corresponding to every divisor d of n.

Lemma 2. *Let F be a field. Then $x^l - 1$ divides $x^m - 1$ in $F[x]$ iff l divides m.*

PROOF. Let $m = ql + r$, where $0 \le r < l$. Then we have

$$\frac{x^m - 1}{x^l - 1} = x^r \frac{x^{ql} - 1}{x^l - 1} + \frac{x^r - 1}{x^l - 1}.$$

Since $(x^{ql} - 1)/(x^l - 1) = (x^l)^{q-1} + (x^l)^{q-2} + \cdots + x^l + 1$, the right-hand side of the above equation is a polynomial iff $(x^r - 1)/(x^l - 1)$ is a polynomial. This is easily seen to be the case iff $r = 0$. The result follows.

Lemma 3. *If a is a positive integer, then $a^l - 1$ divides $a^m - 1$ iff l divides m.*

PROOF. The proof is analogous to that of Lemma 2 with the number a playing the role of x. We leave the details to the reader.   □

Proposition 7.1.5. *Let F be a finite field of dimension n over $\mathbb{Z}/p\mathbb{Z}$. The subfields of F are in one-to-one correspondence with the divisors of n.*

PROOF. Suppose that E is a subfield of F and let d be its dimension over $\mathbb{Z}/p\mathbb{Z}$. We shall show that $d \mid n$.

Since E^* has $p^d - 1$ elements all satisfying $x^{p^d-1} - 1$, we have that $x^{p^d-1} - 1$ divides $x^{p^n-1} - 1$. By Lemma 2, $p^d - 1$ divides $p^n - 1$ and consequently, by Lemma 3, d divides n.

Now suppose that $d \mid n$. Let $E = \{\alpha \in F \mid \alpha^{p^d} = \alpha\}$. We claim that E is a field. For if $\alpha, \beta \in E$, then

(a) $(\alpha + \beta)^{p^d} = \alpha^{p^d} + \beta^{p^d} = \alpha + \beta$.
(b) $(\alpha\beta)^{p^d} = \alpha^{p^d}\beta^{p^d} = \alpha\beta$.
(c) $(\alpha^{-1})^{p^d} = (\alpha^{p^d})^{-1} = \alpha^{-1}$ for $\alpha \neq 0$.

In step (a) we made use of Proposition 7.1.4.

Now E is the set of solutions to $x^{p^d} - x = 0$. Since $d \mid n$, we have $p^d - 1 \mid p^n - 1$ and $x^{p^d-1} - 1 \mid x^{p^n-1} - 1$ by Lemmas 2 and 3. Thus $x^{p^d} - x$ divides $x^{p^n} - x$, and by Corollary 2 to Proposition 7.1.1, it follows that E has p^d elements and so has dimension d over $\mathbb{Z}/p\mathbb{Z}$.

Finally, if E' is another subfield of F of dimension d over $\mathbb{Z}/p\mathbb{Z}$, then the elements of E' must satisfy $x^{p^d} - x = 0$; i.e., E' must coincide with E.   □

Let F_q denote a finite field with q elements. To illustrate Proposition 7.1.5, consider F_{4096} (we shall show in Section 2 the existence of such a field). Since $4096 = 2^{12}$ we have the following lattice diagram:

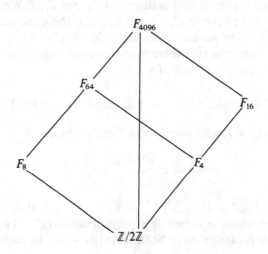

§2 The Existence of Finite Fields

In Section 1 we proved that the number of elements in a finite field has the form p^n, where p is a prime. We shall now show that given a number p^n there exists a finite field with p^n elements. To do this we shall need some results from the theory of fields that connect our problem with the existence of irreducible polynomials. Then we shall prove a theorem going back to Gauss (again!) that shows that $\mathbb{Z}/p\mathbb{Z}[x]$ contains irreducible polynomials of every degree.

Let k be an arbitrary field and $f(x)$ be an irreducible polynomial in $k[x]$. We then have

Proposition 7.2.1. *There exists a field K containing k and an element $\alpha \in K$ such that $f(\alpha) = 0$.*

PROOF. We proved in Chapter 1 that $k[x]$ is a principal ideal domain. It follows that $(f(x))$ is a maximal ideal and thus $k[x]/(f(x))$ is a field. Let $K' = k[x]/(f(x))$ and let ϕ be the homomorphism that maps $k[x]$ onto K' by taking an element to its coset modulo $(f(x))$. We have the diagram

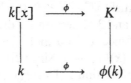

$\phi(k)$ is a subfield of K'. We claim that it is isomorphic to k. It is enough to show that ϕ restricted to k is one to one. Let $a \in k$. If $\phi(a) = 0$, then $a \in (f(x))$. If $a \neq 0$, it is a unit and cannot be an element of a proper ideal. Thus $a = 0$, as was to be shown.

Since ϕ is an isomorphism of k we may identify k with $\phi(k)$. When this is done we relabel K' as K.

Let α be the coset of x in K. Then $0 = \phi(f(x)) = f(\phi(x)) = f(\alpha)$; i.e., α is a root of $f(x)$ in K.   □

We denote the field K constructed in the proposition by $k(\alpha)$. The following proposition about $k(\alpha)$ will be useful.

Proposition 7.2.2. *The elements $1, \alpha, \alpha^2, \ldots, \alpha^{n-1}$ are a vector space basis for $k(\alpha)$ over k, where n is the degree of $f(x)$.*

The proof of this proposition is the same as that of Proposition 6.1.8 and its corollary. One replaces $\mathbb{Q}$ by k and the complex number α of that proposition by the above α.

To turn the matter around, the proposition shows that if we want to find a field extension K of k of degree n, then it is enough to produce an irreducible polynomial $f(x) \in k[x]$ of degree n.

In $\mathbb{Z}/p\mathbb{Z}[x]$ there are finitely many polynomials of a given degree. Let $F_d(x)$ be the product of the monic irreducible polynomials in $\mathbb{Z}/p\mathbb{Z}[x]$ of degree d.

Theorem 2

$$x^{p^n} - x = \prod_{d|n} F_d(x).$$

PROOF. First notice that if $f(x)$ divides $x^{p^n} - x$, then $f(x)^2$ does not divide $x^{p^n} - x$. This follows since if $x^{p^n} - x = f(x)^2 g(x)$ we obtain

$$-1 = 2f(x)f'(x)g(x) + f(x)^2 g'(x)$$

by formal differentiation. This is impossible since it implies that $f(x)$ divides 1.

It remains to prove that if $f(x)$ is a monic irreducible polynomial of degree d, then $f(x)|x^{p^n} - x$ iff $d|n$.

Consider $K = \mathbb{Z}/p\mathbb{Z}(\alpha)$, where α is a root of $f(x)$, as in Proposition 7.2.2. It has dimension d over $\mathbb{Z}/p\mathbb{Z}$ and thus p^d elements. The elements of K satisfy $x^{p^d} - x = 0$.

Assume that $x^{p^n} - x = f(x)g(x)$. Then $\alpha^{p^n} = \alpha$. If $b_1 \alpha^{d-1} + b_2 \alpha^{d-2} + \cdots + b_d$ is an arbitrary element of K, then

$$(b_1 \alpha^{d-1} + \cdots + b_d)^{p^n} = b_1(\alpha^{p^n})^{d-1} + \cdots + b_d = b_1 \alpha^{d-1} + \cdots + b_d.$$

Hence the elements of K satisfy $x^{p^n} - x = 0$. It follows that $x^{p^d} - x$ divides $x^{p^n} - x$, and by Lemmas 2 and 3 of Section 1 d divides n.

Assume now that $d|n$. Since $\alpha^{p^d} = \alpha$ and $f(x)$ is the monic irreducible polynomial for α, we have $f(x)|x^{p^d} - x$. Since $d|n$ we have $x^{p^d} - x|x^{p^n} - x$ again by Lemmas 2 and 3 of Section 1. Thus $f(x)|x^{p^n} - x$. $\square$

Let N_d be the number of monic irreducible polynomials of degree d in $\mathbb{Z}/p\mathbb{Z}[x]$. Equating the degrees on both sides of the identity in the theorem yields

Corollary 1. $p^n = \sum_{d|n} dN_d$.

Corollary 2. $N_n = n^{-1} \sum_{d|n} \mu(n/d)p^d$.

PROOF. Apply the Möbius inversion formula (Theorem 2 of Chapter 2) to the equation in Corollary 1. $\square$

Corollary 3. *For each integer $n \geq 1$, there exists an irreducible polynomial of degree n in $\mathbb{Z}/p\mathbb{Z}[x]$.*

PROOF. $N_n = n^{-1}(p^n - \cdots + p\mu(n))$ by Corollary 2. The term in parentheses cannot be zero since it is the sum of distinct powers of p with coefficients 1 and -1. $\square$

Summarizing, we have

Theorem 3. *Let $n \geq 1$ be an integer and p be a prime. Then there exists a finite field with p^n elements.*

§3 An Application to Quadratic Residues

In Chapter 6 we proved the law of quadratic reciprocity using Gauss sums and the elements of the theory of algebraic numbers. We shall now give an exceptionally short proof along the same lines using the theory of finite fields.

Let p and q be distinct odd primes. Since $(p, q) = 1$ there is an integer n (for example, $p - 1$) such that $q^n \equiv 1 \ (p)$. Let F be a finite field of dimension n over $\mathbb{Z}/q\mathbb{Z}$. Then F^* is cyclic of order $q^n - 1$. Let γ be a generator of F^* and set $\lambda = \gamma^{(q^n - 1)/p}$. Then λ has order p. Define $\tau_a = \sum_{t=0}^{p-1} (t/p)\lambda^{at}$, where $a \in \mathbb{Z}$. The element τ_a of F is an analog of the quadratic Gauss sums introduced in Chapter 6. Set $\tau_1 = \tau$. Then the proofs of Propositions 6.3.1 and 6.3.2 can be used to show that

(1) $\tau_a = (a/p)\tau$.
(2) $\tau^2 = (-1)^{(p-1)/2}\bar{p}$.

In relation 2, $\bar{p}$ is the coset of p in $\mathbb{Z}/q\mathbb{Z}$. Let $p^* = (-1)^{(p-1)/2}p$. Then relation 2 can be written as $\tau^2 = \overline{p^*}$. This relation implies that $(p^*/q) = 1$ iff $\tau \in \mathbb{Z}/q\mathbb{Z}$. By Corollary 1 to Proposition 7.1.1, this is true iff $\tau^q = \tau$. Now,

$$\tau^q = \left(\sum_t \left(\frac{t}{p}\right)\lambda^t\right)^q = \sum_t \left(\frac{t}{p}\right)\lambda^{qt} = \tau_q.$$

By relation 1 we have $\tau_q = (q/p)\tau$. Thus $\tau^q = \tau$ iff $(q/p) = 1$.
We have proved that

$$\left(\frac{p^*}{q}\right) = 1 \quad \text{iff} \quad \left(\frac{q}{p}\right) = 1.$$

This is the law of quadratic reciprocity.

A proof that $(2/q) = (-1)^{(q^2-1)/8}$ can be given using the same technique. In Chapter 6 we gave Euler's proof that $(2/q) = 1$ if $q \equiv 1 \ (8)$. If $q \not\equiv 1 \ (8)$, it is nevertheless true that $q^2 \equiv 1 \ (8)$. In this case one can carry through the proof working in a finite field F of dimension 2 over $\mathbb{Z}/q\mathbb{Z}$. We leave the details to the reader.

NOTES

The first systematic account of the theory of finite fields is found in Dickson [25], although E. Galois had axiomatically developed a number of their properties much earlier in his note "Sur la théorie des nombres" [33]. As the existence of a finite field with p^n elements is equivalent to the existence of an irreducible polynomial of degree n in the ring $F[x]$ we must include Gauss once again as a founder. In his paper "Die Lehre von den Reste" he derives the formula we have given for the number of irreducibles of degree n (see [34]).

The use of finite fields to give a proof of quadratic reciprocity has been observed by a number of mathematicians, e.g., Hausner [43] and Holzer [45, pp. 76–78].

Our treatment of finite fields throughout this book is much more elementary than is usual in modern times. Most treatments first develop the full Galois theory of fields and apply the general results of that theory to the special case of finite fields. This is done in A. Albert's compact book [1]. The advantage of Albert's book for those readers already familiar with the theory of fields is that he discusses finite fields extensively in his last chapter and provides a very long bibliography on the subject. Many interesting references are provided.

EXERCISES

1. Use the method of Theorem 1 to show that a finite subgroup of the multiplicative group of a field is cyclic.

2. Let R and C be the real and complex numbers, respectively. Find the finite subgroups of R^* and C^* and show directly that they are cyclic.

3. Let F be a field with q elements and suppose that $q \equiv 1$ (n). Show that for $\alpha \in F^*$ the equation $x^n = \alpha$ has either no solutions or n solutions.

4. (continuation)   Show that the set of $\alpha \in F^*$ such that $x^n = \alpha$ is solvable is a subgroup with $(q - 1)/n$ elements.

5. (continuation)   Let K be a field containing F such that $[K : F] = n$. For all $\alpha \in F^*$ show that the equation $x^n = \alpha$ has n solutions in K. [*Hint*: Show that $q^n - 1$ is divisible by $n(q - 1)$ and use the fact that $\alpha^{q-1} = 1$.]

6. Let $K \supset F$ be finite fields with $[K : F] = 3$. Show that if $\alpha \in F$ is not a square in F, it is not a square in K.

7. Generalize Exericse 6 by showing that if α is not a square in F, it is not a square in any extension of odd degree and is a square in every extension of even degree.

8. In a field with 2^n elements what is the subgroup of squares?

9. If $K \supset F$ are finite fields, $|F| = q$, $\alpha \in F$, $q \equiv 1$ (n), and $x^n = \alpha$ is not solvable in F, show that $x^n = \alpha$ is not solvable in K if $(n, [K : F]) = 1$.

10. Let $K \supset F$ be finite fields and $[K : F] = 2$. For $\beta \in K$ show that $\beta^{1+q} \in F$ and moreover that every element in F is of the form β^{1+q} for some $\beta \in K$.

11. With the situation being that of Exercise 10 suppose that $\alpha \in F$ has order $q - 1$. Show that there is a $\beta \in K$ with order $q^2 - 1$ such that $\beta^{1+q} = \alpha$.

12. Use Proposition 7.2.1 to show that given a field k and a polynomial $f(x) \in k[x]$ there is a field $K \supset k$ such that $[K : k]$ is finite and $f(x) = (x - \alpha_1)(x - \alpha_2) \cdots (x - \alpha_n)$ in $K[x]$.

13. Apply Exercise 12 to $k = \mathbb{Z}/p\mathbb{Z}$ and $f(x) = x^{p^n} - x$ to obtain another proof of Theorem 2.

14. Let F be a field with q elements and n a positive integer. Show that there exist irreducible polynomials in $F[x]$ of degree n.

15. Let $x^n - 1 \in F[x]$, where F is a finite field with q elements. Suppose that $(q, n) = 1$. Show that $x^n - 1$ splits into linear factors in some extension field and that the least degree of such a field is the smallest integer f such that $q^f \equiv 1 \ (n)$.

16. Calculate the monic irreducible polynomials of degree 4 in $\mathbb{Z}/2\mathbb{Z}[x]$.

17. Let q and p be distinct odd primes. Show that the number of monic irreducibles of degree q in $\mathbb{Z}/p\mathbb{Z}[x]$ is $q^{-1}(p^q - p)$.

18. Let p be a prime with $p \equiv 3 \ (4)$. Show that the residue classes modulo p in $\mathbb{Z}[i]$ form a field with p^2 elements.

19. Let F be a finite field with q elements. If $f(x) \in F[x]$ has degree t, put $|f| = q^t$. Verify the formal identity $\sum_f |f|^{-s} = (1 - q^{1-s})^{-1}$. The sum is over all monic polynomials.

20. With the notation of Exercise 19 let $d(f)$ be the number of monic divisors of f and $\sigma(f) = \sum_{g|f} |g|$, where the sum is over the monic divisors of f. Verify the following identities:
(a) $\sum_f d(f)|f|^{-s} = (1 - q^{1-s})^{-2}$.
(b) $\sum_f \sigma(f)|f|^{-s} = (1 - q^{1-s})^{-1}(1 - q^{2-s})^{-1}$.

21. Let F be a field with $q = p^n$ elements. For $\alpha \in F$ set $f(x) = (x - \alpha)(x - \alpha^p) \times (x - \alpha^{p^2}) \cdots (x - \alpha^{p^{n-1}})$. Show that $f(x) \in \mathbb{Z}/p\mathbb{Z}[x]$. In particular, $\alpha + \alpha^p + \cdots + \alpha^{p^{n-1}}$ and $\alpha \alpha^p \alpha^{p^2} \cdots \alpha^{p^{n-1}}$ are in $\mathbb{Z}/p\mathbb{Z}$.

22. (continuation)   Set $\operatorname{tr}(\alpha) = \alpha + \alpha^p + \cdots + \alpha^{p^{n-1}}$. Prove that
(a) $\operatorname{tr}(\alpha) + \operatorname{tr}(\beta) = \operatorname{tr}(\alpha + \beta)$.
(b) $\operatorname{tr}(a\alpha) = a \operatorname{tr}(\alpha)$ for $a \in \mathbb{Z}/p\mathbb{Z}$.
(c) There is an $\alpha \in F$ such that $\operatorname{tr}(\alpha) \neq 0$.

23. (continuation)   For $\alpha \in F$ consider the polynomial $x^p - x - \alpha \in F[x]$. Show that this polynomial is either irreducible or the product of linear factors. Prove that the latter alternative holds iff $\operatorname{tr}(\alpha) = 0$.

24. Suppose that $f(x) \in \mathbb{Z}/p\mathbb{Z}[x]$ has the property that $f(x + y) = f(x) + f(y) \in \mathbb{Z}/p\mathbb{Z}[x, y]$. Show that $f(x)$ must be of the form $a_0 x + a_1 x^p + a_2 x^{p^2} + \cdots + a_m x^{p^m}$.

Chapter 8

Gauss and Jacobi Sums

In Chapter 6 we introduced the notion of a quadratic Gauss sum. In this chapter a more general notion of Gauss sum will be introduced. These sums have many applications. They will be used in Chapter 9 as a tool in the proofs of the laws of cubic and biquadratic reciprocity. Here we shall consider the problem of counting the number of solutions of equations with coefficients in a finite field. In this connection, the notion of a Jacobi sum arises in a natural way. Jacobi sums are interesting in their own right, and we shall develop some of their properties.

To keep matters as simple as possible, we shall confine our attention to the finite field $\mathbb{Z}/p\mathbb{Z} = F_p$ and come back later to the question of associating Gauss sums with an arbitrary finite field.

§1 Multiplicative Characters

A multiplicative character on F_p is a map χ from F_p^* to the nonzero complex numbers that satisfies

$$\chi(ab) = \chi(a)\chi(b) \quad \text{for all } a, b \in F_p^*.$$

The Legendre symbol, (a/p), is an example of such a character if it is regarded as a function of the coset of a modulo p.

Another example is the trivial multiplicative character defined by the relation $\varepsilon(a) = 1$ for all $a \in F_p^*$.

It is often useful to extend to domain of definition of a multiplicative character to all of F_p. If $\chi \neq \varepsilon$, we do this by defining $\chi(0) = 0$. For ε we define $\varepsilon(0) = 1$. The usefulness of these definitions will soon become apparent.

Proposition 8.1.1. *Let χ be a multiplicative character and $a \in F_p^*$. Then*

(a) $\chi(1) = 1$.
(b) $\chi(a)$ *is a* $(p-1)$*st root of unity.*
(c) $\chi(a^{-1}) = \chi(a)^{-1} = \overline{\chi(a)}$.

[*In part* (a) *the* 1 *on the left-hand side is the unit of* F_p, *whereas the* 1 *on the right-hand side is the complex number* 1. *The bar in part* (c) *is complex conjugation.*]

PROOF. $\chi(1) = \chi(1 \cdot 1) = \chi(1)\chi(1)$. Thus $\chi(1) = 1$, since $\chi(1) \neq 0$.

To prove part (b), notice that $a^{p-1} = 1$ implies that $1 = \chi(1) = \chi(a^{p-1}) = \chi(a)^{p-1}$.

To prove part (c), notice that $1 = \chi(1) = \chi(a^{-1}a) = \chi(a^{-1})\chi(a)$. This shows that $\chi(a^{-1}) = \chi(a)^{-1}$. The fact that $\chi(a)^{-1} = \overline{\chi(a)}$ follows from the fact that $\chi(a)$ is a complex number of absolute value 1 by part (b). $\qquad\square$

Proposition 8.1.2. *Let* χ *be a multiplicative character. If* $\chi \neq \varepsilon$, *then* $\sum_t \chi(t) = 0$, *where the sum is over all* $t \in F_p$. *If* $\chi = \varepsilon$, *the value of the sum is* p.

PROOF. The last assertion is obvious, so we may assume that $\chi \neq \varepsilon$. In this case there is an $a \in F_p^*$ such that $\chi(a) \neq 1$. Let $T = \sum_t \chi(t)$. Then

$$\chi(a)T = \sum_t \chi(a)\chi(t) = \sum_t \chi(at) = T.$$

The last equality follows since at runs over all elements of F_p as t does. Since $\chi(a)T = T$ and $\chi(a) \neq 1$ it follows that $T = 0$. $\qquad\square$

The multiplicative characters form a group by means of the following definitions. (We shall drop the use of the word *multiplicative* for the remainder of this chapter.)

(1) If χ and λ are characters, then $\chi\lambda$ is the map that takes $a \in F_p^*$ to $\chi(a)\lambda(a)$.
(2) If χ is a character, χ^{-1} is the map that takes $a \in F_p^*$ to $\chi(a)^{-1}$.

We leave it to the reader to verify that $\chi\lambda$ and χ^{-1} are characters and that these definitions make the set of characters into a group. The identity of this group is, of course, the trivial character ε.

Proposition 8.1.3. *The group of characters is a cyclic group of order* $p - 1$. *If* $a \in F_p^*$ *and* $a \neq 1$, *then there is a character* χ *such that* $\chi(a) \neq 1$.

PROOF. We know that F_p^* is cyclic (see Theorem 1 of Chapter 4). Let $g \in F_p^*$ be a generator. Then every $a \in F_p^*$ is equal to a power of g. If $a = g^l$ and χ is a character, then $\chi(a) = \chi(g)^l$. This shows that χ is completely determined by the value $\chi(g)$. Since $\chi(g)$ is a $(p-1)$st root of unity, and since there are exactly $p - 1$ of these, it follows that the character group has order at most $p - 1$.

Now define a function λ by the equation $\lambda(g^k) = e^{2\pi i(k/(p-1))}$. It is easy to check that λ is well defined and is a character. We claim that $p - 1$ is the smallest integer n such that $\lambda^n = \varepsilon$. If $\lambda^n = \varepsilon$, then $\lambda^n(g) = \varepsilon(g) = 1$. However, $\lambda^n(g) = \lambda(g)^n = e^{2\pi i(n/(p-1))}$. It follows that $p - 1$ divides n. Since $\lambda^{p-1}(a) = \lambda(a)^{p-1} = \lambda(a^{p-1}) = \lambda(1) = 1$ we have $\lambda^{p-1} = \varepsilon$. We have established that the characters $\varepsilon, \lambda, \lambda^2, \ldots, \lambda^{p-2}$ are all distinct. Since by the first part of the proof there are at most $p - 1$ characters, we now have that there are exactly $p - 1$ characters and that the group is cyclic with λ as a generator.

If $a \in F_p^*$ and $a \neq 1$, then $a = g^l$ with $p - 1 \nmid l$. Let us compute $\lambda(a)$. $\lambda(a) = \lambda(g)^l = e^{2\pi i(l/(p-1))} \neq 1$. This concludes the proof. $\qquad\square$

Corollary. *If $a \in F_p^*$ and $a \neq 1$, then $\sum_\chi \chi(a) = 0$, where the summation is over all characters.*

PROOF. Let $S = \sum_\chi \chi(a)$. Since $a \neq 1$ there is, by the theorem, a character λ such that $\lambda(a) \neq 1$. Then

$$\lambda(a)S = \sum_\chi \lambda(a)\chi(a) = \sum_\chi \lambda\chi(a) = S.$$

The final equality holds since $\lambda\chi$ runs over all characters as χ does. It follows that $(\lambda(a) - 1)S = 0$ and thus $S = 0$. $\qquad\square$

Characters are useful in the study of equations. To illustrate this, consider the equation $x^n = a$ for $a \in F_p^*$. By Proposition 4.2.1 we know that solutions exist iff $a^{(p-1)/d} = 1$, where $d = (n, p - 1)$, and that if a solution exists, then there are exactly d solutions. For simplicity, we shall assume that n divides $p - 1$. In this case $d = (n, p - 1) = n$.

We shall now derive a criterion for the solution of $x^n = a$ using characters.

Proposition 8.1.4. *If $a \in F_p^*, n \mid p - 1$, and $x^n = a$ is not solvable, then there is a character χ such that*

(a) $\chi^n = \varepsilon$.
(b) $\chi(a) \neq 1$.

PROOF. Let g and λ be as in Proposition 8.1.3 and set $\chi = \lambda^{(p-1)/n}$. Then $\chi(g) = \lambda^{(p-1)/n}(g) = \lambda(g)^{(p-1)/n} = e^{2\pi i/n}$. Now $a = g^l$ for some l, and since $x^n = a$ is not solvable, we must have $n \nmid l$. Then $\chi(a) = \chi(g)^l = e^{2\pi i(l/n)} \neq 1$. Finally, $\chi^n = \lambda^{p-1} = \varepsilon$. $\qquad\square$

For $a \in F_p$, let $N(x^n = a)$ denote the number of solutions of the equation $x^n = a$. If $n \mid p - 1$, we have

Proposition 8.1.5. $N(x^n = a) = \sum_{\chi^n = \varepsilon} \chi(a)$ *where the sum is over all characters of order dividing n.*

PROOF. We claim first that there are exactly n characters of order dividing n. Since the value of $\chi(g)$ for such a character must be an nth root of unity, there are at most n such characters. In Proposition 8.1.4, we found a character χ such that $\chi(g) = e^{2\pi i/n}$. It follows that $\varepsilon, \chi, \chi^2, \ldots, \chi^{n-1}$ are n distinct characters of order dividing n.

To prove the formula, notice that $x^n = 0$ has one solution, namely, $x = 0$. Now $\sum_{\chi^n = \varepsilon} \chi(0) = 1$, since $\varepsilon(0) = 1$ and $\chi(0) = 0$ for $\chi \neq \varepsilon$.

Now suppose that $a \neq 0$ and that $x^n = a$ is solvable; i.e., there is an element b such that $b^n = a$. If $\chi^n = \varepsilon$, then $\chi(a) = \chi(b^n) = \chi(b)^n = \chi^n(b) = \varepsilon(b) = 1$. Thus $\sum_{\chi^n = \varepsilon} \chi(a) = n$, which is $N(x^n = a)$ in this case.

Finally, suppose that $a \neq 0$ and that $x^n = a$ is not solvable. We must show that $\sum_{\chi^n = \varepsilon} \chi(a) = 0$. Call the sum T. By Proposition 8.1.4, there is a character ρ such that $\rho(a) \neq 1$ and $\rho^n = \varepsilon$. A simple calculation shows that $\rho(a)T = T$ (one uses the obvious fact that the characters of order dividing n form a group). Thus $(\rho(a) - 1)T = 0$ and $T = 0$, as required. $\qquad\square$

As a special case, suppose that p is odd and that $n = 2$. Then the theorem says that $N(x^2 = a) = 1 + (a/p)$, where (a/p) is the Legendre symbol. This equation is easy to check directly.

In Section 3 we shall return to equations over the field F_p.

§2 Gauss Sums

In Chapter 6 we introduced quadratic Gauss sums. The following definition generalizes that notion.

Definition. Let χ be a character on F_p and $a \in F_p$. Set $g_a(\chi) = \sum_t \chi(t)\zeta^{at}$, where the sum is over all t in F_p, and $\zeta = e^{2\pi i/p}$. $g_a(\chi)$ is called a *Gauss sum* on F_p belonging to the character χ.

Proposition 8.2.1. *If $a \neq 0$ and $\chi \neq \varepsilon$, we have $g_a(\chi) = \chi(a^{-1})g_1(\chi)$. If $a \neq 0$ and $\chi = \varepsilon$ we have $g_a(\varepsilon) = 0$. If $a = 0$ and $\chi \neq \varepsilon$, we have $g_0(\chi) = 0$. If $a = 0$ and $\chi = \varepsilon$, we have $g_0(\varepsilon) = p$.*

PROOF. Suppose that $a \neq 0$ and that $\chi \neq \varepsilon$. Then

$$\chi(a)g_a(\chi) = \chi(a) \sum_t \chi(t)\zeta^{at} = \sum_t \chi(at)\zeta^{at} = g_1(\chi).$$

This proves the first assertion.

If $a \neq 0$, then

$$g_a(\varepsilon) = \sum_t \varepsilon(t)\zeta^{at} = \sum_t \zeta^{at} = 0.$$

We have used Lemma 1 of Chapter 6.

To finish the proof notice that $g_0(\chi) = \sum_t \chi(t)\zeta^{0t} = \sum_t \chi(t)$. If $\chi = \varepsilon$, the result is p; if $\chi \neq \varepsilon$, the result is zero by Proposition 8.1.2. $\qquad\square$

From now on we shall denote $g_1(\chi)$ by $g(\chi)$. We wish to determine the absolute value of $g(\chi)$. This can be done fairly easily by imitating the proof of Proposition 6.3.2.

Proposition 8.2.2. *If $\chi \neq \varepsilon$, then $|g(\chi)| = \sqrt{p}$.*

PROOF. The idea is to evaluate the sum $\sum_a g_a(\chi)\overline{g_a(\chi)}$ in two ways.

If $a \neq 0$, then by Proposition 8.2.1, $\overline{g_a(\chi)} = \overline{\chi(a^{-1})g(\chi)} = \chi(a)\overline{g(\chi)}$ and $g_a(\chi) = \chi(a^{-1})g(\chi)$. Thus $g_a(\chi)\overline{g_a(\chi)} = \chi(a^{-1})\chi(a)g(\chi)\overline{g(\chi)} = |g(\chi)|^2$. Since $g_0(\chi) = 0$ our sum has the value $(p-1)|g(\chi)|^2$.

On the other hand,

$$g_a(\chi)\overline{g_a(\chi)} = \sum_x \sum_y \chi(x)\overline{\chi(y)}\zeta^{ax-ay}.$$

Summing both sides over a and using the corollary to Lemma 1 of Chapter 6 yields

$$\sum_a g_a(\chi)\overline{g_a(x)} = \sum_x \sum_y \chi(x)\overline{\chi(y)}\delta(x, y)p = (p-1)p.$$

Thus $(p-1)|g(\chi)|^2 = (p-1)p$ and the result follows. $\qquad\square$

The relation of the above result to Proposition 6.3.2 is made clearer by the following considerations.

What is the relation between $\overline{g(\chi)}$ and $g(\bar\chi)$ ($\bar\chi$ is the character that takes a to $\overline{\chi(a)}$; i.e., it coincide with the character χ^{-1})?

$$\overline{g(\chi)} = \sum_t \overline{\chi(t)}\zeta^{-t} = \chi(-1)\sum_t \overline{\chi(-t)}\zeta^{-t} = \chi(-1)g(\bar\chi).$$

We have used the fact that $\overline{\chi(-1)} = \chi(-1)$, which is obvious since $\chi(-1) = \pm 1$. Thus the fact that $|g(\chi)|^2 = p$ can be written as $g(\chi)g(\bar\chi) = \chi(-1)p$. If χ is the Legendre symbol, this relation is precisely the result in Proposition 6.3.2.

§3 Jacobi Sums

Consider the equation $x^2 + y^2 = 1$ over the field F_p. Since F_p is finite, the equation has only finitely many solutions. Let $N(x^2 + y^2 = 1)$ be that number. We would like to determine this value explicitly.

Notice that

$$N(x^2 + y^2 = 1) = \sum_{a+b=1} N(x^2 = a)N(y^2 = b),$$

where the sum is over all pairs $a, b \in F_p$ such that $a + b = 1$. Since $N(x^2 = a) = 1 + (a/p)$, we obtain by substitution that

$$N(x^2 + y^2 = 1) = p + \sum_a \left(\frac{a}{p}\right) + \sum_b \left(\frac{b}{p}\right) + \sum_{a+b=1} \left(\frac{a}{p}\right)\left(\frac{b}{p}\right).$$

The first two sums are zero, so we are left with the task of evaluating the last sum. We shall see shortly that its value is $-(-1)^{(p-1)/2}$. Thus $N(x^2 + y^2 = 1)$ is $p - 1$ if $p \equiv 1$ (4) and $p + 1$ if $p \equiv 3$ (4). The reader is invited to check this result numerically for the first few primes.

Let us go a step further and try to evaluate $N(x^3 + y^3 = 1)$. As before we have

$$N(x^3 + y^3 = 1) = \sum_{a+b=1} N(x^3 = a)N(y^3 = b).$$

If $p \equiv 2$ (3), then $N(x^3 = a) = 1$ for all a since $(3, p - 1) = 1$. It follows that $N(x^3 + y^3 = 1) = p$ in this case. Assume now that $p \equiv 1$ (3). Let $\chi \neq \varepsilon$ be a character of order 3. Then χ^2 is a character of order 3 and $\chi^2 \neq \varepsilon$. Thus ε, χ, and χ^2 are all the characters of order 3, henceforth called cubic characters. By Proposition 8.1.5 we have $N(x^3 = a) = 1 + \chi(a) + \chi^2(a)$. Thus

$$N(x^3 + y^3 = 1) = \sum_{a+b=1} \sum_{i=0}^{2} \chi^i(a) \sum_{j=0}^{2} \chi^j(b)$$

$$= \sum_{i} \sum_{j} \left(\sum_{a+b=1} \chi^i(a)\chi^j(b) \right).$$

The inner sums are similar to the sum that occurred in the analysis of $N(x^2 + y^2 = 1)$.

Definition. Let χ and λ be characters of F_p and set $J(\chi, \lambda) = \sum_{a+b=1} \chi(a)\lambda(b)$. $J(\chi, \lambda)$ is called a *Jacobi sum*.

To complete the analysis of $N(x^2 + y^2 = 1)$ and $N(x^3 + y^3 = 1)$ we need to obtain information on the value of Jacobi sums. The following theorem not only supplies this information, but shows as well a surprising connection between Jacobi sums and Gauss sums.

Theorem 1. *Let χ and λ be nontrivial characters. Then*

(a) $J(\varepsilon, \varepsilon) = p$.
(b) $J(\varepsilon, \chi) = 0$.
(c) $J(\chi, \chi^{-1}) = -\chi(-1)$.
(d) *If $\chi\lambda \neq \varepsilon$, then*

$$J(\chi, \lambda) = \frac{g(\chi)g(\lambda)}{g(\chi\lambda)}.$$

PROOF. Part (a) is immediate, and part (b) is an immediate consequence of Proposition 8.1.2.

To prove part (c), notice that

$$J(\chi, \chi^{-1}) = \sum_{\substack{a+b=1}} \chi(a)\chi^{-1}(b) = \sum_{\substack{a+b=1 \\ b \neq 0}} \chi\left(\frac{a}{b}\right) = \sum_{a \neq 1} \chi\left(\frac{a}{1-a}\right).$$

Set $a/(1-a) = c$. If $c \neq -1$, then $a = c/(1+c)$. It follows that as a varies over F_p, less the element 1, that c varies over F_p, less the element -1. Thus

$$J(\chi, \chi^{-1}) = \sum_{c \neq -1} \chi(c) = -\chi(-1).$$

To prove part (d), notice that

$$g(\chi)g(\lambda) = \left(\sum_x \chi(x)\zeta^x\right)\left(\sum_y \lambda(y)\zeta^y\right)$$

$$= \sum_{x,y} \chi(x)\lambda(y)\zeta^{x+y}$$

$$= \sum_t \left(\sum_{x+y=t} \chi(x)\lambda(y)\right)\zeta^t. \tag{1}$$

If $t = 0$, then $\sum_{x+y=0} \chi(x)\lambda(y) = \sum_x \chi(x)\lambda(-x) = \lambda(-1)\sum_x \chi\lambda(x) = 0$, since $\chi\lambda \neq \varepsilon$ by assumption.

If $t \neq 0$, define x' and y' by $x = tx'$ and $y = ty'$. If $x + y = t$, then $x' + y' = 1$. It follows that

$$\sum_{x+y=t} \chi(x)\lambda(y) = \sum_{x'+y'=1} \chi(tx')\lambda(ty') = \chi\lambda(t)J(\chi, \lambda).$$

Substituting into Equation (1) yields

$$g(\chi)g(\lambda) = \sum_t \chi\lambda(t)J(\chi, \lambda)\zeta^t = J(\chi, \lambda)g(\chi\lambda). \qquad \square$$

Corollary. *If χ, λ, and $\chi\lambda$ are not equal to ε, then $|J(\chi, \lambda)| = \sqrt{p}$.*

PROOF. Take the absolute value of both sides of the equation in part (d) and use Proposition 8.2.2. $\square$

We now return to the analysis of $N(x^2 + y^2 = 1)$ and $N(x^3 + y^3 = 1)$. In the former case, it was necessary to evaluate the sum $\sum_{a+b=1} (a/p) \times (b/p)$. Case (c) of Theorem 1 is applicable and gives the result $-(-1/p) = -(-1)^{(p-1)/2}$, as was stated earlier.

In the case of $N(x^3 + y^3 = 1)$ we had to evaluate the sums $\sum_{a+b=1} \chi^i(a)\chi^j(b)$, where χ is a cubic character. Applying the theorem leads to the result

$$N(x^3 + y^3 = 1) = p - \chi(-1) - \chi^2(-1) + J(\chi, \chi) + J(\chi^2, \chi^2).$$

Since $-1 = (-1)^3$ we have $\chi(-1) = \chi^3(-1) = 1$. Also notice that $\chi^2 = \chi^{-1} = \bar{\chi}$. Thus

$$N(x^3 + y^3 = 1) = p - 2 + 2 \operatorname{Re} J(\chi, \chi).$$

This result is not as nice as the result for $N(x^2 + y^2 = 1)$, since we do not know $J(\chi, \chi)$ explicitly. Nevertheless, by the corollary to Theorem 1 we know that $|J(\chi, \chi)| = \sqrt{p}$ so we have the estimate

$$|N(x^3 + y^3 = 1) - p + 2| \le 2\sqrt{p}.$$

If we write N_p for the number of solutions to $x^3 + y^3 = 1$ in the field F_p, then the estimate says that N_p is approximately equal to $p - 2$ with an "error term" $2\sqrt{p}$. This shows that for large primes p there are always many solutions.

If $p \equiv 1 \ (3)$, there are always at least six solutions since $x^3 = 1$ and $y^3 = 1$ have three solutions each and we can write $1 + 0 = 1$ and $0 + 1 = 1$. For $p = 7$ and 13 these are the only solutions. For $p = 19$ other solutions exist; e.g., $3^3 + 10^3 \equiv 1 \ (19)$. These "nontrivial" solutions exist for all primes $p \ge 19$ since it follows from the estimate that $N_p \ge p - 2 - 2\sqrt{p} > 6$ for $p \ge 19$.

Using Jacobi sums we can easily extend our analysis to equations of the form $ax^n + by^n = 1$, but we shall not go more deeply into this matter now.

The corollary to Theorem 1 has two immediate consequences of considerable interest.

Proposition 8.3.1. *If $p \equiv 1 \ (4)$, then there exist integers a and b such that $a^2 + b^2 = p$.*

If $p \equiv 1 \ (3)$, then there exist integers a and b such that $a^2 - ab + b^2 = p$.

PROOF. If $p \equiv 1 \ (4)$, there is a character χ of order 4 (if λ has order $p - 1$, let $\chi = \lambda^{(p-1)/4}$). The values of χ are in the set $\{1, -1, i, -i\}$, where $i = \sqrt{-1}$. Thus $J(\chi, \chi) = \sum_{s+t=1} \chi(s)\chi(t) \in \mathbb{Z}[i]$ (see Chapter 1, Section 4). It follows that $J(\chi, \chi) = a + bi$, where $a, b \in \mathbb{Z}$; thus $p = |J(\chi, \chi)|^2 = a^2 + b^2$.

If $p \equiv 1 \ (3)$, there is a character χ of order 3. The values of χ are in the set $\{1, \omega, \omega^2\}$, where $\omega = e^{2\pi i/3} = (-1 + \sqrt{-3})/2$. Thus $J(\chi, \chi) \in \mathbb{Z}[\omega]$. As above, we have $J(\chi, \chi) = a + b\omega$, where $a, b \in \mathbb{Z}$ and $p = |J(\chi, \chi)|^2 = |a + b\omega|^2 = a^2 - ab + b^2$. $\square$

The fact that primes $p \equiv 1 \ (4)$ can be written as the sum of two squares was discovered by Fermat. It is not hard to prove that if $a, b > 0$, a is odd and b is even, then the representation $p = a^2 + b^2$ is unique.

If $p \equiv 1 \ (3)$, the representation $p = a^2 - ab + b^2$ is not unique even if we assume that $a, b > 0$. This can be seen from the equations

$$a^2 - ab + b^2 = (b - a)^2 - (b - a)b + b^2 = a^2 - a(a - b) + (a - b)^2.$$

However, we can reformulate things so that the result is unique. If $p = a^2 - ab + b^2$, then $4p = (2a - b)^2 + 3b^2 = (2b - a)^2 + 3a^2 = (a + b)^2 + 3(a - b)^2$.

We claim that 3 divides either a, b, or $a - b$. Suppose that $3 \nmid a$ and that $3 \nmid b$. If $a \equiv 1$ (3) and $b \equiv 2$ (3), or $a \equiv 2$ (3) and $b \equiv 1$ (3), then $a^2 - ab + b^2 \equiv 0$ (3), which implies that $3|p$, a contradiction. Thus $3|a - b$, and we have

Proposition 8.3.2. *If $p \equiv 1$ (3), then there are integers A and B such that $4p = A^2 + 27B^2$. In this representation of $4p$, A and B are uniquely determined up to sign.*

PROOF. The proof of the uniqueness is left to the Exercises.   □

Theorem 1 together with a simple argument leads to a further interesting relation between Gauss sums and Jacobi sums.

Proposition 8.3.3. *Suppose that $p \equiv 1$ (n) and that χ is a character of order $n > 2$. Then*

$$g(\chi)^n = \chi(-1)pJ(\chi, \chi)J(\chi, \chi^2) \cdots J(\chi, \chi^{n-2}).$$

PROOF. Using part (d) of Theorem 1 we have $g(\chi)^2 = J(\chi, \chi)g(\chi^2)$. Multiply both sides by $g(\chi)$ and we get $g(\chi)^3 = J(\chi, \chi)J(\chi, \chi^2)g(\chi^3)$. Continuing in this way shows that

$$g(\chi)^{n-1} = J(\chi, \chi)J(\chi, \chi^2) \cdots J(\chi, \chi^{n-2})g(\chi^{n-1}). \qquad (2)$$

Now $\chi^{n-1} = \chi^{-1} = \bar{\chi}$. Thus, as we have seen, $g(\chi)g(\chi^{n-1}) = g(\chi)g(\bar{\chi}) = \chi(-1)p$. The result follows upon multiplying both sides of Equation (2) by $g(\chi)$.   □

Corollary. *If χ is a cubic character, then*

$$g(\chi)^3 = pJ(\chi, \chi).$$

PROOF. This is simply a special case of the proposition and the fact that $\chi(-1) = \chi((-1)^3) = 1$.   □

Using this corollary, we are in a position to analyze more fully the complex number $J(\chi, \chi)$ that occurred in the discussion of $N(x^3 + y^3 = 1)$. We have seen that $J(\chi, \chi) = a + b\omega$, where $a, b \in Z$ and $\omega = e^{2\pi i/3} = (-1 + \sqrt{-3})/2$.

Proposition 8.3.4. *Suppose that $p \equiv 1$ (3) and that χ is a cubic character. Set $J(\chi, \chi) = a + b\omega$ as above. Then*

(a) $b \equiv 0$ (3).
(b) $a \equiv -1$ (3).

PROOF. We shall work with congruences in the ring of algebraic integers as in Chapter 6:

$$g(\chi)^3 = \left(\sum_t \chi(t)\zeta^t \right)^3 \equiv \sum_t \chi(t)^3 \zeta^{3t} \ (3).$$

Since $\chi(0) = 0$ and $\chi(t)^3 = 1$ for $t \neq 0$ we have $\sum_t \chi(t)^3 \zeta^{3t} = \sum_{t \neq 0} \zeta^{3t}$ $= -1$. Thus

$$g(\chi)^3 = pJ(\chi, \chi) \equiv a + b\omega \equiv -1 \ (3).$$

Working with $\bar{\chi}$ instead of χ and remembering that $\overline{g(\chi)} = g(\bar{\chi})$ we find that

$$g(\bar{\chi})^3 = pJ(\bar{\chi}, \bar{\chi}) \equiv a + b\bar{\omega} \equiv -1 \ (3).$$

Subtracting yields $b(\omega - \bar{\omega}) \equiv 0 \ (3)$, or $b\sqrt{-3} \equiv 0 \ (3)$. Thus $-3b^2 \equiv 0 \ (9)$ and it follows that $3 | b$. Since $3 | b$ and $a + b\omega \equiv -1 \ (3)$, we must have $a \equiv -1 \ (3)$, which completes the proof. $\square$

Corollary. *Let $A = 2a - b$ and $B = b/3$. Then $A \equiv 1 \ (3)$ and*

$$4p = A^2 + 27B^2.$$

PROOF. Since $J(\chi, \chi) = a + b\omega$ and $|J(\chi, \chi)|^2 = p$ we have $p = a^2 - ab + b^2$. Thus $4p = (2a - b)^2 + 3b^2$ and $4p = A^2 + 27B^2$.

By Proposition 8.3.4, $3 | b$ and $a \equiv -1 \ (3)$. Therefore, $A = 2a - b \equiv 1 \ (3)$. $\square$

We are now ready to prove the following beautiful theorem due to Gauss.

Theorem 2. *Suppose that $p \equiv 1 \ (3)$. Then there are integers A and B such that $4p = A^2 + 27B^2$. If we require that $A \equiv 1 \ (3)$, A is uniquely determined, and*

$$N(x^3 + y^3 = 1) = p - 2 + A.$$

PROOF. We have already shown that $N(x^3 + y^3 = 1) = p - 2 + 2 \, \text{Re} \, J(\chi, \chi)$. Since $J(\chi, \chi) = a + b\omega$ as above, we have $\text{Re} \, J(\chi, \chi) = (2a - b)/2$. Thus $2 \, \text{Re} \, J(\chi, \chi) = 2a - b = A \equiv 1 \ (3)$. Uniqueness is left as an exercise. $\square$

Let us illustrate this result with two examples, $p = 61$ and $p = 67$.
$4 \cdot 61 = 1^2 + 27 \cdot 3^2$. Thus the number of solutions to $x^3 + y^3 = 1$ in F_{61} is $61 - 2 + 1 = 60$.

Now, $4 \cdot 67 = 5^2 + 27 \cdot 3^2$. We must be careful here; since $5 \not\equiv 1 \ (3)$ we must choose $A = -5$. The answer is thus $67 - 2 - 5 = 60$, which by coincidence (?) is the same as for $p = 61$.

§4 The Equation $x^n + y^n = 1$ in F_p

We shall assume that $p \equiv 1 \ (n)$ and investigate the number of solutions to the equation $x^n + y^n = 1$ over the field F_p. The methods of Section 3 are directly applicable.

We have

$$N(x^n + y^n = 1) = \sum_{a+b=1} N(x^n = a)N(y^n = b).$$

Let χ be a character of order n. By Proposition 8.1.5

$$N(x^n = a) = \sum_{i=0}^{n-1} \chi^i(a).$$

Combining these results yields

$$N(x^n + y^n = 1) = \sum_{j=0}^{n-1} \sum_{i=0}^{n-1} J(\chi^j, \chi^i).$$

Theorem 1 can be used to estimate this sum. When $i = j = 0$ we have $J(\chi^0, \chi^0) = J(\varepsilon, \varepsilon) = p$. When $j + i = n$, $\chi^j = (\chi^i)^{-1}$ so that $J(\chi^j, \chi^i) = -\chi^j(-1)$. The sum of these terms is $-\sum_{j=1}^{n-1} \chi^j(-1)$. Notice that $\sum_{j=0}^{n-1} \chi^j(-1)$ is n when -1 is an nth power and zero otherwise. Thus the contribution of these terms is $1 - \delta_n(-1)n$, where $\delta_n(-1)$ has the obvious meaning. Finally, if $i = 0$ and $j \neq 0$ or $i \neq 0$ and $j = 0$, then $J(\chi^i, \chi^j) = 0$. Thus

$$N(x^n + y^n = 1) = p + 1 - \delta_n(-1)n + \sum_{i,j} J(\chi^i, \chi^j).$$

The sum is over indices i and j between 1 and $n - 1$ subject to the condition that $i + j \neq n$. There are $(n-1)^2 - (n-1) = (n-1)(n-2)$ such terms and they all have absolute value $\sqrt{p}$. Thus

Proposition 8.4.1.

$$|N(x^n + y^n = 1) + \delta_n(-1)n - (p+1)| \leq (n-1)(n-2)\sqrt{p}.$$

The term $\delta_n(-1)n$ will be interpreted later as the number of points "at infinity" on the curve $x^n + y^n = 1$.

For large p the above estimate shows the existence of many nontrivial solutions.

§5 More on Jacobi Sums

Theorem 1 can be generalized in a very fruitful manner. First we need a definition.

Definition. Let $\chi_1, \chi_2, \ldots, \chi_l$ be characters on F_p. A Jacobi sum is defined by the formula

$$J(\chi_1, \chi_2, \ldots, \chi_l) = \sum_{t_1 + \cdots + t_l = 1} \chi_1(t_1)\chi_2(t_2) \cdots \chi_l(t_l).$$

Notice that when $l = 2$ this reduces to our former definition of Jacobi sum.

It is useful to define another sum, which will be left unnamed:

$$J_0(\chi_1, \ldots, \chi_l) = \sum_{t_1 + \cdots + t_l = 0} \chi_1(t_1)\chi_2(t_2) \cdots \chi_l(t_l).$$

Proposition 8.5.1.

(a) $J_0(\varepsilon, \varepsilon, \ldots, \varepsilon) = J(\varepsilon, \varepsilon, \ldots, \varepsilon) = p^{l-1}$.

(b) *If some but not all of the χ_i are trivial, then* $J_0(\chi_1, \chi_2, \ldots, \chi_l) = J(\chi_1, \chi_2, \ldots, \chi_l) = 0$.

(c) *Assume that $\chi_l \neq \varepsilon$. Then*

$$J_0(\chi_1, \chi_2, \ldots, \chi_l) = \begin{cases} 0, & \text{if } \chi_1\chi_2 \cdots \chi_l \neq \varepsilon, \\ \chi_l(-1)(p-1)J(\chi_1, \chi_2, \ldots, \chi_{l-1}), & \text{otherwise.} \end{cases}$$

PROOF. If $t_1, t_2, \ldots, t_{l-1}$ are chosen (arbitrarily) in F_p, then t_l is uniquely determined by the condition $t_1 + t_2 + \cdots + t_{l-1} + t_l = 0$. Thus $J_0(\varepsilon, \varepsilon, \ldots, \varepsilon) = p^{l-1}$. Similarly for $J(\varepsilon, \varepsilon, \ldots, \varepsilon)$.

To prove part (b), assume that $\chi_1, \chi_2, \ldots, \chi_s$ are nontrivial and that $\chi_{s+1} = \chi_{s+2} = \cdots = \chi_l = \varepsilon$. Then

$$\sum_{t_1 + \cdots + t_l = 0} \chi_1(t_1)\chi_2(t_2) \cdots \chi_l(t_l)$$

$$= \sum_{t_1, t_2, \ldots, t_{l-1}} \chi_1(t_1)\chi_2(t_2) \cdots \chi_s(t_s)$$

$$= p^{l-s-1} \left(\sum_{t_1} \chi_1(t_1) \right) \left(\sum_{t_2} \chi_2(t_2) \right) \cdots \left(\sum_{t_s} \chi_s(t_s) \right) = 0.$$

We have used Proposition 8.1.2. Thus $J_0(\chi_1, \chi_2, \ldots, \chi_l) = 0$. Similarly for $J(\chi_1, \ldots, \chi_l)$.

To prove part (c), notice that

$$J_0(\chi_1, \chi_2, \ldots, \chi_l) = \sum_s \left(\sum_{t_1 + \cdots + t_{l-1} = -s} \chi_1(t_1) \cdots \chi_{l-1}(t_{l-1}) \right) \chi_l(s)$$

Since $\chi_l \neq \varepsilon$, $\chi_l(0) = 0$, so we may assume that $s \neq 0$ in the above sum. If $s \neq 0$, define t_i' by $t_i = -st_i'$. Then

$$\sum_{t_1 + \cdots + t_{l-1} = -s} \chi_1(t_1) \cdots \chi_{l-1}(t_{l-1})$$

$$= \chi_1\chi_2 \cdots \chi_{l-1}(-s) \sum_{t_1' + \cdots + t_{l-1}' = 1} \chi_1(t_1') \cdots \chi_{l-1}(t_{l-1}')$$

$$= \chi_1\chi_2 \cdots \chi_{l-1}(-s)J(\chi_1, \ldots, \chi_{l-1}).$$

Combining these results yields

$$J_0(\chi_1, \chi_2, \ldots, \chi_l) = \chi_1\chi_2 \cdots \chi_{l-1}(-1)J(\chi_1, \ldots, \chi_{l-1}) \sum_{s \neq 0} \chi_1\chi_2 \cdots \chi_l(s).$$

The main result follows since the sum is zero if $\chi_1\chi_2\cdots\chi_l \neq \varepsilon$ and $p-1$ if $\chi_1\chi_2\cdots\chi_l = \varepsilon$. $\qquad\qquad\qquad\qquad\qquad\qquad\qquad\qquad\qquad\qquad\qquad\square$

Parts (a) and (b) of Proposition 8.5.1 generalize parts (a) and (b) of Theorem 1. Part (d) of Theorem 1 can be generalized as follows.

Theorem 3. *Assume that* $\chi_1, \chi_2, \ldots, \chi_r$ *are nontrivial and also that* $\chi_1\chi_2\cdots\chi_r$ *is nontrivial. Then*

$$g(\chi_1)g(\chi_2)\cdots g(\chi_r) = J(\chi_1, \chi_2, \ldots, \chi_r)g(\chi_1\chi_2\cdots\chi_r).$$

PROOF. Let $\psi: F_p \to \mathbb{C}$ be defined by $\psi(t) = \zeta^t$. Then $\psi(t_1 + t_2) = \psi(t_1)\psi(t_2)$, and $g(\chi) = \sum \chi(t)\psi(t)$. The introduction of ψ is for notational convenience.

$$g(\chi_1)g(\chi_2)\cdots g(\chi_r)$$

$$= \left(\sum_{t_1} \chi_1(t_1)\psi(t_1)\right)\cdots\left(\sum_{t_r} \chi_r(t_r)\psi(t_r)\right)$$

$$= \sum_s \left(\sum_{t_1+t_2+\cdots+t_r=s} \chi_1(t_1)\chi_2(t_2)\cdots\chi_r(t_r)\right)\psi(s).$$

If $s = 0$, then by part (c) of Proposition 8.5.1 and the assumption that $\chi_1\cdots\chi_r \neq \varepsilon$

$$\sum_{t_1+\cdots+t_r=0} \chi_1(t_1)\cdots\chi_r(t_r) = 0.$$

If $s \neq 0$, the substitution $t_i = st_i'$ shows that

$$\sum_{t_1+\cdots+t_r=s} \chi_1(t_1)\cdots\chi_r(t_r) = \chi_1\chi_2\cdots\chi_r(s)J(\chi_1, \chi_2, \ldots, \chi_r).$$

Putting these remarks together, we have

$$g(\chi_1)\cdots g(\chi_r) = J(\chi_1, \chi_2, \ldots, \chi_r)\sum_{s\neq 0} \chi_1\chi_2\cdots\chi_r(s)\psi(s)$$

$$= J(\chi_1, \chi_2, \ldots, \chi_r)g(\chi_1\chi_2\cdots\chi_r). \qquad\qquad\square$$

Corollary 1. *Suppose that* $\chi_1, \chi_2, \ldots, \chi_r$ *are nontrivial and that* $\chi_1\chi_2\cdots\chi_r$ *is trivial. Then*

$$g(\chi_1)g(\chi_2)\cdots g(\chi_r) = \chi_r(-1)pJ(\chi_1, \chi_2, \ldots, \chi_{r-1}).$$

PROOF. $g(\chi_1)g(\chi_2)\cdots g(\chi_{r-1}) = J(\chi_1, \ldots, \chi_{r-1})g(\chi_1\chi_2\cdots\chi_{r-1})$ by Theorem 3. Multiply both sides by $g(\chi_r)$. Since $\chi_1\chi_2\cdots\chi_{r-1} = \chi_r^{-1}$ we have

$$g(\chi_1\cdots\chi_{r-1})g(\chi_r) = g(\chi_r^{-1})g(\chi_r) = \chi_r(-1)p. \qquad\qquad\square$$

Corollary 2. *Let the hypotheses be as in Corollary 1. Then*

$$J(\chi_1, \ldots, \chi_r) = -\chi_r(-1)J(\chi_1, \chi_2, \ldots, \chi_{r-1}).$$

[If $r = 2$, *we set* $J(\chi_1) = 1$.]

PROOF. If $r = 2$, this is the assertion of part (c) of Theorem 1.

Suppose that $r > 2$. In the proof of Theorem 3 use the hypothesis that $\chi_1\chi_2 \cdots \chi_r = \varepsilon$. This yields

$$g(\chi_1)g(\chi_2) \cdots g(\chi_r) = J_0(\chi_1, \chi_2, \ldots, \chi_r) + J(\chi_1, \ldots, \chi_r) \sum_{s \neq 0} \psi(s).$$

Since $\sum_s \psi(s) = 0$, the sum in the formula is equal to -1. By part (c) of Proposition 8.5.1, we have $J_0(\chi_1, \ldots, \chi_r) = \chi_r(-1)(p - 1)J(\chi_1, \ldots, \chi_{r-1})$. By Corollary 1, $g(\chi_1) \cdots g(\chi_r) = \chi_r(-1)pJ(\chi_1, \chi_2, \ldots, \chi_{r-1})$. Putting these results together proves the corollary. □

Theorem 4. *Assume that* $\chi_1, \chi_2, \ldots, \chi_r$ *are nontrivial.*

(a) *If* $\chi_1\chi_2 \cdots \chi_r \neq \varepsilon$, *then*

$$|J(\chi_1, \chi_2, \ldots, \chi_r)| = p^{(r-1)/2}.$$

(b) *If* $\chi_1\chi_2 \cdots \chi_r = \varepsilon$, *then*

$$|J_0(\chi_1, \chi_2, \ldots, \chi_r)| = (p - 1)p^{(r/2)-1}$$

and

$$|J(\chi_1, \chi_2, \ldots, \chi_r)| = p^{(r/2)-1}.$$

PROOF. If χ is nontrivial, $|g(\chi)| = \sqrt{p}$. Part (a) follows directly from Theorem 3.

Part (b) follows similarly from part (c) of Proposition 8.5.1 and from Corollary 2 to Theorem 3. □

§6 Applications

Earlier in this chapter we investigated the number of solutions of the equation $x^2 + y^2 = 1$ in the field F_p. It is natural to ask the same question about the equation $x_1^2 + x_2^2 + \cdots + x_r^2 = 1$. The answer can easily be found using the results of Section 5.

Let χ be a character of order 2 ($\chi(a) = (a/p)$ in our earlier notation). Then $N(x^2 = a) = 1 + \chi(a)$. Thus

$$N(x_1^2 + \cdots + x_r^2 = 1) = \sum N(x_1^2 = a_1)N(x_2^2 = a_2) \cdots N(x_r^2 = a_r),$$

where the sum is over all r-tuples $(a_1, \ldots, a_r)$ such that $a_1 + a_2 + \cdots + a_r = 1$. Multiplying out, and using Proposition 8.5.1, yields

$$N(x_1^2 + \cdots + x_r^2 = 1) = p^{r-1} + J(\chi, \chi, \ldots, \chi).$$

If r is odd, $\chi' = \chi$, and if r is even, $\chi' = \varepsilon$.

Suppose that r is odd. Then Theorem 3 applies and we have $J(\chi, \ldots, \chi) = g(\chi)^{r-1}$. Since $g(\chi)^2 = \chi(-1)p$ it follows that $J(\chi, \ldots, \chi) = \chi(-1)^{(r-1)/2}p^{(r-1)/2}$.

If r is even, we use Corollary 2 to Theorem 3 and find that $J(\chi, \chi, \ldots, \chi) = -\chi(-1)^{r/2}p^{(r-2)/2}$. Finally, remember that $\chi(-1) = (-1)^{(p-1)/2}$. Thus

Proposition 8.6.1. *If r is odd, then*

$$N(x_1^2 + x_2^2 + \cdots + x_r^2 = 1) = p^{r-1} + (-1)^{((r-1)/2)((p-1)/2)}p^{(r-1)/2}.$$

If r is even, then

$$N(x_1^2 + x_2^2 + \cdots + x_r^2 = 1) = p^{r-1} - (-1)^{(r/2)((p-1)/2)}p^{(r/2)-1}.$$

The most general equation that can be treated by these methods has the form $a_1x_1^{l_1} + a_2x_2^{l_2} + \cdots + a_rx_r^{l_r} = b$, where $a_1, \ldots, a_r$, $b \in F_p$, and $l_1, l_2, \ldots, l_r$ are positive integers. We shall return to this subject in Section 7. For now, we shall use Jacobi sums to give yet another proof of the law of quadratic reciprocity.

Let q be an odd prime not equal to p, and χ the character of order 2 on F_p. Then by Corollary 1 to Theorem 3

$$g(\chi)^{q+1} = (-1)^{(p-1)/2}pJ(\chi, \chi, \ldots, \chi),$$

where there are q components in the Jacobi sum.

Since $q + 1$ is even $g(\chi)^{q+1} = (g(\chi)^2)^{(q+1)/2} = (-1)^{((p-1)/2)((q+1)/2)} \cdot p^{(q+1)/2}$. Substituting into the formula we find that

$$(-1)^{((p-1)/2)((q-1)/2)}p^{(q-1)/2} = J(\chi, \chi, \ldots, \chi).$$

Now, $J(\chi, \chi, \ldots, \chi) = \sum \chi(t_1)\chi(t_2) \cdots \chi(t_q)$, where the sum is over all $(t_1, t_2, \ldots, t_q)$ with $t_1 + t_2 + \cdots + t_q = 1$. If $t = t_1 = t_2 = \cdots = t_q$, then $t = 1/q$, and the corresponding term of the sum has value $\chi(1/q)^q = \chi(q)^{-q} = \chi(q)$. If not all the t_i are equal, then there are q different q-tuples obtained from $(t_1, t_2, \ldots, t_q)$ by cyclic permutation. The corresponding terms of the sum all have the same value. Thus

$$(-1)^{((p-1)/2)((q-1)/2)}p^{(q-1)/2} \equiv \chi(q) \; (q).$$

Since $\chi(q) = (q/p)$ and $p^{(q-1)/2} \equiv (p/q) \; (q)$ we have

$$(-1)^{((p-1)/2)((q-1)/2)}\left(\frac{p}{q}\right) \equiv \left(\frac{q}{p}\right) \; (q)$$

and thus

$$(-1)^{((p-1)/2)((q-1)/2)}\left(\frac{p}{q}\right) = \left(\frac{q}{p}\right).$$

§7 A General Theorem

All the equations we have considered up to now are special cases of

$$a_1x_1^{l_1} + a_2x_2^{l_2} + \cdots + a_rx_r^{l_r} = b, \tag{3}$$

where $a_1, a_2, \ldots, a_r, \in F_p^*$ and $b \in F_p$. Let N be the number of solutions. Our object is to give a formula for N and an estimate for N. The methods to be used are identical with those already developed in the previous sections.

To begin with, we have

$$N = \sum N(x_1^{l_1} = u_1)N(x_2^{l_2} = u_2) \cdots N(x_r^{l_r} = u_r), \tag{4}$$

where the sum is over all r-tuples $(u_1, u_2, \ldots, u_r)$ such that $\sum_{i=1}^r a_i u_i = b$.

We shall assume that $l_1, l_2, \ldots, l_r$ are divisors of $p - 1$, although this is not necessary (see the Exercises). Let χ_i vary over the characters of order dividing l_i. Then

$$N(x_i^{l_i} = u_i) = \sum_{\chi_i} \chi_i(u_i).$$

Substituting into Equation (4) we get

$$N = \sum_{\chi_1, \chi_2, \ldots, \chi_r} \sum_{\sum a_i u_i = b} \chi_1(u_1)\chi_2(u_2) \cdots \chi_r(u_r). \tag{5}$$

The inner sum is closely related to the Jacobi sums that we have considered.

It is necessary to treat the cases $b = 0$ and $b \neq 0$ separately.

If $b = 0$, let $t_i = a_i u_i$. Then the inner sum becomes

$$\chi_1(a_1^{-1})\chi_2(a_2^{-1}) \cdots \chi_r(a_r^{-1})J_0(\chi_1, \chi_2, \ldots, \chi_r).$$

If $b \neq 0$, let $t_i = b^{-1}a_i u_i$. The inner sum becomes

$$\chi_1\chi_2 \cdots \chi_r(b)\chi_1(a_1^{-1}) \cdots \chi_r(a_r^{-1})J(\chi_1, \chi_2, \ldots, \chi_r).$$

In both cases, if $\chi_1 = \chi_2 = \cdots = \chi_r = \varepsilon$, the term has the value p^{r-1} since $J_0(\varepsilon, \ldots, \varepsilon) = J(\varepsilon, \varepsilon, \ldots, \varepsilon) = p^{r-1}$. If some but not all the χ_i are equal to ε, then the term has the value zero. In the first case the value is zero unless $\chi_1\chi_2 \cdots \chi_r = \varepsilon$. All this is a consequence of Proposition 8.5.1.

Putting this together with Theorem 4 we obtain

Theorem 5. *If $b = 0$, then*

$$N = p^{r-1} + \sum \chi_1(a_1^{-1})\chi_2(a_2^{-1}) \cdots \chi_r(a_r^{-1})J_0(\chi_1, \chi_2, \ldots, \chi_r).$$

The sum is over all r-tuples of characters $\chi_1, \chi_2, \ldots, \chi_r$, where $\chi_i^{l_i} = \varepsilon$, $\chi_i \neq \varepsilon$ for $i = 1, \ldots, r$, and $\chi_1\chi_2 \cdots \chi_r = \varepsilon$. If M is the number of such r-tuples, then

$$|N - p^{r-1}| \leq M(p - 1)p^{(r/2)-1}.$$

If $b \neq 0$, then

$$N = p^{r-1} + \sum \chi_1\chi_2 \cdots \chi_r(b)\chi_1(a_1^{-1}) \cdots \chi_r(a_r^{-1})J(\chi_1, \chi_2, \ldots, \chi_r).$$

The summation is over all r-tuples of characters $\chi_1, \ldots, \chi_r$, where $\chi_i^{l_i} = \varepsilon$ and $\chi_i \neq \varepsilon$ for $i = 1, \ldots, r$. If M_0 is the number of such r-tuples with $\chi_1\chi_2 \cdots \chi_r = \varepsilon$, and M_1 is the number of such r-tuples with $\chi_1\chi_2 \cdots \chi_r \neq \varepsilon$, then

$$|N - p^{r-1}| \leq M_0 p^{(r/2)-1} + M_1 p^{(r-1)/2}.$$

An immediate consequence of Theorem 5 is worth noting. Let $a_1, a_2, \ldots, a_r$ and $b \in \mathbb{Z}$ and consider the congruence

$$a_1 x_1^{l_1} + a_2 x_2^{l_2} + \cdots + a_r x_r^{l_r} \equiv b\ (p).$$

Then if p is sufficiently large, the congruence has many solutions. In fact, the number of solutions tends to infinity as p is taken larger and larger.

NOTES

The inspiration for this chapter is the famous paper of A. Weil [80]. The basic relationship between Gauss sums, also known as Lagrange resolvents, and Jacobi sums was known to Gauss [34] (unpublished), Jacobi [47], Eisenstein [27], and Cauchy. Complete proofs of the fundamental relations given in Proposition 8.3.3 and Theorem 1 were published by Eisenstein in his paper "Beiträge zur Kreistheilung" in 1844. Eisenstein also introduced generalized Jacobi sums (Section 5) to obtain a proof of the law of biquadratic reciprocity (see Chapter 9).

Aside from its usefulness in obtaining the Weil–Riemann hypothesis for certain hypersurfaces over finite fields (see Chapter 11), the generalized Jacobi sum is of importance in the theory of cyclotomy and difference sets. For an introduction to this material, see Storer [74]. See also the difficult but important continuation of [80] by Weil [81].

Material on Gauss and Jacobi sums is scattered throughout the treatise of Hasse [41]. He gives a systematic presentation in his last chapter where in addition to developing many interesting results he shows how both types of sum arise naturally in the theory of cyclotomic number fields. Much of the theory in that chapter is distilled from the paper of Davenport and Hasse [23]. The latter paper is well worth close study, but it is unfortunately of an advanced nature and is probably inaccessible to a beginner. Somewhat less difficult are the more recent papers of K. Yamamoto [82] and A. Yokoyama [83]. One should also consult the classical treatise of P. Bachman [5].

More recently B. C. Berndt and R. J. Evans have studied Gauss, Jacobi, and other classical character sums attached to characters of order 6, 8, 12, 24. For their interesting results and extensive bibliography the reader should consult [92] and [95]. See also Leonard and Williams [177].

Theorem 2 is proved by Gauss in §358 of *Disquisitiones Arithmeticae*. He does not really state the theorem explicitly. It comes out as a by-product of another investigation. What he does, in fact, is to use the theorem to help find the algebraic equation satisfied by certain Gauss sums. We have done the reverse, using the theory of Gauss sums to derive the theorem. Gauss derived other results of this type in his first memoir on biquadratic reciprocity [34]. For further historical remarks about this subject, see the introduction to the paper of Weil [80].

The estimates given in Theorem 5 are derived in the first chapter of Borevich and Shafarevich [9]. They use a somewhat different method which we have outlined in the Exercises. In the special case of quadratic forms, i.e., when all the $l = 2$, the result goes back at least to Dickson [25].

The technique of counting solutions by means of characters lends itself naturally to the problem of finding sequences of integers of prescribed length having prescribed kth power character modulo p. This problem is dealt with to some extent in Hasse [41]. In an interesting, and elementary paper, Davenport [21] shows that the number of sequences of four successive quadratic residues between 1 and p satisfies the inequality $|R - p/8| < Kp^{3/4}$, where K is a constant independent of p. Better estimates can be obtained using the results of Weil. For another paper along the same lines, see Graham [36].

One final remark on Theorem 5. It is due originally to Weil and independently (and almost simultaneously) to L. K. Hua and H. S. Vandiver (*Proc. Nat. Acad. Sci. U.S.A.*, **35** (1949), 94–99). With a few simplifications and addenda we have essentially followed Weil's presentation.

EXERCISES

1. Let p be a prime and $d = (m, p - 1)$. Prove that $N(x^m = a) = \sum \chi(a)$, the sum being over all χ such that $\chi^d = \varepsilon$.

2. With the notation of Exercise 1 show that $N(x^m = a) = N(x^d = a)$ and conclude that if $d_i = (m_i, p - 1)$, then $\sum_i a_i x^{m_i} = b$ and $\sum_i a_i x^{d_i} = b$ have the same number of solutions.

3. Let χ be a nontrivial multiplicative character of F_p and ρ be the character of order 2. Show that $\sum_t \chi(1 - t^2) = J(\chi, \rho)$. [*Hint*: Evaluate $J(\chi, \rho)$ using the relation $N(x^2 = a) = 1 + \rho(a)$.]

4. Show, if $k \in F_p$, $k \neq 0$, that $\sum_t \chi(t(k - t)) = \chi(k^2/2^2)J(\chi, \rho)$.

5. If $\chi^2 \neq \varepsilon$, show that $g(\chi)^2 = \chi(2)^{-2}J(\chi, \rho)g(\chi^2)$. [*Hint*: Write out $g(\chi)^2$ explicitly and use Exercise 4.]

6. (continuation) Show that $J(\chi, \chi) = \chi(2)^{-2}J(\chi, \rho)$.

7. Suppose that $p \equiv 1\ (4)$ and that χ is a character of order 4. Then $\chi^2 = \rho$ and $J(\chi, \chi) = \chi(-1)J(\chi, \rho)$. [*Hint*: Evaluate $g(\chi)^4$ in two ways.]

8. Generalize Exercise 3 in the following way. Suppose that p is a prime, $\sum_t \chi(1 - t^m) = \sum_\lambda J(\chi, \lambda)$, where λ varies over all characters such that $\lambda^m = \varepsilon$. Conclude that $|\sum_t \chi(1 - t^m)| \leq (m - 1)p^{1/2}$.

9. Suppose that $p \equiv 1\ (3)$ and that χ is a character of order 3. Prove (using Exercise 5) that $g(\chi)^3 = p\pi$, where $\pi = \chi(2)J(\chi, \rho)$.

10. (continuation) Show that $\chi\rho$ is a character of order 6 and that $g(\chi\rho)^6 = (-1)^{(p-1)/2}p\bar{\pi}^4$.

11. Use Gauss' theorem to find the number of solutions to $x^3 + y^3 = 1$ in F for $p = 13$, 19, 37, and 97.

12. If $p \equiv 1$ (4), then we have seen that $p = a^2 + b^2$ with $a, b \in \mathbb{Z}$. If we require that a
and b be positive, that a be odd, and that b be even, show that a and b are uniquely
determined. (*Hint*: Use the fact that unique factorization holds in $\mathbb{Z}[i]$ and that if
$p = a^2 + b^2$ then $a + bi$ is a prime in $\mathbb{Z}[i]$.)

13. If $p \equiv 1$ (3), we have seen that $4p = A^2 + 27B^2$ with $A, B \in \mathbb{Z}$. If we require that
$A \equiv 1$ (3), show that A is uniquely determined. (*Hint*: Use the fact that unique
factorization holds in $\mathbb{Z}[\omega]$. This proof is a little trickier than that for Exercise 12.)

14. Suppose that $p \equiv 1$ (n) and that χ is a character of order n. Show that $g(\chi)^n \in \mathbb{Z}[\zeta]$,
where $\zeta = e^{2\pi i/n}$.

15. Suppose that $p \equiv 1$ (6) and let χ and ρ be characters of order 3 and 2, respectively.
Show that the number of solutions to $y^2 = x^3 + D$ in F is $p + \pi + \bar{\pi}$, where
$\pi = \chi\rho(D)J(\chi, \rho)$. If $\chi(2) = 1$, show that the number of solutions to $y^2 = x^3 + 1$
is $p + A$, where $4p = A^2 + 27B^2$ and $A \equiv 1$ (3). Verify this result numerically
when $p = 31$.

16. Suppose that $p \equiv 1$ (4) and that χ is a character of order 4. Let N be the number of
solutions to $x^4 + y^4 = 1$ in F_p. Show that $N = p + 1 - \delta_4(-1)4 + 2\,\mathrm{Re}\,J(\chi, \chi) +$
$4\,\mathrm{Re}\,J(\chi, \rho)$.

17. (continuation)   By Exercise 7, $J(\chi, \chi) = \chi(-1)J(\chi, \rho)$. Let $\pi = -J(\chi, \rho)$. Show that
    (a) $N = p - 3 - 6\,\mathrm{Re}\,\pi$ if $p \equiv 1$ (8).
    (b) $N = p + 1 - 2\,\mathrm{Re}\,\pi$ if $p \equiv 5$ (8).

18. (continuation)   Let $\pi = a + bi$. One can show (see Chapter 11, Section 5) that a is
odd, b is even, and $a \equiv 1$ (4) if $4 | b$ and $a \equiv -1$ (4) if $4 \nmid b$. Let $p = A^2 + B^2$ and
fix A by requiring that $A \equiv 1$ (4). Then show that
    (a) $N = p - 3 - 6A$ if $p \equiv 1$ (8).
    (b) $N = p + 1 + 2A$ if $p \equiv 5$ (8).

19. Find a formula for the number of solutions to $x_1^2 + x_2^2 + \cdots + x_r^2 = 0$ in F_p.

20. Generalize Proposition 8.6.1 by finding an explicit formula for the number of
solutions to $a_1 x_1^2 + a_2 x_2^2 + \cdots + a_r x_r^2 = 1$ in F_p.

21. Suppose that $p \equiv 1$ (d), $\zeta = e^{2\pi i/p}$, and consider $\sum_x \zeta^{ax^d}$. Show that $\sum_x \zeta^{ax^d} =$
$\sum_r m(r)\zeta^{ar}$, where $m(r) = N(x^d = r)$.

22. (continuation)   Prove that $\sum_x \zeta^{ax^d} = \sum_\chi g_a(\chi)$, where the sum is over all χ such
that $\chi^d = \varepsilon, \chi \neq \varepsilon$. Assume that $p \nmid a$.

23. Let $f(x_1, x_2, \ldots, x_n) \in F_p[x_1, x_2, \ldots, x_n]$. Let N be the number of zeros of f
in F_p. Show that $N = p^{n-1} + p^{-1} \sum_{a \neq 0} \left(\sum_{x_1, \ldots, x_n} \zeta^{af(x_1, \ldots, x_n)} \right)$.

24. (continuation)   Let $f(x_1, x_2, \ldots, x_n) = a_1 x_1^{m_1} + a_2 x_2^{m_2} + \cdots + a_n x_n^{m_n}$. Let $d_i =$
$(m_i, p - 1)$. Show that $N = p^{n-1} + p^{-1} \sum_{a \neq 0} \prod_{i=1}^n \sum_{\chi_i} g_{aa_i}(\chi_i)$, where χ_i runs over
all characters such that $\chi_i^{d_i} = \varepsilon$ and $\chi_i \neq \varepsilon$.

25. Deduce from Exercise 24 that $|N - p^{n-1}| \leq (p - 1)(d_1 - 1) \cdots (d_n - 1)p^{(n/2)-1}$.

26. Let p be a prime, $p \equiv 1$ (4), χ a multiplicative character of order 4 on F_p, and ρ the
Legendre symbol. Put $J(\chi, \rho) = a + bi$. Show
    (a) $N(y^2 + x^4 = 1) = p - 1 + 2a$.
    (b) $N(y^2 = 1 - x^4) = p + \sum \rho(1 - x^4)$.

(c) $2a \equiv -(-1)^{(p-1)/4}\binom{2m}{m} (p)$ where $m = (p-1)/4$.

(d) Verify (c) for $p = 13, 17, 29$.

27. Let $p \equiv 1$ (3), χ a character of order 3, ρ the Legendre symbol. Show

(a) $N(y^2 = 1 - x^3) = p + \sum \rho(1 - x^3)$.

(b) $N(y^2 + x^3 = 1) = p + 2 \operatorname{Re} J(\chi, \rho)$.

(c) $2a - b \equiv -\binom{(p-1)/2}{(p-1)/3} (p)$ where $J(\chi, \rho) = a + b\omega$.

28. Let $p \equiv 3$ (4) and χ the quadratic character defined on $\mathbb{Z}/p\mathbb{Z}$. Show

(a) $\sum_{x=1}^{p-1} x\chi(x) = 2 \sum_{x=1}^{(p-1)/2} x\chi(x) - p \sum_{x=1}^{(p-1)/2} \chi(x)$.

(b) $\sum_{x=1}^{p-1} x\chi(x) = 4\chi(2) \sum_{x=1}^{(p-1)/2} x\chi(x) - p\chi(2) \sum_{x=1}^{(p-1)/2} \chi(x)$.

(c) If $p \equiv 3$ (8) then $\sum_{x=1}^{p-1} x\chi(x)/p = \frac{1}{3} \sum_{x=1}^{(p-1)/2} \chi(x)$.

(d) If $p \equiv 7$ (8) then $\sum_{x=1}^{p-1} x\chi(x)/p = \sum_{x=1}^{(p-1)/2} \chi(x)$.

Chapter 9

Cubic and Biquadratic Reciprocity

In Chapter 5 we saw that the law of quadratic reciprocity provided the answer to the question. For which primes p is the congruence $x^2 \equiv a\ (p)$ solvable? Here a is a fixed integer. If the same question is considered for congruences $x^n \equiv a\ (p)$, n a fixed positive integer, we are led into the realm of the higher reciprocity laws. When $n = 3$ and 4 we speak of cubic and biquadratic reciprocity.

In the introduction to his famous pair of papers, "Theorie der biquadratischen Reste I, II" [34], Gauss claims that the theory of quadratic residues had been brought to such a state of perfection that nothing more could be wished. On the other hand, "The theory of cubic and biquadratic residues is by far more difficult." He had only been able to deal with certain special cases for which the proofs had been so difficult that he soon came to the realization that ". . . the previously accepted principles of arithmetic are in no way sufficient for the foundations of a general theory, that rather such a theory necessarily demands that to a certain extent the domain of higher arithmetic needs to be endlessly enlarged . . . ." In modern language, he is calling for the establishment of a theory of algebraic numbers. As a first step, because this is what is needed for discussing biquadratic residues, he investigated in detail the arithmetic of the ring $\mathbb{Z}[\sqrt{-1}]$, which we now refer to as the ring of Gaussian integers.

Curiously, although Gauss formulated and discovered the law of biquadratic reciprocity, he did not prove it completely. The first complete published proofs of cubic and biquadratic reciprocity are due to G. Eisenstein.

In this chapter we shall formulate and prove the laws of cubic and biquadratic reciprocity. We shall give two proofs to the law of cubic reciprocity. The first is due to Eisenstein and is similar in every way to the proof of the law of quadratic reciprocity given in Chapter 6. The second proof uses Jacobi sums and is analogous to the proof of

108

quadratic reciprocity given in Chapter 8, Section 6. Our proof of biquadratic reciprocity is also due to Eisenstein.

In Section 10 we establish a "rational" reciprocity law for biquadratic residues. This elegant result, discovered by K. Burde in 1969 answers the following problem. If $p \equiv 1$ (4) and $q \equiv 1$ (4) are primes and p is a fourth power modulo q give necessary and sufficient conditions that q is a fourth power modulo p.

In Section 11 we establish, with the use of Jacobi sums, Gauss' criterion for the constructibility of a regular polygon.

The chapter concludes with a short discussion of Kummer's problem concerning the distribution of cubic Gauss sums.

§1 The Ring $\mathbb{Z}[\omega]$

Let $\omega = (-1 + \sqrt{-3})/2$. The ring $\mathbb{Z}[\omega]$ was defined and discussed in Chapter 1, Section 4. Its elements are complex numbers of the form $a + b\omega$, $a, b \in \mathbb{Z}$. If $\alpha = a + b\omega \in \mathbb{Z}[\omega]$, define the norm of α, $N\alpha$, by the formula $N\alpha = \alpha\bar{\alpha} = a^2 - ab + b^2$. Here $\bar{\alpha}$ means the complex conjugate of α. In Chapter 1 we used the notation $\lambda(\alpha)$ instead of $N\alpha$. The change is merely a matter of conforming to standard notation. For notational convenience we shall set $D = \mathbb{Z}[\omega]$.

We have proved earlier that D is a unique factorization domain. Our first task here is to discover the units and the prime elements in D.

Proposition 9.1.1. $\alpha \in D$ is a unit iff $N\alpha = 1$. The units in D are $1, -1, \omega,$ $-\omega, \omega^2,$ and $-\omega^2$.

PROOF. If $N\alpha = 1$, $\alpha\bar{\alpha} = 1$, which implies that α is a unit since $\bar{\alpha} \in D$.

If α is a unit, there is a $\beta \in D$ such that $\alpha\beta = 1$. Thus $N\alpha N\beta = 1$. Since $N\alpha$ and $N\beta$ are positive integers this implies that $N\alpha = 1$.

Now suppose that $\alpha = a + b\omega$ is a unit. Then $1 = a^2 - ab + b^2$ or $4 = (2a - b)^2 + 3b^2$. There are two possibilities:

(a) $2a - b = \pm 1, b = \pm 1$.
(b) $2a - b = \pm 2, b = 0$.

Solving these six pairs of equations yields the result $1, -1, \omega, -\omega,$ $-1 - \omega$ and $1 + \omega$. Since $\omega^2 + \omega + 1 = 0$ the last two elements are ω^2 and $-\omega^2$. We are done. $\qquad\square$

To investigate primes in D it is important to realize that primes in $\mathbb{Z}$ need not be prime in D. For example, $7 = (3 + \omega)(2 - \omega)$. For this reason we shall speak of primes in $\mathbb{Z}$ as rational primes and refer to primes in D simply as primes.

Proposition 9.1.2. *If π is a prime in D, then there is a rational prime p such that $N\pi = p$ or p^2. In the former case π is not associate to a rational prime; in the latter case π is associate to p.*

PROOF. We have $N\pi = n > 1$, or $\pi\bar{\pi} = n$. n is a product of rational primes. Thus $\pi | p$ for some rational prime p. If $p = \pi\gamma$, $\gamma \in D$, then $N\pi N\gamma = Np = p^2$. Thus either $N\pi = p^2$ and $N\gamma = 1$ or $N\pi = p$. In the former case γ is a unit and therefore π is associate to p. In the latter case if $\pi = uq$, u a unit and q a rational prime, then $p = N\pi = NuNq = q^2$, which is nonsense. Thus π is not associate to a rational prime. $\square$

Proposition 9.1.3. *If $\pi \in D$ is such that $N\pi = p$, a rational prime, then π is a prime in D.*

PROOF. If π were not prime in D, then we could write $\pi = \rho\gamma$ with $N\rho$, $N\gamma > 1$. Then $p = N\pi = N\rho N\gamma$, which cannot be true since p is prime in $\mathbb{Z}$. Thus π is a prime in D. $\square$

The following result classifies primes in D.

Proposition 9.1.4. *Suppose that p and q are rational primes. If $q \equiv 2$ (3), then q is prime in D. If $p \equiv 1$ (3), then $p = \pi\bar{\pi}$, where π is prime in D. Finally $3 = -\omega^2(1 - \omega)^2$, and $1 - \omega$ is prime in D.*

PROOF. Suppose that p were not a prime. Then $p = \pi\gamma$, with $N\pi > 1$, $N\gamma > 1$. Thus $p^2 = N\pi N\gamma$ and $N\pi = p$. Let $\pi = a + b\omega$. Then $p = a^2 - ab + b^2$ or $4p = (2a - b)^2 + 3b^2$, yielding $p \equiv (2a - b)^2$ (3). If $3 \nmid p$ we have $p \equiv 1$ (3) for 1 is the only nonzero square mod 3. It follows immediately that if $q \equiv 2$ (3), it is a prime in D.

Now, suppose that $p \equiv 1$ (3). By quadratic reciprocity we have

$$\left(\frac{-3}{p}\right) = \left(\frac{-1}{p}\right)\left(\frac{3}{p}\right) = (-1)^{(p-1)/2}\left(\frac{p}{3}\right)(-1)^{((p-1)/2)((3-1)/2)}$$

$$= \left(\frac{p}{3}\right) = \left(\frac{1}{3}\right) = 1.$$

Hence, there is an $a \in \mathbb{Z}$ such that $a^2 \equiv -3$ (p) or $pb = a^2 + 3$ for some $b \in \mathbb{Z}$. Thus p divides $(a + \sqrt{-3})(a - \sqrt{-3}) = (a + 1 + 2\omega) \times (a - 1 - 2\omega)$. If p were a prime in D, it would have to divide one of the factors but this cannot happen since $p \neq 2$ and $2/p \notin \mathbb{Z}$. Thus $p = \pi\gamma$ with π and γ nonunits. Taking norms we see that $p^2 = N\pi N\gamma$ and that $p = N\pi = \pi\bar{\pi}$.

The last case is handled as follows; $x^3 - 1 = (x - 1)(x - \omega)(x - \omega^2)$ implies that $x^2 + x + 1 = (x - \omega)(x - \omega^2)$. Setting $x = 1$ yields $3 =$

$(1 - \omega)(1 - \omega^2) = (1 + \omega)(1 - \omega)^2 = -\omega^2(1 - \omega)^2$. Taking norms we see that $9 = N(1 - \omega)^2$ and so $3 = N(1 - \omega)$. Thus $1 - \omega$ is a prime. $\square$

As a matter of notation q will be a positive rational prime congruent to 2 modulo 3 and π a complex prime whose norm, $N\pi = p$, is a rational prime congruent to 1 modulo 3. Occasionally π will refer to an arbitrary prime of D. The context should make the usage clear.

§2 Residue Class Rings

Just as in the ring $\mathbb{Z}$ and in the ring of all algebraic integers, the notion of congruence is extremely useful in D. If α, β, $\gamma \in D$ and $\gamma \neq 0$ is a nonunit, we say that $\alpha \equiv \beta$ (γ) if γ divides $\alpha - \beta$. Just as in $\mathbb{Z}$ the congruence classes modulo γ may be made into a ring $D/\gamma D$, called the residue class ring modulo γ.

Proposition 9.2.1. *Let $\pi \in D$ be a prime. Then $D/\pi D$ is a finite field with $N\pi$ elements.*

PROOF. We first show that $D/\pi D$ is a field. Let $\alpha \in D$ be such that $\alpha \not\equiv 0$ (π). By Corollary 1 to Proposition 1.3.2 there exist elements β, $\gamma \in D$ such that $\beta\alpha + \gamma\pi = 1$. Thus $\beta\alpha \equiv 1$ (π), which shows that the residue class of α is a unit in $D/\pi D$.

To show that $D/\pi D$ has $N\pi$ elements we must consider separately the cases in Proposition 9.1.4.

Suppose that $\pi = q$ is a rational prime congruent to 2 modulo 3. We claim that $\{a + b\omega | 0 \leq a < q$ and $0 \leq b < q\}$ is a complete set of coset representatives. This will show that D/qD has $q^2 = Nq$ elements. Let $\mu = m + n\omega \in D$. Then $m = qs + a$ and $n = qt + b$, where $s, t, a, b \in \mathbb{Z}$ and $0 \leq a, b < q$. Clearly $\mu \equiv a + b\omega$ (q). Next, suppose that $a + b\omega \equiv a' + b'\omega$ (q), where $0 \leq a, b, a', b' < q$. Then $((a - a')/q) + ((b - b')/q)\omega \in D$, implying that $(a - a')/q$ and $(b - b')/q$ are in $\mathbb{Z}$. This is possible only if $a = a'$ and $b = b'$.

Now suppose that $p \equiv 1$ (3) is a rational prime and $\pi\bar{\pi} = N\pi = p$. We claim that $\{0, 1, \ldots, p - 1\}$ is a complete set of coset representatives. This will show that $D/\pi D$ has $p = N\pi$ elements. Let $\pi = a + b\omega$. Since $p = a^2 - ab + b^2$ it follows that $p \nmid b$. Let $\mu = m + n\omega$. There is an integer c such that $cb \equiv n$ (p). Then $\mu - c\pi \equiv m - ca$ (p) and so $\mu \equiv m - ca$ (π). Every element of D is congruent to a rational integer modulo π. If $l \in \mathbb{Z}$, $l = sp + r$, where $s, r \in \mathbb{Z}$ and $0 \leq r < p$. Thus $l \equiv r$ (p) and *a fortiori* $l \equiv r$ (π). We have shown that every element of D is congruent to an element of $\{0, 1, 2, \ldots, p - 1\}$ modulo π. If $r \equiv r'$ (π) with $r, r' \in \mathbb{Z}$ and $0 \leq r, r' < p$, then $r - r' = \pi\gamma$ and $(r - r')^2 = pN\gamma$, implying that $p | r - r'$. Thus $r = r'$ and we are done.

We leave the case of the prime $1 - \omega$ as an exercise. $\square$

§3 Cubic Residue Character

Let π be a prime. Then the multiplicative group of $D/\pi D$ has order $N\pi - 1$. Hence we have an analog of Fermat's Little Theorem.

Proposition 9.3.1. *If $\pi \nmid \alpha$, then*

$$\alpha^{N\pi - 1} \equiv 1 \ (\pi).$$

If the norm of π is different from 3, then the residue classes of 1, ω, and ω^2 are distinct in $D/\pi D$. To see this, suppose, for example, that $\omega \equiv 1 \ (\pi)$. Then $\pi | (1 - \omega)$, and since $1 - \omega$ is prime, π and $1 - \omega$ are associate. Thus $N\pi = N(1 - \omega) = 3$, a contradiction. The other cases are handled in the same way.

Since $\{1, \omega, \omega^2\}$ is a cyclic group of order 3 it follows that 3 divides the order of $(D/\pi D)^*$; i.e., $3 | N\pi - 1$. This can be seen in another way using Proposition 9.1.3. If $\pi = q$, a rational prime, then $N\pi = q^2 \equiv 1 \ (3)$. If π is such that $N\pi = p$, then $p \equiv 1 \ (3)$.

Proposition 9.3.2. *Suppose that π is a prime such that $N\pi \neq 3$ and that $\pi \nmid \alpha$. Then there is a unique integer $m = 0, 1, or 2$ such that $\alpha^{(N\pi - 1)/3} \equiv \omega^m \ (\pi)$.*

PROOF. We know that π divides $\alpha^{N\pi - 1} - 1$. Now,

$$\alpha^{N\pi - 1} - 1 = (\alpha^{(N\pi - 1)/3} - 1)(\alpha^{(N\pi - 1)/3} - \omega)(\alpha^{(N\pi - 1)/3} - \omega^2).$$

Since π is prime it must divide one of the three factors on the right. By the preceding remarks it can divide at most one factor, since if it divided two factors it would divide the difference. This proves the proposition. □

On the basis of this result we can make the following definition.

Definition. If $N\pi \neq 3$, the *cubic residue character* of α modulo π is given by

(a) $(\alpha/\pi)_3 = 0$ if $\pi | \alpha$.
(b) $\alpha^{(N\pi - 1)/3} \equiv (\alpha/\pi)_3 \ (\pi)$, with $(\alpha/\pi)_3$ equal to 1, ω, or ω^2.

This character plays the same role in the theory of cubic residues as the Legendre symbol plays in the theory of quadratic residues.

Proposition 9.3.3.

(a) $(\alpha/\pi)_3 = 1$ *iff $x^3 \equiv \alpha \ (\pi)$ is solvable, i.e., iff α is a cubic residue.*
(b) $\alpha^{(N\pi - 1)/3} \equiv (\alpha/\pi)_3 \ (\pi)$.
(c) $(\alpha\beta/\pi)_3 = (\alpha/\pi)_3(\beta/\pi)_3$.
(d) *If $\alpha \equiv \beta \ (\pi)$, then $(\alpha/\pi)_3 = (\beta/\pi)_3$.*

PROOF. Part (a) is a special case of Proposition 7.1.2. Take $F = D/\pi D$, $q = N\pi$, and $n = 3$ in that proposition.

Part (b) is immediate from the definition.

Part (c): $(\alpha\beta/\pi)_3 \equiv (\alpha\beta)^{(N\pi-1)/3} \equiv \alpha^{(N\pi-1)/3}\beta^{(N\pi-1)/3} \equiv (\alpha/\pi)_3(\beta/\pi)_3 \ (\pi)$. The result follows.

Part (d): If $\alpha \equiv \beta \ (\pi)$, then $(\alpha/\pi)_3 \equiv \alpha^{(N\pi-1)/3} \equiv \beta^{(N\pi-1)/3} \equiv (\beta/\pi)_3 \ (\pi)$, and so $(\alpha/\pi)_3 = (\beta/\pi)_3$. □

Since we shall be dealing only with cubic characters in this section the notation $\chi_\pi(\alpha) = (\alpha/\pi)_3$ will be convenient.

It is useful to study the behavior of characters under complex conjugation.

Proposition 9.3.4.

(a) $\overline{\chi_\pi(\alpha)} = \chi_\pi(\alpha)^2 = \chi_\pi(\alpha^2)$.
(b) $\overline{\chi_\pi(\alpha)} = \chi_{\bar\pi}(\bar\alpha)$.

PROOF.

(a) $\chi_\pi(\alpha)$ is by definition 1, ω, or ω^2, and each of these numbers squared is equal to its conjugate.

(b)
$$\alpha^{(N\pi-1)/3} \equiv \chi_\pi(\alpha)\,(\pi),$$

we get

$$\bar\alpha^{(N\pi-1)/3} \equiv \overline{\chi_\pi(\alpha)}\,(\bar\pi).$$

Since $N\bar\pi = N\pi$ this shows that $\chi_{\bar\pi}(\bar\alpha) \equiv \overline{\chi_\pi(\alpha)}\,(\bar\pi)$ and thus that $\chi_{\bar\pi}(\bar\alpha) = \overline{\chi_\pi(\alpha)}$. □

Corollary. $\chi_q(\bar\alpha) = \chi_q(\alpha^2)$ and $\chi_q(n) = 1$ if n is a rational integer prime to q.

PROOF. Since $\bar q = q$ we have $\chi_q(\bar\alpha) = \chi_{\bar q}(\bar\alpha) = \overline{\chi_q(\alpha)} = \chi_q(\alpha^2)$. This gives the first relation.

Since $\bar n = n$ we have $\chi_q(n) = \overline{\chi_q(n)} = \chi_q(n)^2$. Since $\chi_q(n) \neq 0$ it follows that $\chi_q(n) = 1$. □

The corollary states that n is a cubic residue modulo q. Thus, if $q_1 \neq q_2$ are two primes congruent to 2 modulo 3, then we have (trivially) $\chi_{q_1}(q_2) = \chi_{q_2}(q_1)$. This is a special case of the law of cubic reciprocity. To formulate the general law we need to introduce the idea of a "primary" prime.

Definition. If π is a prime in D, we say that π is *primary* if $\pi \equiv 2\ (3)$.

If $\pi = q$ is rational, this is nothing new. If $\pi = a + b\omega$ is a complex prime, the definition is equivalent to $a \equiv 2\ (3)$ and $b \equiv 0\ (3)$.

We need a notion such as "primary" to eliminate the ambiguity caused by the fact that every nonzero element of D has six associates.

Proposition 9.3.5. *Suppose that $N\pi = p \equiv 1\ (3)$. Among the associates of π exactly one is primary.*

PROOF. Write $\pi = a + b\omega$. The associates of π are $\pi, \omega\pi, \omega^2\pi, -\pi, -\omega\pi$, and $-\omega^2\pi$. In terms of a and b these elements can be expressed as

(a) $a + b\omega$.
(b) $-b + (a - b)\omega$.
(c) $(b - a) - a\omega$.
(d) $-a - b\omega$.
(e) $b + (b - a)\omega$.
(f) $(a - b) + a\omega$.

Since $p = a^2 - ab + b^2$, not both a and b are divisible by 3. By looking at parts (a) and (b) it is clear that we can assume that $3 \nmid a$. Considering parts (a) and (d) we can assume further that $a \equiv 2$ (3). Under this assumption $p = a^2 - ab + b^2$ leads to $1 \equiv 4 - 2b + b^2$ (3) or $b(b - 2) \equiv 0$ (3). If $3 | b$, then $a + b\omega$ is primary. If $b \equiv 2$ (3), then $b + (b - a)\omega$ is primary.

To show uniqueness, assume that $a + b\omega$ is primary. By considering the congruence class of the first term in part (b) to part (e) we see that none of these expressions is primary. Neither is the expression in part (f) since the coefficient of ω, a, is not divisible by 3.   □

For example, $3 + \omega$ is prime since $N(3 + \omega) = 7$, and $-\omega^2(3 + \omega) = 2 + 3\omega$ is the primary prime associated to it.

We can now state

Theorem 1 (The Law of Cubic Reciprocity). *Let π_1 and π_2 be primary, $N\pi_1$, $N\pi_2 \neq 3$, and $N\pi_1 \neq N\pi_2$. Then*

$$\chi_{\pi_1}(\pi_2) = \chi_{\pi_2}(\pi_1).$$

A proof will be given in Section 4, but first a few remarks are in order.

(a) There are three cases to consider. Namely, both π_1 and π_2 are rational, π_1 is rational and π_2 is complex, and both π_1 and π_2 are complex. The first case is, as we have seen, trivial.

(b) The cubic character of the units can be dealt with as follows. Since $-1 = (-1)^3$ we have $\chi_\pi(-1) = 1$ for all primes π.

   In $N\pi \neq 3$, then it follows from Proposition 9.3.3, part (b), that $\chi_\pi(\omega) = \omega^{(N\pi - 1)/3}$. Thus $\chi_\pi(\omega) = 1$, ω, or ω^2 according to whether $N\pi \equiv 1, 4$, or 7 modulo 9.

(c) The prime $1 - \omega$ causes particular difficulty. If $N\pi \neq 3$, we would like to evaluate $\chi_\pi(1 - \omega)$. This is done by Eisenstein in [29] by a highly ingenious argument. An elegant proof due to K. Williams is given in the Exercises.

Theorem 1' (Supplement to the Cubic Reciprocity Law). *Suppose that $N\pi \neq 3$. If $\pi = q$ is rational, write $q = 3m - 1$. If $\pi = a + b\omega$ is a primary complex prime, write $a = 3m - 1$. Then*

$$\chi_\pi(1 - \omega) = \omega^{2m}.$$

We give a proof for the case of a rational prime q. Since $(1 - \omega)^2 = -3\omega$ we have

$$\chi_q(1 - \omega)^2 = \chi_q(-3)\chi_q(\omega).$$

By the corollary to Proposition 9.3.4 we know that $\chi_q(-3) = 1$. By remark (b) $\chi_q(\omega) = \omega^{(Nq-1)/3} = \omega^{(q^2-1)/3}$. Thus $\chi_q(1 - \omega)^2 = \omega^{(q^2-1)/3}$. Squaring both sides yields

$$\chi_q(1 - \omega) = \omega^{(2/3)(q^2-1)}.$$

Now, $q^2 - 1 = 9m^2 - 6m$ so that $\frac{2}{3}(q^2 - 1) \equiv -4m \equiv 2m$ (3). The result follows. For extensions of these results to primary elements see exercises 17 to 20 on page 135.

§4 Proof of the Law of Cubic Reciprocity

Let π be a complex prime such that $N\pi = p \equiv 1$ (3). Since $D/\pi D$ is a finite field of characteristic p it contains a copy of $\mathbb{Z}/p\mathbb{Z}$. Both $D/\pi D$ and $\mathbb{Z}/p\mathbb{Z}$ have p elements. Thus we may identify the two fields. More explicitly the identification is given by sending the coset of n in $\mathbb{Z}/p\mathbb{Z}$ to the coset of n in $D/\pi D$.

This identification allows us to consider χ_π as a cubic character on $\mathbb{Z}/p\mathbb{Z}$ in the sense of Chapter 8 [see Proposition 9.3.3, parts (c) and (d)]. Thus we may work with the Gauss sums $g_a(\chi_\pi)$ and the Jacobi sum $J(\chi_\pi, \chi_\pi)$.

If χ is any cubic character, we have proved (see the corollary to Proposition 8.3.3 and Proposition 8.3.4) that

(a) $g(\chi)^3 = pJ(\chi, \chi)$.
(b) If $J(\chi, \chi) = a + b\omega$, then $a \equiv -1$ (3) and $b \equiv 0$ (3).

Since $J(\chi, \chi)\overline{J(\chi, \chi)} = p$, the second assertion says that $J(\chi, \chi)$ is a primary prime in D of norm p.

We need a lemma. Assume π is primary.

Lemma 1. $J(\chi_\pi, \chi_\pi) = \pi$.

PROOF. Let $J(\chi_\pi, \chi_\pi) = \pi'$. Since $\pi\bar{\pi} = p = \pi'\bar{\pi}'$ we have $\pi | \pi'$ or $\pi | \bar{\pi}'$.

Since all the primes involved are primary we must have $\pi = \pi'$ or $\pi = \bar{\pi}'$. We wish to eliminate the latter possibility.

From the definitions,

$$J(\chi_\pi, \chi_\pi) = \sum_x \chi_\pi(x)\chi_\pi(1 - x) \equiv \sum_x x^{(p-1)/3}(1 - x)^{(p-1)/3} \ (\pi),$$

where the sum is over $\mathbb{Z}/p\mathbb{Z}$. The polynomial $x^{(p-1)/3}(1 - x)^{(p-1)/3}$ is of degree $\frac{2}{3}(p - 1) < p - 1$. By Exercise 11 of Chapter 4 it follows that $\sum_x x^{(p-1)/3}(1 - x)^{(p-1)/3} \equiv 0$ (p). This shows that $J(\chi_\pi, \chi_\pi) \equiv 0$ (π); i.e., $\pi | \pi'$ and therefore $\pi = \pi'$.   □

Corollary. $g(\chi_\pi)^3 = p\pi$.

We can now prove the law of cubic reciprocity. We first consider the case where $\pi_1 = q \equiv 2 \ (3)$ and $\pi_2 = \pi$ with $N\pi = p$.

Raise both sides of the relation $g(\chi_\pi)^3 = p\pi$ to the $(q^2 - 1)/3$ power. This gives $g(\chi_\pi)^{q^2-1} = (p\pi)^{(q^2-1)/3}$. Taking congruences modulo q we see that

$$g(\chi_\pi)^{q^2-1} \equiv \chi_q(p\pi) \ (q).$$

Since $\chi_q(p) = 1$ this leads to

$$g(\chi_\pi)^{q^2} \equiv \chi_q(\pi)g(\chi_\pi) \ (q). \tag{1}$$

We now analyze the left-hand side:

$$g(\chi_\pi)^{q^2} = \left(\sum \chi_\pi(t)\zeta^t\right)^{q^2} \equiv \sum \chi_\pi(t)^{q^2}\zeta^{q^2 t} \ (q).$$

Since $q^2 \equiv 1 \ (3)$ and $\chi_\pi(t)$ is a cube root of 1 we have

$$g(\chi_\pi)^{q^2} \equiv g_{q^2}(\chi_\pi) \ (q). \tag{2}$$

By Proposition 8.2.1 $g_{q^2}(\chi_\pi) = \chi_\pi(q^{-2})g(\chi_\pi) = \chi_\pi(q)g(\chi_\pi)$. Thus, combining Equations (1) and (2)

$$\chi_\pi(q)g(\chi_\pi) \equiv \chi_q(\pi)g(\chi_\pi) \ (q).$$

Multiply both sides of this congruence by $g(\overline{\chi_\pi})$. Since $g(\chi_\pi)g(\overline{\chi_\pi}) = p$,

$$\chi_\pi(q)p \equiv \chi_q(\pi)p \ (q)$$

or

$$\chi_\pi(q) \equiv \chi_q(\pi) \ (q),$$

implying that

$$\chi_\pi(q) = \chi_q(\pi).$$

It remains to consider the case of two complex primes π_1 and π_2, where $N\pi_1 = p_1 \equiv 1 \ (3)$ and $N\pi_2 = p_2 \equiv 1 \ (3)$. This case is handled by essentially the same technique, but it is a little trickier.

Let $\gamma_1 = \bar{\pi}_1$ and $\gamma_2 = \bar{\pi}_2$. Then γ_1 and γ_2 are primary and $p_1 = \pi_1\gamma_1$ and $p_2 = \pi_2\gamma_2$.

Starting from the relation $g(\chi_{\gamma_1})^3 = p_1\gamma_1$, raising to the $(N\pi_2 - 1)/3 = (p_2 - 1)/3$ power, and taking congruences modulo π_2, we obtain by the same method as above the relation

$$\chi_{\gamma_1}(p_2^2) = \chi_{\pi_2}(p_1\gamma_1). \tag{3}$$

Similarly, starting from $g(\chi_{\pi_2})^3 = p_2\pi_2$, raising to the $(p_1 - 1)/3$ power, and taking congruences modulo π_1, we obtain

$$\chi_{\pi_2}(p_1^2) = \chi_{\pi_1}(p_2\pi_2). \tag{4}$$

We also need the relation $\chi_{\gamma_1}(p_2^2) = \chi_{\pi_1}(p_2)$, which follows from Proposition 9.3.4 since $\gamma_1 = \bar{\pi}_1$ and $\bar{p}_2 = p_2$. Now we calculate

$$\chi_{\pi_1}(\pi_2)\chi_{\pi_2}(p_1\gamma_1) = \chi_{\pi_1}(\pi_2)\chi_{\gamma_1}(p_2^2) \qquad \text{by Equation (3)}$$

$$= \chi_{\pi_1}(\pi_2)\chi_{\pi_1}(p_2) = \chi_{\pi_1}(p_2\pi_2) \qquad \text{by above remark}$$

$$= \chi_{\pi_2}(p_1^2) = \chi_{\pi_2}(p_1\pi_1\gamma_1) \qquad \text{by Equation (4)}$$

$$= \chi_{\pi_2}(\pi_1)\chi_{\pi_2}(p_1\gamma_1).$$

Equating the first and last terms and canceling $\chi_{\pi_2}(p_1\gamma_1)$ gives the sought for result:

$$\chi_{\pi_1}(\pi_2) = \chi_{\pi_2}(\pi_1).$$

§5 Another Proof of the Law of Cubic Reciprocity

We present a proof of cubic reciprocity using Jacobi sums. This proof is somewhat shorter and more elegant than the one given in Section 4. It should be noticed, however, that more background material is used.

Consider the case $\pi_1 = q$, $\pi_2 = \pi$. Let $\chi_\pi = \chi$, and consider the Jacobi sum $J(\chi, \chi, \ldots, \chi)$ with q terms. Since $3 | q + 1$ we have by Corollary 1 to Theorem 3 of Chapter 8,

$$g(\chi)^{q+1} = pJ(\chi, \chi, \ldots, \chi). \qquad (5)$$

Since $g(\chi)^3 = p\pi$,

$$g(\chi)^{q+1} = (p\pi)^{(q+1)/3}. \qquad (6)$$

Now, recall that

$$J(\chi, \chi, \ldots, \chi) = \sum \chi(x_1)\chi(x_2)\cdots\chi(x_q),$$

where the sum is over all $x_1, x_2, \ldots, x_q \in Z/pZ$ such that $x_1 + x_2 + \cdots + x_q = 1$. Consider the term for which $x_1 = x_2 = \cdots = x_q$. Then $qx_1 = 1$ and $\chi(q)\chi(x_1) = 1$. Raising both sides to the qth power, and recalling that $q \equiv 2$ (3), yields $\chi(q)^2\chi(x_1)^q = 1$ and so $\chi(x_1)^q = \chi(q)$. Thus the "diagonal" term of $J(\chi, \chi, \ldots, \chi)$ has the value $\chi(q)$. If not all the x_i are equal, there are q different q-tuples obtained from $(x_1, x_2, \ldots, x_q)$ by cyclic permutation. The corresponding terms of $J(\chi, \chi, \ldots, \chi)$ all have the same value. Thus

$$J(\chi, \chi, \ldots, \chi) \equiv \chi(q)(q). \qquad (7)$$

Combining Equations (5), (6), and (7) we obtain

$$(p\pi)^{(q+1)/3} \equiv p\chi(q)(q)$$

or

$$p^{(q-2)/3}\pi^{(q+1)/3} \equiv \chi(q)(q).$$

Raising both sides to the $q - 1$ power (remember that $q - 1 \equiv 1\ (3)$)

$$p^{((q-2)/3)(q-1)}\pi^{(q^2-1)/3} \equiv \chi(q)^{q-1} \equiv \chi(q)\ (q)$$

Since $p^{((q-2)/3)(q-1)} \equiv 1\ (q)$ by Fermat's theorem and $\pi^{(q^2-1)/3} \equiv \chi_q(\pi)\ (q)$ it follows that

$$\chi_q(\pi) \equiv \chi_\pi(q)\ (q)$$

and

$$\chi_q(\pi) = \chi_\pi(q).$$

Now consider the case of two primary complex primes π_1 and π_2. Let $\gamma_1 = \bar{\pi}_1, \gamma_2 = \bar{\pi}_2, p_1 = \pi_1\gamma_1,$ and $p_2 = \pi_2\gamma_2$. Then $p_1, p_2 \equiv 1\ (3)$. By Theorem 3 of Chapter 8 we have

$$g(\chi_{\gamma_1})^{p_2} = J(\chi_{\gamma_1}, \ldots, \chi_{\gamma_1})g(\chi_{\gamma_1}^{p_2}).$$

There are p_2 terms in the Jacobi sum. Since $p_2 \equiv 1\ (3)$, $\chi_{\gamma_1}^{p_2} = \chi_{\gamma_1}$. Thus

$$[g(\chi_{\gamma_1})^3]^{(p_2-1)/3} = J(\chi_{\gamma_1}, \ldots, \chi_{\gamma_1}). \tag{8}$$

By isolating the diagonal term of the Jacobi sum (as we have done a number of times by now) we find that

$$J(\chi_{\gamma_1}, \ldots, \chi_{\gamma_1}) \equiv \chi_{\gamma_1}(p_2^{-1}) \equiv \chi_{\gamma_1}(p_2^2)\ (p_2).$$

Using this and the fact that $g(\chi_{\gamma_1})^3 = p_1\gamma_1$, we obtain from Equation (8) the congruence

$$\chi_{\pi_2}(p_1\gamma_1) \equiv \chi_{\gamma_1}(p_2^2)\ (\pi_2)$$

and therefore

$$\chi_{\pi_2}(p_1\gamma_1) = \chi_{\gamma_1}(p_2^2). \tag{9}$$

Similarly one proves that

$$\chi_{\pi_1}(p_2\pi_2) = \chi_{\pi_2}(p_1^2). \tag{10}$$

Equations (9) and (10) are the basic relations. From here on one proceeds exactly as in Section 4 to the desired conclusion $\chi_{\pi_1}(\pi_2) = \chi_{\pi_2}(\pi_1)$.

§6 The Cubic Character of 2

The law of cubic reciprocity can be used to develop the theory of cubic residues in the same manner as the law of quadratic reciprocity led to the results of Chapter 5, Section 2. We shall forego a development of the general theory in favor of a discussion of an illuminating special case. Namely, we shall ask for all primes π in D for which 2 is a cubic residue.

To begin with, notice that $x^3 \equiv 2\ (\pi)$ is solvable iff $x^3 \equiv 2\ (\pi')$ is solvable for any associate of π. Thus we may assume that π is primary. If $\pi = q$ is a rational prime, then $\chi_q(2) = 1$ and so 2 is a cubic residue for all such primes. We assume from now on that $\pi = a + b\omega$ is a primary complex prime. By cubic reciprocity $\chi_\pi(2) = \chi_2(\pi)$. The norm of 2 is $2^2 = 4$. Thus

$$\pi \equiv \pi^{(4-1)/3} \equiv \chi_2(\pi) \ (2).$$

It follows that $\chi_\pi(2) = 1$ iff $\pi \equiv 1$ (2). We have proved

Proposition 9.6.1. $x^3 \equiv 2 \ (\pi)$ *is solvable iff* $\pi \equiv 1$ (2), *i.e., iff* $a \equiv 1$ (2) *and* $b \equiv 0$ (2).

It is possible to formulate this proposition in another way. Let $\pi = a + b\omega$ be a primary complex prime and $p = N\pi = a^2 - ab + b^2$. Then $4p = (2a - b)^2 + 3b^2$. If we set $A = 2a - b$ and $B = b/3$, then $4p = A^2 + 27B^2$. According to Proposition 8.3.2 the integers A and B are uniquely determined up to sign.

Proposition 9.6.2. *If* $p \equiv 1$ (3), *then* $x^3 \equiv 2 \ (p)$ *is solvable iff there are integers* C *and* D *such that* $p = C^2 + 27D^2$.

PROOF. If $x^3 \equiv 2 \ (p)$ is solvable, so is $x^3 \equiv 2 \ (\pi)$ and thus $\pi \equiv 1$ (2) by Proposition 9.6.1. We have

$$4p = A^2 + 27B^2, \quad \text{where } A = 2a - b, B = \frac{b}{3}.$$

Since b is even, so are B and A. Let $D = B/2$ and $C = A/2$. Then $p = C^2 + 27D^2$.

Suppose, conversely, that $p = C^2 + 27D^2$. Then $4p = (2C)^2 + 27(2D)^2$. By uniqueness $B = \pm 2D$; i.e., B is even and thus so is b. It follows that $\pi = a + b\omega \equiv 1$ (2), and $x^3 \equiv 2 \ (\pi)$ is solvable. Since $D/\pi D$ has $p = N\pi$ elements there is an integer h such that $h^3 \equiv 2 \ (\pi)$. It is now easy to show that $h^3 \equiv 2 \ (p)$. If $\pi | h^3 - 2$, then $\bar{\pi} | h^3 - 2$ and $\pi\bar{\pi} = p | (h^3 - 2)^2$. Consequently, $p | h^3 - 2$ and we are done. ☐

As an example take $p = 7$. Then $x^3 \equiv 2 \ (7)$ is not solvable since there are clearly no integers C and D such that $7 = C^2 + 27D^2$.

On the other hand, $p = 31 = 2^2 + 27 \cdot 1^2$. Thus $x^3 \equiv 2 \ (31)$ is solvable. Indeed, $4^3 \equiv 2 \ (31)$.

§7 Biquadratic Reciprocity: Preliminaries

In his second memoir (1832) on biquadratic residues, Gauss stated, without proof, the law of biquadratic reciprocity. The proof, he asserted, belonged to the mysteries of the higher arithmetic. The details were to be published in a third memoir, which unfortunately never appeared. Subsequently Eisenstein published several proofs (1844), using Jacobi and Gauss sums. The basic idea is the same as in the cubic case, although the details are more extensive. The use of Gauss sums to prove reciprocity laws is due to Gauss himself, who utilized them essentially in his sixth proof of quadratic reciprocity.

Throughout the following three sections D denotes the ring $\mathbb{Z}[i]$ of Gaussian integers. If $\alpha \in D$ then $(\alpha) = \alpha D$ is the principal ideal generated by α. By a prime will always be meant a positive prime of $\mathbb{Z}$. Recall from Chapter 1 that D is a Euclidean ring. Thus if π is irreducible and $\pi | \alpha\beta$ then either $\pi | \alpha$ or $\pi | \beta$. If $N(\alpha) = \alpha\bar{\alpha}$ is the norm of α then by Exercise 32 of Chapter 1, $N(\alpha) = 1$ iff α is a unit. From this, one sees that the units of D are ± 1, $\pm i$.

Lemma 1. *If π is irreducible then there is a prime $p \in \mathbb{Z}$ such that $\pi | p$.*

PROOF. $N(\pi) = \pi\bar{\pi} = n = p_1 \cdots p_s$, p_i prime, $p_i \in \mathbb{Z}$. Thus $\pi | p_i$ for some i. $\square$

Thus the irreducibles are found by decomposing in D all primes in $\mathbb{Z}$. The following lemma is useful.

Lemma 2. *If $\alpha \in D$, and $N(\alpha)$ is prime then α is irreducible.*

PROOF. If $\alpha = \mu\lambda$ then $N(\alpha) = N(\mu)N(\lambda)$. Since $N(\alpha)$ is prime it follows that $N(\mu) = 1$ or $N(\lambda) = 1$. Thus either μ or λ is a unit.

Lemma 3. $1 + i$ *is irreducible and* $2 = -i(1 + i)^2$ *is the prime factorization of 2 in D.*

PROOF. $N(1 + i) = 2$ and so the first assertion follows from Lemma 2. The second assertion results from a direct calculation.

Lemma 4. *If $q \equiv 3$ (4) is a prime in $\mathbb{Z}$, then q is irreducible considered as an element of D.*

PROOF. If q were not irreducible in D, then $q = \alpha\beta$ with $N(\alpha) > 1$ and $N(\beta) > 1$. Taking norms we find $q^2 = N(\alpha)N(\beta)$. It follows that $q = N(\alpha)$. If $\alpha = a + bi$ with a, $b \in \mathbb{Z}$, then $q = a^2 + b^2$. This is a contradiction since a sum of two squares in $\mathbb{Z}$ is congruent to 0 or 1 modulo 4, and q is congruent to 3 modulo 4.

Lemma 5. *If p is prime, $p \equiv 1$ (4) then there is an irreducible π such that $p = \pi\bar{\pi}$. Furthermore $(\pi) \neq (\bar{\pi})$.*

PROOF. The first statement is part (a) of Proposition 8.3.1. Another proof not using Jacobi sums is the following. Since $p \equiv 1$ (4) there is, by Proposition 5.1.2, an integer a with $a^2 \equiv -1(p)$. Thus $p | a^2 + 1 = (a + i)(a - i)$. If p were irreducible then $p | a + i$ which is absurd. Thus $p = \alpha\beta$, $N(\alpha) > 1$, $N(\beta) > 1$. Taking norms enables one to conclude that $p = N(\alpha)$. Since $N(\alpha)$ is prime it follows by Lemma 2 that α is irreducible. The fact that $(\alpha) \neq (\bar{\alpha})$ is left as an exercise. $\square$

This completes the description of the irreducibles in D.

Definition. A nonunit $\alpha \in D$ is primary if $\alpha \equiv 1 \ (1 + i)^3$.

Lemma 6. *A nonunit α is primary iff either $a \equiv 1 \ (4)$, $b \equiv 0 \ (4)$ or $a \equiv 3 \ (4)$, $b \equiv 2 \ (4)$.*

PROOF. Since $(1 + i)^3 = 2i(1 + i)$ it follows that $a + bi$ is primary iff

$$\frac{(a - 1) + bi}{2 + 2i} = \frac{a + b - 1}{4} + \frac{b - a + 1}{4} i \in D.$$

This is equivalent to the congruences $a + b \equiv 1 \ (4)$, $a - b \equiv 1 \ (4)$. The result follows easily from this. $\qquad\square$

We note that any nonunit $\alpha \equiv 1 \ (4)$ in D is primary. Furthermore if α is primary then $(1 + i) \nmid \alpha$. If q is a real prime, $q \equiv 3 \ (4)$ then $-q$ is a primary irreducible. As for the irreducibles arising from primes $p \equiv 1 \ (4)$ one has the following important result.

Lemma 7. *Let $\alpha \in D$ be a nonunit, $(1 + i) \nmid \alpha$. Then there is a unique unit u such that $u\alpha$ is primary.*

PROOF. There is a unit ε such that $\varepsilon\alpha = a + bi$ where a is odd and b is even. Multiplying if necessary by -1, Lemma 6 shows that α has a primary associate. If u_1 and u_2 are units such that $u_1\alpha$ and $u_2\alpha$ are primary then since $(1 + i) \nmid \alpha$ it follows that $u_1 \equiv u_2 \ (1 + i)^3$. An examination of cases shows easily that this implies $u_1 = u_2$.

Lemma 8. *A primary element can be written as the product of primary irreducibles.*

PROOF. Let $\alpha \in D$ be primary. Then there are rational primes $q_i \equiv 3 \ (4)$, primary irreducibles π_i, $N(\pi_i) \equiv 1 \ (4)$ and a unit u such that $\alpha = u\pi_1 \cdots \pi_t(-q_1) \cdots (-q_s)$. Reduction modulo $(1 + i)^3$ shows that $1 \equiv u \ (1 + i)^3$. This implies that $u = 1$. $\qquad\square$

§8 The Quartic Residue Symbol

Consider an irreducible π in D.

Proposition 9.8.1. *The residue class ring $D/\pi D$ is a finite field with $N(\pi)$ elements.*

PROOF. The proof proceeds in exactly the same way as Proposition 9.2.1, replacing the classification of irreducibles in $\mathbb{Z}[\omega]$ by the corresponding classification in $D = \mathbb{Z}[i]$. $\qquad\square$

Corollary. *If $\pi \nmid \alpha$ then $\alpha^{N\pi - 1} \equiv 1\ (\pi)$.*

Proposition 9.8.2. *If $\pi \nmid \alpha$, $(\pi) \neq (1 + i)$ there exists a unique integer j, $0 \leq j \leq 3$ such that*

$$\alpha^{(N(\pi) - 1)/4} \equiv i^j\ (\pi).$$

PROOF. It is easy to see that the residue classes of $1, -1, i, -i$ are distinct. They are the roots of $x^4 \equiv 1\ (\pi)$. However the residue class of $\alpha^{(N(\pi) - 1)/4}$ is also a solution to $x^4 \equiv 1\ (\pi)$ by the above corollary. The result follows from this.
□

Definition. If π is an irreducible, $N(\pi) \neq 2$, then the biquadratic (or quartic) residue character of α, for $\pi \nmid \alpha$, is defined by $\chi_\pi(\alpha) = i^j$ where j is determined by Proposition 9.8.2. If $\pi \mid \alpha$ then $\chi_\pi(\alpha) = 0$.

Proposition 9.8.3.

(a) *If $\pi \nmid \alpha$ then $\chi_\pi(\alpha) = 1 \Leftrightarrow x^4 \equiv \alpha\ (\pi)$ has a solution in D.*
(b) *$\chi_\pi(\alpha\beta) = \chi_\pi(\alpha) \cdot \chi_\pi(\beta)$.*
(c) *$\overline{\chi_\pi(\alpha)} = \chi_\pi(\bar{\alpha})$.*
(d) *If π is a primary irreducible then $\chi_\pi(-1) = (-1)^{(a-1)/2}$, where $\pi = a + bi$.*
(e) *If $\alpha \equiv \beta\ (\pi)$ then $\chi_\pi(\alpha) = \chi_\pi(\beta)$.*
(f) *$\chi_\pi(\alpha) = \chi_\lambda(\alpha)$ if $(\pi) = (\lambda)$.*

PROOF. Part (a) follows from Proposition 7.1.2. Parts (b), (c), (e), and (f) follow immediately from the definition. Part (d) follows from Lemma 6 (see Exercise 38).
□

Proposition 9.8.4. *Let q be prime, $q \equiv 3\ (4)$. Then $\chi_q(a) = 1$ for $a \in \mathbb{Z}, q \nmid a$.*

PROOF. $N(q) = q^2$. Thus

$$\chi_q(a) \equiv a^{(q^2 - 1)/4} = (a^{q - 1})^{(q + 1)/4} \equiv 1\ (q),$$

by Fermat's Little Theorem.
□

The quartic residue character is generalized as follows.

Definition. Let $\alpha \in D$ be a nonunit such that $(1 + i) \nmid \alpha$, and $\beta \in D$. Write $\alpha = \prod_i \lambda_i$ where λ_i is irreducible. If $(\alpha, \beta) = 1$ define $\chi_\alpha(\beta)$ by

$$\chi_\alpha(\beta) = \prod_i \chi_{\lambda_i}(\beta).$$

This is well defined by Proposition 9.8.3(f). By part (e) of that proposition one sees that if $\beta \equiv \gamma\ (\alpha)$ then $\chi_\alpha(\beta) = \chi_\alpha(\gamma)$.

Proposition 9.8.5. *Let $\alpha \in \mathbb{Z}, \alpha \neq 0$, and $a \in \mathbb{Z}$ be an odd nonunit. If $(a, \alpha) = 1$, then*

$$\chi_a(\alpha) = 1.$$

PROOF. We may assume $a > 0$. Write $a = \prod p_i \prod q_i$ where p_i, q_i are prime, $p_i \equiv 1$ (4) and $q_i \equiv 3$ (4). By Proposition 9.8.4 we need only verify that $\chi_{p_i}(\alpha) = 1$. If $p_i = \pi\bar{\pi}$ where π is irreducible then $\chi_{p_i}(\alpha) = \chi_\pi(\alpha)\chi_{\bar\pi}(\alpha) = \overline{\chi_\pi(\alpha)\chi_\pi(\alpha)} = 1$ by Proposition 9.8.3(c).

Proposition 9.8.6. *If $n \neq 1$ is an integer $n \equiv 1$ (4), then $\chi_n(i) = (-1)^{(n-1)/4}$.*

PROOF. Note that n may be negative. If n is a positive prime $p \equiv 1$ (4) then writing $p = \pi\bar{\pi}$ one has

$$\chi_p(i) = \chi_\pi(i)\chi_{\bar\pi}(i) = (i^{(p-1)/4})^2 = (-1)^{(p-1)/4}.$$

If on the other hand $n = -q$, $q \equiv 3$ (4) and prime, then $\chi_{-q}(i) = i^{(q^2-1)/4} = (i^{q-1})^{(q+1)/4} = (-1)^{(-q-1)/4}$. If $n \equiv 1$ (4) is arbitrary then one may write $n = p_1 \cdots p_t(-q_1)\cdots(-q_s)$, $p_i \equiv 1$ (4), $q_i \equiv 3$ (4). The result then follows from Exercise 44. $\qquad\square$

§9 The Law of Biquadratic Reciprocity

The general law of biquadratic reciprocity may be stated as follows. Let λ and π be relatively prime primary elements of D. Then

Theorem 2. $\chi_\pi(\lambda) = \chi_\lambda(\pi)(-1)^{((N(\lambda)-1)/4)((N(\pi)-1)/4)}$.

If λ and π are primary, where $\lambda = c + di$ and $\pi = a + bi$, it is simple to see that $((N(\lambda) - 1)/4)((N(\pi) - 1)/4)$ and $((a-1)/2)((c-1)/2)$ have the same parity, so one may write

$$\chi_\pi(\lambda) = \chi_\lambda(\pi)(-1)^{((a-1)/2)((c-1)/2)}.$$

In other words if either π or λ is congruent to 1 modulo 4 then π and λ have the same biquadratic character. If however both are congruent to $3 + 2i$ (see Lemma 6) then π and λ have "opposite" character in the sense that $\chi_\pi(\lambda) = -\chi_\lambda(\pi)$.

Consider a primary irreducible π with $N(\pi) = p \equiv 1$ (4) and let χ_π be the associated quartic residue character. Then χ_π may be viewed as a multiplicative character on the finite field $D/\pi D = F$. Recall that F is a finite field with p elements consisting of the residue classes of $0, 1, \ldots, p - 1$. If $\zeta = e^{2\pi i/p}$ let $g(\chi_\pi) = \sum_{j \in F} \chi_\pi(j)\zeta^j$ be the Gauss sum belonging to χ_π. If $\psi = \chi_\pi^2$ then ψ is the nontrivial character of order 2 on F and thus is the Legendre symbol.

Proposition 9.9.1. $J(\chi_\pi, \chi_\pi) = \chi_\pi(-1)J(\chi_\pi, \psi)$.

PROOF. By Theorem 1, Chapter 8, one has $J(\chi_\pi, \chi_\pi) = g(\chi_\pi)^2/g(\psi)$. Thus

$$J(\chi_\pi, \chi_\pi)^2 = \frac{g(\chi_\pi)^4}{g(\psi)^2} = \chi_\pi(-1)J(\chi_\pi, \chi_\pi)J(\chi_\pi, \psi)$$

using Propositions 6.3.2 and 8.3.3. This gives the result. $\qquad\square$

Proposition 9.9.2. $g(\chi_\pi)^4 = pJ(\chi_\pi, \chi_\pi)^2$.

PROOF. This follows immediately from Propositions 9.9.1 and 8.3.3. $\qquad\square$

Proposition 9.9.3. $-\chi_\pi(-1)J(\chi_\pi, \chi_\pi)$ is primary.

PROOF. Clearly

$$J(\chi_\pi, \chi_\pi) = 2 \sum_{t=2}^{(p-1)/2} \chi_\pi(t)\chi_\pi(1-t) + \chi_\pi\left(\frac{p+1}{2}\right)^2.$$

But any unit in D is congruent to 1 modulo $1 + i$. Also $p \equiv 1 \, (2 + 2i)$. Finally $\chi_\pi((p+1)/2)^2 = (\chi_\pi(2^{-1}))^2 = \chi_\pi(2)^{-2} = \chi_\pi(2)^2 = \chi_\pi(-i(1+i)^2)^2 = \chi_\pi(-i)^2 = \chi_\pi(-1)$. Thus

$$J(\chi_\pi, \chi_\pi) \equiv 2\left(\frac{p-3}{2}\right) + \chi_\pi(-1)\,(2 + 2i)$$

$$\equiv -2 + \chi_\pi(-1)\,(2 + 2i).$$

Thus

$$-\chi_\pi(-1)J(\chi, \chi) \equiv 2\chi_\pi(-1) - 1\,(2 + 2i)$$
$$\equiv 1\,(2 + 2i),$$

since $\chi_\pi(-1) = \pm 1$. $\qquad\square$

The next proposition identifies the primary element $-\chi_\pi(-1)J(\chi_\pi, \chi_\pi)$.

Proposition 9.9.4. $-\chi_\pi(-1)J(\chi_\pi, \chi_\pi) = \pi$.

PROOF. By Lemma 7 of Section 7 it is enough to show that the left- and right-hand sides differ by a unit. Now $J(\chi_\pi, \chi_\pi) \equiv \sum_{t=1}^{p-1} t^{(p-1)/4}(1-t)^{(p-1)/4} \, (\pi)$. By Exercise 11 of Chapter 4 it follows that $J(\chi_\pi, \chi_\pi) \equiv 0 \, (\pi)$. By the corollary to Theorem 1 of Chapter 8, $N(J(\chi_\pi, \chi_\pi)) = p$. Thus $J(\chi_\pi, \chi_\pi)$ is irreducible and the proposition is complete. $\qquad\square$

Combining Proposition 9.9.4 with Proposition 9.9.2 gives the factorization of $g(\chi)^4$ in D.

Proposition 9.9.5. $g(\chi_\pi)^4 = \pi^3 \bar{\pi}$.

We will now prove two particular cases of the law of biquadratic reciprocity. The general statement will then be a formal, if somewhat tedious, consequence.

Proposition 9.9.6. Let $q > 0$ be a real irreducible in D. Then

$$\chi_\pi(-q) = \chi_q(\pi).$$

PROOF. Since $q \equiv 3 \,(4)$ one has

$$g(\chi_\pi)^q \equiv \sum_{j=1}^{p-1} \chi_\pi(j)^q \zeta^{qj} \equiv \sum \chi_\pi^3(j) \zeta^{qj} \; (q)$$

$$\equiv \chi_\pi(q) g(\bar{\chi}_\pi) \; (q).$$

Thus

$$(g(\chi_\pi)^4)^{(q+1)/4} = g(\chi_\pi)^{q+1} \equiv \chi_\pi(q) g(\chi_\pi) \cdot g(\bar{\chi}_\pi) \; (q).$$

By the observation following Proposition 8.2.2 and noting (see Exercise 45) that $\bar{\pi} \equiv \pi^q \; (q)$ one has, by Proposition 9.9.5

$$\pi^{[(q+3)(q+1)]/4} \equiv \chi_\pi(-1) \chi_\pi(q) \pi^{q+1} \; (q)$$

or

$$\pi^{(q^2-1)/4} \equiv \chi_\pi(-q) \; (q).$$

But $\pi^{(q^2-1)/4} \equiv \chi_q(\pi) \; (q)$. Thus

$$\chi_q(\pi) \equiv \chi_\pi(-q) \; (q),$$

which implies, since both sides are units, that

$$\chi_q(\pi) = \chi_\pi(-q).$$

This completes the proof. □

Notice that $-q$ is a *primary* irreducible and $(N(q) - 1)/4 = (q^2 - 1)/4$ is even. Thus Proposition 9.9.6 is indeed a special case of biquadratic reciprocity.

Proposition 9.9.7. *Let q be prime $q \equiv 1 \,(4)$. Then $\chi_\pi(q) = \chi_q(\pi)$.*

PROOF. Since $q \equiv 1 \,(4)$

$$g(\chi_\pi)^q \equiv \sum \chi_\pi(j)^q \zeta^{qj} \equiv \sum \chi_\pi(j) \zeta^{qj} \equiv \bar{\chi}_\pi(q) g(\chi_\pi) \; (q).$$

Thus

$$g(\chi_\pi)^{q+3} \equiv \bar{\chi}_\pi(q) g(\chi_\pi)^4 \; (q).$$

By Proposition 9.9.5 this becomes

$$(\pi^3 \bar{\pi})^{(q+3)/4} \equiv \bar{\chi}_\pi(q) \pi^3 \bar{\pi} \; (q).$$

Both sides of this congruence belong to D and $(q, \pi) = (q, \bar{\pi}) = 1$. Thus we may divide to obtain

$$(\pi^3)^{(q-1)/4} (\bar{\pi})^{(q-1)/4} \equiv \bar{\chi}_\pi(q) \; (q).$$

If $q = \lambda \bar{\lambda}$ where λ is a irreducible in D then this implies

$$\chi_\lambda(\pi^3) \chi_\lambda(\bar{\pi}) \equiv \bar{\chi}_\pi(q) \; (\lambda).$$

As in the previous case we conclude that

$$\chi_\lambda(\pi^3)\chi_\lambda(\bar{\pi}) = \bar{\chi}_\pi(q).$$

This may be written as

$$\overline{\chi_\lambda(\pi)}\chi_\lambda(\bar{\pi}) = \overline{\chi_\pi}(q)$$

or

$$\chi_{\bar{\lambda}}(\bar{\pi})\chi_\lambda(\bar{\pi}) = \overline{\chi_\pi}(q)$$

which gives, by definition

$$\chi_q(\bar{\pi}) = \bar{\chi}_\pi(q).$$

Taking conjugates completes the proof.   □

The reader should notice that in Proposition 9.9.7, q is not irreducible and that the left-hand side is the generalized biquadratic residue symbol.

The following proposition is a formal exercise using Lemma 8 of Section 7, and Propositions 9.8.6, 9.9.6, and 9.9.7.

Proposition 9.9.8. *Let a be real and $a \equiv 1$ (4) and λ be primary, $(\lambda, a) = 1$. Then $\chi_a(\lambda) = \chi_\lambda(a)$.*

Suppose now that $\pi = a + bi$ and $\lambda = c + di$ are primary and relatively prime. We do not assume that $N(\pi) \neq N(\lambda)$, or that they are irreducible.

Proposition 9.9.9. *If $(a, b) = 1$, $(c, d) = 1$ then*

$$\chi_\pi(\lambda) = \chi_\lambda(\pi)(-1)^{((a-1)/2((c-1)/2)}.$$

PROOF. The hypothesis implies that $(a, \pi) = (b, \pi) = (c, \lambda) = (d, \lambda) = 1$. The relation $c\pi \equiv ac + bd$ (λ) implies $(ac + bd, \lambda) = (ac + bd, \pi) = 1$. Furthermore

$$\chi_\lambda(c)\chi_\lambda(\pi) = \chi_\lambda(ac + bd). \tag{1}$$

Similarly

$$\chi_\pi(a)\chi_\pi(\lambda) = \chi_\pi(ac + bd). \tag{2}$$

Taking the conjugate of (2) and multiplying by (1) one obtains the relation

$$\chi_\lambda(c)\chi_{\bar{\pi}}(a)\chi_\lambda(\pi)\overline{\chi_\pi(\lambda)} = \chi_{\lambda\bar{\pi}}(ac + bd).$$

Thus we have shown, using Proposition 9.8.3(c)

$$\chi_\lambda(\pi)\overline{\chi_\pi(\lambda)} = \chi_\lambda(c)\chi_\pi(a)\chi_{\lambda\bar{\pi}}(ac + bd). \tag{3}$$

Assume that c, a, and $ac + bd$ are nonunits. The three terms on the right-hand side are easily computed. For an odd integer n put $\varepsilon(n) = (-1)^{(n-1)/2}$. Then $\varepsilon(n)n \equiv 1$ (4) and $\varepsilon(ac + bd) = \varepsilon(a)\varepsilon(c)$ since $bd \equiv 0$ (4). Writing

$\chi_a(x) = \chi_a(\varepsilon(x))\chi_a(\varepsilon(x)x)$ for each term on the right-hand side of (3) one obtains, noting that $\chi_a(\varepsilon(x)) = \chi_{\bar{a}}(\varepsilon(x))$ and using Proposition 9.9.8 and 9.8.3(b)

$$\chi_\lambda(\pi)\overline{\chi_\pi(\lambda)} = \chi_c(\bar{\lambda})\chi_a(\pi)\chi_{ac+bd}(\lambda\bar{\pi}). \tag{4}$$

As for the last three terms one computes, using Proposition 9.8.5

$$\chi_c(\bar{\lambda}) = \chi_c(c - di) = \chi_c(-di) = \chi_c(i),$$

$$\chi_a(\pi) = \chi_a(a + bi) = \chi_a(bi) = \chi_a(i),$$

$$\chi_{ac+bd}(\bar{\pi}\lambda) = \chi_{ac+bd}((ad - bc)i) = \chi_{ac+bd}(i).$$

Thus we have the relation

$$\chi_\lambda(\pi)\overline{\chi_\pi(\lambda)} = \chi_{(ac+bd)ac}(i)$$
$$= (-1)^{((ac+bd)ac - 1)/4}$$
$$= (-1)^{((a-1)/2)((c-1)/2)}. \quad \text{(Proposition 9.8.6)} \tag{5}$$

The last equality is a simple exercise using Lemma 6 of Section 7. We leave to the reader the simple task of carrying through the situation in which one of a, c, or $ac + bd$ is a unit. $\qquad\square$

The general law of biquadratic reciprocity follows easily from Proposition 9.9.9. For write $\pi = m(a + bi)$, $\lambda = n(c + di)$, $(\pi, \lambda) = 1$ where $m \equiv n \equiv 1$ (4), $(a, b) = 1$, $(c, d) = 1$. By Proposition 9.9.8, $\chi_\pi(n) = \chi_n(\pi)$ and $\chi_\lambda(m) = \chi_m(\lambda)$. Also $\chi_m(n) = \chi_n(m) = 1$ by Proposition 9.8.5. Then, since $a + bi$ and $c + di$ are primary,

$$\chi_\lambda(\pi) = \chi_\lambda(m)\chi_\lambda(a + bi)$$
$$= \chi_m(\lambda)\chi_n(a + bi)\chi_{c+di}(a + bi)$$
$$= \chi_m(\lambda)\chi_{a+bi}(n)\chi_{a+bi}(c + di)(-1)^{((a-1)/2)((c-1)/2)}$$
$$= \chi_\pi(\lambda)(-1)^{((a-1)/2)((c-1)/2)}$$
$$= \chi_\pi(\lambda)(-1)^{((N(\pi)-1)/4)((N(\lambda)-1)/4)},$$

where in the last line we have used the fact that $m \equiv n \equiv 1$ (4). This completes the proof, a monument to ingenuity and persistence! $\qquad\square$

§10 Rational Biquadratic Reciprocity

Throughout this section p and q denote distinct primes congruent to 1 modulo 4. Then the multiplicative group $(\mathbb{Z}/p\mathbb{Z})^*$ has a unique subgroup of order $(p - 1)/4$ consisting of the residues of fourth powers of integers. Consider the biquadratic residue character χ_π defined by means of an irreducible π in $\mathbb{Z}[i]$ dividing p. By Proposition 9.8.3 $\chi_\pi(q) = 1$ iff $x^4 \equiv q$ (π) has a solution with $x \in \mathbb{Z}[i]$.

Lemma 1. $\chi_\pi(q) = 1$ iff $x^4 \equiv q \ (p)$ has a solution with $x \in \mathbb{Z}$.

PROOF. By Proposition 9.8.1 the integers $0, 1, 2, \ldots, p - 1$ form a complete set of residues for the residue classes of $\mathbb{Z}[i]$ modulo π. Thus $\chi_\pi(q) = 1$ iff $x^4 \equiv q \ (\pi)$ has a solution with $x \in \mathbb{Z}$. It follows that $x^4 \equiv q \ (\bar{\pi})$. However, $(\pi, \bar{\pi}) = 1$. Thus $p = \pi\bar{\pi} | x^4 - q$. $\qquad \square$

Let ψ_p denote the quadratic residue character.

Lemma 2. If $\psi_p(q) = 1$ then $\chi_\pi(q) = \pm 1$.

PROOF. Since $q^{(p-1)/2} \equiv 1 \ (p)$ it follows that $\chi_\pi^2(q) \equiv (q^{(p-1)/4})^2 \equiv q^{(p-1)/2} \equiv 1 \ (\pi)$. Thus $\chi_\pi^2(q) = 1$. $\qquad \square$

Thus, assuming that q is a square modulo p, $\chi_\pi(q)$ is $+1$ or -1 according as q is or is not a fourth power modulo p. By the law of quadratic reciprocity $\psi_q(p) = +1$. Notice that the value $\chi_\pi(q)$ depends only on p and q and not on the choice of the irreducible π. Contrary to what one might expect the relationship between the two integers $\chi_\pi(q)$ and $\chi_\lambda(p)$ where λ is an irreducible dividing q is not a simple consequence of the law of biquadratic reciprocity. In 1969 K. Burde [102] discovered the following remarkable reciprocity law. Since p and q are congruent to 1 modulo 4 we may write $p = a^2 + b^2$, $q = c^2 + d^2$, where $a \equiv c \equiv 1 \ (2)$ and $b \equiv d \equiv 0 \ (2)$. Throughout the following we assume $\psi_q(p) = 1$.

Theorem 3. $\chi_\pi(q)\chi_\lambda(p) = (-1)^{(q-1)/4}\psi_q(ad - bc)$.

The following elegant proof is due to K. Williams [244]. The law of biquadratic reciprocity is not assumed. However the value of the quadratic Gauss sum is used (Chapter 6, Section 4). The following proposition is of interest in itself. (See the comment at the end of Section 12).

Proposition 9.10.1. Let π be the primary irreducible dividing p. Then

$$g(\chi_\pi)^2 = -(-1)^{(p-1)/4}\sqrt{p}\,\pi$$

where $\sqrt{p}$ denotes the positive square root.

PROOF. By Proposition 9.9.4 and Theorem 1, Chapter 8 we have

$$J(\chi_\pi, \chi_\pi) = -\chi_\pi(-1)\pi = \frac{g(\chi_\pi)^2}{g(\psi_p)}.$$

The proposition follows from Theorem 1, Chapter 6 and the observation that $\chi_\pi(-1) = (-1)^{(p-1)/4}$. $\qquad \square$

Proposition 9.10.2. If π is a primary irreducible dividing p then $\chi_\pi(q)\chi_\lambda(p) \equiv \pi^{(q-1)/2} \ (q)$.

PROOF. We have, in the ring of all algebraic integers,

$$g(\chi_\pi)^q = (\sum \chi_\pi(j)\zeta^j)^q$$
$$\equiv \sum \chi_\pi(j)\zeta^{qj} \ (q)$$
$$\equiv \chi_\pi(q^{-1})g(\chi_\pi) \ (q)$$
$$\equiv \chi_\pi(q)g(\chi_\pi) \ (q).$$

The last congruence follows because

$$\chi_\pi(q^{-1}) = \chi_\pi^3(q) = \chi_\pi^2(q)\chi_\pi(q) = \chi_\pi(q).$$

Thus, multiplying by $g(\chi_\pi)^3$

$$g(\chi_\pi)^4(g(\chi_\pi))^{q-1} \equiv \chi_\pi(q)g(\chi_\pi)^4 \ (q).$$

The two terms on the left-hand side are in $\mathbb{Z}[i]$ by Proposition 8.3.3; and by Proposition 8.2.2 $N(g(\chi_\pi)^4) = p^4$. Thus one may cancel $g(\chi_\pi)^4$ to obtain

$$g(\chi_\pi)^{q-1} \equiv \chi_\pi(q) \ (q).$$

Using Proposition 9.10.1 one obtains

$$(g(\chi_\pi)^2)^{(q-1)/2} = p^{(q-1)/4}\pi^{(q-1)/2} \equiv \chi_\pi(q) \ (q).$$

But $p^{(q-1)/4} \equiv \chi_\lambda(p) \ (\lambda)$ and since both sides of this congruence are real it follows, taking conjugates and noting $(\lambda, \bar\lambda) = 1$, that this congruence holds modulo q. This completes the proof. $\square$

In the following proposition π is not assumed to be primary. Write $\pi = a + bi$ and $\lambda = c + di$.

Proposition 9.10.3. $\pi^{(q-1)/2} \equiv \psi_q(d)\psi_q(ad - bc) \ (q).$

PROOF. Since $d\pi \equiv ad - bc \ (\lambda)$ one has

$$(d\pi)^{(q-1)/2} \equiv (ad - bc)^{(q-1)/2} \ (\lambda).$$

Thus

$$\psi_q(d)\pi^{(q-1)/2} \equiv \psi_q(ad - bc) \ (\lambda).$$

Similarly $d\pi \equiv (ad + bc) \ (\bar\lambda)$ implies

$$\psi_q(d)\pi^{(q-1)/2} \equiv \psi_q(ad + bc) \ (\bar\lambda).$$

The proof now follows from the following lemma. $\square$

Lemma 3. $\psi_q(ad - bc) = \psi_q(ad + bc).$

PROOF. Since $c^2 \equiv -d^2 \ (q)$ one has

$$\psi_q(ad - bc)\psi_q(ad + bc) = \psi_q(a^2d^2 - b^2c^2) = \psi_q(d^2p) = \psi_q(p) = 1. \quad \square$$

Note furthermore that since $\psi_q(-1) = 1$ one has as a consequence of the above lemma $\psi_q(ad - bc) = \psi_q(-ad + bc) = \psi_q(-ad - bc)$. Thus in the statement of Theorem 3 there is no loss of generality in assuming that π is primary. With this assumption one concludes from Propositions 9.10.2 and 9.10.3 that

$$\chi_\pi(q)\chi_\lambda(p) = \psi_q(d)\psi_q(ad - bc).$$

The proof of Theorem 3 is completed by the following lemma.

Lemma 4. *If $q = c^2 + d^2, c > 0, c \equiv 1$ (2) then $\psi_q(d) = (-1)^{(q-1)/4}$.*

PROOF. Let ψ_c denote the Jacobi symbol. Then by Proposition 5.2.2 one has $\psi_q(c) = \psi_c(q) = \psi_c(d^2) = 1$. (Cf. Exercise 26, Chapter 5). But $c^2 \equiv -d^2$ (q) implies $c^{(q-1)/2} \equiv (-1)^{(q-1)/4}d^{(q-1)/2}$ (q). Thus $\psi_q(c) = 1 = (-1)^{(q-1)/4}\psi_q(d)$. $\qquad\square$

§11 The Constructibility of Regular Polygons

On March 30, 1796 C. Gauss, then almost 19 years old, began a diary in which he recorded his mathematical discoveries. The first entry reads "Principia quibus innitur sectio circuli, ac divisibilitas eiusdem geometrica in septemdecim partes, etc.," a rough translation of which is "Principles upon which the division of a circle into 17 parts depend, etc....". More generally in his *Disquisitiones Arithmeticae*, §365, Gauss proves, using "cyclotomic periods" that if p is a prime of the form $2^n + 1$ then a regular polygon with p sides is constructible by ruler and compass.

In this section we give a short proof of this result using Gauss and Jacobi sums.

Generally speaking the constructible complex numbers in our context are those numbers that may be obtained from $\mathbb{Q}$ by a finite sequence of rational operations and the formation of square roots. More precisely

Definition. A complex number $\alpha \in \mathbb{C}$ is constructible if there exist subfields of $\mathbb{C}$, $\mathbb{Q} = K_0 \subset K_1 \subset K_2 \subset \cdots \subset K_n$ such that $\alpha \in K_n$ and $K_i = K_{i-1}(\sqrt{\alpha_{i-1}})$ for some $\alpha_i \in K_i$, $i = 1, \ldots, n$.

Here $K(\sqrt{\beta})$ denotes the field of all complex numbers $a + b\sqrt{\beta}, a, b \in K$ (see Exercise 6, Chapter 6). It is easy to see that α is constructible iff the real and imaginary parts of α are constructible. Furthermore if α is constructible then $\sqrt{\alpha}$ is constructible. Let, as usual, $\zeta_t = e^{2\pi i/t}$.

Lemma 1. *ζ_{2^n} is constructible, $n = 1, 2, \ldots$.*

PROOF. Since $(\zeta_{2^n})^2 = \zeta_{2^{n-1}}$ the result follows by induction (ζ_2 is certainly constructible!).

Lemma 2.

$$\sum_\chi \chi(t) = \begin{cases} 1, & \text{if } t = 0, \\ p - 1, & \text{if } t = 1, \\ 0, & \text{if } t \neq 0, 1, \end{cases}$$

the sum being over all characters of F_p^*.

PROOF. If $\chi = \varepsilon$, the trivial character then $\varepsilon(0) = 1$. Thus the result holds for $t = 0$. It is true when $t = 1$ by Proposition 8.1.3 while the remaining case is the corollary to Proposition 8.1.3. □

Recall that a Fermat prime is a prime of the form $2^n + 1$.

Theorem 4. *If p is a Fermat prime then ζ_p is constructible.*

PROOF. If $g(\chi) = \sum_{t=0}^{p-1} \chi(t)\zeta_p^t$ is the Gauss sum associated with χ then

$$\sum_\chi g(\chi) = \sum_{t=0}^{p-1} \left(\sum_\chi \chi(t) \right) \zeta_p^t$$

$$= 1 + (p - 1)\zeta_p.$$

Thus $\zeta_p = (p - 1)^{-1}(-1 + \sum_\chi g(\chi))$ and therefore ζ_p is constructible if each $g(\chi)$ is.

However $p - 1 = 2^n$ and since the characters form a group of order $p - 1$ we see that the order of χ is 2^m for some m. Then using Proposition 8.3.3 we have $g(\chi)^{2^m} = \chi(-1)pJ(\chi, \chi)J(\chi, \chi^2)\ldots J(\chi, \chi^l)$ where $l = 2^m - 2$. But $J(\chi, \chi^j) \in \mathbb{Z}[\zeta_{2^n}]$ so that by Lemma 1 $g(\chi)^{2^m}$ is constructible. It follows that $g(\chi)$ is constructible and the proof is complete. □

§12 Cubic Gauss Sums and the Problem of Kummer

If p is a prime $p \equiv 1 \ (4)$ then the simple argument of Proposition 6.3.2 showed that $g(\chi)^2 = p$ where

$$g(\chi) = \sum_{t=1}^{p-1} \left(\frac{t}{p} \right) \zeta_p^t = \sum_{t=0}^{p-1} \zeta_p^{t^2} = \sum_{t=0}^{p-1} \cos \frac{2\pi t^2}{p}$$

is the classical quadratic Gauss sum. Thus with little effort $g(\chi)$ was shown to be one of the real roots of $x^2 - p = 0$. Using a more sophisticated argument, we have shown in Section 6, Chapter 6 that actually $g(\chi)$ is always the largest root, that is to say $g(\chi) = \sqrt{p}$.

In the case of cubic Gauss sums the matter is more subtle. Let p be a prime $p \equiv 1 \ (3)$ and consider $\sum_{t=0}^{p-1} \cos(2\pi t^3/p) = G$. Write $p = \pi\bar{\pi}$ where π is a complex primary prime in $\mathbb{Z}[\omega]$ and let χ_π be the cubic character associated with π as defined in Section 3.

Lemma 1. $G = g(\chi_\pi) + \overline{g(\chi_\pi)}$.

PROOF. If $\zeta = e^{2\pi i/p}$ then since G is real, and $-1 = (-1)^3$

$$G = \sum_{t=0}^{p-1} \zeta^{t^3} = \sum_{t=0}^{p-1} \zeta^t(1 + \chi_\pi(t) + \chi_\pi(t^2))$$

$$= g(\chi_\pi) + g(\chi_\pi^2)$$

$$= g(\chi_\pi) + g(\bar{\chi}_\pi)$$

$$= g(\chi_\pi) + \chi_\pi(-1)\overline{g(\chi_\pi)}$$

$$= g(\chi_\pi) + \overline{g(\chi_\pi)}. \qquad \square$$

Notice that in the above proof χ can be any character of order 3. However in the following lemma the choice of χ_π is essential. Write $\pi = a + b\omega$.

Lemma 2. G is a real root of $x^3 - 3px - (2a - b)p = 0$.

PROOF. By Lemma 1, writing χ for χ_π,

$$G^3 = g(\chi)^3 + \overline{g(\chi)}^3 + 3g(\chi)\overline{g(\chi)}(g(\chi) + \overline{g(\chi)})$$

$$= p\pi + p\bar{\pi} + 3pG$$

$$= 3pG + p(2a - b).$$

In the second step we have used the corollary to Lemma 1, Section 4. $\qquad \square$

Corollary. G is a root of $x^3 - 3px - Ap = 0$ where $4p = A^2 + 27B^2$, $A \equiv 1 \ (3)$.

PROOF. This is simply the corollary to Proposition 8.3.4. $\qquad \square$

Thus G is twice the real part of $g(\chi_\pi)$ and is a root of the polynomial $x^3 - 3px - Ap$. In the same manner as above we see that the other roots are $2\operatorname{Re}(\omega g(\chi_\pi))$ and $2\operatorname{Re}(\omega^2 g(\chi_\pi))$. Using the fact that $|g(\chi_\pi)| = p^{1/2}$ it is a simple matter to see that each of the intervals $(-2\sqrt{p}, -\sqrt{p}), (-\sqrt{p}, \sqrt{p})$, and $(\sqrt{p}, 2\sqrt{p})$ contains precisely one of the roots (see Exercise 43). By the corollary to Lemma 1, Section 4, the value of $g(\chi_\pi)$ is determined up to 1, ω, or ω^2. Unable to find an expression for this root of unity for general p, Kummer proposed a statistical study of the distribution of those primes for which G, say, is the largest root of $x^3 - 3px - Ap$. He found, for example, that among the primes less than 500, G was in the interval $(\sqrt{p}, 2\sqrt{p})$ for 24 primes. The interval $(-2\sqrt{p}, -\sqrt{p})$ contained 7 primes and the middle interval 14 primes. (See [164], Vol. 1, pp. 50, 296, 353.) Putting $I_1 = (-2\sqrt{p}, -\sqrt{p})$, $I_2 = (-\sqrt{p}, \sqrt{p})$, $I_3 = (\sqrt{p}, 2\sqrt{p})$ and letting $N_j(B)$ be the number of primes less than B such that G is in I_j he noted that the ratio $N_1(500) : N_2(500) : N_3(500)$ is roughly $1 : 2 : 3$.

However in 1953, J. von Neumann and H. H. Goldstine considering all primes ($\equiv 1$ (3)) less than 9973 arrived at a ratio of roughly $2 : 3 : 4$ [197]. They found $N_1(10^4) = 138$, $N_2(10^4) = 201$, $N_3(10^4) = 272$. They stated, "These results would seem to indicate a significant departure from the conjectured densities and a trend toward randomness." Emma Lehmer extended the calculations to include the first 1000 primes, $p \equiv 1$ (3), and discovered a ratio approximately $3 : 4 : 5$. [176]. Thus the suspicion arose that indeed the values of G are asymptotically uniformly distributed in the three intervals. That this is indeed the case was established in 1978 by D. R. Heath-Brown and S. J. Patterson in their paper, "The distribution of Kummer sums at prime arguments" [147].

We mention that J. W. S. Cassels [108], conjectured a precise expression for $g(\chi_\pi)$ involving elliptic functions. This conjecture was established by C. R. Matthews [186]. Furthermore an explicit elementary expression has been obtained for the biquadratic Gauss sum by Matthews [186]. The result of Matthews is as follows. Let p be prime $p \equiv 1$ (4) and write $p = \pi\bar{\pi}$, π primary, $\pi = a + bi$. Define $\beta = \pm i$ by $((p - 1)/2)! \equiv \beta \ (\pi)$. If $g(\chi_\pi)$ is the biquadratic Gauss sum attached to χ_π then by Proposition 9.10.1, $g(\chi_\pi)^2 = (-1)^{(p-1)/4}\pi\sqrt{p}$. Thus $g(\chi_\pi) = \varepsilon\sqrt{(-1)^{(p-1)/4}\pi\sqrt{p}}$ where the square root has positive real part. Matthews proved that $\varepsilon = -\beta\chi_\pi(2i)(2|b|/a)$ where $(2|b|/a)$ is the Jacobi symbol. See also J. H. Loxton [182], and B. C. Berndt and R. J. Evans [93].

NOTES

For the early history of cubic and biquadratic reciprocity we note that Euler, during the years 1748–1750, conjectured Proposition 9.6.2 concerning the cubic character of 2, as well as similar results for the integers 3, 5, and 7. He also conjectured that 2 is a fourth power modulo p, $p \equiv 1$ (4) iff $p = a^2 + 64b^2$ (Exercise 6, Chapter 5) and stated similar results for the primes 3 and 5. All of Euler's conjectures concerning these special cases of reciprocity were correct, a remarkable example of his "inductive" ability. The general biquadratic character of 2 (Exercise 37) was established by Gauss in his first memoir on biquadratic residues (1828) while the general law of biquadratic reciprocity was stated in his second memoir on the same subject (1832). For further historical comments on the history of these results see the paper by M. J. Collision [116].

Gauss wrote to Alexander von Humboldt in 1846 that Eisenstein's mathematical talent was such as nature confers upon few in each century. In 1844, at the age of twenty-one, Eisenstein published a total of 25 papers in

Crelle's journal. The proofs of cubic and biquadratic reciprocity given in this chapter as well as the proofs of quadratic reciprocity given in Chapter 6 are among them (see [28], [130], [131]). The collected works of this remarkable genius, dead at 29, are now available. An informative and charming account of Eisenstein's life and research has been given by A. Weil in his review of the collected works [239]. One should also read the beautiful paper by Weil, "La Cyclotomie, jadis et naguère" [238]. In a later chapter we shall prove a generalization of these reciprocity laws, the celebrated Eisenstein reciprocity law. A discussion of Eisenstein's other proofs of biquadratic reciprocity is contained in H. Smith's report [72]. As far as cubic reciprocity is concerned Jacobi claims to have given the proof in his lectures of 1837 but the first published proof is definitely due to Eisenstein in 1844. The dispute over priority appears to have been quite bitter.

For the actual construction of a 17-sided polygon see Hardy and Wright [40], p. 61. Gauss' treatment of cyclotomy is contained in §7 of his *Disquisitiones Arithmeticae* [136]. In §335 he mentions that the techniques developed there extend to other transcendental functions such as those connected with $\int dx/\sqrt{1-x^4}$, the integral arising from arc length on a lemniscate. Gauss recorded in his diary on March 21, 1797 that he has succeeded in dividing the arc of the lemniscate into five equal parts. In 1827 Abel was able to show that, as in the case of a circle, the arc of a lemniscate can be divided into p equal parts with ruler and compass when p is a Fermat prime. For an examination of Abel's proof from a modern point of view see the article by M. Rosen [212].

In recent times there has been a renewed interest in rational reciprocity laws. The interested reader should consult the survey article by E. Lehmer [175] as well as the paper by H. von Lienen [181].

EXERCISES

1. If $\alpha \in \mathbb{Z}[\omega]$, show that α is congruent to either 0, 1, or -1 modulo $1 - \omega$.

2. From now on we shall set $D = \mathbb{Z}[\omega]$ and $\lambda = 1 - \omega$. For μ in D show that we can write $\mu = (-1)^a \omega^b \lambda^c \pi_1^{a_1} \pi_2^{a_2} \cdots \pi_t^{a_t}$, where a, b, c, and the a_i are nonnegative integers and the π_i are primary primes.

3. Let γ be a primary prime. To evaluate $\chi_\gamma(\mu)$ we see, by Exercise 2, that it is enough to evaluate $\chi_\gamma(-1), \chi_\gamma(\omega), \chi_\gamma(\lambda)$, and $\chi_\gamma(\pi)$, where π is a primary prime. Since $-1 = (-1)^3$ we have $\chi_\gamma(-1) = 1$. We now consider $\chi_\gamma(\omega)$. Let $\gamma = a + b\omega$ and set $a = 3m - 1$ and $b = 3n$. Show that $\chi_\gamma(\omega) = \omega^{m+n}$.

4. (continuation) Show that $\chi_\gamma(\omega) = 1, \omega$, or ω^2 according to whether γ is congruent to 8, 2, or 5 modulo 3λ. In particular, if q is a rational prime, $q \equiv 2$ (3), then $\chi_q(\omega) = 1$, ω, or ω^2 according to whether $q \equiv 8$, 2, or 5 (9). [*Hint*: $\gamma = a + b\omega = -1 + 3(m + n\omega)$, and so $\gamma \equiv -1 + 3(m + n)$ (3λ).]

5. In the text we stated Eisenstein's result $\chi_\gamma(\lambda) = \omega^{2m}$. Show that $\chi_\gamma(3) = \omega^{2n}$.

6. Prove that
   (a) $\chi_\gamma(\lambda) = 1$ for $\gamma \equiv 8, 8 + 3\omega, 8 + 6\omega$ (9).
   (b) $\chi_\gamma(\lambda) = \omega$ for $\gamma \equiv 5, 5 + 3\omega, 5 + 6\omega$ (9).
   (c) $\chi_\gamma(\lambda) = \omega^2$ for $\gamma \equiv 2, 2 + 3\omega, 2 + 6\omega$ (9).

7. Find primary primes associate to $1 - 2\omega$, $-7 - 3\omega$, and $3 - \omega$.

8. Factor the following numbers into primes in D: 7, 21, 45, 22, and 143.

9. Show that $\bar{\alpha}$, the residue class of α, is a cube in the field $D/\pi D$ iff $\alpha^{(N\pi - 1)/3} \equiv 1$ (π). Conclude that there are $(N\pi - 1)/3$ cubes in $D/\pi D$.

10. What is the factorization of $x^{24} - 1$ in $D/5D$?

11. How many cubes are there in $D/5D$?

12. Show that $\omega\lambda$ has order 8 in $D/5D$ and that $\omega^2\lambda$ has order 24. [*Hint*: Show first that $(\omega\lambda)^2$ has order 4.]

13. Show that π is a cube in $D/5D$ iff $\pi \equiv 1, 2, 3, 4, 1 + 2\omega, 2 + 4\omega, 3 + \omega$, or $4 + 3\omega$ (5).

14. For which primes $\pi \in D$ is $x^3 \equiv 5$ (π) solvable?

15. Suppose that $p \equiv 1$ (3) and that $p = \pi\bar{\pi}$, where π is a primary prime in D. Show that $x^3 \equiv a$ (p) is solvable in $\mathbb{Z}$ iff $\chi_\pi(a) = 1$. We assume that $a \in \mathbb{Z}$.

16. Is $x^3 \equiv 2 - 3\omega$ (11) solvable? Since $D/11D$ has 121 elements this is hard to resolve by straightforward checking. Fill in the details of the following proof that it is not solvable. $\chi_\pi(2 - 3\omega) = \chi_{2 - 3\omega}(11)$ and so we shall have a solution iff $x^3 \equiv 11$ $(2 - 3\omega)$ is solvable. This congruence is solvable iff $x^3 = 11$ (7) is solvable in $\mathbb{Z}$. However, $x^3 \equiv a$ (7) is solvable in $\mathbb{Z}$ iff $a \equiv 1$ or 6 (7).

17. An element $\gamma \in D$ is called primary if $\gamma \equiv 2$ (3). If γ and ρ are primary, show that $-\gamma\rho$ is primary. If γ is primary, show that $\gamma = \pm\gamma_1\gamma_2\cdots\gamma_t$, where the γ_i are (not necessarily distinct) primary primes.

18. (continuation) If $\gamma = \pm\gamma_1\gamma_2\cdots\gamma_t$ is a primary decomposition of the primary element γ, define $\chi_\gamma(\alpha) = \chi_{\gamma_1}(\alpha)\chi_{\gamma_2}(\alpha)\cdots\chi_{\gamma_t}(\alpha)$. Prove that $\chi_\gamma(\alpha) = \chi_\gamma(\beta)$ if $\alpha \equiv \beta$ (γ) and $\chi_\gamma(\alpha\beta) = \chi_\gamma(\alpha)\chi_\gamma(\beta)$. If ρ is primary, show that $\chi_\rho(\alpha)\chi_\gamma(\alpha) = \chi_{-\rho\gamma}(\alpha)$.

19. Suppose that $\gamma = A + B\omega$ is primary and that $A = 3M - 1$ and $B = 3N$. Prove that $\chi_\gamma(\omega) = \omega^{M+N}$ and that $\chi_\gamma(\lambda) = \omega^{2M}$.

20. If γ and ρ are primary, show that $\chi_\gamma(\rho) = \chi_\rho(\gamma)$.

21. If γ is primary, show that there are infinitely many primary primes π such that $x^3 \equiv \gamma$ (π) is not solvable. Show also that there are infinitely many primary primes π such that $x^3 \equiv \omega$ (π) is not solvable and the same for $x^3 \equiv \lambda$ (π). (*Hint*: Imitate the proof of Theorem 3 of Chapter 5.)

22. (continuation) Show in general that if $\gamma \in D$ and $x^3 \equiv \gamma$ (π) is solvable for all but finitely many primary primes π, then γ is a cube in D.

23. Suppose that $p \equiv 1$ (3). Use Exercise 5 to show that $x^3 \equiv 3$ (p) is solvable in $\mathbb{Z}$ iff p is of the form $4p = C^2 + 243B^2$.

The following three exercises give K. Williams' elegant proof of the complex case of the supplement to the law of cubic reciprocity [245]. The reader may wish to consult the hints at the end of the book.

24. Let $\pi = a + b\omega$ be a complex primary element of $D = \mathbb{Z}[\omega]$. Put $a = 3m - 1$, $b = 3n$, $p = N(\pi)$.
   (a) $(p - 1)/3 \equiv -2m + n$ (3).
   (b) $(a^2 - 1)/3 \equiv m$ (3).
   (c) $\chi_\pi(a) = \omega^m$.
   (d) $\chi_\pi(a + b) = \omega^{2n}\chi_\pi(1 - \omega)$.

25. Show that $\chi_{a+b}(\pi)$ may be computed as follows.
   (a) $\chi_{a+b}(\pi) = \chi_{a+b}(1 - \omega)$.
   (b) $\chi_{a+b}(\pi) = \omega^{2(m+n)}$.

26. Combine the previous two exercises to conclude that $\chi_\pi(1 - \omega) = \omega^{2m}$.

The following four exercises are taken from Matthews [186].

27. Let $\pi = a + bi$ be a primary irreducible in $\mathbb{Z}[i]$, $b \neq 0$. Show
   (a) $a \equiv (-1)^{(p-1)/4}$ (4), $p = N(\pi)$.
   (b) $b \equiv 1 - (-1)^{(p-1)/4}$ (4).

28. The notation being as in Exercise 27 show $\chi_\pi(\bar{\pi}) = \chi_\pi(2)\chi_\pi(a)$.

29. By Exercise 27, $a(-1)^{(p-1)/4}$ is primary. Use biquadratic reciprocity to show $\chi_\pi(a(-1)^{(p-1)/4}) = (-1)^{(a^2-1)/8}$.

30. Use the preceding two exercises to show $\chi_\pi(\bar{\pi}) = \chi_\pi(-2)(-1)^{(a^2-1)/8}$.

31. Let p be prime, $p \equiv 1$ (4). Show that $p = a^2 + b^2$ where a and b are uniquely determined by the conditions $a \equiv 1$ (4), $b \equiv -((p - 1)/2)! \, a$ (p).

The following five exercises are taken from Eisenstein [130], §9.

32. Let p be prime, $p \equiv 1$ (4) and write $p = \pi\bar{\pi}$, $\pi \in \mathbb{Z}[i]$. Show $\chi_p(1 + i) = i^{(p-1)/4}$.

33. Let q be a positive prime, $q \equiv 3$ (4). Show $\chi_q(1 + i) = i^{(q+1)/4}$. [*Hint*: $(1 + i)^{q-1} \equiv -i$ (q).]

34. Let $\pi = a + bi$ be a primary irreducible, $(a, b) = 1$. Show
   (a) if $\pi \equiv 1$ (4) then $\chi_\pi(a) = i^{(a-1)/2}$.
   (b) if $\pi \equiv 3 + 2i$ (4) then $\chi_\pi(a) = -i^{(-a-1)/2}$.

35. If $\pi = a + bi$ is as in Exercise 34 show $\chi_\pi(a)\chi_\pi(1 + i) = i^{(3(a+b-1))/4}$. [*Hint*: $a(1 + i) = a + b + i(a + bi)$. Generalize Exercises 32 and 33 to any integer $\equiv 1$ (4) and use Proposition 9.9.8. Note $a + b \equiv 1$ (4).]

36. Remove the restriction $(a, b) = 1$ in Exercise 34.

37. Combine Exercises 32, 33, 34, and 35 to show $\chi_\pi(1 + i) = i^{(a-b-b^2-1)/4}$. Show that this result implies Exercise 26 of Chapter 5 (the "biquadratic character of 2").

38. Prove part (d) of Proposition 9.8.3.

39. Let $p \equiv 1$ (6) and write $4p = A^2 + 27B^2$, $A \equiv 1$ (3). Put $m = (p - 1)/6$. Show $\binom{3m}{m} \equiv -A$ $(p) \Leftrightarrow 2|B$.

40. Let $p \equiv 1$ (6), and put $p = \pi\bar{\pi}$ where π is primary. Write $\pi = a + b\omega$ and show
  (a) If $\chi_\pi(2) = \omega$ then $2b - a \equiv -\binom{3m}{m}$ (p).
  (b) If $\chi_\pi(2) = \omega^2$ then $a + b \equiv \binom{3m}{m}$ (p).
  (c) If $\chi_\pi(2) = \omega$ put $A = 2a - b$, $B = b/3$. Show $(A - 9B)/2 \equiv \binom{3m}{m}$ (p).
  (d) If $\chi_\pi(2) = \omega^2$ put $2a - b = A$ and $B = -b/3$. Show $(A - 9B)/2 \equiv \binom{3m}{m}$ (p).
  (e) Show that the "normalization" of B in (c) and (d) is equivalent to $A \equiv B$ (4).
  [Recall $\chi_\pi(2) \equiv \pi$ (2) by cubic reciprocity.]

41. Let $p \equiv 1$ (6), $4p = A^2 + 27B^2$, $A \equiv 1$ (3), A and B odd. Put $\pi = a + b\omega$, $2a - b = A$, $b = 3B$. Let χ_π be the cubic residue character.
  (a) If $\chi_\pi(2) = \omega$ show $N(x^3 + 2y^3 = 1) = p + 1 + 2b - a \equiv 0$ (2).
  (b) If $\chi_\pi(2) = \omega^2$ show $N(x^3 + 2y^3 = 1) = p + 1 - a - b \equiv 0$ (2).
  (c) Show that if $A \equiv B$ (4) then, assuming $\chi_\pi(2) \neq 1$, one has $\chi_\pi(2) = \omega$.
  (d) If $\chi_\pi(2) \neq 1$, $A \equiv B$ (4) then $2^{(p-1)/3} \equiv (-A - 3B)/6B \equiv (A + 9B)/(A - 9B)$ (π).
  (This generalization of Euler's criterion is due to E. Lehmer [174]. See also K. Williams [243].)

42. The notation being as in Section 12 show that the minimal polynomial of $g(\chi_\pi)$ is $x^3 - 3px - Ap$.

43. Find the local maxima and minima of $x^3 - 3px - Ap$ and show that each of the intervals $(-2\sqrt{p}, -\sqrt{p})$, $(-\sqrt{p}, \sqrt{p})$, $(\sqrt{p}, 2\sqrt{p})$ contains exactly one of the values $2\,\mathrm{Re}\,(\omega^k g(\chi_\pi))$, $k = 0, 1, 2$.

44. Let $n \in \mathbb{Z}$, $n = s_1 \cdots s_t$, $n \equiv 1$ (4), $s_i \equiv 1$ (4), $i = 1, \ldots, t$. Show $(n - 1)/4 \equiv \sum_{i=1}^{t} (s_i - 1)/4$ (4).

45. Let $\pi = a + bi \in \mathbb{Z}[i]$ and $q \equiv 3$ (4) a rational prime. Show $\pi^q \equiv \bar{\pi}$ (q).

Chapter 10

Equations over Finite Fields

In this chapter we shall introduce a new point of view. Diophantine problems over finite fields will be put into the context of elementary algebraic geometry. The notions of affine space, projective space, and points at infinity will be defined.

After these problems of language have been dealt with, we shall prove a very general theorem due to C. Chevalley, which states that a polynomial in several variables with no constant term over a finite field always has nontrivial zeros if the number of variables exceeds the degree.

Next, our interest turns to the problem of generalizing the results of Chapter 8 to arbitrary finite fields. This turns out to be relatively easy. These more general results are of interest for their own sake and are crucial to the discussion of the zeta function, which we shall take up in Chapter 11.

§1 Affine Space, Projective Space, and Polynomials

Let F be a field and $A^n(F)$ the set of n-tuples $(a_1, a_2, \ldots, a_n)$ with $a_i \in F$. $A^n(F)$ can be considered as a vector space by defining addition and scalar multiplication in the usual way. We shall be concerned principally with the underlying set, which will be called affine n-space over F. As usual the point $(0, 0, \ldots, 0)$ will be called the origin. If there is no chance of confusion we shall denote the point $(a_1, a_2, \ldots, a_n)$ by the single letter a.

Projective n-space over F, $P^n(F)$, is a somewhat more difficult concept. We first consider $A^{n+1}(F)$, denoting its points by $(a_0, a_1, \ldots, a_n)$. On the set $A^{n+1}(F) - \{(0, 0, \ldots, 0)\}$ (affine $(n + 1)$-space from which the origin has been removed) we define an equivalence relation. $(a_0, a_1, \ldots, a_n)$ is said to be equivalent to $(b_0, b_1, \ldots, b_n)$ if there is a $\gamma \in F^*$ such that $a_0 = \gamma b_0$, $a_1 = \gamma b_1, \ldots, a_n = \gamma b_n$. This is easily seen to be an equivalence relation. The equivalence classes are called points of $P^n(F)$. If $a \in A^{n+1}(F)$ is distinct from the origin, then $[a]$ will denote the equivalence class containing a. a will be called a representative of $[a]$. Geometrically, the points of $P^n(F)$

are in one-to-one correspondence with the lines in $A^{n+1}(F)$ that pass through the origin.

If F is a finite field with q elements, then clearly $A^n(F)$ has q^n elements. $P^n(F)$ has $q^n + q^{n-1} + \cdots + q + 1$ elements. To see this, notice that $A^{n+1}(F) - \{(0, 0, \ldots, 0)\}$ has $q^{n+1} - 1$ elements. Since F^* has $q - 1$ elements each equivalence class has $q - 1$ elements. Thus $P^n(F)$ has $(q^{n+1} - 1)/(q - 1) = q^n + q^{n-1} + \cdots + q + 1$ elements.

In general $P^n(F)$ has more points than $A^n(F)$. This is made more precise as follows. If $[x] \in P^n(F)$ and $x_0 \neq 0$, set $\phi([x]) = (x_1/x_0, x_2/x_0, \ldots, x_n/x_0) \in A^n(F)$. This map is easily seen to be independent of the representative x.

Lemma 1. *Let $\bar{H}$ be the set of $[x] \in P^n(F)$ such that $x_0 = 0$. Then ϕ maps $P^n(F) - \bar{H}$ to $A^n(F)$ and this map is one to one and onto. (If S and T are sets, then $S - T$ is the set of elements in S but not in T.)*

PROOF. If $\phi([x]) = \phi([y])$, then $x_i/x_0 = y_i/y_0$ for $i = 0, 1, \ldots, n$. Let $\gamma = y_0/x_0$. Then $\gamma x_i = y_i$ for $i = 0, 1, \ldots, n$ and so $[x] = [y]$.

If $v = (v_1, v_2, \ldots, v_n) \in A^n(F)$, set $w = (1, v_1, v_2, \ldots, v_n)$. Then $\phi([w]) = v$. $\qquad\square$

The set $\bar{H}$ is called the hyperplane at infinity. It is easy to see that $\bar{H}$ has the structure of $P^{n-1}(F)$. Thus $P^n(F)$ is made up of two pieces, one a copy of $A^n(F)$, called the finite points, and the other a copy of $P^{n-1}(F)$, called the points at infinity.

Notice that $P^0(F)$ consists of just one point. Thus $P^1(F)$ has only one point at infinity. Similarly $P^2(F)$ has a (projective) line at infinity, etc.

Now that affine space and projective space have been defined we take up the subject of polynomials and see how they determine sets called hypersurfaces.

Let $F[x_1, x_2, \ldots, x_n]$ be the ring of polynomials in n variables over F. If $f \in F[x_1, \ldots, x_n]$, then

$$f(x) = \sum_{(i_1, i_2, \ldots, i_n)} a_{i_1 i_2 \cdots i_n} x_1^{i_1} x_2^{i_2} \cdots x_n^{i_n},$$

where the sum is over a finite set of n-tuples of nonnegative integers $(i_1, i_2, \ldots, i_n)$, where $a_{i_1 i_2 \cdots i_n} \neq 0$. A polynomial of the form $x_1^{i_1} x_2^{i_2} \cdots x_n^{i_n}$ is called a monomial. Its total degree is defined to be $i_1 + i_2 + \cdots + i_n$: its degree in the variable x_m is defined as i_m. The degree of $f(x)$ is the maximum of the total degrees of monomials that occur in $f(x)$ with nonzero coefficients. The degree in x_m is the maximum of the degrees in x_m of monomials that occur in $f(x)$ with nonzero coefficients. Call these two numbers $\deg f(x)$ and $\deg_m f(x)$. Then

(a) $\deg f(x)g(x) = \deg f(x) + \deg g(x)$.
(b) $\deg_m f(x)g(x) = \deg_m f(x) + \deg_m g(x)$.

If all the monomials that occur in $f(x)$ have degree l, then $f(x)$ is said to be homogeneous of degree l.

For example, if $f(x) = 1 + x_1 x_2 + x_2 x_3 + x_4^3$, then $\deg f(x) = 3$, $\deg_1 f(x) = \deg_2 f(x) = \deg_3 f(x) = 1$, and $\deg_4 f(x) = 3$. $f(x)$ is not homogeneous, but $h(x) = x_1^3 + x_2^3 + x_3^3 + x_1 x_2 x_3$ is homogeneous of degree 3.

A homogeneous polynomial is sometimes called a form. A form of degree 2 is called a quadratic form, and one of degree 3 is called a cubic form, etc.

Suppose that K is a field containing F. If $f(x) \in F[x_1, x_2, \ldots, x_n]$ and $a \in A^n(K)$, we can substitute a_i for x_i and compute $f(a)$.

This shows that $f(x)$ defines a function from $A^n(K)$ to K by sending a to $f(a)$. A point $a \in A^n(K)$ such that $f(a) = 0$ is called a zero of $f(x)$.

If K is a finite field with q elements, then $x^q - x$ defines the zero function on $A^1(K)$. Thus it may happen that a nonzero polynomial gives rise to the zero function. This cannot happen when K is infinite (see the Exercises).

Let $f(x)$ be a nonzero polynomial and define $H_f(K) = \{a \in A^n(K) \mid f(a) = 0\}$. $H_f(K)$ is called the hypersurface defined by f in $A^n(K)$. When K is a finite field, $H_f(K)$ is a finite set and it is natural to ask for the number of points in $H_f(K)$. In Chapter 8 we dealt with a number of special cases of this problem.

We now wish to define a projective hypersurface. Let $h(x) \in F[x_0, x_1, \ldots, x_n]$ be a nonzero homogeneous polynomial of degree d. As before, K is a field containing F. For $\gamma \in K^*$ we have $h(\gamma x) = \gamma^d h(x)$. It follows that if $a \in A^{n+1}(K)$ and $h(a) = 0$, then $h(\gamma a) = 0$. Thus we may define $\bar{H}_h(K) = \{[a] \in P^n(K) \mid h(a) = 0\}$. This set is called the hypersurface defined by h in $P^n(K)$. Again, if K is finite, we can ask for the number of points in $\bar{H}_h(K)$.

More generally if $f_1, \ldots, f_m$ are polynomials in $F[x_1, \ldots, x_n]$ define $V = \{(a_1, \ldots, a_n) \mid a_i \in F, i = 1, \ldots, n, f_j(a_1, \ldots, a_n) = 0, j = 1, \ldots, m\}$. V is called an algebraic set defined over F. If the ideal defined by $f_1, \ldots, f_m$ in $F[x_1, \ldots, x_n]$ is prime then V is called an algebraic variety. Similarly, the common projective zeros of a finite set of homogeneous polynomials in $F[x_0, \ldots, x_n]$ is called a projective algebraic set.

It turns out that working with projective space leads to more unified results than working with affine space. We shall illustrate this point after defining the projective closure of an affine hypersurface.

Let $f(x) \in F[x_1, x_2, \ldots, x_n]$, and define $\tilde{f}(y) = \tilde{f}(y_0, y_1, \ldots, y_n)$ by

$$\tilde{f}(y) = y_0^{\deg f} f\left(\frac{y_1}{y_0}, \frac{y_2}{y_0}, \ldots, \frac{y_n}{y_0}\right).$$

We shall see in a moment that $\tilde{f}$ is a homogeneous polynomial. It will give rise to a hypersurface in $P^n(K)$. Roughly speaking, the new hypersurface will be obtained from $H_f(K)$ by adding points at infinity.

Lemma 2. $\bar{f}(y)$ *is a homogeneous polynomial of degree equal to* deg f. *Moreover,* $\bar{f}(1, y_1, y_2, \ldots, y_n) = f(y_1, y_2, \ldots, y_n)$.

PROOF. Set $d = \deg f$ and consider a monomial $x_1^{i_1} x_2^{i_2} \cdots x_n^{i_n}$ of degree $l \le d$. Then $y_0^d(y_1/y_0)^{i_1} \cdots (y_n/y_0)^{i_n} = y_0^{d-l} y_1^{i_1} y_2^{i_2} \cdots y_n^{i_n}$, which is of degree d. Thus in $\bar{f}(y)$ all the monomials have degree d, which proves the first statement.

The second statement is immediate from the definition. $\qquad\square$

As examples, if $f(x) = x_1^3 + x_2^3 - 1$, then $\bar{f}(y) = y_1^3 + y_2^3 - y_0^3$: if $f(x) = 1 + 2x_1^3 - 3x_2^2$, then $\bar{f}(y) = y_0^3 + 2y_1^3 - 3y_0 y_2^2$.

Consider the hypersurface $H_f(K) \subset A^n(K)$. $\bar{f}(y)$ is homogeneous in the variables $y_0, y_1, \ldots, y_n$ and so $\bar{f}$ defines a hypersurface $\bar{H}_{\bar{f}}(K)$ in $P^n(K)$. This projective hypersurface is called the projective closure of $H_f(K)$ in $P^n(K)$.

Let $\lambda : A^n(K) \to P^n(K)$ by $\lambda(a_1, a_2, \ldots, a_n) = [1, a_1, a_2, \ldots, a_n]$. λ is one to one and moreover the image of $H_f(K)$ under λ is contained in $\bar{H}_{\bar{f}}(K)$ since clearly $\bar{f}([1, a_1, \ldots, a_n]) = f(a_1, a_2, \ldots, a_n) = 0$ for all $a \in H_f(K)$. In general $\bar{H}_{\bar{f}}(K)$ has more points than $H_f(K)$, namely, the intersection of $\bar{H}_{\bar{f}}(K)$ with the hyperplane at infinity.

All this will become clearer by means of examples, but before giving some we recall the definitions of the maps ϕ and λ and give a diagrammatic picture of $P^n(K)$:

$$\lambda : A^n(K) \to P^n(K) \quad \text{by} \quad \lambda(a_1, a_2, \ldots, a_n) = [1, a_1, a_2, \ldots, a_n],$$

$$\phi : P^n(K) - \bar{H} \to A^n(K) \quad \text{by} \quad \phi([b_0, b_1, \ldots, b_n]) = \left(\frac{b_1}{b_0}, \frac{b_2}{b_0}, \ldots, \frac{b_n}{b_0}\right).$$

$\underline{P^n(K)}$

im $\lambda \approx A^n(K)$	$\bar{H} \approx P^{n-1}(K)$
Finite points	Points at infinity

EXAMPLES

1. $f(x) = x_1^2 + x_2^2 - 1$ over the field $F = \mathbb{Z}/p\mathbb{Z}$.

We have seen in Chapter 8, Section 3, that $f(x) = 0$ has $p - 1$ solutions if $p \equiv 1$ (4) and $p + 1$ solutions if $p \equiv 3$ (4).

$\bar{f}(y) = y_1^2 + y_2^2 - y_0^2$. The solutions $[p_0, p_1, p_2]$, where $p_0 \ne 0$ corresponds to the affine solution $(p_1/p_0, p_2/p_0)$. Suppose that $[0, p_1, p_2]$ is a solution. Then $p_1^2 + p_2^2 = 0$ or $(p_2/p_1)^2 = -1$. If $p \equiv 1$ (4), there is an $a \in F$ such that $a^2 = -1$ and in this case there are two points at infinity, namely, $[0, 1, a]$ and $[0, 1, -a]$. If $p \equiv 3$ (4), there is no $a \in F$ such that $a^2 = -1$ and consequently there are no points at infinity. In both cases, then, the hypersurface $\bar{H}_{\bar{f}}(F)$ has exactly $p + 1$ points.

2. $f(x) = x_1^n + x_2^n - 1$ over $F = \mathbb{Z}/p\mathbb{Z}$ where $p \equiv 1\ (n)$.

We have $\tilde{f}(y) = y_1^n + y_2^n - y_0^n$. Thus the points at infinity on $\bar{H}_f(K)$ are of the form $[0, y_1, y_2]$, where $y_1^n + y_2^n = 0$. If -1 is not an nth power in F, then there are no points at infinity. If $a^n = -1$ for some $a \in F$, then there are n solutions to $x^n = -1$ in F [this follows from Proposition 4.2.1 since $p \equiv 1\ (n)$]. Call these solutions $a_1 = a,\ a_2, \ldots, a_n$. Then $[0, 1, a_1]$, $\ldots, [0, 1, a_n]$ are the points at infinity that are zeros of $\tilde{f}(y)$. In the notation of Chapter 8, Section 4, the number of points at infinity is $\delta_n(-1)n$, and $N(x_1^n + x_2^n = 1) + \delta_n(-1)n$ is the number of points on the projective hypersurface (curve) defined by $y_1^n + y_2^n - y_0^n = 0$. Since the number of points in $P^1(F)$ is $p + 1$ the formula in Proposition 8.4.1 can be interpreted in the following way: The number of points on the projective curve $y_1^n + y_2^n - y_0^n = 0$ over $\mathbb{Z}/p\mathbb{Z}$ differs from the number of points on the projective line by an error term that does not exceed $(n - 1)(n - 2)\sqrt{p}$.

This result is a special case of the so-called Riemann hypothesis for finite fields, which states, roughly, that over a finite field with q elements, the number of points on a projective curve differs from the number of points on the projective line by an error term that does not exceed twice the genus (a number associated with the curve) times $\sqrt{q}$.

Special cases of the result were proved by various authors: Gauss, G. Herglotz, Hasse, and Davenport. The theorem was proved in full generality by Weil.

3. $f(x) = x_1^2 + x_2^2 + \cdots + x_m^2 - 1$ over $F = \mathbb{Z}/p\mathbb{Z}$, where m is even and $p \neq 2$.

The number of finite points is given by $p^{m-1} - (-1)^{(m/2)((p-1)/2)} \cdot p^{(m/2)-1}$ (see Proposition 8.6.1). Since $\tilde{f}(y) = y_1^2 + y_2^2 + \cdots + y_m^2 - y_0^2$ the number of points at infinity is equal to the number of solutions to $y_1^2 + y_2^2 + \cdots + y_m^2 = 0$ in $P^{m-1}(F)$. The number of affine solutions is given by $N = p^{m-1} + (-1)^{(m/2)((p-1)/2)}(p - 1)p^{(m/2)-1}$ (see Exercise 19 in Chapter 8) so the number of projective solutions is

$$\frac{N - 1}{p - 1} = p^{m-2} + p^{m-3} + \cdots + p + 1 + (-1)^{(m/2)((p-1)/2}p^{(m/2)-1}.$$

Adding the number of finite solutions to the solutions at infinity yields

$$p^{m-1} + p^{m-2} + \cdots + p + 1.$$

This result is rather remarkable. It says that the number of points on the projective hypersurface given by $y_1^2 + y_2^2 + \cdots + y_m^2 - y_0^2 = 0$ is exactly equal to the number of points in $P^{m-1}(\mathbb{Z}/p\mathbb{Z})$.

There is a simpler way to achieve this result. Instead of considering the finite and infinite points separately one simply counts the number M of affine solutions to $y_1^2 + y_2^2 + \cdots + y_m^2 - y_0^2 = 0$ in $A^{m+1}(F)$ and then calculates $(M - 1)/(p - 1)$. Since $m + 1$ is odd, the number M is equal to p^m (see Exercise 19 in Chapter 8). Thus $(M - 1)/(p - 1) = p^{m-1} + p^{m-2} + \cdots + p + 1$.

§2 Chevalley's Theorem

In this section F will denote a finite field with q elements.

If q is a prime, i.e., $F = \mathbb{Z}/q\mathbb{Z}$, the equation $x_1^{q-1} + x_2^{q-1} + \cdots + x_{q-1}^{q-1} = 0$ has no solution except $(0, 0, \ldots, 0)$ because a^{q-1} is equal to 1 or zero depending on whether $a \neq 0$ or $a = 0$ for $a \in F$. Thus the values taken on by the above polynomial are $0, 1, 2, \ldots, q - 1$ and it is zero only if $x_1 = x_2 = \cdots = x_{q-1} = 0$. Notice that for this polynomial the number of variables is equal to the degree.

In 1935 E. Artin conjectured the following theorem, which was proved almost immediately by C. Chevalley [16].

Theorem 1. *Let $f(x) \in F[x_1, x_2, \ldots, x_n]$ and suppose that*

(a) $f(0, 0, \ldots, 0) = 0$.
(b) $n > d = \deg f$.

Then f has at least two zeros in $A^n(F)$.

Before giving the proof we shall deduce an immediate corollary.

Corollary. *Let $h(y) \in F[y_0, y_1, \ldots, y_n]$ be a homogeneous polynomial of degree $d > 0$. If $n + 1 > d$. then $\bar{H}_h(F)$ is not empty.*

PROOF. Since h is homogeneous $(0, 0, \ldots, 0)$ is a zero. By Theorem 1 h has another zero, $(a_0, a_1, \ldots, a_n)$. Clearly $[a_0, a_1, \ldots, a_n] \in \bar{H}_h(F)$. □

We shall need the following elementary lemmas.

Lemma 1. *Let $f(x_1, x_2, \ldots, x_n)$ be a polynomial that is of degree less than q in each of its variables. Then if f vanishes on all of $A^n(F)$, it is the zero polynomial.*

PROOF. The proof is by induction on n. If $n = 1$, $f(x)$ is a polynomial in one variable of degree less than q with q distinct roots, namely, all the elements of F. Thus f is identically zero.

Suppose that we have proved the result for $n - 1$ and consider

$$f(x_1, x_2, \ldots, x_n).$$

We can write

$$f(x_1, \ldots, x_n) = \sum_{i=0}^{q-1} g_i(x_1, \ldots, x_{n-1}) x_n^i.$$

Select $a_1, a_2, \ldots, a_{n-1} \in F$. Then $\sum_{i=0}^{q-1} g_i(a_1, a_2, \ldots, a_{n-1}) x_n^i$ has q roots and so $g_i(a_1, a_2, \ldots, a_{n-1}) = 0$. By induction each polynomial g_i is identically zero and hence so is f. □

Remember that $f(x) = x^q - x$ is a nonzero polynomial that vanishes on all of $A^1(F)$, so the hypothesis of the lemma is crucial.

If a polynomial is of degree less than q in each variable, it is said to be reduced. Two polynomials f, g are said to be equivalent if $f(a) = g(a)$ for all $a \in A^n(F)$. We write $f \sim g$.

Lemma 2. *Each polynomial* $f(x) \in F[x_1, \ldots, x_n]$ *is equivalent to a reduced polynomial.*

PROOF. Consider the case of one variable. Clearly $x^q \sim x$. If $m > 0$ is an integer, let l be the least positive integer such that $x^m \sim x^l$. We claim that $l < q$. If not, $l = qs + r$ with $0 \le r < q$ and $s \ne 0$. Then $x^l = (x^q)^s x^r \sim x^{s+r}$. Since $s + r < l$ this contradicts the minimality of l.

In the case of n variables consider the monomial $x_1^{i_1} x_2^{i_2} \cdots x_n^{i_n}$. By what has been said, $x_1^{i_1} x_2^{i_2} \cdots x_n^{i_n} \sim x_1^{j_1} x_2^{j_2} \cdots x_n^{j_n}$, where $j_k < q$ for $k = 1, 2, \ldots, n$. Lemma 2 follows directly from this remark. $\square$

We are now in a position to prove Theorem 1. Suppose that $(0, 0, \ldots, 0)$ is the only zero of f. Then $1 - f^{q-1}$ has the value 1 at $(0, 0, \ldots, 0)$ and the value zero elsewhere. The same is true of the polynomial $(1 - x_1^{q-1})(1 - x_2^{q-1}) \cdots (1 - x_n^{q-1})$. Thus

$$1 - f^{q-1} - (1 - x_1^{q-1})(1 - x_2^{q-1}) \cdots (1 - x_n^{q-1})$$

vanishes on all of $A^n(F)$. Replace $1 - f^{q-1}$ by an equivalent reduced polynomial g. Then

$$g - (1 - x_1^{q-1}) \cdots (1 - x_n^{q-1})$$

is of degree less than q in each of its variables and vanishes on all of $A^n(F)$. By Lemma 1 it vanishes identically. Thus $\deg g = n(q - 1)$. On the other hand, $\deg g \le \deg(1 - f^{q-1}) = d(q - 1)$. Recall that $d = \deg f$. This implies that $n \le d$, which is contrary to the hypothesis. Consequently f must have more than one zero.

We shall give another proof due to Ax [3]. It is based on the following lemma.

Lemma 3. *Let* $i_1, i_2, \ldots, i_n$ *be nonnegative integers. Then unless each* i_j *is nonzero and divisible by* $q - 1$ *we have*

$$\sum_{a \in A^n(F)} a_1^{i_1} a_2^{i_2} \cdots a_n^{i_n} = 0.$$

PROOF. Suppose first that $n = 1$. If $i = 0$, then $\sum_{a \in F} a^0 = q = 0$ in F. Suppose $i \ne 0$. F^* is cyclic. Let b be a generator. If $q - 1 \nmid i$, then

$$\sum_{a \in F} a^i = \sum_{k=0}^{q-2} b^{ki} = \frac{b^{(q-1)i} - 1}{b^i - 1} = 0.$$

In general

$$\sum_{a \in A^n(F)} a_1^{i_1} a_2^{i_2} \cdots a_n^{i_n} = \left(\sum_{a_1 \in F} a_1^{i_1}\right)\left(\sum_{a_2 \in F} a_2^{i_2}\right) \cdots \left(\sum_{a_n \in F} a_n^{i_n}\right).$$

Lemma 3 is now clear.   □

It should be remarked that if $q - 1 | i_j$ and $i_j \neq 0$ for all j, then the value of the above sum is $(q - 1)^n$.

To return to Theorem 1, let N_f be the number of solutions of $f(x) = 0$ in $A^n(F)$. We shall show that $p | N_f$, where p is the characteristic of F. This refinement of Chevalley's theorem was first given by Warning [78].

As we have seen, $1 - f^{q-1}$ has the value 1 at a zero of f and the value zero otherwise. Thus

$$\bar{N}_f = \sum_{a \in A^n(F)} (1 - f(a)^{q-1}),$$

where $\bar{N}_f$ is the residue class of N_f mod p considered as an element of F.

Let $x_1^{i_1} x_2^{i_2} \cdots x_n^{i_n}$ be a monomial occurring in $1 - f(x)^{q-1}$. Since this polynomial has degree $d(q - 1)$ we must have $i_j < q - 1$ for some j since otherwise the degree of the monomial would exceed $n(q - 1)$ and we have assumed that $d < n$. By Lemma 3 $\sum_{a \in A^n(F)} a_1^{i_1} a_2^{i_2} \cdots a_n^{i_n} = 0$. Since $1 - f(x)^{q-1}$ is a linear combination of monomials it follows that $\bar{N}_f = 0$, or $p | N_f$.

Warning was able to prove that $N_f \geq q^{n-d}$. In a somewhat different direction Ax showed that $q^b | N_f$, where b is the largest integer less than n/d. See [78] and [3] for details.

§3 Gauss and Jacobi Sums over Finite Fields

Let $\zeta_p = e^{2\pi i/p}$ and $F_p = \mathbb{Z}/p\mathbb{Z}$. In Chapter 8 the function $\psi: F_p \to \mathbb{C}$ given by $\psi(t) = \zeta_p^t$ played a crucial role. To carry over the principal results of Chapter 8 to an arbitrary finite field F, we need an analog of ψ for F. This is done by means of the trace.

Suppose that F has $q = p^n$ elements. For $\alpha \in F$ define $\text{tr}(\alpha) = \alpha + \alpha^p + \alpha^{p^2} + \cdots + \alpha^{p^{n-1}}$. $\text{tr}(\alpha)$ is called the trace of α.

Proposition 10.3.1. *If $\alpha, \beta \in F$ and $a \in F_p$, then*
(a) $\text{tr}(\alpha) \in F_p$.
(b) $\text{tr}(\alpha + \beta) = \text{tr}(\alpha) + \text{tr}(\beta)$.
(c) $\text{tr}(a\alpha) = a\,\text{tr}(\alpha)$.
(d) tr *maps F onto F_p.*

PROOF.

(a) We have

$$(\alpha + \alpha^p + \cdots + \alpha^{p^{n-1}})^p = \alpha^p + \alpha^{p^2} + \cdots + \alpha^{p^{n-1}} + \alpha^{p^n}.$$

Since $\alpha^{p^n} = \alpha^q = \alpha$ we see that $\mathrm{tr}(\alpha)^p = \mathrm{tr}(\alpha)$. This proves property (a) (see Proposition 7.1.1, Corollary 1).

(b) $\mathrm{tr}(\alpha + \beta) = (\alpha + \beta) + (\alpha + \beta)^p + \cdots + (\alpha + \beta)^{p^{n-1}}$
$\qquad\qquad = (\alpha + \beta) + (\alpha^p + \beta^p) + \cdots + (\alpha^{p^{n-1}} + \beta^{p^{n-1}})$
$\qquad\qquad = (\alpha + \alpha^p + \cdots + \alpha^{p^{n-1}}) + (\beta + \beta^p + \cdots + \beta^{p^{n-1}})$
$\qquad\qquad = \mathrm{tr}(\alpha) + \mathrm{tr}(\beta).$

(c) $\mathrm{tr}(a\alpha) = a\alpha + a^p\alpha^p + \cdots + a^{p^{n-1}}\alpha^{p^{n-1}}$
$\qquad\qquad = a(\alpha + \alpha^p + \cdots + \alpha^{p^{n-1}})$
$\qquad\qquad = a\,\mathrm{tr}(\alpha).$

We have used the fact that $a^p = a$ for $a \in F_p$.

(d) The polynomial $x + x^p + \cdots + x^{p^{n-1}}$ has at most p^{n-1} roots in F. Since F has p^n elements there is an $\alpha \in F$ such that $\mathrm{tr}(\alpha) = c \neq 0$. If $b \in F_p$, then using property (c) we see that $\mathrm{tr}((b/c)\alpha) = (b/c)\,\mathrm{tr}(\alpha) = b$. Thus the trace is onto. $\qquad\square$

We now define $\psi: F \to \mathbb{C}$ by the formula $\psi(\alpha) = \zeta_p^{\mathrm{tr}(\alpha)}$. If $F = F_p$, this coincides with the previous definition.

Proposition 10.3.2. *The function ψ has the following properties:*

(a) $\psi(\alpha + \beta) = \psi(\alpha)\psi(\beta).$
(b) *There is an $\alpha \in F$ such that $\psi(\alpha) \neq 1$.*
(c) $\sum_{a \in F} \psi(\alpha) = 0.$

PROOF.

(a) $\psi(\alpha + \beta) = \zeta_p^{\mathrm{tr}(\alpha + \beta)} = \zeta_p^{\mathrm{tr}(\alpha) + \mathrm{tr}(\beta)} = \zeta_p^{\mathrm{tr}(\alpha)}\zeta_p^{\mathrm{tr}(\beta)} = \psi(\alpha)\psi(\beta).$
(b) tr is onto, so there is an $\alpha \in F$ such that $\mathrm{tr}(\alpha) = 1$. Then $\psi(\alpha) = \zeta_p \neq 1$.
(c) Let $S = \sum_{\alpha \in F} \psi(\alpha)$. Choose β such that $\psi(\beta) \neq 1$. Then $\psi(\beta)S = \sum_{\alpha \in F} \psi(\beta)\psi(\alpha) = \sum_{\alpha \in F} \psi(\beta + \alpha) = S$. It follows that $S = 0$. $\qquad\square$

Proposition 10.3.3. *Let $\alpha, x, y \in F$. Then*

$$\frac{1}{q}\sum_{\alpha \in F} \psi(\alpha(x - y)) = \delta(x, y),$$

where $\delta(x, y) = 1$ if $x = y$ and zero otherwise.

PROOF. If $x = y$, then $\sum_{\alpha \in F} \psi(\alpha(x - y)) = \sum_{\alpha \in F} \psi(0) = q$.

If $x \neq y$, then $x - y \neq 0$ and $\alpha(x - y)$ ranges over all of F as α ranges over all of F. Thus $\sum_{\alpha \in F} \psi(\alpha(x - y)) = \sum_{\beta \in F} \psi(\beta) = 0$ by property (c) of Proposition 10.3.2. $\qquad\square$

Proposition 10.3.3 generalizes the corollary to Lemma 1 of Chapter 6.

In Chapter 7 we proved that the multiplicative group of a finite field is cyclic. On the basis of this fact, one easily see that all the definitions and propositions of Chapter 8, Section 1, can be applied to F as well as to F_p. It is only necessary to replace p by q whenever it occurs. Thus we may assume that the theory of multiplicative characters for F is known.

We are now in a position to define Gauss sums on F.

Definition. Let χ be a character of F and $\alpha \in F^*$. Let $g_\alpha(\chi) = \sum_{t \in F} \chi(t)\psi(\alpha t)$. $g_\alpha(\chi)$ is called a *Gauss sum on F* belonging to the character χ.

If we replace p by q, Propositions 8.2.1 and 8.2.2 can now be proved for the sums $g_\alpha(\chi)$. In the proof of Proposition 8.2.2 one needs Proposition 10.3.3.

In particular, we have $|g_\alpha(\chi)| = q^{1/2}$ and $g_\alpha(\chi)g_\alpha(\chi^{-1}) = \chi(-1)q$ for $\chi \neq \varepsilon$.

The general theory of Jacobi sums and the interrelation between Gauss sums and Jacobi sums that is developed in Chapter 8, Section 5, generalizes with no difficulty (just replace p by q everywhere), and all the results of Chapter 8, Section 7, also hold. The reader may wish to go back to these sections to assure himself that there are indeed no difficulties in generalizing the definitions and results.

As an exercise in working with these new tools, we present a theorem that is really a reformulation of some of our earlier work. This theorem will also be of use in Chapter 11.

Theorem 2. *Suppose that F is a field with q elements and $q \equiv 1$ (m). The homogeneous equation $a_0 y_0^m + a_1 y_1^m + \cdots + a_n y_n^m = 0$, $a_0, a_1, \ldots, a_n \in F^*$, defines a hypersurface in $P^n(F)$. The number of points on this hypersurface is given by*

$$q^{n-1} + q^{n-2} + \cdots + q + 1$$

$$+ \frac{1}{q-1} \sum_{\chi_0, \chi_1, \ldots, \chi_n} \chi_0(a_0^{-1}) \cdots \chi_n(a_n^{-1}) J_0(\chi_0, \chi_1, \ldots, \chi_n), \qquad (1)$$

where $\chi_i^m = \varepsilon$, $\chi_i \neq \varepsilon$, and $\chi_0 \chi_1 \cdots \chi_n = \varepsilon$.

Moreover, under these conditions

$$\frac{1}{q-1} J_0(\chi_0, \chi_1, \ldots, \chi_n) = \frac{1}{q} g(\chi_0)g(\chi_1) \cdots g(\chi_n). \qquad (2)$$

PROOF. The number of points N on the hypersurface in $A^{n+1}(F)$ defined by $a_0 y_0^m + a_1 y_1^m + \cdots + a_n y_n^m = 0$ is given by

$$q^n + \sum_{\chi_0, \chi_1, \ldots, \chi_n} \chi_0(a_0^{-1})\chi_1(a_1^{-1}) \cdots \chi_n(a_n^{-1}) J_0(\chi_0, \chi_1, \ldots, \chi_n),$$

where the characters χ_i are subject to the conditions stated in Theorem 2. This follows from Theorem 5 of Chapter 8. The number we are looking for is $(N - 1)/(q - 1)$ and this yields Equation (1).

By Proposition 8.5.1, part (c), we have

$$J_0(\chi_0, \chi_1, \ldots, \chi_n) = \chi_0(-1)(q-1)J(\chi_1, \chi_2, \ldots, \chi_n). \qquad (3)$$

By Theorem 3 of Chapter 8

$$J(\chi_1, \chi_2, \ldots, \chi_n) = \frac{g(\chi_1)g(\chi_2) \cdots g(\chi_n)}{g(\chi_1\chi_2 \cdots \chi_n)}. \qquad (4)$$

Multiply the numerator and denominator of the right-hand side by $g(\chi_0)$. Since $\chi_0\chi_1 \cdots \chi_n = \varepsilon$, we have $g(\chi_0)g(\chi_1\chi_2 \cdots \chi_n) = \chi_0(-1)q$. Combining this comment, Equations (3) and (4) yield Equation (2). $\square$

NOTES

There is a pleasant introduction to geometry over finite fields in the book *Excursions into Mathematics* [7]. The authors discuss affine, projective, and even hyperbolic geometry. There is also a short but useful bibliography.

Artin's conjecture on polynomials over finite fields was made much earlier by Dickson (On the Representations of Numbers by Modular Forms, *Bull. Am. Math. Soc.*, **15** (1909), 338–347). The first proof we gave is the original proof of Chevally [16]. The second proof is due to J. Ax [3] and is found in M. Greenberg [37] and Samuel [68]. E. Warning's proof of a sharper result can be found in his original paper [78] and in Borevich and Shafarevich [9].

A. Meyer, in 1884, was able to prove that a quadratic form over the rationals in five or more variables always has a rational zero if it has a real zero. Hasse was able to prove that the same result, suitably generalized, holds over any algebraic number field. E. Artin was led by this and other considerations to conjecture that over a certain class of number fields a form of degree d in $n > d^2$ variable always has a nontrivial zero. He also made conjectures of this nature over other types of fields. For a discussion of this area of research, see the paper of S. Lang [53], as well as the book of Greenberg [37], which includes a counterexample to Artin's conjecture for p-adic fields, discovered in 1966 by G. Terjanian. Other counterexamples were provided shortly thereafter by S. Shanuel. There is much left to discover in this area, which is one of the most fascinating in modern number theory.

If one looks at the case where the ground field is the field of rational functions over a finite field, then the Artin conjecture mentioned above has been proved by Carlitz [11]. More precisely, let F be a finite field and $K = F(x)$. Then every form of degree d in more than d^2 variables has a nontrivial zero in K. The proof makes ingenious use of the theorem of Chevalley proved in this chapter. It is a special case of a general result of S. Lang.

Many of the most important advances in number theory demand an extensive knowledge of modern algebraic geometry. For a readable and not too sophisticated introduction to algebraic geometry see W. Fulton [135]. A more extensive introduction is contained in Shafarevich [219]. Finally, for a reader with more background in commutative algebra, see R. Hartshorne [144].

EXERCISES

1. If K is an infinite field and $f(x_1, x_2, \ldots, x_n)$ is a non-zero polynomial with coefficients in K, show that f is not identically zero on $A^n(K)$. (*Hint*: Imitate the proof of Lemma 1 in Section 2.)

2. In Section 1 it was asserted that H, the hyperplane at infinity in $P^n(F)$, has the structure of $P^{n-1}(F)$. Verify this by constructing a one-to-one, onto map from $P^{n-1}(F)$ to H.

3. Suppose that F has q elements. Use the decomposition of $P^n(F)$ into finite points and points at infinity to give another proof of the formula for the number of points in $P^n(F)$.

4. The hypersurface defined by a homogeneous polynomial of degree 1, $a_0 x_0 + a_1 x_1 + a_2 x_2 + \cdots + a_n x_n$, is called a hyperplane. Show that any hyperplane in $P^n(F)$ has the same number of elements as $P^{n-1}(F)$.

5. Let $f(x_0, x_1, x_2)$ be a homogeneous polynomial of degree n in $F[x_0, x_1, x_2]$. Suppose that not every zero of $a_0 x_0 + a_1 x_1 + a_2 x_2$ is a zero of f. Prove that there are at most n common zeros of f and $a_0 x_0 + a_1 x_1 + a_2 x_2$ in $P^2(F)$. In more geometric language this says that a curve of degree n and a line have at most n points in common unless the line is contained in the curve.

6. Let F be a field with q elements. Let $M_n(F)$ be the set of $n \times n$ matrices with coefficients in F. Let $Sl_n(F)$ be the subset of those matrices with determinant equal to one. Show that $Sl_n(F)$ can be considered as a hypersurface in $A^{n^2}(F)$. Find a formula for the number of points on this hypersurface. [*Answer*: $(q - 1)^{-1}(q^n - 1)(q^n - q) \cdots (q^n - q^{n-1})$.]

7. Let $f \in F[x_0, x_1, x_2, \ldots, x_n]$. One can define the partial derivatives $\partial f / \partial x_0$, $\partial f / \partial x_1, \ldots, \partial f / \partial x_n$ in a formal way. Suppose that f is homogeneous of degree m. Prove that $\sum_{i=0}^{n} x_i(\partial f / \partial x_i) = mf$. This result is due to Euler. (*Hint*: Do it first for the case that f is a monomial.)

8. (continuation) If f is homogeneous, a point $\bar{a}$ on the hypersurface defined by f is said to be singular if it is simultaneously a zero of all the partial derivatives of f. If the degree of f is prime to the characteristic, show that a common zero of all the partial derivatives of f is automatically a zero of f.

9. If m is prime to the characteristic of F, show that the hypersurface defined by $a_0 x_0^m + a_1 x_1^m + \cdots + a_n x_n^m$ has no singular points.

10. A point on an affine hypersurface is said to be singular if the corresponding point on the projective closure is singular. Show that this is equivalent to the following definition. Let $f \in F[x_1, x_2, \ldots, x_n]$, not necessarily homogeneous, and $a \in H_f(F)$. Then a is singular iff it is a common zero of $\partial f / \partial x_i$ for $i = 1, 2, \ldots, n$.

11. Show that the origin is a singular point on the curve defined by $y^2 - x^3 = 0$.

12. Show that the affine curve defined by $x^2 + y^2 + x^2y^2 = 0$ has two points at infinity and that both are singular.

13. Suppose that the characteristic of F is not 2, and consider the curve defined by $ax^2 + bxy + cy^2 = 1$, where $a, b, c \in F^*$. If $b^2 - 4ac \notin F^2$, show that there are no points at infinity in $P^2(F)$. If $b^2 - 4ac \in F^2$, show that there are one or two points at infinity depending on whether $b^2 - 4ac$ is zero. If $b^2 - 4ac = 0$, show that the point at infinity is singular.

14. Consider the curve defined by $y^2 = x^3 + ax + b$. Show that it has no singular points (finite or infinite) if $4a^3 + 27b^2 \neq 0$.

15. Let $\mathbb{Q}$ be the field of rational numbers and p a prime. Show that the form $x_0^{n+1} + px_1^{n+1} + p^2x_2^{n+1} + \cdots + p^nx_n^{n+1}$ has no zeros in $P^n(\mathbb{Q})$. (*Hint*: If $\bar{a}$ is a zero, one can assume that the components of a are integers and that they are not all divisible by p.)

16. Show by explicit calculation that every cubic form in two variables over $Z/2Z$ has a nontrivial zero.

17. Show that for each $m > 0$ and finite field F_q there is a form of degree m in m variables with no nontrivial zero. [*Hint*: Let $\omega_1, \omega_2, \ldots, \omega_m$ be a basis for F_{q^m} over F_q and show that $f(x_1, x_2, \ldots, x_m) = \prod_{i=0}^{m-1} (\omega_1^{q^i}x_1 + \cdots + \omega_m^{q^i}x_m)$ has the required properties.]

18. Let $g_1, g_2, \ldots, g_m \in F_q[x_1, x_2, \ldots, x_n]$ be homogeneous polynomials of degree d and assume that $n > md$. Prove that there is nontrivial common zero. [*Hint*: Let f be as in Exercise 17 and consider the polynomial $f(g_1(x_1, \ldots, x_n), \ldots, g_m(x_1, \ldots, x_n))$.]

19. Characterize those extensions F_{p^n} of F_p that are such that the trace is identically zero on F_p.

20. Show that if $\alpha \in F_q$ has trace zero, then $\alpha = \beta - \beta^p$ for some $\beta \in F_q$.

21. Let ψ be a map from F_q to $\mathbb{C}^*$ such that $\psi(\alpha + \beta) = \psi(\alpha)\psi(\beta)$ for all $\alpha, \beta \in F_q$. Show that there is a $\gamma \in F_q$ such that $\psi(x) = \zeta^{\text{tr}(\gamma x)}$ for all $x \in F_q$, where $\zeta = e^{2\pi i/p}$.

22. If $g_\alpha(\chi)$ is a Gauss sum on F, defined in Section 3, show that
    (a) $g_\alpha(\chi) = \overline{\chi(\alpha)}g(\chi)$.
    (b) $g(\chi^{-1}) = g(\bar{\chi}) = \chi(-1)\overline{g(\chi)}$.
    (c) $|g_\alpha(\chi)| = q^{1/2}$.
    (d) $g(\chi)g(\chi^{-1}) = \chi(-1)q$.

23. Suppose that f is a function mapping F to $\mathbb{C}$. Define $\hat{f}(s) = (1/q)\sum_t f(t)\overline{\psi(st)}$ and prove that $f(t) = \sum_s \hat{f}(s)\psi(st)$. The last sum is called the finite Fourier series expansion of f.

24. In Exercise 23 take f to be a nontrivial character χ and show that $\hat{\chi}(s) = (1/q)g_{-s}(\chi)$.

Chapter 11

The Zeta Function

The zeta function of an algebraic variety has played a major role in recent developments in diophantine geometry.

In 1924 E. Artin introduced the notion of a zeta function for a certain class of curves defined over a finite field. By analogy with the classical Riemann zeta function he conjectured that the Riemann hypothesis was valid for the functions he had defined. In special cases he was able to prove this. Remarkably, results of this nature can already be found in the work of Gauss (naturally, Gauss stated his results differently from Artin). Weil was able to prove (in 1948) that the Riemann hypothesis for nonsingular curves over a finite field was true in general.

In 1949 Weil published a paper in the Bulletin of the American Mathematical Society *entitled "Numbers of Solutions of Equations over Finite Fields." In this paper he defined the zeta function of an algebraic variety and announced a number of conjectures. Weil had already proved the validity of his conjectures for curves. Here he establishes the same results for a class of projective hypersurfaces. We shall give an exposition of part of this material. Most of the necessary tools have already been developed. The main new result that is needed is the Hasse–Davenport relation between Gauss sums. Weil gave a proof of this relation that is substantially simpler than the original. We shall give a proof due to P. Monsky that is even simpler than Weil's, although it is far from trivial.*

In 1973 Pierre Deligne succeeded in establishing the validity of the Weil conjectures in all generality. The proof utilizes the most advanced techniques of modern algebraic geometry and represents one of the most remarkable mathematical achievements of this century.

§1 The Zeta Function of a Projective Hypersurface

In Chapter 7, Section 2, we showed that if $F = \mathbb{Z}/p\mathbb{Z}$ and $s \geq 1$ an integer, then there exists a field K containing F with p^s elements. The same result holds true in general. Namely, if F is a finite field with q elements and $s \geq 1$

an integer, then there exists a field F_s containing F with q^s elements (this is F_{q^s} in our former terminology). The proof of the general case is almost identical with that of the special case (see the Exercises to Chapter 7).

Now, let $f(y) \in F[y_0, y_1, \ldots, y_n]$ be a homogeneous polynomial and let N_s be the number of points on the projective hypersurface $\bar{H}_f(F_s) \subset P^n(F_s)$. In less fancy language, N_s is the number of zeros of f in $P^n(F_s)$. We wish to investigate the way in which the numbers N_s depend on s.

At the end of this section we shall prove that the number N_s depends only on s and not on the field F_s. This will follow once we show that any two fields containing F and of the same dimension over F are isomorphic.

To study the numbers N_s we introduce the power series $\sum_{s=1}^{\infty} N_s u^s$. In all that follows it is possible to deal only with formal power series and thus to avoid all questions of convergence. To those who are uncomfortable with that notion, notice that $N_s \leq (q^{s(n+1)} - 1)/(q^s - 1) < (n + 1)q^{sn}$. It follows that our series converges for all complex numbers u such that $|u| < q^{-n}$ and defines an analytic function in that disc.

Let $\exp u = \sum_{s=0}^{\infty} (1/s!)u^s$.

Definition. The *zeta function* of the hypersurface defined by f is the series given by

$$Z_f(u) = \exp\left(\sum_{s=1}^{\infty} \frac{N_s u^s}{s}\right).$$

It is possible to regard $Z_f(u)$ either as a formal power series or as a function of a complex variable defined and analytic on the disc $\{u \in \mathbb{C} \mid |u| < q^{-n}\}$.

It may seem strange to deal with $Z_f(u)$ instead of directly considering the series $\sum_{s=1}^{\infty} N_s u^s$. The reasons are mainly historical, although as we shall see the zeta function is, in fact, easier to handle. See the remarks at the end of this section.

As a first example, consider the hyperplane at infinity. By definition this is the set of points $[a_0, \ldots, a_n] \in P^n(F)$ with $a_0 = 0$. It is defined by the equation $x_0 = 0$. As we pointed out in Chapter 10 it is easy to see that $\bar{H}_{x_0}(F_s)$ has the same number of points as $P^{n-1}(F_s)$; that is,

$$N_s = q^{s(n-1)} + q^{s(n-2)} + \cdots + q^s + 1.$$

It follows that

$$\sum_{s=1}^{\infty} \frac{N_s u^s}{s} = \sum_{m=0}^{n-1}\left(\sum_{s=1}^{\infty} \frac{(q^m u)^s}{s}\right) = -\sum_{m=0}^{n-1} \ln(1 - q^m u). \tag{1}$$

We have used the identity $\sum_{s=1}^{\infty} w^s/s = -\ln(1 - w)$. Exponentiating Equation (1) yields

$$Z_{x_0}(u) = (1 - q^{n-1}u)^{-1}(1 - q^{n-2}u)^{-1} \cdots (1 - qu)^{-1}(1 - u)^{-1}.$$

In particular, we see that $Z_{x_0}(u)$ is a rational function of u.

We shall now compute a somewhat more involved example. Consider the hypersurface defined by $-y_0^2 + y_1^2 + y_2^2 + y_3^2 = 0$. To compute N_1 we use Theorem 2 of Chapter 10. Specializing to our case we find that

$$N_1 = q^2 + q + 1 + \chi(-1)\frac{1}{q}g(\chi)^4,$$

where χ is the character of order 2 on F. We know that $g(\chi)^2 = \chi(-1)q$. Thus

$$N_1 = q^2 + q + 1 + \chi(-1)q.$$

To compute N_s we must replace q by q^s and χ by χ_s, the character of order 2 on F_s. Then

$$N_s = q^{2s} + q^s + 1 + \chi_s(-1)q^s.$$

If -1 is a square in F, then $\chi_s(-1) = 1$ for all s. If -1 is not a square in F, it is not hard to see that $\chi_s(-1) = -1$ for s odd and $\chi_s(-1) = 1$ for s even.

In the first case

$$\sum_{s=1}^{\infty} \frac{N_s u^s}{s} = \sum_{s=1}^{\infty} \frac{(q^2 u)^s}{s} + 2\sum_{s=1}^{\infty} \frac{(qu)^s}{s} + \sum_{s=1}^{\infty} \frac{u^s}{s}$$

and so

$$Z(u) = (1 - q^2 u)^{-1}(1 - qu)^{-2}(1 - u)^{-1}.$$

In the second case the last term gives rise to the sum

$$\sum_{s=1}^{\infty} \frac{(-qu)^s}{s} = -\ln(1 + qu).$$

Thus in this case

$$Z(u) = (1 - q^2 u)^{-1}(1 - qu)^{-1}(1 + qu)^{-1}(1 - u)^{-1}.$$

Notice that in the first case the zeta function has a pole at $u = q^{-1}$ of order 2, whereas in the second case there is a pole at $u = q^{-1}$ of order 1. This is in accordance with a conjecture of John Tate, which relates the order of the pole at $u = q^{-1}$ to certain geometric properties of the hypersurface. We cannot go more deeply into this here.

As a final example, consider the curve $y_0^3 + y_1^3 + y_2^3 = 0$ over $F = \mathbb{Z}/p\mathbb{Z}$, p is a prime congruent to 1 modulo 3.

Specializing Theorem 2 of Chapter 10 once again we find that

$$N_1 = p + 1 + \frac{1}{p}g(\chi)^3 + \frac{1}{p}g(\chi^2)^3.$$

Here χ is a cubic character on $\mathbb{Z}/p\mathbb{Z}$. We know that $g(\chi)^3 = p\pi$, where $\pi = J(\chi, \chi)$, and $\pi\bar{\pi} = p$. Thus

$$N_1 = p + 1 + \pi + \bar{\pi}.$$

It will follow from the Hasse–Davenport relation, to be proved later, that

$$N_s = p^s + 1 - (-\pi)^s - (-\bar{\pi})^s.$$

Calculation now shows that

$$Z_f(u) = \frac{(1 + \pi u)(1 + \bar{\pi} u)}{(1 - u)(1 - pu)}.$$

The numerator can be evaluated explicitly. In Chapter 8 we proved that $\pi + \bar{\pi} = A$, where A is uniquely determined by $4p = A^2 + 27B^2$ and $A \equiv 1\ (3)$.

So our final expression is

$$Z_f(u) = \frac{1 + Au + pu^2}{(1 - u)(1 - pu)}.$$

In this example $Z_f(u)$ is a rational function; the numerator and denominator are polynomials with integer coefficients. The roots of $Z_f(u)$, $-\pi^{-1}$ and $-\bar{\pi}^{-1}$, both have absolute value $p^{-1/2}$.

More generally, let $f(x_0, x_1, x_2) \in F[x_0, x_1, x_2]$ be a nonzero homogeneous polynomial that is nonsingular over every algebraic extension of F. Then, Weil was able to prove that the zeta function of f has the form

$$\frac{P(u)}{(1 - u)(1 - qu)},$$

where $P(u)$ is a polynomial with integer coefficients of degree $(d - 1)(d - 2)$, d being the degree of f. Furthermore, if α is a root of $P(u)$, then $|\alpha| = q^{-1/2}$. The last statement is called the Riemann hypothesis for curves.

[To see the relation with the classical Riemann hypothesis, make the change of variables $u = q^{-s}$ and set $\zeta_f(s) = Z_f(q^{-s})$. $\zeta_f(s)$ is directly analogous to the classical zeta function (see the end of this section). The condition that the roots of $Z_f(u)$ have absolute value $q^{-1/2}$ is equivalent to the condition that the roots of $\zeta_f(s)$ have real part $\frac{1}{2}$.]

In all our examples the zeta function is rational. In 1959 B. Dwork proved that any algebraic set has a rational zeta function [26]. His proof is extremely beautiful, but unfortunately it is based on methods that are beyond the scope of this book.

Our examples suggest another characterization of the condition that the zeta function is rational.

It is immediate from the definition of the zeta function that if it is expanded in a power series about the origin, then the constant term is 1. Consequently, if $Z_f(u) = P(u)/Q(u)$, where $P(u)$ and $Q(u)$ are polynomials, we may assume that $P(0) = Q(0) = 1$ (prove it). With this assumption, the zeta function can be factored as follows:

$$Z_f(u) = \frac{\prod_i (1 - \alpha_i u)}{\prod_j (1 - \beta_j u)}$$

where $\alpha_i, \beta_j \in \mathbb{C}$. We can now prove

Proposition 11.1.1. *The zeta function is rational iff there exist complex numbers α_i and β_j such that*

$$N_s = \sum_j \beta_j^s - \sum_i \alpha_i^s.$$

PROOF. Suppose that the zeta function is rational. Then by the above remarks

$$Z(u) = \frac{\prod_i (1 - \alpha_i u)}{\prod_j (1 - \beta_j u)}$$

with $\alpha_i, \beta_j \in \mathbb{C}$. Taking the logarithmic derivative of both sides:

$$\frac{Z'(u)}{Z(u)} = \sum_i \frac{-\alpha_i}{1 - \alpha_i u} - \sum_j \frac{-\beta_j}{1 - \beta_j u}.$$

Multiply both sides by u and then use the geometric series to expand in a power series. One finds finally that

$$\frac{uZ'(u)}{Z(u)} = \sum_{s=1}^{\infty} \left(\sum_j \beta_j^s - \sum_i \alpha_i^s \right) u^s. \tag{2}$$

We now compute the left-hand side in a different way. From the definition

$$Z(u) = \exp \sum_{s=1}^{\infty} \frac{N_s u^s}{s}.$$

Differentiate logarithmically both sides and then multiply both sides by u. We find that

$$\frac{uZ'(u)}{Z(u)} = \sum_{s=1}^{\infty} N_s u^s. \tag{3}$$

Comparing coefficients of u^s in Equations (2) and (3) we have

$$N_s = \sum_j \beta_j^s - \sum_i \alpha_i^s.$$

The converse is an easy calculation that we have done in special cases. We leave the details to the reader.   □

It remains to prove that the number N_s is independent of the choice of field F_s. The reader may wish to simply accept this fact and proceed to Section 2.

Suppose that E and E' are two fields containing F both with q^s elements.

Proposition 11.1.2. *E and E′ are isomorphic over F; i.e., there exists a map*
$\sigma: E \to E'$ *such that*

(a) σ *is one to one and onto.*
(b) $\sigma(a) = a$ *for all* $a \in F$.
(c) $\sigma(\alpha + \beta) = \sigma(\alpha) + \sigma(\beta)$ *for all* $\alpha, \beta \in E$.
(d) $\sigma(\alpha\beta) = \sigma(\alpha)\sigma(\beta)$ *for all* $\alpha, \beta \in E$.

PROOF. We shall show that both E and E' are isomorphic over F to $F[x]/(f(x))$
for some irreducible polynomial $f(x) \in F[x]$.

To begin with there is an $\alpha' \in E'$ such that $E' = F(\alpha')$ (for example, take
α' to be a primitive $q^s - 1$ root of unity). Let $f(x) \in F[x]$ be the monic
irreducible polynomial for α'. Then $E' \approx F[x]/(f(x))$. Since α' satisfies
$x^{q^s} - x = 0$ we have $f(x) | x^{q^s} - x$.

Since E has q^s elements we have $x^{q^s} - x = \prod_{\alpha \in E} (x - \alpha)$. It follows that
$f(\alpha) = 0$ for some $\alpha \in E$.

Thus $F(\alpha) \approx F[x]/(f(x))$ is a subfield of E with q^s elements. One concludes that $E = F(\alpha) \approx F[x]/(f(x)) \approx F(\alpha') = E'$.   □

We can now use the isomorphism σ to induce a map $\bar{\sigma}$ from $P^n(E)$ to
$P^n(E')$. Namely,

$$\bar{\sigma}([\alpha_0, \ldots, \alpha_n]) = [\sigma(\alpha_0), \ldots, \sigma(\alpha_n)]$$

$\bar{\sigma}$ is one to one and onto. Moreover, if $f(y_0, y_1, \ldots, y_n) \in F[y_0, y_1, \ldots, y_n]$
and we restrict $\bar{\sigma}$ to the projective hypersurface $\bar{H}_f(E)$, it maps onto the
projective hypersurface $\bar{H}_f(E')$. This proves the independence of the numbers
N_s from the choice of field F_s. We leave the details to the reader.

We conclude this section with a discussion of the analogy between the
congruence zeta function and the Riemann zeta function.

The Riemann zeta function $\zeta(s) = \sum_{n=1}^{\infty} n^{-s}$ may be written, by the
fundamental theorem of arithmetic as an infinite product $\prod_p (1 - p^{-s})^{-1}$
the product being over all prime numbers p (see Exercise 25, Chapter 2).
We will establish an analogous infinite product for $Z_f(u)$ the product being
over certain objects called the prime divisors of the underlying algebraic
set. This will be done with a minimum of technical language from algebraic
geometry.

If F is a finite field with q elements consider any algebraic set V in $A^n(F)$.
Then we may define as in Section 1 the zeta function of V over F as

$$\exp\left(\sum_{s=1}^{\infty} \frac{N_s u^s}{s} \right)$$

where N_s is the number of points in $A^n(F_{q^s})$ satisfying the equations defining
V. We consider an affine algebraic set rather than a projective algebraic set to
simplify the discussion. Furthermore it is convenient to have a single field
$K \supset F$ which is algebraic over F and contains an extension of degrees s

over F for every integer $s \geq 1$. It follows easily from Proposition 7.1.1 that K then contains precisely one field with q^s elements. A simple construction for such a field K is given in the Exercises. This field is uniquely determined up to isomorphism and is called an algebraic closure of F. We may then consider $A^n(K)$ and extend V to be an algebraic set still denoted by V in $A^n(K)$ with N_s points whose coordinates are in F_{q^s}.

If $\alpha = (a_1, a_2, \ldots, a_n) \in V$ let F_{q^d} be the smallest field containing F and $a_1, a_2, \ldots, a_n$. We say that α is a point of degree d. Since $a^q = a$ for $a \in F$ it follows that the points $\alpha, \alpha^q, \alpha^{q^2}, \ldots, \alpha^{q^{d-1}}$ are also in V where the exponent denotes raising each coordinate to the indicated power. Furthermore these points are distinct (by say, the corollary to Proposition 7.1.1).

Definition. A prime divisor on V is a set of the form $\{\alpha^{q^j} | j = 0, 1, 2, \ldots, d-1\}$ where α is a point on V of degree d. This is somewhat at variance with common usage. What we call a prime divisor is usually referred to as a prime zero cycle defined over F.

Prime divisors are traditionally denoted by $\mathfrak{P}$. The degree of $\mathfrak{P}$, deg $\mathfrak{P}$, is d.

The prime divisors clearly partition $V \subset A^n(K)$. Furthermore if α is a point on V with coordinates in F_{q^s} then α defines a unique prime divisor $\mathfrak{P}$ of degree d for some $d|s$ by Proposition 7.1.5. This prime divisor contains d points on V each with coordinates in F_{q^s}. If we define a_d to be the number of prime divisors on V of degree d (a number which is finite) then we have by the above the following important result.

Lemma 1. $N_s = \sum_{d|s} d a_d$.

The main result of this section may now be stated.

Proposition 11.1.3. $Z_V(u) = \prod_{\mathfrak{P}} (1/(1 - u^{\deg \mathfrak{P}}))$.

PROOF. The right-hand side is clearly

$$\prod_{n=1}^{\infty} \left(\frac{1}{1 - u^n} \right)^{a_n}.$$

The logarithmic derivative of this expression is

$$\frac{1}{u} \sum_{n=1}^{\infty} \frac{n a_n u^n}{1 - u^n}.$$

Expanding the denominator into a geometric series and computing the coefficient of t^m we obtain

$$\frac{1}{u} \sum_{m=1}^{\infty} \left(\sum_{d|m} d a_d \right) u^m$$

which by Lemma 1 becomes

$$\sum_{m=1}^{\infty} N_m u^{m-1}.$$

Integrating and taking the exponential gives the result.   □

This result shows that $Z(u)$ has integral coefficients. The analogy with the Riemann zeta function becomes even more striking if one introduces a new variable s related to u by $u = q^{-s}$. Then we have

$$Z(q^{-s}) = \prod_{\mathfrak{P}} \left(\frac{1}{1 - q^{-s \deg \mathfrak{P}}} \right)$$

$$= \prod_{\mathfrak{P}} \left(\frac{1}{1 - (1/N(\mathfrak{P})^s)} \right)$$

in perfect analogy with the Riemann zeta function.

§2 Trace and Norm in Finite Fields

In Chapter 10, Section 3, we introduce the notion of trace. Here we shall generalize that notion and also define the norm in finite fields.

Let F be a finite field with q elements and E a field containing F with q^s elements.

Definition. If $\alpha \in E$, the *trace* of α from E to F is given by

$$\text{tr}_{E/F}(\alpha) = \alpha + \alpha^q + \cdots + \alpha^{q^{s-1}}.$$

The norm of α from E to F is given by

$$N_{E/F}(\alpha) = \alpha \cdot \alpha^q \cdots \alpha^{q^{s-1}}.$$

The following two propositions describe the basic properties of trace and norm.

Proposition 11.2.1. *If α, $\beta \in E$ and $a \in F$, then*

(a) $\text{tr}_{E/F}(\alpha) \in F$.
(b) $\text{tr}_{E/F}(\alpha + \beta) = \text{tr}_{E/F}(\alpha) + \text{tr}_{E/F}(\beta)$.
(c) $\text{tr}_{E/F}(a\alpha) = a\, \text{tr}_{E/F}(\alpha)$.
(d) $\text{tr}_{E/F}$ maps E onto F.

Proposition 11.2.2. *If α, $\beta \in E$ and $a \in F$, then*

(a) $N_{E/F}(\alpha) \in F$.
(b) $N_{E/F}(\alpha\beta) = N_{E/F}(\alpha)N_{E/F}(\beta)$.

(c) $N_{E/F}(a\alpha) = a^s N_{E/F}(\alpha)$.
(d) $N_{E/F}$ maps E^* onto F^*.

PROOF. The proof of Proposition 11.2.1 is exactly analogous to that of Proposition 10.3.1 and will be omitted.

To prove Proposition 11.2.2 notice that

$$N_{E/F}(\alpha)^q = (\alpha \cdot \alpha^q \cdot \cdots \cdot \alpha^{q^{s-1}})^q = \alpha^q \cdot \alpha^{q^2} \cdot \cdots \cdot \alpha^{q^s} = N_{E/F}(\alpha).$$

Thus $N_{E/F}(\alpha) \in F$.

Now,

$$\begin{aligned} N_{E/F}(\alpha\beta) &= (\alpha\beta) \cdot (\alpha\beta)^q \cdot \cdots \cdot (\alpha\beta)^{q^{s-1}} \\ &= (\alpha \cdot \alpha^q \cdot \cdots \cdot \alpha^{q^{s-1}}) \cdot (\beta \cdot \beta^q \cdot \cdots \cdot \beta^{q^{s-1}}) \\ &= N_{E/F}(\alpha)N_{E/F}(\beta). \end{aligned}$$

This proves step (b).

To prove step (c) notice that for $a \in F$, $N_{E/F}(a) = a \cdot a^q \cdot \cdots \cdot a^{q^{s-1}} = a^s$ since $a^q = a$. Now apply the result of step (b).

Finally, consider the kernel of the homomorphism $N_{E/F}$, i.e., the set of all $\alpha \in E$ such that $N_{E/F}(\alpha) = 1$. α is in the kernel iff

$$1 = \alpha \cdot \alpha^q \cdot \cdots \cdot \alpha^{q^{s-1}} = \alpha^{1+q+\cdots+q^{s-1}} = \alpha^{(q^s-1)/(q-1)}$$

Since $(q^s - 1)/(q - 1) | q^s - 1$ we have by Proposition 7.1.2 that $x^{(q^s-1)/(q-1)} = 1$ has $(q^s - 1)/(q - 1)$ solutions in E. By elementary group theory it follows that the image, $N_{E/F}(E^*)$, has $q - 1$ elements, but this is exactly the number of elements in F^*. Thus $N_{E/F}$ is onto. $\qquad\square$

Given a tower of fields $F \subset E \subset K$ we have the relation $[K:F] = [K:E][E:F]$. This result is easy to prove in general. If all three fields are finite, we can prove it as follows. Let q be the number of elements in F. Then the number of elements in E and K are $q^{[E:F]}$ and $q^{[K:F]}$, respectively. Considering K as an extension of E we can express the number of elements in K as $(q^{[E:F]})^{[K:E]}$. Thus

$$q^{[K:F]} = q^{[E:F][K:E]}$$

and therefore $[K:F] = [E:F][K:E]$.

We can now prove another simple property of trace and norm that will be useful.

Proposition 11.2.3. *Let $F \subset E \subset K$ be three finite fields and $\alpha \in K$. Then*

(a) $\text{tr}_{K/F}(\alpha) = \text{tr}_{E/F}(\text{tr}_{K/E}(\alpha))$.
(b) $N_{K/F}(\alpha) = N_{E/F}(N_{K/E}(\alpha))$.

PROOF. We shall prove only property (a). The proof of property (b) is similar.

Let $d = [E:F]$, $m = [K:E]$, and $n = [K:F]$. As we have pointed out above, $n = dm$.

The number of elements in E is $q_1 = q^d$. Thus

$$\text{tr}_{K/E}(\alpha) = \alpha + \alpha^{q_1} + \cdots + \alpha^{q_1^{m-1}}$$

and

$$\begin{aligned}
\text{tr}_{E/F}(\text{tr}_{K/E}(\alpha)) &= \sum_{i=0}^{d-1} \text{tr}_{K/E}(\alpha)^{q^i} \\
&= \sum_{i=0}^{d-1} \sum_{j=0}^{m-1} \alpha^{q_1^j q^i} \\
&= \sum_{i=0}^{d-1} \sum_{j=0}^{m-1} \alpha^{q^{dj+i}} \\
&= \sum_{k=0}^{n-1} \alpha^{q^k} \\
&= \text{tr}_{K/F}(\alpha).
\end{aligned}$$

We have used the fact that as j varies from zero to $m - 1$ and i varies from zero to $d - 1$ the quantity $dj + i$ varies from zero to $md - 1 = n - 1$. $\square$

Suppose now that $F \subset K$ are finite fields, $n = [K:F]$, and $\alpha \in K$. Let $E = F(\alpha)$ and $f(x) \in F[x]$ be the minimal polynomial for α over F. By the Proposition 7.2.2 we have $[E:F] = d$, where d is the degree of $f(x)$.

Proposition 11.2.4. *Write* $f(x) = x^d - c_1 x^{d-1} + \cdots + (-1)^d c_d$. *Then*

(a) $f(x) = (x - \alpha)(x - \alpha^q) \cdots (x - \alpha^{q^{d-1}})$.
(b) $\text{tr}_{K/F}(\alpha) = (n/d)c_1$.
(c) $N_{K/F}(\alpha) = c_d^{n/d}$.

PROOF. Since the coefficients of f satisfy $a^q = a$ we have

$$0 = f(\alpha)^q = f(\alpha^q).$$

Thus α^q is a root of f. Similarly,

$$0 = f(\alpha^q)^q = f(\alpha^{q^2}).$$

Thus α^{q^2} is a root of f. Continuing in this manner we see that $\alpha, \alpha^q, \alpha^{q^2}, \ldots,$ $\alpha^{q^{d-1}}$ are all roots of f. If we can show that all these roots are distinct, assertion (a) will follow.

Suppose that $0 \leq i \leq j < d$ and that $\alpha^{q^i} = \alpha^{q^j}$. Set $k = j - i$. We shall show that $k = 0$.

We have

$$\alpha^{q^i} = \alpha^{q^j} = (\alpha^{q^k})^{q^i},$$

which implies that

$$(\alpha - \alpha^{q^k})^{q^i} = 0$$

and so

$$\alpha = \alpha^{q^k}.$$

Since $f(x)$ is the minimal polynomial for α it follows that $f(x)$ divides $x^{q^k} - x$ and so by Theorem 2 of Chapter 7 we have $d|k$. However, $0 \leq k < d$ and so $k = 0$ and we are done.

It follows immediately from assertion (a) that $c_1 = \mathrm{tr}_{E/F}(\alpha)$ and that $c_d = N_{E/F}(\alpha)$.

Since $\alpha \in E = F(\alpha)$ we have $\mathrm{tr}_{K/E}(\alpha) = [K:E]\alpha = (n/d)\alpha$ and $N_{K/E}(\alpha) = \alpha^{n/d}$.

By Proposition 11.2.3,

$$\mathrm{tr}_{K/F}(\alpha) = \mathrm{tr}_{E/F}(\mathrm{tr}_{K/E}(\alpha)) = \mathrm{tr}_{E/F}\left(\frac{n}{d}\alpha\right) = \frac{n}{d}\mathrm{tr}_{E/F}(\alpha) = \frac{n}{d}c_1.$$

Similarly,

$$N_{K/F}(\alpha) = N_{E/F}(N_{K/E}(\alpha)) = N_{E/F}(\alpha^{n/d}) = N_{E/F}(\alpha)^{n/d} = c_d^{n/d}. \qquad \square$$

§3 The Rationality of the Zeta Function Associated to $a_0 x_0^m + a_1 x_1^m + \cdots + a_n x_n^m$

Let $f(x_0, x_1, \ldots, x_n)$ be the polynomial given in the title of this section [notice that this is *not* the $f(x)$ of Section 2]. Suppose that the coefficients are in F, a finite field, with q elements and that $q \equiv 1 \ (m)$. We have to investigate the number N_s of elements in $\bar{H}_f(F_s)$, where $[F_s : F] = s$. Theorem 2 of Chapter 10 shows that N_s is given by

$$q^{s(n-1)} + q^{s(n-2)} + \cdots + q^s + 1$$

$$+ \frac{1}{q^s} \sum_{\chi_0^{(s)}, \ldots, \chi_n^{(s)}} \chi_0^{(s)}(a_0^{-1}) \cdots \chi_n^{(s)}(a_n^{-1}) g(\chi_0^{(s)}) \cdots g(\chi_n^{(s)}), \quad (4)$$

where q^s is the number of elements in F_s, and the $\chi_i^{(s)}$ are multiplicative characters of F_s such that $\chi_i^{(s)m} = \varepsilon$, $\chi_i^{(s)} \neq \varepsilon$, and $\chi_0^{(s)}\chi_1^{(s)} \cdots \chi_n^{(s)} = \varepsilon$.

We must analyze the terms $\chi_i^{(s)}(a_i^{-1})$ and $g(\chi_i^{(s)})$. To do this we first relate characters of F_s to characters of F.

Let χ be a character of F and set $\chi' = \chi \circ N_{F_s/F}$; i.e., for $\alpha \in F_s$, $\chi'(\alpha) = \chi(N_{F_s/F}(\alpha))$. Then one sees, using Proposition 11.2.2, that χ' is a character of F_s, and moreover that

(a) $\chi \neq \rho$ implies that $\chi' \neq \rho'$.
(b) $\chi^m = \varepsilon$ implies that $\chi'^m = \varepsilon$.
(c) $\chi'(a) = \chi(a)^s$ for all $a \in F$.

It follows easily that as χ varies over the characters of F of order dividing m, χ' varies over the characters of F_s of order dividing m.

The sum in Equation (4) can now be rewritten as

$$\sum_{\chi_0, \ldots, \chi_n} \chi_0(a_0^{-1})^s \cdots \chi_n(a_n^{-1})^s g(\chi_0') \cdots g(\chi_n'), \tag{5}$$

where $\chi_0, \ldots, \chi_n$ are characters of F satisfying $\chi_i^m = \varepsilon$, $\chi_i \neq \varepsilon$, and $\chi_0 \chi_1 \cdots \chi_n = \varepsilon$.

It remains to analyze the Gauss sums $g(\chi')$. This is the content of the following theorem of Hasse and Davenport (see [23]).

Theorem 1. $(-g(\chi))^s = -g(\chi')$.

We postpone the proof of this relation. Using Theorem 1 and Equations (4) and (5) we see that N_s is given by

$$\sum_{k=0}^{n-1} q^{ks} + (-1)^{n+1} \sum_{\chi_0, \chi_1, \ldots, \chi_n} \left[\frac{(-1)^{n+1}}{q} \chi_0(a_n^{-1}) \cdots \chi_n(a_n^{-1}) g(\chi_0) \cdots g(\chi_n) \right]^s, \tag{6}$$

where the second sum is restricted by the same conditions as Equation (5).

Applying Proposition 11.1.1 gives us the main result of this chapter.

Theorem 2. *Let $a_0, a_1, \ldots, a_n \in F^*$, where F is a finite field with q elements, and $q \equiv 1 \ (m)$. Let $f(x_0, \ldots, x_n) = a_0 x_0^m + a_1 x_1^m + \cdots + a_n x_n^m$. Then the zeta function $Z_f(u)$ is a rational function of the form*

$$\frac{P(u)^{(-1)^n}}{(1 - u)(1 - qu) \cdots (1 - q^{n-1}u)},$$

where $P(u)$ is the polynomial

$$\prod_{\chi_0, \chi_1, \ldots, \chi_n} \left(1 - (-1)^{n+1} \frac{1}{q} \chi_0(a_0^{-1}) \cdots \chi_n(a_n^{-1}) g(\chi_0) g(\chi_1) \cdots g(\chi_n) u \right),$$

the $(n + 1)$-tuples $\chi_0, \chi_1, \ldots, \chi_n$ being subject to the conditions $\chi_i^m = \varepsilon$, $\chi_i \neq \varepsilon$, and $\chi_0 \chi_1 \cdots \chi_n = \varepsilon$.

A number of remarks are in order:

(1) The degree of $P(u)$ can be computed explicitly. It is

$$m^{-1}[(m - 1)^{n+1} + (-1)^{n+1}(m - 1)].$$

(2) Since $|g(\chi)| = q^{1/2}$ it follows from the explicit expression for $P(u)$ that the zeros of $Z_f(u)$ have absolute value $q^{-((n-1)/2)}$. This is in accord with the general Riemann hypothesis.

(3) If we write $P(u) = \prod (1 - \alpha u)$, then numbers α are algebraic integers. This is not hard to see. Each α has the form

$$\zeta \frac{1}{q} g(\chi_0) \cdots g(\chi_n),$$

where ζ is a root of unity and $\chi_0 \chi_1 \cdots \chi_n = \varepsilon$. Using Corollary 1 to Theorem 3 of Chapter 8 we see that

$$\frac{1}{q} g(\chi_0) g(\chi_1) \cdots g(\chi_n) = \chi_n(-1) J(\chi_0, \chi_1, \ldots, \chi_{n-1}).$$

The Jacobi sum is a sum of roots of unity and so is an algebraic integer. Thus $\alpha = \zeta \chi_n(-1) J(\chi_0, \chi_1, \ldots, \chi_{n-1})$ is an algebraic integer as well.

Let $f(x_0, x_1, \ldots, x_n)$ be a homogeneous form of degree d with coefficients in a finite field F. Assume furthermore that the partial derivatives $f_{x_0}, \ldots, f_{x_n}$ have no common projective zero in any algebraic extension of F. In this case we say that the projective hypersurface defined by f is absolutely non-singular. Then one may consider the zeta function $Z(t)$ of the hypersurface, $f = 0$. In this case the Weil conjectures (now theorems) state the following:

(a) $Z(t)$ is a rational function which can be written as

$$Z(t) = \frac{P(t)^{(-1)^n}}{(1 - t)(1 - qt) \cdots (1 - q^{n-1}t)},$$

where $P(t)$ is a polynomial with integer coefficients.

(b) $P(t) = (1 - a_1 t)(1 - a_2 t) \cdots (1 - a_m t)$. The mapping $a \to q^{n-1}/a$ is a bijection of the set of reciprocal roots $a_1, \ldots, a_m$.

(c) $|a_i| = q^{(n-1)/2}$.

(d) The degree of $P(t)$ is $d^{-1}[(d-1)^{n+1} + (-1)^{n+1}(d-1)]$.

The statement regarding the absolute value of the roots is known as the Riemann hypothesis for the hypersurface. The proof of (a), (b), and (d) for a general hypersurface is due to B. Dwork [26]. The proof of the Riemann hypothesis is due to P. Deligne (1973). For the general statement of the Weil conjectures we refer the reader to Weil [80] and Katz [161].

§4 A Proof of the Hasse–Davenport Relation

Let F be a finite field with q elements and F_s be a field containing F such that $[F_s : F] = s$. Let χ be a nontrivial multiplicative character of F and $\chi' = \chi \circ N_{F_s/F}$. χ' is a character of F_s. We wish to compare the Gauss sums $g(\chi)$ and $g(\chi')$.

Let us recall the definition of $g(\chi)$ (see Chapter 10, Section 3):

$$g(\chi) = \sum_{t \in F} \chi(t)\psi(t),$$

where $\psi(t)$ is equal to $\zeta_p^{\mathrm{tr}(t)}$. The trace function in this definition coincides with the function tr_{F/F_p} introduced in this chapter. Since we are considering more than one field, it is important to attach subscripts to tr. Now,

$$g(\chi') = \sum_{t \in F_s} \chi'(t)\psi'(t),$$

where $\psi'(t) = \zeta_p^{\mathrm{tr}\,F_s/F_p\,(t)}$. Since $\mathrm{tr}_{F_s/F_p}(t) = \mathrm{tr}_{F/F_p}(\mathrm{tr}_{F_s/F}(t))$ it follows that $\psi' = \psi \circ \mathrm{tr}_{F_s/F}$.

For a monic polynomial $f(x) = x^n - c_1 x^{n-1} + \cdots + (-1)^n c_n$ in $F[x]$ define $\lambda(f) = \psi(c_1)\chi(c_n)$.

Lemma 1. $\lambda(fg) = \lambda(f)\lambda(g)$ for all monic $f, g \in F[x]$.

PROOF. If $g(x) = x^m - b_1 x^{m-1} + \cdots + (-1)^m b_m$, then $f(x)g(x) = x^{n+m} - (b_1 + c_1)x^{n+m-1} + \cdots + (-1)^{n+m}b_m c_n$. Thus $\lambda(fg) = \psi(b_1 + c_1) \cdot \chi(b_m c_n) = \psi(b_1)\psi(c_1)\chi(b_m)\chi(c_n) = \psi(b_1)\chi(b_m)\psi(c_1)\chi(c_n) = \lambda(g)\lambda(f)$. $\square$

Lemma 2. Let $\alpha \in F_s$ and $f(x)$ be the monic irreducible polynomial for α over F. Then

$$\lambda(f)^{s/d} = \chi'(\alpha)\psi'(\alpha), \quad where \; d = \deg f.$$

PROOF. This result follows easily from Proposition 11.2.4. Namely, if $f(x) = x^d - c_1 x^{d-1} + \cdots + (-1)^d c_d$, then

$$\mathrm{tr}_{F_s/F}(\alpha) = \frac{s}{d}c_1 \quad \text{and} \quad N_{F_s/F}(\alpha) = c_d^{s/d}.$$

Now, $\lambda(f) = \psi(c_1)\chi(c_d)$, so

$$\lambda(f)^{s/d} = \psi(c_1)^{s/d}\chi(c_d)^{s/d} = \psi\left(\frac{s}{d}c_1\right)\chi(c_d^{s/d})$$

$$= \psi(\mathrm{tr}_{F_s/F}(\alpha))\chi(N_{F_s/F}(\alpha)) = \psi'(\alpha)\chi'(\alpha). \qquad \square$$

Lemma 3. $g(\chi') = \sum (\deg f)\lambda(f)^{s/\deg f}$, where the sum is over all monic irreducible polynomials of $F[x]$ with degree dividing s.

PROOF. According to Theorem 1 of Chapter 7—generalized to F as base field—$x^{q^s} - x$ is the product of all monic irreducible polynomials of degree dividing s. It follows that every such irreducible polynomial has all its roots in F_s and conversely that every element in F_s satisfies such a polynomial.

Let $f(x)$ be monic irreducible of degree $d \mid s$. Let $\alpha_1, \alpha_2, \ldots, \alpha_d \in F_s$ be its roots. Then by Lemma 2

$$\sum_{i=1}^{d} \chi'(\alpha_i)\psi'(\alpha_i) = d\lambda(f)^{s/d}.$$

Summing over all polynomials of the required type yields the result.   □

We are now in a position to prove the Hasse–Davenport relation. The proof is based on the following identity:

$$\sum_{f} \lambda(f)t^{\deg f} = \prod_{f} (1 - \lambda(f)t^{\deg f})^{-1}, \tag{7}$$

where the sum is over all monic polynomials and the product is over all monic irreducible polynomials in $F[x]$.

The identity is proved by expanding each term $(1 - \lambda(f)t^{\deg f})^{-1}$ in a geometric series and using the fact that every monic polynomial can be written as the product of monic irreducible polynomials in a unique way. The details are left as an exercise.

Now,

$$\sum_{f} \lambda(f)t^{\deg f} = \sum_{s=0}^{\infty} \left(\sum_{\deg f = s} \lambda(f) \right) t^s.$$

We define $\lambda(1) = 1$, as this is necessary for Equation (7) to hold. For $s = 1$ we have

$$\sum_{\deg f = 1} \lambda(f) = \sum_{a \in F} \lambda(x - a) = \sum_{a \in F} \chi(a)\psi(a) = g(\chi).$$

For $s > 1$ we have

$$\sum_{\deg f = s} \lambda(f) = \sum_{c_i \in F} \lambda(x^s - c_1 x^{s-1} + \cdots + (-1)^s c_s)$$

$$= q^{s-2} \sum_{c_1, c_s} \chi(c_s)\psi(c_1) = q^{s-2} \left(\sum_{c_s} \chi(c_s) \right) \left(\sum_{c_1} \psi(c_1) \right) = 0.$$

Putting all this together we see that the left-hand side of Equation (7) reduces to $1 + g(\chi)t$. Using this, take the logarithm of both sides of Equation (7), differentiate, and multiply both sides of the result by t. This yields

$$\frac{g(\chi)t}{1 + g(\chi)t} = \sum_{f} \frac{\lambda(f)(\deg f)t^{\deg f}}{1 - \lambda(f)t^{\deg f}}.$$

Expand the denominators in geometric series. Then

$$\sum_{s=1}^{\infty} (-1)^{s-1} g(\chi)^s t^s = \sum_{f} \left(\sum_{r=1}^{\infty} (\deg f)\lambda(f)^r t^{r \deg f} \right).$$

Equating the coefficients of t^s yields

$$(-1)^{s-1}g(\chi)^s = \sum_{\deg f | s} (\deg f)\lambda(f)^{s/\deg f}.$$

By Lemma 3, the right-hand side is $g(\chi')$. This completes the proof.   □

§5 The Last Entry

The last entry of Gauss's mathematical diary is a statement of the following remarkable conjecture:

Suppose that $p \equiv 1$ (4). Then the number of solutions to the congruence $x^2 + y^2 + x^2y^2 \equiv 1$ (p) is $p + 1 - 2a$, where $p = a^2 + b^2$ and $a + bi \equiv 1$ (2 + 2i).

Some explanation is in order. If $p \equiv 1$ (4), then by Proposition 8.3.1 we know that $p = a^2 + b^2$ for some integers a and b. If we choose a odd and b even, then a and b are uniquely determined up to sign. The congruence $a + bi \equiv 1$ (2 + 2i) determines the sign of a. We shall give a simpler formulation of this.

Lemma. *If* $p \equiv 1$ (4), $p = a^2 + b^2$, *and* $a + bi \equiv 1$ (2 + 2i), *then a is odd and b is even. Moreover, if* $4|b$, *then* $a \equiv 1$ (4), *and if* $4 \nmid b$, *then* $a \equiv -1$ (4).

PROOF. $a + bi \equiv 1$ (2 + 2i) implies that $a + bi \equiv 1$ (2) and so a is odd and b even.

Since $4 = -2(i - 1)(i + 1)$ it follows that if $4|b$, then $a + bi \equiv a \equiv 1$ (2 + 2i). Taking conjugates $a \equiv 1$ (2 − 2i). Thus $(2 + 2i)(2 - 2i) = 8|(a - 1)^2$ and $a \equiv 1$ (4).

If $4 \nmid b$, then $b = 4k + 2$ for some k. Thus $a + bi \equiv a + 2i \equiv 1$ (2 + 2i). Since $2i \equiv -2$ (2 + 2i) we have $a \equiv 3 \equiv -1$ (2 + 2i). As before $8|(a + 1)^2$ and so $a \equiv -1$ (4).   □

Theorem. *Consider the curve C determined by* $x^2t^2 + y^2t^2 + x^2y^2 - t^4$ *over* F_p, *where* $p \equiv 1$ (4). *Write* $p = a^2 + b^2$ *with a odd and b even. If* $4|b$, *choose* $a \equiv 1$ (4); *if* $4 \nmid b$, *choose* $a \equiv -1$ (4). *Then the number of points on C in* $P^2(F_p)$ *is* $p - 1 - 2a$.

The zeta function of C is

$$Z(u) = \frac{1 - 2au + pu^2}{1 - pu}(1 - u).$$

Before giving the proof a few remarks are in order.

The answer $p - 1 - 2a$ differs from Gauss' $p + 1 - 2a$. The difficulty is that Gauss counts four points at infinity, whereas a simple calculation shows that [0, 1, 0] and [0, 0, 1] are the only points at infinity according to our definition. Thus our answer differs from his by 2.

Since there are two points at infinity independently of p it suffices to count the number of finite points, i.e., the solutions to $x^2 + y^2 + x^2y^2 = 1$.

As an example take $p = 5$. Since $5 = 1^2 + 2^2$ we have $4 \nmid b$ so we must take $a = -1$. The formula $p - 1 - 2a$ gives the answer 6 in this case. Indeed, in addition to the two points at infinity, $(1, 0)$, $(-1, 0)$, $(0, 1)$, and $(0, -1)$ are the other points on the curve in F_p.

The form of the zeta function may be surprising. The explanation is that the two points at infinity are singular. Thus the form of this zeta function is not in contradiction to our earlier observations.

We now proceed to prove the theorem. Denote by C_1 the curve given by $x^2 + y^2 + x^2y^2 = 1$ and by C_2 the curve given by $w^2 = 1 - z^4$. We shall construct maps from C_1 to C_2 and from C_2 to C_1.

Notice that

$$x^2 + y^2 + x^2y^2 = 1$$

implies that

$$(1 + x^2)y^2 = 1 - x^2$$

and

$$[(1 + x^2)y]^2 = 1 - x^4.$$

Thus, if (a, b) is on C_1, then $(a, (1 + a^2)b)$ is on C_2. Let

$$\lambda(x, y) = (x, (1 + x^2)y),$$

λ maps C_1 to C_2. It is easy to see that this map is one to one.

Now let

$$\mu(z, w) = \left(z, \frac{w}{1 + z^2}\right),$$

μ is not always defined. If $\alpha \in F_p$ is such that $\alpha^2 = -1$, then $(\alpha, 0)$ and $(-\alpha, 0)$ are on C_2 but μ is undefined at these points. μ is defined at all other points of C_2 and maps these points to C_1. It is easy to check that μ is inverse to λ where it is defined. Thus

$$N_1 = N_2 - 2,$$

where N_1 and N_2 are the number of finite points in F_p on C_1 and C_2, respectively.

We can compute N_2 by using Theorem 5 of Chapter 8. Specializing Theorem 5 to $w^2 + z^4 = 1$ we see that

$$N_2 = p + J(\rho, \chi) + J(\rho, \chi^2) + J(\rho, \chi^3),$$

where ρ is the character of order 2 and χ is a character of order 4.

Since $\chi^2 = \rho$, we have $J(\rho, \chi^2) = J(\rho, \rho) = -\rho(-1) = -1$. Also, since $\chi^4 = \varepsilon$ we have $\chi^3 = \bar{\chi}$ so that $J(\rho, \chi^3) = J(\rho, \bar{\chi}) = \overline{J(\rho, \chi)}$.

Let $\pi = -J(\rho, \chi)$. Then

$$N_2 = p - 1 - \pi - \bar{\pi},$$

ρ takes on the values ± 1 and χ takes on the values ± 1, $\pm i$. Thus $\pi = a + bi$, where a, $b \in Z$. Moreover $|J(\rho, \chi)|^2 = p$ so that $a^2 + b^2 = \pi\bar{\pi} = p$. It follows that $N_2 = p - 1 - 2a$ and $N_1 = p - 3 - 2a$. Since C_1 has two points at infinity, the total number of points on C_1 in F_p is given by

$$N = p - 1 - 2a.$$

By the lemma it suffices to prove that $\pi \equiv 1$ $(2 + 2i)$ in order to complete the proof of the first part of the theorem. This is accomplished by means of the following pretty calculation given in Hasse–Davenport [23].

Notice that $\rho(a) - 1 \equiv 0$ (2) and that $\chi(a) - 1 \equiv 0$ $(1 + i)$ for all $a \neq 0$ in F_p. The first assertion is obvious; the second follows from $1 - 1 = 0$, $-1 - 1 = -(1 - i)(1 + i)$, $-i - 1 = -(1 + i)$, and $i - 1 = i(1 + i)$. Thus if $a \neq 0$ and $b \neq 0$, $(\rho(a) - 1)(\chi(b) - 1) \equiv 0$ $(2 + 2i)$. This congruence is trivially true for the pairs $a = 0, b = 1$ and $a = 1, b = 0$. Therefore,

$$\sum_{a+b=1} (\rho(a) - 1)(\chi(b) - 1) \equiv 0 \ (2 + 2i).$$

Expanding we see that

$$-\pi - \sum_b \chi(b) - \sum_a \rho(a) + p \equiv 0 \ (2 + 2i).$$

The second and third terms are zero. Thus

$$\pi \equiv p \equiv 1 \ (2 + 2i).$$

The last step follows because $p \equiv 1$ (4) by hypothesis, and $2 + 2i$ divides 4; indeed $4 = (1 - i)(2 + 2i)$.

To calculate the zeta function it suffices to notice that by the Hasse–Davenport relation the number of points on $x^2t^2 + y^2t^2 + x^2y^2 - t^4$ in $P^2(F_{p^s})$ is given by

$$p^s - 1 - (-J(\rho, \chi))^s - (-J(\overline{\rho, \chi}))^s = p^s - 1 - \pi^s - \bar{\pi}^s.$$

Thus

$$Z(u) = \frac{(1 - \pi u)(1 - \bar{\pi}u)}{(1 - pu)}(1 - u)$$

$$= \frac{1 - 2au + pu^2}{(1 - pu)}(1 - u).$$

NOTES

As we have mentioned, in his thesis E. Artin [2] introduced the congruence zeta function. In that work he establishes the analog of the Riemann hypothesis for about 40 curves of the type $y^2 = f(x)$, where f is a cubic or quartic polynomial. In 1934 Hasse proved that the result held in general for nonsingular cubics (the case of elliptic curves). The Riemann hypothesis for arbitrary nonsingular curves was established in full generality by Weil in 1948. His proof is far from elementary and uses deep techniques in algebraic geometry.

Weil's conjecture that the zeta function of any algebraic set is rational was proved in 1959 by B. Dwork using methods of p-adic analysis [26].

In 1969 S. A. Stepanov succeeded in giving an elementary proof of the Riemann hypothesis for curves [222]. A complete account of Stepanov's method is given in the book by W. M. Schmidt, *Equations over Finite Fields: An Elementary Approach* [218]. This method was simplified further by E. Bombieri, who, using the Riemann–Roch theorem, gives a complete proof in five pages [98]. Sharper estimates in special cases have been obtained by H. Stark [221]. For an analysis of Deligne's proof and an historical discussion of the entire issue the reader should consult N. Katz's "Overview of Deligne's proof..." [161]. This paper also contains an extensive bibliography of the subject. See also the survey [248]. The discovery of these remarkable theorems is discussed by Weil in the first volume of his *Collected Papers*, [241], pp. 568–569. Finally we mention the paper by J. R. Joly, "Equations et varietés algébriques sur un corps fini" [160].

Section 5 on Gauss' conjecture is logically out of place since it could have been given in Chapter 8. We felt it was appropriate at this point since the relation between this conjecture and Weil's Riemann hypothesis reveals once again the remarkable acuity of Gauss' insight and how his imposing presence continues to make itself felt to this very day.

A new edition of the mathematical journal of Gauss, translated from Latin to German, with an historical review by K. Biermann and comments by H. Wussing is now available [137]. This important historical document records the major discoveries of Gauss between the years 1796 and 1814. It is interesting to note that both the first entry (Section 11 of chapter 9) and the last entry are concerned with cyclotomy. For more biographical information on Gauss see T. Hall [143] and the recent biography by W. K. Bühler [101].

EXERCISES

1. Suppose that we may write the power series $1 + a_1 u + a_2 u^2 + \cdots$ as the quotient of two polynomials $P(u)/Q(u)$. Show that we may assume that $P(0) = Q(0) = 1$.

2. Prove the converse to Proposition 11.1.1.

3. Give the details of the proof that N_s is independent of the field F_s (see the concluding paragraph to Section 1).

4. Calculate the zeta function of $x_0 x_1 - x_2 x_3 = 0$ over F_p.

5. Calculate as explicitly as possible the zeta function of $a_0 x_0^2 + a_1 x_1^2 + \cdots + a_n x_n^2$ over F_q, where q is odd. The answer will depend on whether n is odd or even and whether $q \equiv 1 \ (4)$ or $q \equiv 3 \ (4)$.

6. Consider $x_0^3 + x_1^3 + x_2^3 = 0$ as an equation over F_4, the field with four elements. Show that there are nine points on the curve in $P^2(F_4)$. Calculate the zeta function. [*Answer*: $(1 + 2u)^2/((1 - u)(1 - 4u))$.]

7. Try this exercise if you know a little projective geometry. Let N_s be the number of lines in $P^n(F_{p^s})$. Find N_s and calculate $\sum_{s=1}^{\infty} N_s u^s/s$. (The set of lines in projective space form an algebraic variety called a Grassmannian variety. So do the set of planes, three-dimensional linear subspaces, etc.)

8. If f is a nonhomogeneous polynomial, we can consider the zeta function of the projective closure of the hypersurface defined by f (see Chapter 10). One way to calculate this is to count the number of points on $H_f(F_q)$ and then add to it the number of points at infinity. For example, consider $y^2 = x^3$ over F_{p^s}. Show that there is one point at infinity. The origin $(0, 0)$ is clearly on this curve. If $x \neq 0$, write $(y/x)^2 = x$ and show that there are $p^s - 1$ more points on this curve. Altogether we have p^s points and the zeta function over F_p is $(1 - pu)^{-1}$.

9. Calculate the zeta function of $y^2 = x^3 + x^2$ over F_p.

10. If $A \neq 0$ in F_q and $q \equiv 1$ (3), show that the zeta function of $y^2 = x^3 + A$ over F_q has the form $Z(u) = (1 + au + qu^2)/(1 - u)(1 - qu)$, where $a \in Z$ and $|a| \leq 2q^{1/2}$.

11. Consider the curve $y^2 = x^3 - Dx$ over F_p, where $D \neq 0$. Call this curve C_1. Show that the substitution $x = \frac{1}{2}(u + v^2)$ and $y = \frac{1}{2}v(u + v^2)$ transforms C_1 into the curve C_2 given by $u^2 - v^4 = 4D$. Show that in any given finite field the number of finite points on C_1 is one more than the number of finite points on C_2.

12. (continuation) If $p \equiv 3$ (4), show that the number of projective points on C_1 is just $p + 1$. If $p \equiv 1$ (4), show that the answer is $p + 1 + \chi(D)J(\chi, \chi^2) + \overline{\chi(D)J(\chi, \chi^2)}$, where χ is a character of order 4 on F_p.

13. (continuation) If $p \equiv 1$ (4), calculate the zeta function of $y^2 = x^3 - Dx$ over F in terms of π and $\chi(D)$, where $\pi = -J(\chi, \chi^2)$. This calculation in somewhat sharpened form is contained in [23]. The result has played a key role in recent empirical work of B. J. Birch and H. P. F. Swinnerton-Dyer on elliptic curves.

14. Suppose that $p \equiv 1$ (4) and consider the curve $x^4 + y^4 = 1$ over F_p. Let χ be a character of order 4 and $\pi = -J(\chi, \chi^2)$. Give a formula for the number of projective points over F_p and calculate the zeta function. Both answers should depend only on π. (*Hint*: See Exercises 7 and 16 of Chapter 8, but be careful since there we were counting only finite points.)

15. Find the number of points on $x^2 + y^2 + x^2y^2 = 1$ for $p = 13$ and $p = 17$. Do it both by means of the formula in Section 5 and by direct calculation.

16. Let F be a field with q elements and F_s an extension of degree s. If χ is a character of F, let $\chi' = \chi \circ N_{F_s/F}$. Show that
    (a) χ' is a character of F_s.
    (b) $\chi \neq \rho$ implies that $\chi' \neq \rho'$.
    (c) $\chi^m = \varepsilon$ implies that $\chi'^m = \varepsilon$.
    (d) $\chi'(a) = \chi(a)^s$ for $a \in F$.
    (e) As χ varies over all characters of F with dividing m, χ' varies over all characters of F_s with order dividing m. Here we are assuming that $q \equiv 1$ (m).

17. In Theorem 2 show that the order of the numerator of the zeta function, $P(u)$ has degree $m^{-1}((m-1)^{n+1} + (-1)^{n+1}(m-1))$.

18. Let the notation be as in Exercise 16. Use the Hasse–Davenport relation to show that $J(\chi_1', \chi_2', \ldots, \chi_n') = (-1)^{(s-1)(n-1)}J(\chi_1, \chi_2, \ldots, \chi_n)^s$, where the χ_i are nontrivial characters of F and $\chi_1\chi_2 \cdots \chi_n \neq \varepsilon$.

19. Prove the identity $\sum \lambda(f)t^{\deg f} = \prod (1 - \lambda(f)t^{\deg f})^{-1}$, where the sum is over all monic polynomials in $F[t]$ and the product is over all monic irreducibles in $F[t]$. λ is defined in Section 4.

20. If in Theorem 2 we keep f fixed but consider the base field to be F_s instead of F, we get a different zeta function, $Z_f^{(s)}(u)$. Show that $Z_f^{(s)}(u)$ and $Z_f(u)$ are related by the equation $Z_f^{(s)}(u^s) = Z_f(u)Z_f(\rho u) \cdots Z_f(\rho^{s-1}u)$, where $\rho = e^{2\pi i/s}$.

21. In Exercise 6 we considered the equation $x_0^3 + x_1^3 + x_2^3 = 0$ over the field with four elements. Consider the same equation over the field with two elements. The trouble here is that $2 \not\equiv 1$ (3) and so our usual calculations do not work. Prove that in every extension of $Z/2Z$ of odd degree every element is a cube and that in every extension of even degree, 3 divides the order of the multiplicative group. Use this information to calculate the zeta function over $Z/2Z$. [*Answer:* $(1 + 2u^2)/(1 - u)(1 - 2u)$.]

22. Use the ideas developed in Exercise 21 to show that Theorem 2 continues to hold (in a suitable sense) even when the hypothesis $q \equiv 1$ (m) is removed.

23. Let $p_1 < p_2 < p_3 < \cdots$ denote the positive prime numbers arranged in order. Let $N_m = p_1^m p_2^m \cdots p_m^m$ and let E_m denote the field with q^{N_m} elements. Show that E_m can be considered as a subfield of E_{m+1} and that $E = \bigcup E_m$ is an extension of $E_0 = F$, a finite field with q elements, with the following property; for every positive integer n, E contains one and only one subfield F_n with q^n elements.

Chapter 12

Algebraic Number Theory

In this chapter we shall introduce the concept of an algebraic number field and develop its basic properties. Our treatment will be classical, developing directly only those aspects that will be needed in subsequent chapters. The study of these fields, and their interaction with other branches of mathematics forms a vast area of current research. Our objective is to develop as much of the general theory as is needed to study higher-power reciprocity. The reader who is interested in a more systematic treatment of these fields should consult any one of the standard texts on this subject, e.g., Ribenboim [207], Lang [168], Goldstein [140], Marcus [183].

We will assume that the reader has some familiarity with the theory of separable field extensions as can be found, for example, in Herstein's Topics in Algebra *[150]. Some of the results assumed are given in the Exercises.*

§1 Algebraic Preliminaries

In this section we will recall some facts from field theory and prove some results about discriminants.

Let L/K be a finite algebraic extension of fields. The dimension of L/K, $[L:K]$, will be denoted by n.

Suppose $\alpha_1, \alpha_2, \ldots, \alpha_n$ is a basis for L/K and $\alpha \in L$. Then $\alpha\alpha_i = \sum_j a_{ij}\alpha_j$, with $a_{ij} \in K$.

Definition. The norm of α, $N_{L/K}(\alpha)$, is $\det(a_{ij})$. The trace of α, $t_{L/K}(\alpha)$, is $a_{11} + a_{22} + \cdots + a_{nn}$.

It is easy to check that this definition is independent of the choice of a basis. In what follows, norm and trace will be denoted by N and t since the extension L/K will be fixed.

If $\alpha, \beta \in L$ and $a \in K$ then $N(\alpha\beta) = N(\alpha)N(\beta)$, $t(\alpha + \beta) = t(\alpha) + t(\beta)$, $N(a\beta) = a^n N(\beta)$, and $t(a\alpha) = at(\alpha)$. If $\alpha \neq 0$ then $N(\alpha)N(\alpha^{-1}) = N(\alpha\alpha^{-1}) = N(1) = 1$. Thus, if $\alpha \neq 0$, $N(\alpha) \neq 0$, and $N(\alpha^{-1}) = N(\alpha)^{-1}$. If L/K is separable, then t is not identically zero. If char $K = 0$ this is easy to see since then $t(1) =$

$n \neq 0$. The only fields of characteristic $p > 0$ that we will consider are finite fields and in this case the result follows from Proposition 11.2.1(d).

Suppose L/K is separable and let $\sigma_1, \sigma_2, \ldots, \sigma_n$ be the distinct isomorphisms of L into a fixed algebraic closure of K which leave K fixed. For $\alpha \in L$ denote $\sigma_j(\alpha)$ by $\alpha^{(j)}$. The elements $\alpha^{(j)}$ are called the conjugates of α. Here $\alpha^{(1)}$ is α.

One can show using linear algebra (see Exercises 21–23). $t(\alpha) = \alpha^{(1)} + \alpha^{(2)} + \cdots + \alpha^{(n)}$ and that $N(\alpha) = \alpha^{(1)}\alpha^{(2)}\cdots\alpha^{(n)}$. If $\alpha \in L$ consider $f(x) = (x - \alpha^{(1)})(x - \alpha^{(2)})\cdots(x - \alpha^{(n)})$. Then $f(x) \in K[x]$. The coefficient of x^{n-1} is $-t(\alpha)$ and the constant term is $(-1)^n N(\alpha)$. The reader should verify that our definitions of norm and trace generalize those of Chapter 11, Section 2.

Definition. If $\alpha_1, \alpha_2, \ldots, \alpha_n$ is an n-tuple of elements of L we define the discriminant $\Delta(\alpha_1, \ldots, \alpha_n)$ to be $\det(t(\alpha_i \alpha_j))$.

Proposition 12.1.1. *If $\Delta(\alpha_1, \ldots, \alpha_n) \neq 0$ then $\alpha_1, \ldots, \alpha_n$ is a basis for L/K. If L/K is separable and $\alpha_1, \ldots, \alpha_n$ is a basis for L/K then $\Delta(\alpha_1, \ldots, \alpha_n) \neq 0$.*

PROOF. Suppose $\alpha_1, \ldots, \alpha_n$ are linearly dependent. Then there exist $a_1, \ldots, a_n \in K$, not all zero, such that $\sum a_i \alpha_i = 0$. Multiply this equation by α_j and take the trace. One finds

$$\sum_i a_i t(\alpha_i \alpha_j) = 0, \qquad j = 1, 2, \ldots, n.$$

This shows that the matrix $(t(\alpha_i \alpha_j))$ is singular and so its determinant is zero.

Now suppose $\alpha_1, \ldots, \alpha_n$ is a basis and $\Delta(\alpha_1, \ldots, \alpha_n) = 0$. Then the system of linear equations

$$\sum_i x_i t(\alpha_i \alpha_j) = 0, \qquad j = 1, \ldots, n,$$

has a nontrivial solution $x_i = a_i \in K$, $i = 1, \ldots, n$. Let $\alpha = \sum a_i \alpha_i \neq 0$. Then, $t(\alpha \alpha_j) = 0$ for $j = 1, 2, \ldots, n$, and since $\alpha_1, \ldots, \alpha_n$ is a basis it follows that $t(\alpha \beta) = 0$ for all $\beta \in L$. This implies t is identically zero which it is not since L/K is separable. This establishes the second assertion. $\qquad \square$

Proposition 12.1.2. *Suppose $\alpha_1, \ldots, \alpha_n$ and $\beta_1, \ldots, \beta_n$ are bases for L/K. Let $\alpha_i = \sum_j a_{ij} \beta_j$, $a_{ij} \in K$. Then $\Delta(\alpha_1, \ldots, \alpha_n) = \det(a_{ij})^2 \Delta(\beta_1, \ldots, \beta_n)$.*

PROOF. Take the trace of both sides of the identity $\alpha_i \alpha_k = \sum_j \sum_l a_{ij} a_{kl} \beta_j \beta_l$. Let $A = (t(\alpha_i \alpha_j))$, $B = (t(\beta_j \beta_l))$, and $C = (a_{ij})$. Then we find the matrix identity, $A = C'BC$, where C' is the transpose of C. Taking the determinant of both sides of this matrix identity and noting that $\det C = \det C'$ gives the result.

Proposition 12.1.3. *For $\alpha_1, \alpha_2, \ldots, \alpha_n \in L$ and L/K separable we have*

$$\Delta(\alpha_1, \ldots, \alpha_n) = \det(\alpha_i^{(j)})^2.$$

PROOF. $t(\alpha_i \alpha_j) = \alpha_i^{(1)} \alpha_j^{(1)} + \alpha_i^{(2)} \alpha_j^{(2)} + \cdots + \alpha_i^{(n)} \alpha_j^{(n)}$. Let $A = (t(\alpha_i \alpha_j))$ and $B = (\alpha_i^{(k)})$. Then $A = BB'$. Taking determinants of both sides of this matrix equation gives the result. $\qquad\square$

Proposition 12.1.4. *Suppose* $1, \beta, \ldots, \beta^{n-1}$ *are in* L *and linearly independent over* K. *Let* $f(x) \in K[x]$ *be the minimal polynomial for* β *over* K. *If* L/K *is separable then*

$$\Delta(1, \beta, \ldots, \beta^{n-1}) = (-1)^{(n(n-1))/2} N(f'(\beta))$$

where $f'(x)$ *is the formal derivative of* $f(x)$.

PROOF. The matrix $((\beta^{(j)})^i)$ where $j = 1, \ldots, n$ and $i = 0, \ldots, n - 1$ is of Vandermonde type and so its determinant is

$$\prod_{i < j} (\beta^{(j)} - \beta^{(i)}).$$

Thus we have

$$\Delta(1, \beta, \ldots, \beta^{n-1}) = (-1)^{(n(n-1))/2} \prod_{i \neq j} (\beta^{(j)} - \beta^{(i)}).$$

Now, $f(x) = \prod_i (x - \beta^{(i)})$, so $f'(\beta^{(j)}) = \prod_i (\beta^{(j)} - \beta^{(i)})$ with $i \neq j$. Since $f'(\beta^{(j)}) = (f'(\beta))^{(j)}$ the result follows by taking the product over j. $\qquad\square$

§2 Unique Factorization in Algebraic Number Fields

Elementary number theory is concerned with the properties of the natural numbers $1, 2, 3, \ldots$. In the course of studying these properties it became necessary to take into account the ring of integers $\mathbb{Z}$ and then the field of rational numbers $\mathbb{Q}$. In his attempt to understand biquadratic reciprocity Gauss introduced the ring $\mathbb{Z}[i]$. Likewise to study higher reciprocity laws and Fermat's Last Theorem (see Chapter 14) other rings were introduced. Eventually a general definition of an algebraic number fields and rings of algebraic integers emerged, principally through the efforts of E. Kummer and R. Dedekind.

Definition. A subfield F of the complex numbers is called an algebraic number field if $[F : \mathbb{Q}]$ is finite. If F is such a field, the subset of F consisting of algebraic integers forms a ring D, called the ring of algebraic integers in F.

Proposition 6.1.2 shows that an algebraic number field consists of algebraic numbers (just take $V = F$ and choose $\gamma_1, \ldots, \gamma_n$ to be a basis for F over $\mathbb{Q}$).

Let Ω be the set of all algebraic integers. Then Proposition 6.1.5 shows Ω

is a ring. Since $D = \Omega \cap F$. D is also a ring. We will often refer to D simply as the ring of integers in F.

It turns out that in general D is not a unique factorization domain (Exercise 7). However D does have a property which is almost as good. Namely, every nonzero ideal can be written uniquely as a product of prime ideals. An integral domain with this property is called a Dedekind ring. In this section we will prove that D is a Dedekind ring following a method due to A. Hurwitz [154] (pp. 236–243).

Throughout the discussion the word ideal will mean nonzero ideal. Hopefully, this will not cause confusion.

Lemma 1. *Suppose $\beta \in F$. There is a $b \in \mathbb{Z}$, $b \neq 0$, such that $b\beta \in D$.*

PROOF. β satisfies an equation $a_0\beta^n + a_1\beta^{n-1} + \cdots + a_n = 0$ with the $a_i \in \mathbb{Z}$, $a_0 \neq 0$. Multiply both sides by a_0^{n-1} and notice that $(a_0\beta)^n + a_1(a_0\beta)^{n-1} + \cdots + a_n a_0^{n-1} = 0$. This shows $a_0\beta$ is an algebraic integer since for all i, $a_i a_0^{i-1} \in \mathbb{Z}$. $\qquad\square$

Proposition 12.2.1. *Every ideal A of D contains a basis for F over $\mathbb{Q}$.*

PROOF. Let $\beta_1, \ldots, \beta_n$ be a basis for F over $\mathbb{Q}$. By the preceding lemma there is a $b \in \mathbb{Z}$, $b \neq 0$, such that $b\beta_1, \ldots, b\beta_n \in D$. Choose $\alpha \in A$, $\alpha \neq 0$. Then the elements $b\beta_1\alpha, \ldots, b\beta_n\alpha$ are in A and are a basis for F over $\mathbb{Q}$. $\qquad\square$

In the first section we considered a field extension L/K and considered the trace, norm, and discriminant of a basis. Here we fix the extension $F/\mathbb{Q}$ and consider all these concepts with respect to this extension.

If $\alpha \in D$ we claim $N(\alpha)$ and $t(\alpha)$ are in $\mathbb{Z}$. To see this notice that if α satisfies a monic polynomial with coefficients in $\mathbb{Z}$ so do the conjugates of α. Thus $N(\alpha)$ and $t(\alpha)$ which are respectively the product and sum of the conjugates of α are algebraic integers. They are also in $\mathbb{Q}$ so by Proposition 6.1.1 they are in $\mathbb{Z}$. The fact that the trace has this property shows that if $\alpha_1, \ldots, \alpha_n$ is a basis for F over $\mathbb{Q}$ and all the $\alpha_i \in D$ then $\Delta(\alpha_1, \ldots, \alpha_n) \in \mathbb{Z}$.

Before proceeding we remark that the discriminant of a basis can be negative. For example, let $i = \sqrt{-1}$ and consider the basis $1, i$ for $\mathbb{Q}(i)/\mathbb{Q}$. A simple calculation shows $\Delta(1, i) = -4$.

Proposition 12.2.2. *Let A be an ideal in D and suppose $\alpha_1, \ldots, \alpha_n \in A$ is a basis for $F/\mathbb{Q}$ with $|\Delta(\alpha_1, \ldots, \alpha_n)|$ minimal. Then $A = \mathbb{Z}\alpha_1 + \mathbb{Z}\alpha_2 + \cdots + \mathbb{Z}\alpha_n$.*

PROOF. Since the absolute value of the discriminant of a basis in A is a positive integer, there is such a basis with $|\Delta(\alpha_1, \ldots, \alpha_n)|$ minimal.

Suppose $\alpha \in A$ and write $\alpha = \gamma_1\alpha_1 + \gamma_2\alpha_2 + \cdots + \gamma_n\alpha_n$ with $\gamma_i \in \mathbb{Q}$. We need to show that the γ_i are in $\mathbb{Z}$. Suppose not. Then some $\gamma_i \notin \mathbb{Z}$ and by relabeling if necessary we can assume $\gamma_1 \notin \mathbb{Z}$. Write $\gamma_1 = m + \theta$ where $m \in \mathbb{Z}$ and $0 < \theta < 1$. Let $\beta_1 = \alpha - m\alpha_1$, $\beta_2 = \alpha_2, \ldots, \beta_n = \alpha_n$. Then $\beta_1, \beta_2, \ldots,$

$\beta_n \in A$ and is a basis for $F/\mathbb{Q}$. Since $\beta_1 = \theta\alpha_1 + \gamma_2\alpha_2 + \cdots + \gamma_n\alpha_n$ the matrix of transition between these two bases is

$$\begin{pmatrix} \theta & \gamma_2 & \gamma_3 & \cdots & \gamma_n \\ 0 & 1 & 0 & \cdots & 0 \\ 0 & 0 & 1 & \cdots & 0 \\ 0 & 0 & 0 & \cdots & 1 \end{pmatrix}$$

By Proposition 12.1.2 we find $\Delta(\beta_1, \ldots, \beta_n) = \theta^2 \Delta(\alpha_1, \ldots, \alpha_n)$ which contradicts the minimality of $|\Delta(\alpha_1, \ldots, \alpha_n)|$ since $0 < \theta < 1$. Thus all the $\gamma_i \in \mathbb{Z}$ and $A = \mathbb{Z}\alpha_1 + \cdots + \mathbb{Z}\alpha_n$ as asserted. $\qquad\square$

If $\alpha_1, \alpha_2, \ldots, \alpha_n \in A$ is a basis for F over $\mathbb{Q}$ and $A = \mathbb{Z}\alpha_1 + \cdots + \mathbb{Z}\alpha_n$ we say that $\alpha_1, \ldots, \alpha_n$ is an integral basis for A. It follows from Proposition 12.1.2 that the discriminants of any two integral bases for A are equal. This common value is called the discriminant of A, written $\Delta(A)$. The discriminant of D is particularly important and, by "abuse of language," $\delta_F = \Delta(D)$ is called the discriminant of $F/\mathbb{Q}$.

We now apply the last proposition to deduce some important properties of the ring D. Recall our convention that all ideals are nonzero ideals.

Lemma 2. *If $A \subset D$ is an ideal then $A \cap \mathbb{Z} \neq 0$.*

PROOF. Let $\alpha \in A$, $\alpha \neq 0$. There exist $a_i \in \mathbb{Z}$ such that $\alpha^m + a_1\alpha^{m-1} + \cdots + a_m = 0$. Since we are working in a field we may assume $a_m \neq 0$. But then, $0 \neq a_m \in A \cap \mathbb{Z}$. $\qquad\square$

Proposition 12.2.3. *For any ideal A, D/A is finite.*

PROOF. By the lemma there is an $a \in A \cap \mathbb{Z}$, $a \neq 0$. Let (a) be the principal ideal generated by a in D. Since $D/(a)$ maps onto D/A it is enough to show $D/(a)$ is finite. In fact we will show it has precisely a^n elements.

By Proposition 12.2.2 we may write $D = \mathbb{Z}\omega_1 + \mathbb{Z}\omega_2 + \cdots + \mathbb{Z}\omega_n$. Let $S = \{\sum \gamma_i\omega_i | 0 \leq \gamma_i < a\}$. We claim S is a set of coset representatives for $D/(a)$. Suppose $\omega = \sum m_i\omega_i \in D$. Write $m_i = q_i a + \gamma_i$ with $0 \leq \gamma_i < a$. Then clearly $\omega \equiv \sum \gamma_i\omega_i$ (a). Thus every coset of A contains an element of S. If $\sum \gamma_i\omega_i$ and $\sum \gamma_i'\omega_i$ are in S and in the same coset modulo (a) then using the linear independence of the ω_i we see $\gamma_i - \gamma_i'$ is divisible by a in $\mathbb{Z}$. Since $0 \leq \gamma_i, \gamma_i' < a$ it follows that $\gamma_i = \gamma_i'$. Thus S is a set of coset representatives and $D/(a)$ has a^n elements as claimed. $\qquad\square$

Corollary 1. *D is a Noetherian ring, i.e., every ascending chain of ideals $A_1 \subset A_2 \subset A_3 \subset \cdots$ terminates. In other words, there is an $N > 0$ such that $A_m = A_{m+1}$ for all $m \geq N$.*

PROOF. Since D/A_1 is finite there are only finitely many ideals containing A_1. $\qquad\square$

Corollary 2. *Every prime ideal of D is maximal.*

PROOF. If P is a prime ideal then D/P is a finite integral domain. Such a ring is necessarily a field (see Exercise 19). Thus D/P is a field and so P is maximal. □

The ring D is also integrally closed. This means that if $\alpha \in F$ satisfies a monic polynomial with coefficients in D then $\alpha \in D$. This is not too hard to establish using Proposition 6.1.4. In standard algebra texts it is shown that if an integral domain is Noetherian, integrally closed, and every nonzero prime ideal is maximal then every ideal is a product of prime ideals in a unique way, i.e., such a ring is a Dedekind domain. We will establish the fact that D is a Dedekind domain in a different way using a very important property of number fields, namely that the class number of D is finite (see below).

Our initial goal is to prove the following two results:

(i) If A, B, and C are ideals and $AB = AC$, then $B = C$.
(ii) If A and B are ideals and $A \subset B$, then there is an ideal C such that $A = BC$.

These will be proved later. We begin by establishing a special case of (i).

Lemma 3. *Let $A \subset D$ be an ideal. If $\beta \in F$ is such that $\beta A \subset A$ then $\beta \in D$.*

PROOF. By Proposition 12.2.2 A is a finitely generated $\mathbb{Z}$ module so the result follows from Proposition 6.1.4. □

Lemma 4. *If A and B are ideals in D and $A = AB$ then $B = D$.*

PROOF. Let $\alpha_1, \alpha_2, \ldots, \alpha_n$ be an integral basis for A. Since $A = AB$ we can find elements $b_{ij} \in B$ such that $\alpha_i = \sum_j b_{ij}\alpha_j$. It follows that the determinant of the matrix $(b_{ij} - \delta_{ij})$ is zero. Writing this out shows $1 \in B$, i.e., $B = D$. □

Proposition 12.2.4. *Let A, $B \subset D$ be ideals and suppose $\omega \in D$ is such that $(\omega)A = BA$. Then $(\omega) = B$.*

PROOF. If $\beta \in B$ we see $(\beta/\omega)A \subset A$ so by Lemma 3, $\beta/\omega \in D$. It follows that $B \subset (\omega)$ and so $\omega^{-1}B \subset D$ is an ideal. Since $A = \omega^{-1}BA$, Lemma 4 shows $\omega^{-1}B = D$ and so $B = (\omega)$ as required. □

The following definition plays a major role in algebraic number theory.

Definition. Two ideals A, $B \subset D$ are said to be equivalent, $A \sim B$, if there exist nonzero α, $\beta \in D$ such that $(\alpha)A = (\beta)B$. This is an equivalence relation. The equivalence classes are called ideal classes. The number of ideal classes, h_F, is called the class number of F. (We will see that h_F is finite.)

We leave the easy verification that $A \sim B$ is an equivalence relation to the reader.

It is worthwhile to point out that $h_F = 1$ if and only if D is a principal ideal domain (PID). To see this suppose $h_F = 1$ and let A be an ideal. Since $A \sim D$ there are nonzero $\alpha, \beta \in D$ such that $(\alpha)A = (\beta)D = (\beta)$. Thus $\beta/\alpha \in A$ and $A = (\beta/\alpha)$. Every ideal is principal. On the other hand it is obvious that if D is a PID then $h_F = 1$.

Thus we see that the class number measures, in some sense, how far D is from being a PID (see Exercises 15, 16 and Masley [184]).

The following lemma is due to A. Hurwitz [154], p. 237. We will use it to show h_F is finite. It is to be noticed that the lemma is a (weak) generalization of the Euclidean algorithm to an arbitrary number field.

Lemma 5. *There exists a positive integer M depending only on F with the following property. Given $\alpha, \beta \in D$, $\beta \neq 0$, there is an integer t, $1 \leq t \leq M$, and an element $\omega \in D$ such that $|N(t\alpha - \omega\beta)| < |N(\beta)|$.*

PROOF. We first reformulate the statement slightly. Let $\gamma = \alpha/\beta \in F$. Then it is sufficient to show that for all $\gamma \in F$ there is an M such that $|N(t\gamma - \omega)| < 1$ for some $1 \leq t \leq M$ and $\omega \in D$.

Let $\omega_1, \omega_2, \ldots, \omega_n$ be an integral basis for D. For $\gamma \in F$, $\gamma = \sum_{i=1}^{n} \gamma_i \omega_i$ with $\gamma_i \in \mathbb{Q}$. Notice that

$$|N(\gamma)| = \left| \prod_j \left(\sum_i \gamma_i \omega_i^{(j)} \right) \right| \leq C \left(\max_i |\gamma_i| \right)^n,$$

where $C = \prod_j (\sum_i |\omega_i^{(j)}|)$. Choose an integer $m > \sqrt[n]{C}$ and set $M = m^n$.

For $\gamma \in F$, $\gamma = \sum_{i=1}^{n} \gamma_i \omega_i$, write $\gamma_i = a_i + b_i$ where $a_i \in \mathbb{Z}$ and $0 \leq b_i < 1$. Let $[\gamma] = \sum_{i=1}^{n} a_i \omega_i$ and $\{\gamma\} = \sum_{i=1}^{n} b_i \omega_i$. Then $\gamma = [\gamma] + \{\gamma\}$ where $[\gamma] \in D$ and $\{\gamma\}$ has coordinates between 0 and 1.

Map F to Euclidean n-space $\mathbb{R}^n$ by $\phi(\sum_{i=1}^{n} \gamma_i \omega_i) = (\gamma_1, \gamma_2, \ldots, \gamma_n)$. For any $\gamma \in F$, $\phi(\{\gamma\})$ lies in the unit cube. Partition the unit cube into m^n subcubes of side $1/m$. Consider the points $\phi(\{k\gamma\})$ for $1 \leq k \leq m^n + 1$. By the pigeonhole principle two of them, at least, must lie in the same subcube, say those corresponding to $h\gamma$ and $l\gamma$. If we write $h\gamma = [h\gamma] + \{h\gamma\}$ and $l\gamma = [l\gamma] + \{l\gamma\}$ and subtract we find $t\gamma = \omega + \delta$ where (assuming $h > l$) $t = h - l \leq m^n = M$, $\omega \in D$, and the coordinates of δ have absolute value less than or equal to $1/m$.

By our previous remark, $N(\delta) \leq C(1/m)^n = C/m^n < 1$. □

Theorem 1. *The class number of F is finite.*

PROOF. Let A be an ideal in D. For $\alpha \in A$, $\alpha \neq 0$, $|N(\alpha)|$ is a positive integer. Choose $\beta \in A$, $\beta \neq 0$, so that $|N(\beta)|$ is minimal. For any $\alpha \in A$ there is a t, $1 \leq t \leq M$, such that $|N(t\alpha - \omega\beta)| < |N(\beta)|$ with $\omega \in D$. Since $t\alpha - \omega\beta \in A$ we must have $t\alpha - \omega\beta = 0$. It follows that $M! A \subset (\beta)$. Let $B = (1/\beta)M! A \subset D$. B is an ideal and $M! A = (\beta)B$. Since $\beta \in A$, $M! \beta \in (\beta)B$ and so $M! \in B$. By Proposition 12.2.3 $M!$ can be contained in at most finitely many ideals. We have shown $A \sim B$ where B is one of at most finitely many ideals. Thus h_F is finite, as asserted. $\qquad\square$

An interesting and significant application of this theorem is the following proposition.

Proposition 12.2.5. *For any ideal $A \subset D$ there is an integer k, $1 \le k \le h_F$, such that A^k is principal.*

PROOF. Consider the set of ideals $\{A^i | 1 \le i \le h_F + 1\}$. At least two of these ideals must lie in the same class, say $A^i \sim A^j$ with $i < j$. There exist $\alpha, \beta \in D$ such that $(\alpha)A^i = (\beta)A^j$. Let $k = j - i$ and $B = A^k$. We will show that B is principal.

Since, clearly, $(\alpha)A^i = (\beta)BA^i$ we see $(\alpha/\beta)A^i \subset A^i$ so $\alpha/\beta \in D$. Let $\omega = \alpha/\beta$. Then $(\omega)A^i = BA^i$. By Proposition 12.2.4, $(\omega) = B$. $\qquad\square$

We remark that the set of ideal classes can be made into a group. Let $\overline{A}$ denote the class of A. We define the product of $\overline{A}$ and $\overline{B}$ to be $\overline{AB}$. One can check without trouble that this is well defined, i.e., if $\overline{A} = \overline{A}_1$ and $\overline{B} = \overline{B}_1$ then $\overline{AB} = \overline{A_1 B_1}$. Associativity follows from the fact that ideal multiplication is associative. The class of D serves as an identity element. Finally, the last proposition shows that an inverse to $\overline{A}$ is the class $\overline{A^{k-1}}$. The structure of the class group has been a major research problem ever since the concept was invented.

One consequence of the fact that the ideal classes form a group is that A^{h_F} is principal for all ideals A. This will not be needed in the remainder of this chapter.

We can now give proofs for the two results mentioned earlier (before Lemma 3).

Proposition 12.2.6. *If A, B, and C are ideals, and $AB = AC$, then $B = C$.*

PROOF. By the last proposition, there is a $k > 0$ such that $A^k = (\alpha)$. Multiply $AB = AC$ on both sides by A^{k-1}. We find $(\alpha)B = (\alpha)C$. It follows that $B = C$. $\qquad\square$

Proposition 12.2.7. *If A and B are ideals, such that $B \supset A$, then there is an ideal C such that $A = BC$.*

PROOF. As above there is a $k > 0$ such that $B^k = (\beta)$.

Now, since $A \subset B$ we have $B^{k-1}A \subset B^k = (\beta)$ so $C = (1/\beta)B^{k-1}A \subset D$ is an ideal.

Thus, $BC = (1/\beta)B^k A = (1/\beta)(\beta)A = A$. $\qquad\square$

This proposition can be phrased "to contain is to divide."

We now have all the tools we need to establish unique factorization into prime ideals.

Proposition 12.2.8. *Every ideal in D can be written as a product of prime ideals.*

PROOF. Let A be a proper ideal. Since D/A is finite, A is contained in a maximal ideal P_1 (using Zorn's lemma one can show that in an arbitrary commutative ring with identity a proper ideal is contained in a maximal ideal). By the last proposition $A = P_1 B_1$ for some ideal B_1. If $B_1 \neq D$ then B_1 is contained in a maximal ideal P_2 and so $A = P_1 P_2 B_2$. If $B_2 \neq D$ we can continue the process. Notice that $A \subset B_1 \subset B_2 \cdots$ is a proper ascending chain of ideals. By Corollary 1 to Proposition 12.2.3 we see that in finitely many steps $B_t = D$. Thus $A = P_1 P_2 \cdots P_t$. $\square$

Let P be a prime ideal. The descending chain $P \supset P^2 \supset P^3 \cdots$ is proper since if $P^i = P^{i+1}$ for some i then $PP^i = P^i$ and so $P = D$ by Lemma 4. This fact is the basis of the following definition.

Definition. Let P be a prime ideal and A an ideal. Then $\text{ord}_P A$ is defined to be the unique nonnegative integer t such that $P^t \supset A$ and $P^{t+1} \not\supset A$.

Proposition 12.2.9. *Let P be a prime ideal and A and B ideals. Then*

(i) $\text{ord}_P P = 1$
(ii) *If* $P' \neq P$ *is prime* $\text{ord}_P P' = 0$
(iii) $\text{ord}_P AB = \text{ord}_P A + \text{ord}_P B$

PROOF. The first assertion is clear. As for (ii) assume $\text{ord}_P P' > 0$. Then $P \supset P'$. Since prime ideals are maximal $P = P'$ contradicting the assumption.

Let $t = \text{ord}_P A$ and $s = \text{ord}_P B$. By Proposition 12.2.7 we have $A = P^t A_1$ and $B = P^s B_1$. By the same proposition we must have $P \not\supset A_1$ and $P \not\supset B_1$.

Now, $AB = P^{s+t} A_1 B_1$. If $P^{s+t+1} \supset AB$ then $AB = P^{s+t+1} C$ and so by Proposition 12.2.6, $PC = A_1 B_1$. This implies $P \supset A_1 B_1$ and since P is prime that $P \supset A_1$ or $P \supset B_1$. This is a contradiction.

Thus $\text{ord}_P AB = t + s = \text{ord}_P A + \text{ord}_P B$. $\square$

Theorem 2. *Let* $A \subset D$ *be an ideal. Then* $A = \prod P^{a(P)}$ *where the product is over the distinct prime ideals of D, and the a(P) are nonnegative integers all but finitely many of which are zero. Finally, the integers a(P) are uniquely determined by* $a(P) = \text{ord}_P A$.

PROOF. The product representation follows from Proposition 12.2.8.

Let P_0 be a prime ideal and apply ord_{P_0} to both sides of the product given in the theorem. Using Proposition 12.2.9 we see

$$\text{ord}_{P_0} A = \sum_P a(P) \, \text{ord}_{P_0}(P) = a(P_0).$$ $\square$

§3 Ramification and Degree

Let P be a prime ideal of D. By Lemma 2, $P \cap \mathbb{Z}$ is not zero. Since it is clearly a prime ideal of $\mathbb{Z}$ it must be generated by a prime number p.

Definition. The number $e = \mathrm{ord}_P(p)$ is called the ramification index of P (here (p) is the principal ideal generated by p in D).

D/P is a finite field containing $\mathbb{Z}/p\mathbb{Z}$. Thus the number of elements in D/P is of the form p^f for some $f \geq 1$. The number f is called the degree of P.

Let $p \in \mathbb{Z}$ be a prime number and let $P_1, P_2, \ldots, P_g$ be the primes in D containing (p). Let e_i and f_i be ramification index and degree of P_i. By Theorem 2, $(p) = P_1^{e_1} P_2^{e_2} \cdots P_g^{e_g}$.

There exists a remarkable relation among the numbers e_i, f_i, and n.

Theorem 3. $\sum_{i=1}^{g} e_i f_i = n$.

We postpone the proof until we have developed some necessary background.

Proposition 12.3.1. *Let R be a commutative ring with identity. Suppose A_1, $A_2, \ldots, A_g$ are ideals such that $A_i + A_j = R$ for $i \neq j$. Let $A = A_1 A_2 \cdots A_g$. Then*

$$R/A \approx R/A_1 \oplus R/A_2 \oplus \cdots \oplus R/A_g.$$

PROOF. Let ψ_i be the natural map from R to R/A_i and define $\psi : R \to R/A_1 \oplus \cdots \oplus R/A_g$ by $\psi(\gamma) = (\psi_1(\gamma), \psi_2(\gamma), \ldots, \psi_g(\gamma))$. We will show ψ is onto and the kernel is A.

To show ψ is onto, it is sufficient to show that for any $\gamma_1, \gamma_2, \ldots, \gamma_g \in R$ the set of simultaneous congruences $x \equiv \gamma_i (A_i)$, $i = 1, \ldots, g$ is solvable.

Expanding the product $(A_1 + A_2)(A_1 + A_3) \cdots (A_1 + A_g) = R$ we see that all the summands, except the last, are in A_1. Thus $A_1 + A_2 A_3 \cdots A_g = R$. There exist elements $v_1 \in A_1$ and $u_1 \in A_2 \cdots A_g$ such that $u_1 + v_1 = 1$. Then $u_1 \equiv 1 (A_1)$ and $u_1 \equiv 0 (A_i)$ for $i \neq 1$. Similarly, for each j there is a u_j such that $u_j \equiv 1 (A_j)$ and $u_j \equiv 0 (A_i)$ for $i \neq j$. It is then clear that $x = \gamma_1 u_1 + \gamma_2 u_2 + \cdots + \gamma_g u_g$ is a solution to our set of congruences.

Having shown that ψ is onto, we now investigate the kernel. Clearly, $\ker \psi = A_1 \cap A_2 \cap \cdots \cap A_g$. We must show that under the hypotheses the intersection is equal to the product. This can be done by induction on g. Suppose $g = 2$. Then, since $A_1 + A_2 = R$, there exist $a_1 \in A_1$ and $a_2 \in A_2$ such that $a_1 + a_2 = 1$. If $a \in A_1 \cap A_2$ then $a = aa_1 + aa_2 \in A_1 A_2$. This shows $A_1 \cap A_2 \subset A_1 A_2$. The reverse inclusion is obvious so the result follows for

$g = 2$. Now suppose $g > 2$ and we know the result for $g - 1$. Then $A_1 \cap A_2 \cap \cdots \cap A_g = A_1 \cap A_2 A_3 \cdots A_g$. However, $A_1 + A_2 A_3 \cdots A_g = R$ by the first part of the proof. Thus, $A_1 \cap A_2 A_3 \cdots A_g = A_1 A_2 \cdots A_g$ and the proof is complete.   □

This proposition is called the Chinese Remainder Theorem for rings. We return from a general commutative ring R to D.

Proposition 12.3.2. *Let $P \subset D$ be a prime ideal and let p^f be the number of elements in D/P. The number of elements in D/P^e is p^{ef}.*

PROOF. The assertion is true for $e = 1$. If $e > 1$ then D/P^e has P^{e-1}/P^e as a subgroup and the quotient is isomorphic to D/P^{e-1} (second law of isomorphism). If we can show P^{e-1}/P^e has p^f elements then the result will follow by induction.

Since $P^e \subset P^{e-1}$ properly we can find an $\alpha \in P^{e-1}$ such that $\alpha \notin P^e$. We claim $(\alpha) + P^e = P^{e-1}$. Since $P^e \subset (\alpha) + P^e$ the latter ideal must be a power of P. Since $(\alpha) + P^e \subset P^{e-1}$ we must have $(\alpha) + P^e = P^{e-1}$.

Map D to P^{e-1}/P^e by $\gamma \to \gamma\alpha + P^e$. This is easily seen to be a homomorphism onto. An element γ is in the kernel if and only if $\gamma\alpha \in P^e$, i.e., iff $\operatorname{ord}_P(\gamma\alpha) \geq e$. Now, $\operatorname{ord}_P(\gamma\alpha) = \operatorname{ord}_P(\gamma) + \operatorname{ord}_P \alpha = \operatorname{ord}_P(\gamma) + e - 1$. Thus γ is in the kernel iff $\operatorname{ord}_P(\gamma) \geq 1$ which is equivalent to saying $\gamma \in P$. Thus $D/P \approx P^{e-1}/P^e$ and so the latter group has p^f elements.   □

We can now prove Theorem 3. Remember $(p) = P_1^{e_1} P_2^{e_2} \cdots P_g^{e_g}$. It is not hard to see that $P_i^{e_i} + P_j^{e_j} = D$ for $i \neq j$ (see Exercise 25). By Proposition 12.3.1

$$D/(p) \approx D/P_1^{e_1} \oplus D/P_2^{e_2} \oplus \cdots \oplus D/P_g^{e_g}.$$

The proof of Proposition 12.2.3 shows $|D/(p)| = p^n$. On the other hand Proposition 12.3.2 shows $|D/P_i^{e_i}|$ has $p^{e_i f_i}$ elements. Thus

$$p^n = p^{e_1 f_1} p^{e_2 f_2} \cdots p^{e_g f_g}.$$

It follows that $n = e_1 f_1 + e_2 f_2 + \cdots + e_g f_g$ as asserted.   □

When $F/\mathbb{Q}$ is a Galois, that is, when all the isomorphisms of F into $\mathbb{C}$ are actually automorphisms, Theorem 3 can be strengthened. Suppose $F/\mathbb{Q}$ is Galois and let G be the Galois group. If A is an ideal and $\sigma \in G$ let $\sigma A = \{\sigma\alpha | \alpha \in A\}$. One easily checks that σA is again an ideal. Also, $\sigma D = D$. Thus $D/\sigma A = \sigma D/\sigma A \approx D/A$. In particular this shows that if P is a prime ideal, then σP is also a prime ideal.

Proposition 12.3.3. *Let $p \in \mathbb{Z}$ be a prime number. Suppose P_i and P_j are prime ideals of D containing p. Then there is a $\sigma \in G$ such that $\sigma P_i = P_j$.*

PROOF. Suppose there is a prime ideal P_0 containing p and not in the set $\{\sigma P_i | \sigma \in G\}$. By Proposition 12.3.1 we can find an $\alpha \in D$ such that $\alpha \equiv 0 \ (P_0)$ and $\alpha \equiv 1 \ (\sigma P_i)$ for all $\sigma \in G$.

Then $N(\alpha) = \prod_{\sigma \in G} \sigma \alpha \in P_0 \cap \mathbb{Z} = p\mathbb{Z}$. It follows that $N(\alpha) \in P_i$ and so $\sigma \alpha \in P_i$ for some σ since P_i is prime. But then $\alpha \in \sigma^{-1} P_i$ contradicting $\alpha \equiv 1 \ (\sigma^{-1} P_i)$. $\qquad \square$

Theorem 3'. *Suppose $F/\mathbb{Q}$ is a Galois extension. Let $p \in \mathbb{Z}$ be a prime number and write $(p) = P_1^{e_1} P_2^{e_2} \cdots P_g^{e_g}$. Then $e_1 = e_2 = \cdots = e_g$ and $f_1 = f_2 = \cdots = f_g$. If e and f denote these common values, then $efg = n$.*

PROOF. For a given index i there is a $\sigma \in G$ such that $\sigma P_1 = P_i$. Since $D/P_1 \approx D/\sigma P_1 = D/P_i$ we find $f_1 = f_i$. Thus all the f_i's are equal.

Apply σ to both sides of $(p) = P_1^{e_1} P_2^{e_2} \cdots P_g^{e_g}$. Since $p \in \mathbb{Z}$ it is clear that $\sigma(p) = (p)$. Thus

$$(p) = (\sigma P_1)^{e_1} (\sigma P_2)^{e_2} \cdots (\sigma P_g)^{e_g}.$$

In this product we see the exponent of $P_i = \sigma P_1$ is e_1. In the first expression the exponent of P_i is e_i. By uniqueness of prime factorization we must have $e_1 = e_i$ and so all the e_i's are equal.

Finally, since $\sum e_i f_i = n$ we see immediately that $efg = n$. $\qquad \square$

We conclude this section by discussing, without proofs, some important facts about number fields. In our applications we will be able to do without this general theory.

Let $P \subset D$ be a prime ideal with ramification index e. Let $P \cap \mathbb{Z} = p\mathbb{Z}$. We say that P is a ramified prime if $e > 1$. One can show that P is ramified only if p divides $\delta_F = \Delta(D)$, the discriminant of F. In particular, only finitely many primes are ramified. If $p \nmid \delta_F$ then (p) is a product of distinct prime ideals in D. An important result of Minkowski asserts that if $[F : \mathbb{Q}] > 1$ then $|\delta_F| > 1$. In fact Minkowski found a more precise result, namely an explicit lower bound for $|\delta_F|$. An important consequence is that every number field strictly bigger than $\mathbb{Q}$ contains ramified primes.

Now suppose $F/\mathbb{Q}$ is a Galois extension with group G. Associate with a prime ideal P the group $G(P) = \{\sigma \in G | \sigma P = P\}$. $G(P)$ is called the decomposition group of P. D/P is a finite field containing $\mathbb{Z}/p\mathbb{Z}$. The field D/P is a Galois extension of $\mathbb{Z}/p\mathbb{Z}$. Call the Galois group $\bar{G}$. There is a homomorphism from $G(P)$ to $\bar{G}$ given as follows. If $\sigma \in G(P)$ and $\bar{\alpha}$ denotes the residue class of α in D/P define $\bar{\sigma}$ by the equation $\bar{\sigma}(\bar{\alpha}) = \overline{\sigma \alpha}$. This is well defined, $\bar{\sigma} \in \bar{G}$, and $\sigma \to \bar{\sigma}$ is a homomorphism. One can show this homomorphism is onto (Exercise 26). Let $T(P)$ be the kernel. $T(P)$ is called the inertia group of P. We have

$$G(P)/T(P) \approx \bar{G}.$$

It is not hard to see that $|\bar{G}| = f$ and $|G(P)| = n/g = ef$. It follows that $|T(P)| = e$. Thus, if P is unramified $G(P) \approx \bar{G}$.

From the theory of finite fields $\bar{G}$ is a cyclic group generated by the automorphism ϕ_p which takes $\bar{\alpha}$ to $\bar{\alpha}^p$. If P is unramified there is a unique $\sigma_P \in G(P)$ such that $\bar{\sigma}_P = \phi_p$. This automorphism σ_P is called the Frobenius automorphism associated to P. Notice that the order of σ_P is equal to the order of ϕ_P which is f, the degree of P. As it turns out, a large part of the arithmetic theory of algebraic number fields centers around the properties of the Frobenius automorphism. We will see illustrations of this in the next chapter.

NOTES

The fact that the ring of integers in an algebraic number field form a Dedekind ring is due to R. Dedekind and appears in the eleventh supplement to Dirichlet's *Vorlesungen uber Zahlentheorie* [127]. This result was subsequently also proven by Kronecker, Hilbert, and Hurwitz. The inertia and decomposition groups were introduced by Hilbert (1894) in his "Grundzüge einer Theorie des Galoisschen Zahlkörpers" (see also §39 of Hilbert's "Zahlbericht" [151] and Dedekind [121], Vol. 2, pp. 43–49).

It can be shown more generally that if D is a Dedekind ring with field of fractions k and K is a finite separable extension of k the integral closure of D in K (Exercise 27) is a Dedekind ring. This follows from a theorem of E. Noether characterizing Dedekind rings as Noetherian domains which are integrally closed and in which every nonzero prime ideal is maximal. For this approach see Samuel–Zariski [214]. In our approach, as in other classical approaches, essential use is made of the fact that the residue class ring modulo a nonzero ideal is finite. The idea of deriving the Dedekind property from the finiteness of the class number is due to Hurwitz. It will be noticed that in our approach no use is made of the fact that the number of elements in the residue class ring is a multiplicative function of the ideal. Butts and Wade [103] have shown that the multiplicativity of this map implies the Dedekind property. The usual classical approach is to show by a suitable generalization of Gauss' lemma (Exercise 4, Chapter 6) that the ideal classes form a group.

Recently the characterization of fields F with class number 2 due to Carlitz (see Exercises 15 and 16) has been generalized by A. Czogala [117]. He proves, among other things, that a number field has an ideal class group which is cyclic of order 2, cyclic of order 3, or the Klein four group iff the product of two irreducibles may be rewritten as the product of at most three other irreducibles.

A deep result conjectured by Hilbert and proved by Furtwängler asserts the existence, for each number field F, of an extension E satisfying the following conditions. First of all the degree of E over F is equal to the class number of F. Every prime ideal $\mathfrak{P}$ of F decomposes into the product of h_F/f distinct prime ideals in E where f is the order of the ideal class of $\mathfrak{P}$ in the class group. Every ideal of F becomes principal in E. Finally the ideal class group of F is isomorphic to the Galois group of E over F. The field E is unique and is called the Hilbert class field of F. The existence of the Hilbert class field is a valuable tool in studying the structure of the ideal class group.

The actual calculation of the class number is a difficult matter. Even for quadratic number fields of small discriminant the calculation requires estimates (due to Minkowski) which we have omitted. These matters are discussed in most standard texts on algebraic number theory. We recommend the treatment in D. Marcus [183]. This book contains a large number of interesting exercises.

In more recent texts it is customary to describe the ideal class group in terms of fractional ideals. If D is an integral domain with field of fractions F, a fractional ideal A is a D submodule of F for which there exists an element d in D with $dA \subset D$. Fractional ideals can be multiplied in the obvious way. It can be shown that D is a Dedekind ring iff the (nonzero) fractional ideals form a group [214]. The subgroup of fractional ideals of the form fD with f in F are the principal fractional ideals. It is not difficult to show that the ideal class group of an algebraic number field is isomorphic to the quotient group of the group of fractional ideals by the subgroup of principal fractional ideals.

EXERCISES

1. Find the minimal polynomial for $\sqrt{3} + \sqrt{7}$.

2. Compute the discriminant of $\mathbb{Q}(\sqrt{2} + \sqrt{5})$.

3. Describe the units in $\mathbb{Q}(\sqrt{5})$.

4. Let D be the ring of integers in $\mathbb{Q}(\sqrt{d})$. Show that, given $N > 0$, there are at most finitely many integers $\alpha \in D$ with $\max(|\alpha|, |\alpha'|) \leq N$, where α' is the conjugate of α.

5. Generalize Exercise 4 to an arbitrary number field.

6. If D is the ring of integers in an algebraic number field and $\mathfrak{P}$ is a prime ideal such that $\mathfrak{P} = (\alpha)$ then show that α is irreducible.

7. Show that the class number of $\mathbb{Q}(\sqrt{-5})$ is greater than one.

8. Let F be a number field. Show that the discriminant δ_F is congruent to 0 or 1 modulo 4. This is one of Stickelberger's theorems. The proof is tricky (cf. [207], p. 97).

9. Compute the discriminant $\Delta(1, \alpha, \alpha^2)$, relative to $\mathbb{Q}(\alpha)$, where α is a root of the reducible cubic $x^3 + px + q$, $p, q \in \mathbb{Q}$.

10. If $R \subset S$ are integral domains $\alpha \in S$ is said to be integral over R if $\alpha^m + b_1\alpha^{m-1} + \cdots + b_m = 0$ for suitable m; $b_1, \ldots, b_m \in R$. S is called integral over R if every element of S is integral over R. Prove that if S is integral over R then S is a field iff R is a field.

11. Let $\alpha_1, \ldots, \alpha_n \in D$, the ring of integers in a number field F, $\Delta(\alpha_1, \ldots, \alpha_n) \neq 0$. Show that if $\Delta(\alpha_1, \ldots, \alpha_n)$ is a product of distinct primes (i.e., Δ is square free) then $\alpha_1, \ldots, \alpha_n$ is an integral basis. Conclude that if d is square free $d \equiv 1$ (4) then $(1 + \sqrt{d})/2, 1$ form an integral basis for the ring of integers in $\mathbb{Q}(\sqrt{d})$.

12. Show that $\sin(\pi/12)$ is an algebraic number.

13. Show that $(3, 1 + \sqrt{-5})$ is a proper ideal in $\mathbb{Z}[\sqrt{-5}]$. Is it prime?

14. Construct an irreducible cubic polynomial over $\mathbb{Q}$ with only real roots.

15. Let F be an algebraic number field, D its ring of integers. Suppose the class number of F is 2. Show that if π is an irreducible such that (π) is not prime then $(\pi) = \mathfrak{P}_1\mathfrak{P}_2$ where $\mathfrak{P}_1, \mathfrak{P}_2$ are (not necessarily distinct) prime ideals.

16. (L. Carlitz) Let F, D be as in Exercise 15. Show that if $\alpha \in D$, $\alpha = \pi_1, \ldots, \pi_t = \lambda_1, \ldots, \lambda_s$ are two decompositions of α into the product or irreducibles then $s = t$. [Note: The converse is also true! (cf. Carlitz [106]).]

17. Let $f(x), g(x)$ be the respective minimal polynomials of α and β of respective degrees n and m. Let the roots in $\mathbb{C}$ of $f(x)$ and $g(x)$ respectively be $\alpha = \alpha_1, \alpha_2, \ldots, \alpha_n$ and $\beta = \beta_1, \beta_2, \ldots, \beta_m$. Recall by Exercise 16, Chapter 6, there are no repeated roots. Choose $t \in \mathbb{Q}$ so that $\alpha_i + t\beta_j \neq \alpha + t\beta, j \neq 1$, all i. Put $\gamma = \alpha + t\beta$. Show that
    (a) $f(\gamma - tx), g(x)$ have greatest common divisor (in $\mathbb{C}[x]$) $x - \beta$.
    (b) (on the other hand) the greatest common divisor of $f(\gamma - tx)$ and $g(x)$ is in $\mathbb{Q}(\gamma)[x]$.
    (c) $\beta \in \mathbb{Q}(\gamma)$, $\alpha \in \mathbb{Q}(\gamma)$.

18. (Theorem on the primitive element.) If F is an algebraic number field show that there exists an element $\gamma \in F$ such that $\mathbb{Q}(\gamma) = F$.

19. Show that a finite integral domain is a field.

20. Let $K = F_2(x)$ and $L = K(\sqrt{x})$. Show that the trace map is identically zero. (Recall, F_2 is the finite field with two elements.)

21. Let F be an algebraic number field of degree n. If $\alpha \in F$, let T be the linear transformation defined by $T(y) = \alpha y$. Show that $\det(xI - T) = f(x)^t$ where $t = n/\deg(f)$, and $f(x)$ is the minimal polynomial of α.

22. Let $F \subset E$ be algebraic number fields. Show that any isomorphism of F into $\mathbb{C}$ extends in exactly $[E:F]$ ways to an isomorphism of E into $\mathbb{C}$.

23. Let F be an algebraic number field of degree n and let $\sigma_1, \ldots, \sigma_n$ be the distinct isomorphisms of F into $\mathbb{C}$. Show that, for $\alpha \in F$, the notation being as in Exercise 21, $f(x)^t = \prod_{i=1}^n (x - \sigma_i(\alpha))$.

24. The notation being as in Exercise 23 show that
$$N_{F/\mathbb{Q}}(\alpha) = \prod_{i=1}^n \sigma_i(\alpha) \quad \text{and} \quad t_{F/\mathbb{Q}}(\alpha) = \sum_{i=1}^n \sigma_i(\alpha).$$

25. Let F be an algebraic number field with ring of integers D. Show that if P and Q are distinct prime ideals then $(P^a, Q^b) = D$, where a and b are positive integers.

26. Let P be a prime ideal in the ring of integers D of an algebraic number field F. If F is Galois show that the natural map from the decomposition group of P to the Galois group of the residue class field is onto.

27. If k is a field containing a ring D the set of all elements in k which are integral over D (Exercise 10) is called the integral closure of D in k. Show that the integral closure is a ring and that it is integrally closed.

28. Let D be the ring of integers in a number field F. Suppose $(p) = P^2 A$ for p prime in $\mathbb{Z}$ and a prime ideal P. Show
    (a) There exists $\alpha \in PA$, $\alpha \notin P^2 A$.
    (b) $(\alpha\beta)^p \in pD$ all $\beta \in D$.
    (c) $(\text{tr}(\alpha\beta))^p \equiv \text{tr}((\alpha\beta)^p) \ (pD)$.
    (d) $p \mid \text{tr}(\alpha\beta)$ all $\beta \in D$.
    (e) $p \mid \Delta$, the discriminant of F.
    (Be sure to use the fact that $\alpha \notin pD$.)

29. Let F be a Galois extension of $\mathbb{Q}$ with abelian Galois group. Show that if $p \in \mathbb{Q}$ is unramified in F then $\sigma_P = \sigma_{P'}$ for prime ideals P and P' dividing p in F, where σ_P denotes the Frobenius automorphism.

30. Let p be an odd prime and consider $\mathbb{Q}(\sqrt{p})$. If $q \neq p$ is prime show that $\sigma_q(\sqrt{p}) = (p/q)\sqrt{p}$ where σ_q is the Frobenius automorphism at a prime ideal in $\mathbb{Q}(\sqrt{p})$ lying above q.

31. Let F be an algebraic number field and $\mathfrak{A}$ an ideal in the ring of integers of F. Show that there is a finite extension L of F with ring of integers S such that $\mathfrak{A}S$ is principal.

32. Let P be a prime ideal in the ring of integers D of a number field F. If $a \equiv b \ (P^t)$ and $\text{ord}_P b < t$ show that $\text{ord}_P a = \text{ord}_P b$.

33. Let $K \subset L$ be number fields with rings of integers R and S respectively. If A and B are ideals in R such that AS divides BS then show that A divides B.

34. The notation being as in Exercise 33 show that $AS \cap R = A$.

Chapter 13

Quadratic and Cyclotomic Fields

*In the last chapter we discussed the general theory of
algebraic number fields and their rings of integers. We
now consider in greater detail two important classes of
these fields which were studied first in the nineteenth
century by Gauss, Eisenstein, Kummer, Dirichlet, and
others in connection with the theory of quadratic forms,
higher reciprocity laws and Fermat's Last Theorem. The
reader who is interested in the historical development of
this subject should consult the book by H. Edwards [128]
as well as the classical treatise by H. Smith [72].*

*We will develop in this chapter only those results that
will be needed for the applications in later chapters. The
fundamental result describes the manner in which rational
primes decompose into a product of prime ideals. However,
we could not resist giving yet another proof of the law of
quadratic reciprocity based on the decomposition laws of
these fields.*

§1 Quadratic Number Fields

An algebraic number field F will be called a quadratic number field if
$[F : \mathbb{Q}] = 2$. Let $D \subset F$ be, as usual, the rings of integers in F. Our first goal
will be to find an explicit integral basis for D.

Let $F = \mathbb{Q}(\alpha)$. The element α must satisfy a quadratic equation $ax^2 +
bx + c = 0$ with $a, b, c \in \mathbb{Z}$. Thus

$$\alpha = \frac{-b \pm \sqrt{b^2 - 4ac}}{2a}.$$

Let $A = b^2 - 4ac$. Then, clearly, $F = \mathbb{Q}(\sqrt{A})$. Let $A = A_1^2 A_2$ where
$A_1, A_2 \in \mathbb{Z}$ and A_2 is square-free. Then $F = \mathbb{Q}(\sqrt{A_2})$. Changing notation, we
have shown that every quadratic number field has the form $\mathbb{Q}(\sqrt{d})$ where d
is a square-free integer.

If σ is any isomorphism of $F/\mathbb{Q}$ into $\mathbb{C}$ we apply σ to $(\sqrt{d})^2 = d$ and
find $(\sigma\sqrt{d})^2 = d$. Thus $\sigma\sqrt{d} = \pm\sqrt{d}$. It follows that $F/\mathbb{Q}$ is a Galois

extension. The Galois group has two elements, the identity and an automorphism taking $\sqrt{d}$ to $-\sqrt{d}$.

Every element of F has the form $\alpha = r + s\sqrt{d}$ with $r, s \in \mathbb{Q}$. The nontrivial automorphism takes α to $\alpha' = r - s\sqrt{d}$. Thus, $t(\alpha) = \alpha + \alpha' = 2r$ and $N(\alpha) = \alpha\alpha' = r^2 - ds^2$.

If $\gamma \in D$ then $t(\gamma)$ and $N(\gamma) \in \mathbb{Z}$. Conversely, if these conditions hold then γ satisfies $0 = (x - \gamma)(x - \gamma') = x^2 - t(\gamma)x + N(\gamma) \in \mathbb{Z}[x]$ showing that $\gamma \in D$. Thus $\gamma \in D$ iff $t(\gamma)$ and $N(\gamma) \in \mathbb{Z}$.

Proposition 13.1.1. *If $d \equiv 2, 3\ (4)$ then $D = \mathbb{Z} + \mathbb{Z}\sqrt{d}$.*
If $d \equiv 1\ (4)$ then $D = \mathbb{Z} + \mathbb{Z}((-1 + \sqrt{d})/2)$.

PROOF. Suppose $\gamma = r + s\sqrt{d}$, $r, s \in \mathbb{Q}$. Then $\gamma \in D$ iff $2r$ and $r^2 - s^2 d \in \mathbb{Z}$. Since $2r \in \mathbb{Z}$ it follows from the second condition that $4s^2 d \in \mathbb{Z}$. Since d is square-free it follows that $2s \in \mathbb{Z}$. Set $2r = m$ and $2s = n$. Then, $r^2 - ds^2 \in \mathbb{Z}$ implies $m^2 - dn^2 \equiv 0\ (4)$.

Recall that a square is congruent to either 0 or 1 modulo 4.

If $d \equiv 2, 3\ (4)$ then $m^2 - dn^2 \equiv m^2 + 2n^2$ or $m^2 + n^2\ (4)$. The only way that $m^2 + 2n^2$ or $m^2 + n^2$ can be divisible by 4 is for both m and n to be even. This is the case iff r and s are in $\mathbb{Z}$. This establishes the first assertion.

If $d \equiv 1\ (4)$ then $m^2 - dn^2$ is congruent to $m^2 - n^2$ modulo 4. But $m^2 - n^2 \equiv 0\ (4)$ iff m and n have the same parity, i.e., they are either both odd or both even. Thus $D = \{(m + n\sqrt{d})/2 \mid m \equiv n\ (2)\}$. Notice

$$\frac{m + n\sqrt{d}}{2} = \frac{m + n}{2} + n\left(\frac{-1 + \sqrt{d}}{2}\right).$$

Since $m \equiv n\ (2)$, $(m + n)/2 \in \mathbb{Z}$. Thus $D \subset \mathbb{Z} + \mathbb{Z}(-1 + \sqrt{d})/2$. To establish the reverse inequality we simply notice that $(-1 + \sqrt{d})/2 \in D$ since $d \equiv 1\ (4)$. $\square$

We can now calculate the discriminant of quadratic number fields.

Proposition 13.1.2. *Let δ_F denote the discriminant of F.*
If $d \equiv 2, 3\ (4)$ then $\delta_F = 4d$.
If $d \equiv 1\ (4)$ then $\delta_F = d$.

PROOF. If $d \equiv 2, 3\ (4)$ set $\omega_1 = 1$ and $\omega_2 = \sqrt{d}$. Then

$$(t(\omega_i\omega_j)) = \begin{pmatrix} 2 & 0 \\ 0 & 2d \end{pmatrix}.$$

Thus $\delta_F = \det(t(\omega_i\omega_j)) = 4d$.

If $d \equiv 1\ (4)$ set $\omega_1 = 1$ and $\omega_2 = (-1 + \sqrt{d})/2$. Then

$$(t(\omega_i\omega_j)) = \begin{pmatrix} 2 & -1 \\ -1 & (1 + d)/2 \end{pmatrix}.$$

Thus $\delta_F = \det(t(\omega_i\omega_j)) = d$. $\square$

Having investigated D and δ_F we now want to determine how rational primes $p \in \mathbb{Z}$ split in D. From Theorem 3' of Chapter 12 we know $efg = 2$, so we have three cases; $e = 2, f = 1, g = 1$ or $e = 1, f = 1, g = 2$, or $e = 1$, $f = 2, g = 1$. We say, respectively, that p ramifies, splits (decomposes), or is inertial (remains prime).

If p is a prime in $\mathbb{Z}$ let P be a prime ideal in D containing p. Let $P' = \{\gamma' | \gamma \in P\}$.

Proposition 13.1.3. *Suppose p is odd.*
 (i) *If $p \nmid \delta_F$ and $x^2 \equiv d \, (p)$ is solvable in $\mathbb{Z}$ then $(p) = PP', P \neq P'$.*
 (ii) *If $p \nmid \delta_F$ and $x^2 \equiv d \, (p)$ is not solvable in $\mathbb{Z}$ then $(p) = P$.*
 (iii) *If $p | \delta_F$ then $(p) = P^2$.*

PROOF. In case (i) suppose $a^2 \equiv d \, (p)$ with $a \in \mathbb{Z}$. We claim that $(p) = (p, a + \sqrt{d})(p, a - \sqrt{d})$. In fact, $(p, a + \sqrt{d})(p, a - \sqrt{d}) = (p)(p, a + \sqrt{d}, a - \sqrt{d}, (a^2 - d)/p)$. The latter ideal is D since it contains p and $2a$ and these two numbers are relatively prime. We claim $(p, a + \sqrt{d}) \neq (p, a - \sqrt{d})$. If equality held then the ideal would contain p and $2a$ and so would equal D and it would follow that $(p) = D$. Thus p splits as asserted.

In case (ii) we claim P has degree 2. If degree P is 1 then D/P has p elements. Since $\mathbb{Z}/p\mathbb{Z}$ injects into D/P it would follow that every coset of D/P is represented by a rational integer. Let $a \in \mathbb{Z}$ be such that $a \equiv \sqrt{d} \, (P)$. Then $a^2 \equiv d \, (P)$ and $a^2 \equiv d \, (p)$ contrary to assumption. Thus p remains prime as asserted.

Finally, in case (iii) we claim $(p) = (p, \sqrt{d})^2$. In fact, $(p, \sqrt{d})^2 = (p)$ $(p, \sqrt{d}, d/p)$. The latter ideal is D since p and d/p are relatively prime (remember that d is square-free). Thus p ramifies as asserted. $\square$

We now discuss the decomposition of the prime $p = 2$. Remember that by Proposition 13.1.2 we have $2 \nmid \delta_F$ if and only if $d \equiv 1 \, (4)$.

Proposition 13.1.4. *Suppose $p = 2$.*

 (i) *If $2 \nmid \delta_F$ and $d \equiv 1 \, (8)$ then $(2) = PP'$ and $P \neq P'$.*
 (ii) *If $2 \nmid \delta_F$ and $d \equiv 5 \, (8)$ then $(2) = P$.*
 (iii) *If $2 | \delta_F$ then $(2) = P^2$.*

PROOF. If $d \equiv 1 \, (8)$ we claim that $(2) = (2, (1 + \sqrt{d})/2)(2, (1 - \sqrt{d})/2)$. In fact $(2, (1 + \sqrt{d})/2)(2, (1 - \sqrt{d})/2) = (2)(2, (1 + \sqrt{d})/2, (1 - \sqrt{d})/2, (1 - d)/8)$. The latter ideal is D since it contains $1 = (1 + \sqrt{d})/2 + (1 - \sqrt{d})/2$. Moreover, $(2, (1 + \sqrt{d})/2) \neq (2, (1 - \sqrt{d})/2)$ since otherwise the ideal contains 1 and it would follow that $(2) = D$.

If $d \equiv 5 \, (8)$ we claim P has degree 2. If not (as in part (ii) of the last proposition) there is an integer $a \in \mathbb{Z}$ such that $a \equiv (1 + \sqrt{d})/2 \, (P)$. Since $(1 + \sqrt{d})/2$ satisfies $x^2 - x + (1 - d)/4 = 0$ we would have $a^2 - a +$

$(1 - d)/4 \equiv 0 \ (P)$ and so $a^2 - a + (1 - d)/4 \equiv 0 \ (2)$. For all $a \in \mathbb{Z}$, $a^2 - a$ is even. It follows that $(1 - d)/4 \equiv 0 \ (2)$ or $d \equiv 1 \ (8)$ contrary to assumption.

Now suppose $2|\delta_F$. We must have $d \equiv 2, 3 \ (4)$. If $d \equiv 2 \ (4)$ then $(2) = (2, \sqrt{d})^2$ and if $d \equiv 3 \ (4)$ then $(2) = (2, 1 + \sqrt{d})^2$. We leave the simple verification to the reader. $\qquad\square$

We note that we can state the decomposition law for odd primes in a succinct manner using the Legendre symbol. Namely, if $(\delta_F/p) = 1$ then p splits, if $(\delta_F/p) = -1$ then p remains prime, and if $(\delta_F/p) = 0$ then p ramifies. Furthermore the decomposition of p, p odd, depends only on the residue class of p modulo δ_F. For if $d \equiv 2$ or 3 modulo 4 then $\delta_F = 4d$ and the result follows from Proposition 5.3.3 and Exercise 37 of Chapter 5. If $d \equiv 1 \ (4)$ then we may argue as follows. Since $d \equiv 1 \ (4)$ we have $\delta_F = d$. Thus

$$\left(\frac{\delta_F}{p}\right) = (-1)^{((p-1)/2)((\delta_F - 1)/2)}\left(\frac{p}{\delta_F}\right) = \left(\frac{p}{\delta_F}\right).$$

The value of (p/δ_F) depends only on the residue class of p modulo δ_F.

Next we determine the structure of the group of units in D. It is simple to see that α is a unit iff $N(\alpha) = \pm 1$. Consider first the case of an imaginary quadratic field, so that $d < 0$. Let U_d denote the group of units in D.

Proposition 13.1.5. *If $d < 0$ and square free then*
(a) $U_{-1} = \{1, i, -1, -i\}$.
(b) $U_{-3} = \{\pm 1, \pm \omega, \pm \omega^2\}$, where $\omega = (-1 + \sqrt{-3})/2$.
(c) $U_d = \{1, -1\}$ for $d < -3$, or $d = -2$.

PROOF. If $d \equiv 2$ or 3 (4) then any unit may be written in the form $x + \sqrt{d}y$, $x, y \in \mathbb{Z}$. Thus $N(\alpha) = \pm 1$ is equivalent to $x^2 + |d|y^2 = 1$. If $d = -1$ we obtain (a). If $|d| > 1$ then clearly $U_d = \{+1, -1\}$.

If $d \equiv 1 \ (4)$ write $= (x + \sqrt{d}y)/2$ where $x \equiv y \ (2)$. Then $N(\alpha) = \pm 1$ is equivalent to $x^2 + |d|y^2 = 4$. If $d = -3$ the solutions to $x^2 + 3y^2 = 4$ give part (b) while if $|d| > 3$ the equation $x^2 + |d|y^2 = 4$ clearly gives $U_d = \{+1, -1\}$. This completes the proof. $\qquad\square$

Thus the determination of the unit group is quite simple in the imaginary case. The case of a real quadratic field is considerably more difficult.

If $d > 0$ and square-free the equation $x^2 - dy^2 = 1$ is called Pell's equation. In Chapter 17, Section 5 it is shown that this equation has a solution in nonzero integers x, y. The proof is elementary. Assuming this result we describe the units in D in the real quadratic case.

Proposition 13.1.6. *If D is the ring of integers in $\mathbb{Q}(\sqrt{d})$, $d > 0$ then there exists a unit $u > 1$ such that every unit is of the form $\pm u^m$, $m \in \mathbb{Z}$.*

PROOF. By Proposition 17.5.2 there exist positive nonzero integers x, y such that $x^2 - dy^2 = +1$. Thus $x + \sqrt{d}y = u$ is a unit in D, $u > 1$. Let M be a fixed real number, $M > u$. By Exercise 4, Chapter 12 there are at most a finite number of $\alpha \in D$ with $|\alpha| < M$, $|\alpha'| < M$ where α' is the conjugate of α. If β is a unit $1 < \beta < M$ then $N(\beta) = \beta\beta' = \pm 1$. If $\beta' = -1/\beta$ then $-M < -1/\beta < M$ and if $\beta' = 1/\beta$ then also $-M < 1/\beta < M$. Thus here are only finitely many units β with $1 < \beta < M$ and there is at least one, viz., u. Let ε be the smallest positive unit $\varepsilon > 1$. If τ is any positive unit then there is a unique integer s (not necessarily positive) with $\varepsilon^s \leq \tau < \varepsilon^{s+1}$. Then $1 \leq \tau\varepsilon^{-s} < \varepsilon$ and since $\tau\varepsilon^{-s}$ is a unit we have $\tau\varepsilon^{-s} = 1$. If τ is negative then $-\tau$ is positive and $-\tau = \varepsilon^s$. This completes the proof.   □

The unique unit ε defined in Proposition 13.1.6 is called the fundamental unit of $\mathbb{Q}(\sqrt{d})$. The set of $d > 0$ for which the norm of ε is -1 has not been determined. However there are many interesting results in that direction (see [196], pp. 124–126). It has been conjectured that for $d = p$, $p \equiv 1$ (4) and prime, and $\varepsilon = (u + v\sqrt{p})/2$ that $p \nmid v$ [86]. The fundamental unit, even for small discriminants, can be difficult to compute. For example, the fundamental unit of $Q(\sqrt{94})$ is $2143295 + 221064\sqrt{94}$.

These results on units are special cases of the important Dirichlet unit theorem which gives the structure of the group of units in an arbitrary number field. This theorem states that the group of units modulo the subgroup of roots of unity in the field is a finitely generated group with $r + s - 1$ generators, where s is the number of pairs of complex conjugate roots and r is the number of real roots of a generator for the field. In the case of quadratic fields this number is clearly 0 or 1 according as the field is imaginary or real, which agrees with the above results.

As regards the class number there is an exceedingly rich theory for quadratic number fields. In fact there exist explicit formulas, discovered by Dirichlet. We give a particularly elegant special case. Suppose $q > 3$ is a prime and $q \equiv 3$ (4). Let $F = \mathbb{Q}(\sqrt{-q})$. Let V and R represent the sum of the quadratic nonresidues and quadratic residues modulo q, respectively, among the numbers $1, 2, 3, \ldots, q - 1$. Then $h_F = (1/q)(V - R)$.

For example, let $q = 7$. Then $V = 3 + 5 + 6 = 14$ and $R = 1 + 2 + 4 = 7$. Thus $h_F = \frac{1}{7}(14 - 7) = 1$.

If we restrict our attention to $d < 0$ then C. L. Siegel proved that $\ln h_F / \ln |\delta_F|^{1/2} \to 1$ as $|\delta_F| \to \infty$. It follows that there are at most finitely many $d < 0$ for which $\mathbb{Q}(\sqrt{-d})$ has class number below a fixed bound.

Gauss conjectured that the only d for which the class number of $\mathbb{Q}(\sqrt{-d})$ is 1 are $d = -1, -2, -3, -7, -11, -19, -43, -67$, and -163. The first generally accepted proof was provided by H. Stark. In essence a proof had been given earlier by K. Heegner, but because of obscurities in the exposition his proof was at first not thought to be valid.

For positive d, Gauss conjectured that infinitely many of the fields $\mathbb{Q}(\sqrt{d})$ have class number 1. This, however, remains an open problem.

A beautiful formula that determines the class number of a real quadratic field of discriminant p, p a prime congruent to 1 modulo 4, is $\varepsilon^h = \prod (\sin(\pi j/p))^{-\chi(j)}$ where ε is the fundamental unit, χ is the Legendre symbol, and the product is over the numbers $j = 1, \ldots, (p - 1)/2$. A similar formula holds for arbitrary discriminant. For these results and their proofs see Borevich and Shafarevich [9], Chapter 5.

We conclude this section by mentioning several other results whose proofs are beyond the scope of an elementary treatment. Consider an imaginary quadratic field of discriminant d. Then the class number of this field is divisible by 2^{t-1} where t is the number of distinct prime divisors of d. Thus the class number of $\mathbb{Q}(\sqrt{-210})$ is divisible by 8. It turns out that the class number is exactly 8. A similar result holds for real quadratic number fields.

The following most remarkable fact has been discovered by F. Hirzebruch. Let p be a prime congruent to 3 modulo 4 and assume that the class number of $\mathbb{Q}(\sqrt{p})$ is one. Then the class number of the imaginary quadratic field $\mathbb{Q}(\sqrt{-p})$ is one third of the alternating sum $a_s - a_{s-1} + a_{s-2} - \cdots \pm a_1$, where the continued fraction of $\sqrt{p}$ is, in the standard notation, $(a_0, \overline{a_1, a_2, \ldots, a_s})$, (see Stark [73], Chapter 7). For example, both $\mathbb{Q}(\sqrt{67})$ and $\mathbb{Q}(\sqrt{-67})$ have class number one and

$$\sqrt{67} = (8, \overline{5, 2, 1, 1, 7, 1, 1, 2, 5, 16}).$$

§2 Cyclotomic Fields

Let m be a positive integer and $\zeta_m = e^{2\pi i/m}$. The number ζ_m satisfied $x^m - 1 = 0$ as do all the powers of ζ_m. Thus, we have $x^m - 1 = (x - 1)(x - \zeta_m) \cdots (x - \zeta_m^{m-1})$. It follows that the field $F = \mathbb{Q}(\zeta_m)$ is the splitting field of the polynomial $x^m - 1$. Thus $F/\mathbb{Q}$ is a Galois extension.

We call $F = \mathbb{Q}(\zeta_m)$ the cyclotomic field of mth roots of unity. It was first studied by Gauss in connection with his investigations into the constructability of regular polygons (see Chapter 9, Section 11).

Proposition 13.2.1. *Let G be the Galois group of $F/\mathbb{Q}$. There is a monomorphism* $\theta: G \to U(\mathbb{Z}/m\mathbb{Z})$ *such that for $\sigma \in G$*

$$\sigma\zeta_m = \zeta_m^{\theta(\sigma)}.$$

PROOF. Since $\zeta_m^m = 1$ we have $(\sigma\zeta_m)^m = 1$. Thus $\sigma\zeta_m = \zeta_m^{\theta(\sigma)}$ where $\theta(\sigma)$ is an integer modulo m. If $\tau = \sigma^{-1}$ then $\zeta_m = \tau\sigma\zeta_m = \tau(\zeta_m^{\theta(\sigma)}) = \zeta_m^{\theta(\tau)\theta(\sigma)}$. Thus $\theta(\tau)\theta(\sigma) = \overline{1}$ (where $\overline{1}$ is the coset of 1 in $\mathbb{Z}/m\mathbb{Z}$). Thus $\theta: G \to U(\mathbb{Z}/m\mathbb{Z})$. It is easily checked that θ is a homomorphism. Finally, if $\theta(\sigma) = \overline{1}$ then $\sigma\zeta_m = \zeta_m$ implying σ is the identity of G since ζ_m generates F over $\mathbb{Q}$. □

Corollary. $[\mathbb{Q}(\zeta_m) : \mathbb{Q}]$ *divides $\phi(m)$.*

We will show later that in fact $[\mathbb{Q}(\zeta_m) : \mathbb{Q}] = \phi(m)$.

Definition. Let $\Phi_m(x) = \prod_{(a,m)=1} (x - \zeta_m^a)$ where $1 \leq a < m$. This polynomial is called the mth cyclotomic polynomial.

The roots of $\Phi_m(x)$ are precisely the primitive mth roots of unity, i.e., those mth roots of unity of order m. Clearly the degree of $\Phi_m(x)$ is $\phi(m)$.

Proposition 13.2.2. $x^m - 1 = \prod_{d/m} \Phi_d(x)$.

PROOF.

$$x^m - 1 = \prod_{i=0}^{m-1} (x - \zeta_m^i) = \prod_{d/m} \prod_{(i,m)=d} (x - \zeta_m^i).$$

We claim $\prod_{(i,m)=d} (x - \zeta_m^i) = \Phi_{m/d}(x)$. The proposition will follow from this.

If $(i, m) = d$, let $i = dj$. Then $\zeta_m^i = \zeta_m^{dj} = \zeta_{m/d}^j$. Moreover, $(j, m/d) = 1$. Thus

$$\prod_{(i,m)=d} (x - \zeta_m^i) = \prod_{(j,m/d)=1} (x - \zeta_{m/d}^j) = \Phi_{m/d}(x). \qquad \square$$

Corollary. $\Phi_m(x) \in \mathbb{Z}[x]$.

PROOF. We proceed by induction on m. $\Phi_1(x) = x - 1$. Now suppose the corollary has been established for integers less than m. By the proposition, $\Phi_m(x) = (x^m - 1)/f(x)$, where $f(x)$ is a monic polynomial which by the induction hypothesis is in $\mathbb{Z}[x]$. It follows by "long division" that $\Phi_m(x) \in \mathbb{Z}[x]$. $\qquad \square$

An alternate proof of the corollary goes as follows. Every $\sigma \in G$ permutes the primitive mth roots of unity. Thus the coefficients of $\Phi_m(x)$ are left fixed by G and so are in $\mathbb{Q}$. Since they are clearly algebraic integers they must be in $\mathbb{Z}$.

From now on we write $\zeta_m = \zeta$, $F = \mathbb{Q}(\zeta)$, and D for the ring of integers in F.

Proposition 13.2.3. *Suppose p is a rational prime and $p \nmid m$. Let P be a prime ideal in D containing p. Then the cosets of $1, \zeta, \zeta^2, \ldots, \zeta^{m-1}$ in D/P are all distinct. If f denotes the degree of P then $p^f \equiv 1 \ (m)$.*

PROOF. For $w \in D$ let $\bar{w}$ denote its coset in D/P.

Divide both sides of $x^m - 1 = \prod (x - \zeta^i)$ by $x - 1$. We find

$$1 + x + \cdots + x^{m-1} = \prod_{i=1}^{m-1} (x - \zeta^i).$$

Let $x = 1$ in this identity. We find $m = \prod (1 - \zeta^i)$ where $1 \leq i \leq m - 1$. Thus $\bar{m} = \prod \overline{(1 - \zeta^i)}$. Since $\bar{m} \neq \bar{0}$ it follows that $\bar{\zeta}^i \neq \bar{1}$ for $1 \leq i \leq m - 1$, and so $\bar{\zeta}^i \neq \bar{\zeta}^j$ for $0 \leq i, j \leq m - 1$.

The elements $\{\bar{\zeta}^i \mid 0 \leq i \leq m - 1\}$ form a subgroup of order m in the multiplicative group of D/P. The latter group has order $p^f - 1$. Therefore $p^f \equiv 1 \ (m)$. $\qquad \square$

Theorem 1. *The* m*th cyclotomic polynomial,* $\Phi_m(x)$, *is irreducible in* $\mathbb{Z}[x]$.

PROOF. Let $f(x) \in \mathbb{Z}[x]$ be the monic irreducible polynomial for ζ. The fact that $f(x)$ has coefficients in $\mathbb{Z}$ follows from the fact that ζ is an algebraic integer (Exercise 16, Chapter 6). If $p \nmid m$ is a prime we will show that ζ^p is also a root of $f(x)$. If $a \in \mathbb{Z}$, and $(a, m) = 1$, then by factoring a into a product of primes it will follow that ζ^a is a root of $f(x)$. Thus deg $f(x) \geq \phi(m)$. On the other hand, since $\Phi_m(\zeta) = 0$, $f(x)$ divides $\Phi_m(x)$ which has degree $\phi(m)$. It will then follow that $f(x) = \Phi_m(x)$.

Now, let p be a prime, $p \nmid m$, and let P be a prime ideal of D containing p. As usual, if $w \in D$ then $\bar{w}$ will denote the residue class of w in D/P. We have $x^m - 1 = f(x)g(x)$ and so $x^m - \bar{1} = \bar{f}(x)\bar{g}(x)$ in $\mathbb{Z}/p\mathbb{Z}[x]$. By the last proposition $x^m - \bar{1}$ has distinct roots in D/P. It follows that $\bar{f}(x)$ and $\bar{g}(x)$ have no common root. Suppose $f(\zeta^p) \neq 0$. Then $g(\zeta^p) = 0$ and $\bar{g}(\bar{\zeta}^p) = 0$. The coefficients of $\bar{g}(x)$ are in $\mathbb{Z}/p\mathbb{Z}$ and are thus equal to their own pth power. From this we see $\bar{0} = \bar{g}(\bar{\zeta}^p) = \bar{g}(\bar{\zeta})^p$ and so $\bar{0} = \bar{g}(\bar{\zeta})$. It follows that $\bar{f}(\bar{\zeta}) \neq \bar{0}$ which is not true because $f(\zeta) = 0$. One concludes $f(\zeta^p) = 0$ as asserted.. $\qquad\square$

Corollary 1. $[\mathbb{Q}(\zeta_m) : \mathbb{Q}] = \phi(m)$.

Corollary 2. *The map* θ *of Proposition* 13.2.1 *is an isomorphism of* G *onto* $U(\mathbb{Z}/m\mathbb{Z})$.

PROOF. Both G and $U(\mathbb{Z}/m\mathbb{Z})$ have $\phi(m)$ elements. Since θ is one-to-one it must be onto. $\qquad\square$

By Corollary 2 we see that for every $a \in \mathbb{Z}$ with $(a, m) = 1$ there is a $\sigma_a \in G$ such that $\sigma_a \zeta = \zeta^a$. The map $a \to \sigma_a$ gives rise to a homomorphism from $U(\mathbb{Z}/m\mathbb{Z})$ to G which is inverse to θ.

If p is a prime, $p \nmid m$, we wish to study more closely the automorphism σ_p. Before we do so, some preliminary work is needed.

Lemma 1. *Let* $F/\mathbb{Q}$ *be an algebraic number field of degree* n. *Let* $D \subset F$ *be the ring of integers and* $\alpha_1, \alpha_2, \ldots, \alpha_n \in D$ *a field basis for* $F/\mathbb{Q}$. *Let* $\Delta = \Delta(\alpha_1, \alpha_2, \ldots, \alpha_n)$ *be the discriminant of this basis. Then* $\Delta D \subset \mathbb{Z}\alpha_1 + \mathbb{Z}\alpha_2 + \cdots + \mathbb{Z}\alpha_n$.

PROOF. Let $w \in D$. We have $w = \sum r_i \alpha_i$ with $r_i \in \mathbb{Q}$. Multiply both sides by α_j and take the trace. We find $t(w\alpha_j) = \sum r_i t(\alpha_i \alpha_j)$. The elements $t(w\alpha_j)$ and $t(\alpha_i \alpha_j)$ are all in $\mathbb{Z}$ since they are traces of algebraic integers. Using Cramer's rule to solve for the r_i we see that each r_i is an integer divided by Δ. The result follows. $\qquad\square$

Lemma 2. *The discriminant* $\Delta = \Delta(1, \zeta, \ldots, \zeta^{\phi(m)-1})$ *divides* $m^{\phi(m)}$.

PROOF. Differentiate both sides of $x^m - 1 = \Phi_m(x)g(x)$. We find $mx^{m-1} = \Phi_m'(x)g(x) + \Phi_m(x)g'(x)$. Substitute $x = \zeta$. The result is $m\zeta^{m-1} = \Phi_m'(\zeta)g(\zeta)$.

Now take the norm of both sides. Using Proposition 12.1.4 and the fact that $N(\zeta) = \pm 1$ we find $\pm m^{\phi(m)} = \Delta N(g(\zeta))$. We note by Theorem 1, that, 1, $\zeta, \ldots, \zeta^{\phi(m)-1}$ is a field basis for $\mathbb{Q}(\zeta)/\mathbb{Q}$ so that $\Delta(1, \zeta, \ldots, \zeta^{\phi(m)-1}) \neq 0$. $\square$

Proposition 13.2.4. *Let $p \in \mathbb{Z}$ be a prime such that $p \nmid m$. Let $w \in D$ the ring of integers in $\mathbb{Q}(\zeta)$. There is an element $\sum a_i \zeta^i \in \mathbb{Z}[\zeta]$ such that $w \equiv \sum a_i \zeta^i (p)$.*

PROOF. Let $\Delta = \Delta(1, \zeta, \ldots, \zeta^{\phi(m)-1})$. By Lemma 2, $p \nmid \Delta$. Thus there is a $\Delta' \in \mathbb{Z}$ such that $\Delta'\Delta \equiv 1\ (p)$. Thus $w \equiv \Delta'\Delta w\ (p)$. By Lemma 1, $\Delta w \in \mathbb{Z}[\zeta]$. Thus the result. $\square$

We remark that in fact $D = \mathbb{Z}[\zeta]$ but this is not so easy to prove for general m. When m is a prime however, the proof is reasonably easy (see Proposition 13.2.10).

Corollary. *Suppose $p \nmid m$ and $n > 0$ is such that $p^n \equiv 1\ (m)$. Then, for $w \in D$ we have $w^{p^n} \equiv w\ (p)$.*

PROOF. By the proposition, $w \equiv \sum a_i \zeta^i\ (p)$ with the $a_i \in \mathbb{Z}$. Since $a_i^p \equiv a_i\ (p)$ we must have $w^p \equiv \sum a_i \zeta^{pi}\ (p)$. Repeating this process n times and using the fact that $p^n \equiv 1\ (m)$ implies $\zeta^{p^n} = \zeta$ yields the result. $\square$

Proposition 13.2.5. *If p is a prime and $p \nmid m$ the every prime ideal P in D containing p is unramified.*

PROOF. Assume P is ramified. Then $(p) \subset P^2$. Let w be an element of P not in P^2. By the above corollary $w^{p^n} \equiv w\ (p)$ and so $w^{p^n} \equiv w\ (P^2)$. Since $p^n \geq 2$ it follows that $w \in P^2$, a contradiction. $\square$

We will see later that the converse of this proposition is "almost" true. See Proposition 13.2.8.

Recall that, for p prime, $p \nmid m$ the automorphism σ_p sends ζ to ζ^p.

Proposition 13.2.6. *For all $w \in D$ we have $\sigma_p w \equiv w^p\ (p)$.*

PROOF. By Proposition 13.2.4 we have $w \equiv \sum a_i \zeta^i\ (p)$. Apply σ_p to both sides. We find that $\sigma_p w \equiv \sum a_i \zeta^{pi}\ (p)$. Since the $a_i \in \mathbb{Z}$ we have $\sum a_i \zeta^{pi} \equiv \sum a_i^p \zeta^{pi} \equiv (\sum a_i \zeta^i)^p\ (p)$. Thus $\sigma_p w \equiv w^p\ (p)$ as asserted. $\square$

Corollary. *Let P be a prime ideal of D containing p. Then $\sigma_p P = P$.*

PROOF. If $w \in P$ then $\sigma_p w \equiv w^p \equiv 0\ (P)$ and so $\sigma_p P \subset P$. Since $\sigma_p P$ is a maximal ideal we have equality. $\square$

Theorem 2. *Let p be a prime, $p \nmid m$. Let f be the smallest positive integer such that $p^f \equiv 1\ (m)$. Then in $D \subset \mathbb{Q}(\zeta)$ we have*

$$(p) = P_1 P_2 \cdots P_g,$$

where each P_i has degree f and $g = \phi(m)/f$.

PROOF. We first observe that it follows directly from the definition that f is the order of the automorphism σ_p.

Now, $p^{f_1} = |D/P_1|$ where f_1 is the degree of P_1. Since D/P_1 is a finite field we have $w^{p^{f_1}} \equiv w \ (P_1)$ for all $w \in D$ and f_1 is the smallest positive integer with this property.

By the last proposition, we have $w \equiv \sigma_p^f(w) \equiv w^{p^f}(P_1)$ for all $w \in D$. It follows that $f_1 \leq f$.

On the other hand, $\zeta^{p^{f_1}} \equiv \zeta \ (P_1)$ implies $\zeta^{p^{f_1}} = \zeta$ by Proposition 13.2.3. Thus $p^{f_1} \equiv 1 \ (m)$ and it follows that $f \leq f_1$.

We now see $f = f_1 =$ degree of P_1. All the P_i have degree f. By Proposition 13.2.5 all the P_i are unramified. Using the relation $efg = \phi(m)$ we conclude $g = \phi(m)/f$.   □

Corollary. *With the notation of the theorem, let P be one of the P_i. Define $G(P) = \{\sigma \in G | \sigma P = P\}$. Then $G(P)$ is a cyclic group generated by σ_p.*

PROOF. By the corollary to Proposition 13.2.6 we know $\sigma_p \in G(P)$. Let $\langle \sigma_p \rangle$ be the cyclic group generated by σ_p. Then $\langle \sigma_p \rangle \subset G(P)$. By Proposition 12.3.3 we have $g|G(P)| = \phi(m)$. Thus $|G(P)| = \phi(m)/g = f = |\langle \sigma_p \rangle|$ and we are done.   □

Theorem 2 is a very satisfactory result on the decomposition of primes which do not divide m. One can also find the decomposition of those primes which do divide m. We content ourselves with the following important special case.

Proposition 13.2.7. *Let l be a prime in $\mathbb{Z}$. Then, in $\mathbb{Q}(\zeta_l)$, l ramifies completely. More precisely, let $L = (1 - \zeta_l)$. Then L is a prime ideal and $(l) = L^{l-1}$. Moreover L has degree 1.*

PROOF. As in the proof of Proposition 13.2.3 we have $l = \prod (1 - \zeta_l^i)$ where the product is over $1 \leq i \leq l - 1$.

Let $u_i = (1 - \zeta^i)/(1 - \zeta) = 1 + \zeta + \cdots + \zeta^{i-1}$. We claim that u_i is a unit. Since $l \nmid i$ there is a $j \in \mathbb{Z}$ such that $ij \equiv 1 \ (l)$. Thus, $u_i^{-1} = (1 - \zeta)/(1 - \zeta^i) = (1 - \zeta^{ij})/(1 - \zeta^i) = 1 + \zeta^i + \cdots + (\zeta^i)^{j-1}$ is an algebraic integer which proves the claim.

It follows that $l = \prod (1 - \zeta^i) = (1 - \zeta)^{l-1} \prod u_i$ and so $(l) = L^{l-1}$. Using the relation $efg = \phi(l) = l - 1$ we see L must be prime, $e = l - 1$, $g = 1$, and $f = 1$.   □

Proposition 13.2.8. *Let P be a prime ideal in $\mathbb{Q}(\zeta_m)$ and set $P \cap \mathbb{Z} = p\mathbb{Z}$. If p is odd then P is ramified iff $p|m$. If $p = 2$ then P is ramified iff $4|m$.*

PROOF. By Proposition 13.2.5 we know that $p \nmid m$ implies P is unramified.

Suppose p is odd and $p|m$. Then $\mathbb{Q}(\zeta_p) \subset \mathbb{Q}(\zeta_m)$. Let D_p and D_m be the rings of integers in $\mathbb{Q}(\zeta_p)$ and $\mathbb{Q}(\zeta_m)$ respectively. By the last proposition $pD_p = (1 - \zeta_p)^{p-1}$. Write $(1 - \zeta_p)D_m = P_1 P_2 \cdots P_t$ where the P_i are, not necessarily distinct, prime ideals in D_m. Then $pD_m = (P_1 P_2 \cdots P_t)^{p-1}$. Since $p - 1 > 1$ all the primes in D_m containing p are ramified.

Now suppose $p = 2$. If $2 | m$ but $4 \nmid m$ then $m = 2m_0$, with m_0 odd. In this case, $-\zeta_{m_0}$ is a primitive mth root of unity so $\mathbb{Q}(\zeta_m) = \mathbb{Q}(\zeta_{m_0})$. Since $2 \nmid m_0$, P is unramified.

Finally, suppose $p = 2$ and $4 | m$. Then $\zeta_4 = \sqrt{-1} = i \in \mathbb{Q}(\zeta_m)$. Since $(1 - i)^2 = -2i$ we see $2D_m = ((1 - i)D_m)^2$ and it follows, as before, that all the primes in D_m containing 2 are ramified. $\square$

Suppose p is a prime and $p \nmid m$. For later use (in the next chapter) we need to know how p decomposes in the field $\mathbb{Q}(\zeta_p, \zeta_m)$.

Lemma 3. *If* $(m, n) = 1$ *then* $\mathbb{Q}(\zeta_m, \zeta_n) = \mathbb{Q}(\zeta_{mn})$.

PROOF. Since $\zeta_{mn}^m = \zeta_n$ and $\zeta_{mn}^n = \zeta_m$ we have $\mathbb{Q}(\zeta_m, \zeta_n) \subset \mathbb{Q}(\zeta_{mn})$.

On the other hand, since $(m, n) = 1$ there exist integers u and v such that $um + vn = 1$. Thus $\zeta_{mn} = \zeta_{mn}^{um} \zeta_{mn}^{vn} = \zeta_n^u \zeta_m^v \in \mathbb{Q}(\zeta_m, \zeta_n)$. $\square$

Proposition 13.2.9. *Let* p *be a prime such that* $p \nmid m$. *Let* D *be the ring of integers in* $\mathbb{Q}(\zeta_p, \zeta_m)$. *Then*

$$pD = (P_1 P_2 \cdots P_g)^{p-1},$$

where the P_i *are distinct prime ideals of degree* f *and* $g = \phi(m)/f$. *The integer* f *is the least positive integer such that* $p^f \equiv 1 \ (m)$.

PROOF. Since $\mathbb{Q}(\zeta_p) \subset \mathbb{Q}(\zeta_p, \zeta_m)$ we see, as in the proof of the last proposition, that all the ramification indices of primes in D containing p are divisible by $p - 1$. Thus

$$pD = (P_1 P_2 \cdots P_{g'})^{e'(p-1)} \qquad (*)$$

where the P_i are distinct prime ideals of degree f', say, and $e' \geq 1$ is some integer.

Let D_m be the ring of integers in $\mathbb{Q}(\zeta_m)$. By Theorem 2

$$pD_m = \tilde{P}_1 \tilde{P}_2 \cdots \tilde{P}_g$$

where the $\tilde{P}_i$ are prime ideals in D_m of degree f and $g = \phi(m)/f$.

By considering the prime decomposition of $\tilde{P}_i D$ and comparing with equation $(*)$ we see $f' \geq f$ and $g' \geq g$.

From equation $(*)$ and Lemma 3 we see

$$(p - 1)\phi(m) = \phi(pm) = e'(p - 1)f'g' \geq e'(p - 1)f \frac{\phi(m)}{f}.$$

It follows that $\phi(m) \geq e'\phi(m)$. Thus $e' = 1$ and all the inequalities are equalities, i.e., $f' = f$ and $g' = g = \phi(m)/f$. This concludes the proof. $\square$

We conclude this section by showing that $D = \mathbb{Z}[\zeta_l]$ when l is prime. This result holds even when l is not prime but the proof is more difficult (see, for example, pp. 265–268, [207]). The case when l is prime will be needed in Chapter 17 where a special case of Fermat's conjecture is discussed.

Proposition 13.2.10. *If l is prime then $D = \mathbb{Z}[\zeta_l]$.*

PROOF. Clearly $\mathbb{Z}[\zeta_l] \subset D$. If $\alpha \in D$ there exist $a_0, a_1, \ldots, a_{l-2}$ rational numbers such that $\alpha = a_0 + a_1\zeta + \cdots + a_{l-2}\zeta^{l-2}$. We show first of all that $la_i \in \mathbb{Z}, i = 0, \ldots, l - 2$. For if tr denotes the trace map from $\mathbb{Q}(\zeta)$ to $\mathbb{Q}$ then one computes easily tr $\zeta^j = -1$ if $l \nmid j$, using say, Corollary 1 of Theorem 1. Thus one sees that $\text{tr}(\alpha\zeta^{-s}) = -a_0 - a_1 - \cdots - a_{s-1} + (l - 1)a_s - a_{s+1} - \cdots - a_{l-2}$. Therefore $\text{tr}(\alpha\zeta^{-s} - \alpha\zeta) = la_s, s = 0, \ldots, l - 2$. Since $\alpha\zeta^{-s} - \alpha\zeta \in D$ it follows that $la_s \in \mathbb{Z}$. If $\lambda = 1 - \zeta$ then by Proposition 13.2.7 one has $(\lambda)^{l-1} = (l)$. By the above there exist $b_0, \ldots, b_{l-2}$ in $\mathbb{Z}$ such that $l\alpha = b_0 + b_1\lambda + \cdots + b_{l-2}\lambda^{l-2}$. Thus $\lambda | b_0$ and taking norms shows that $l | b_0$. Thus $\lambda^{l-1} | b_0$ and reduction modulo λ^2 given $\lambda^2 | b_1\lambda$ so that $\lambda | b_1$. Again this implies $l | b_1$. Clearly, successive reduction modulo higher powers of λ leads to $l | b_j, j = 0, \ldots, l - 2$ and division by l then shows that $\alpha \in \mathbb{Z}[\zeta_l]$. □

§3 Quadratic Reciprocity Revisited

As an application of some of the theory developed in this chapter we give yet another proof of quadratic reciprocity. The idea for this proof goes back, in essence, to Kronecker.

Let p be an odd prime and consider the field $\mathbb{Q}(\zeta_p)$. We claim that this field contains the square root of $(-1)^{(p-1)/2}p = p^*$. This follows from Proposition 6.3.2. However, in order to make our present considerations independent of the theory of Gauss sums, we give a direct proof using the relation

$$p = \prod_{i=1}^{p-1}(1 - \zeta^i).$$

We combine the terms corresponding to i and $p - i$ as follows

$$(1 - \zeta^i)(1 - \zeta^{p-i}) = (1 - \zeta^i)(1 - \zeta^{-i}) = -\zeta^{-i}(1 - \zeta^i)^2.$$

Thus

$$p = (-1)^{(p-1)/2}\zeta^b \prod_{i=1}^{(p-1)/2}(1 - \zeta^i)^2 \quad \text{where } b = -1 - 2 - \cdots - \frac{(p-1)}{2}.$$

Let $c \in \mathbb{Z}$ be such that $2c \equiv 1 \ (p)$. Then $\zeta^b = (\zeta^{bc})^2$. It follows that p^* is a square in $\mathbb{Q}(\zeta)$ as asserted. Let $\tau^2 = p^*$.

Now suppose q is an odd prime $q \neq p$. Consider the automorphism σ_q. Then $\sigma_q\tau = \pm\tau$ with the plus sign holding iff σ_q is in the Galois group of $\mathbb{Q}(\zeta_p)/\mathbb{Q}(\tau)$. Since the Galois group G of $\mathbb{Q}(\zeta)/\mathbb{Q}$ is isomorphic via θ to

$U(\mathbb{Z}/p\mathbb{Z})$ and the latter group is cyclic of order $p - 1$ we see $\sigma_q \tau = \tau$ iff σ_q is a square in G and this is so iff q is a square in $U(\mathbb{Z}/p\mathbb{Z})$. In other words

$$\sigma_q \tau = \left(\frac{q}{p}\right)\tau.$$

Let Q be a prime ideal in $D \subset \mathbb{Q}(\zeta)$ containing q. By Proposition 13.2.6 we have

$$\sigma_q \tau \equiv \tau^q \ (Q).$$

Thus $(q/p)\tau \equiv \tau^q \ (Q)$ implying $(p^*/q) \equiv p^{*(q-1)/2} \equiv \tau^{q-1} \equiv (q/p) \ (Q)$.

This latter congruence implies $(p^*/q) = (q/p)$ since Q does not contain 2.

It may be thought that this proof, pretty as it is, is much more complicated than the previous proofs and so does not add much. This is not the case, because the ideas involved provide the key to studying higher reciprocity laws.

Notes

There is an introduction to the arithmetic of quadratic number fields in J. Sommer's *Introduction à la Théorie des Nombres Algebriques* (Hermann: Paris, 1911). This book is based upon D. Hilbert's lectures in 1897–1898. See also F. Châtelet [111], W. Adams and L. Goldstein [84], and H. Stark [73].

As mentioned earlier all imaginary quadratic fields whose ring of integers form a unique factorization domain have been determined. The imaginary quadratic fields of class number two have also been determined. There are 18 such fields, the one with smallest discriminant being $\mathbb{Q}(\sqrt{-427})$.

In the case of cyclotomic fields Masley has shown that if m is a positive integer, $m \not\equiv 2 \ (4)$, then there are exactly 29 values of m for which $\mathbb{Q}(\zeta_m)$ has class number one. Furthermore, the prime cyclotomic fields $\mathbb{Q}(\zeta_p)$ of class number one are given by $p = 3, 5, 7, 11, 13, 17, 19$ a result due to Uchida and Montgomery. For more details see the surveys by Masley [184], [185].

For a more thorough treatment of the arithmetic of quadratic and cyclotomic number fields the reader should consult the treatise of Borevich and Shafarevich [9].

In Section 3, we saw that $\mathbb{Q}(\sqrt{(-1)^{(p-1)/2}p})$ is a subfield of $\mathbb{Q}(\zeta_p)$. More generally, according to a theorem of Kronecker and Weber any algebraic number field which is Galois with an abelian Galois group is a subfield of $\mathbb{Q}(\zeta_m)$ for some m. For a proof of this difficult theorem see P. Ribenboim [207].

Exercises

1. Show that an algebraic number field of odd degree cannot contain a primitive nth root of unity $n > 2$.

2. Let F be a real quadratic field. Show that if F has an element of norm -1 then no prime $p \equiv 3 \ (4)$ is ramified.

3. Prove that if F is an algebraic number field such that $e^{2\pi i/n} \in F$ for some $n \geq 3$ then the norm of any nonzero element of F is positive.

4. Find the fundamental unit for $Q(\sqrt{5}), Q(\sqrt{15}), Q(\sqrt{2}), Q(\sqrt{3}), Q(\sqrt{624})$.

5. Show that a quadratic number field cannot contain $\sqrt{p}$ and $\sqrt{q}$ for two distinct primes p and q.

6. List the subfields of $Q(\zeta_8)$.

7. Let F be a real quadratic field. Show that there are algebraic integers in F arbitrarily close to 1 and distant from 1.

8. Show that the class number of $Q(\sqrt{10})$ is not 1.

9. Let p be an odd prime and consider $Q(\zeta_p)$.
   (a) Show that $N(1 + \zeta) = 1$ where N denotes the norm from $Q(\zeta_p)$ to Q.
   (b) Show that $\prod (1 + \zeta^s) = A$, the product being over the squares modulo p, is in $Q(\sqrt{p})$.
   (c) If $p \equiv 1$ (4), show that $A = (t + u\sqrt{p})/2$ with $t \equiv u$ (2).
   (d) Conclude from (a) that $((t^2 - pu^2)/4)^{(p-1)/2} = +1$ so that
   (e) $t^2 - pu^2 = \pm 4$.
   (f) Show that $A \neq -1$ by showing that $A > 0$ (compare Exercise 3).

   Now let $p \equiv 5$ (8).

   (g) Show that $A \neq 1$ by considering the polynomial $\prod_s (1 + x^s) - 1$, $s = 1^2, 2^2, \ldots, ((p - 1)/2)^2$. (See also Exercise 9, Chapter 16.) This exercise is adapted from Hartung [145].

10. For which d does $Q(\sqrt{d})$ have an integral basis of the form α, α' where α' is the conjugate of α?

11. Show that $-(\zeta^3 + \zeta^2)$ is a unit in $Q(\zeta)$, $\zeta = e^{2\pi i/5}$. What is the relation between this unit and the units in $Q(\sqrt{5})$?

12. Show that $\sin(\pi j/p)/\sin(\pi/p)$ is a unit in $Q(\zeta_p)$, $1 \leq j \leq p - 1$.

13. Show that if $p \equiv 1$ (4), p prime, then the ring of integers in $Q(\zeta_p)$ always contains an infinite number of units.

14. Let p be prime. Show that the discriminant Δ of $Q(\zeta_p)$ is $\prod_{i<j} (\zeta^i - \zeta^j)^2$, $1 \leq i$, $j \leq p - 1$.

15. The notation being as in Exercise 14 show
   (a) $-p\zeta^{-j}/(1 - \zeta^j) = \prod (\zeta^j - \zeta^i)$, the product over all $i, j, i \neq j, 1 \leq i, j \leq p - 1$.
   (b) Multiply for $j = 1, 2, \ldots, p - 1$ to obtain $\Delta = (-1)^{(p-1)/2} p^{p-2}$.

16. Use Proposition 13.2.8 to show that $i \notin Q(\zeta_p)$, p odd.

17. Use Propositions 13.2.7 and 13.2.8 to show that $\zeta_q \notin Q(\zeta_p)$ if p and q are odd primes $p \neq q$.

18. Show that if p is a prime congruent to 3 modulo 4 then $Q(\sqrt{p})$ is contained in the cyclotomic field $Q(\zeta_{4p})$.

19. Show that any quadratic number field is contained in a cyclotomic field.

20. Show that the fundamental unit of the real quadratic field $\mathbb{Q}(\sqrt{10})$ is $3 + \sqrt{10}$ and using the formula given in the text determine the class number of the field.

21. Let $a \in \mathbb{Z}$, a not a square, $a \equiv 0$ (4) or $a \equiv 1$ (4). Define the Kronecker symbol χ_a as follows: If $p \mid a$, $\chi_a(p) = 0$. If $p \nmid a$, is an odd prime then $\chi_a(p) = (a/p)$ the Legendre symbol; $\chi_a(2) = 1$ if $a \equiv 1$ (8), $\chi_a(2) = -1$ if $a \equiv 5$ (8). Finally $\chi_a(b) = \prod_{i=1}^{t} \chi_a(p_i)$ if $\pm b = p_1 \cdots p_t$. Show
    (a) For b odd χ_a coincides with the Jacobi symbol.
    (b) If $b > 0$, $(a, b) = 1$, $a = 2^t c$ with c odd then $\chi_a(b) = \chi_2(b)^t \chi_b(c)(-1)^{((c-1)/2)((b-1)/2)}$
    (c) $\chi_a(x) = \chi_a(y)$ if $x \equiv y$ (a).

22. Let K be a quadratic number field with discriminant d, and let χ_d be the Kronecker symbol. Show, for p *any* prime,
    (a) p splits in K iff $\chi_d(p) = 1$.
    (b) p is inertial iff $\chi_d(p) = -1$.
    (c) p ramifies iff $\chi_d(p) = 0$.

23. Using the table in Stark [73], p. 340, along with the tables in Borevich–Shafarevich [9], pp. 422–425 verify the Hirzebruch formula stated at the end of Section 1 for the primes 7, 19, 23, 31, 43, 47, 67, 83. Furthermore check the class numbers for the imaginary quadratic fields using Dirichlet's formula. Show that, knowing the class number of $\mathbb{Q}(\sqrt{-91})$ to be 2, $\mathbb{Q}(\sqrt{91})$ is not a principal ideal ring.

24. Let K be the field of pth roots of unity, p an odd prime. Show, without using Gauss sums, that the unique quadratic subfield of K has discriminant $(-1)^{(p-1)/2}p$.

25. The situation being as in the preceding problem, let f be the order of q modulo p, p, for an odd prime $q \neq p$. If E denotes the quadratic subfield of K show that q splits in E iff E is contained in the subfield D of degree $(p - 1)/f$. Show furthermore that this is the case iff q is a square modulo p. Using the preceding exercise derive a new proof of the law of quadratic reciprocity.

26. Count the number of proofs to the law of quadratic reciprocity given thus far in this book and devise another one.

27. Show that there are no primes which remain prime in $\mathbb{Q}(\zeta_8)$. Can you generalize?

Chapter 14

The Stickelberger Relation and the Eisenstein Reciprocity Law

Having developed the basic properties of cyclotomic fields we will prove two beautiful and important theorems which play a fundamental role in the further development of the theory of these fields.

The Eisenstein reciprocity law generalizes some of our previous work on quadratic and cubic reciprocity. It lies midway between these special cases and the more general reciprocity laws investigated by Kummer and Hilbert, proven first by Furtwängler and then in full generality by Artin and Hasse. In the last section of this chapter we will give two interesting applications of Eisenstein's result. The first concerns Fermat's Last Theorem and the second the theory of power residues.

The Stickelberger relation is the basis for the proof we give of Eisenstein reciprocity. Its importance goes far beyond that. In recent years the theory of cyclotomic fields has been dramatically advanced principally due to the efforts of K. Iwasawa. In his work the Stickelberger relation occupies a central position. It has also turned out to be of importance in arithmetic algebraic geometry.

§1 The Norm of an Ideal

We will need a few more results from the general theory of algebraic number fields.

Let $K/\mathbb{Q}$ be an algebraic number field, D the ring of integers in K, and A an ideal. We define $N(A)$, the norm of A, to be the number of elements in D/A. We continue to assume that ideals are nonzero.

Proposition 14.1.1. *If A, $B \subset D$ are ideals, then $N(AB) = N(A)N(B)$.*

PROOF. If A and B are relatively prime, then $D/AB \approx D/A \oplus D/B$ so the assertion is clear in this case.

Let $A = P_1^{a_1} P_2^{a_2} \cdots P_t^{a_t}$ be the prime decomposition of A. We claim $N(A) = (N(P_1))^{a_1}(N(P_2))^{a_2} \cdots (N(P_t))^{a_t}$. On the basis of what has been said it will be sufficient to prove $N(P^a) = (N(P))^a$ for any prime ideal P. This, however, is just a reformulation of Proposition 12.3.2.

Now, in the general case, decompose A and B into a product of prime ideals, multiply, apply the above result, and rearrange terms. The result follows. $\square$

Proposition 14.1.2. *Suppose $K/\mathbb{Q}$ is a Galois extension with group G. Then*

$$\prod_{\sigma \in G} \sigma(A) = (N(A)).$$

PROOF. Since both sides are multiplicative in A it suffices to prove the result when A is a prime ideal P.

Let $P_1, P_2, \ldots, P_g$ be the distinct prime ideals in the set $\{\sigma(P) | \sigma \in G\}$. Then $|G| = g|G(P)|$ where $G(P) = \{\sigma \in G | \sigma(P) = P\}$. Since $efg = n = [K : \mathbb{Q}] = |G|$ we see $|G(P)| = ef$. Thus, using Proposition 12.3.3 and Theorem 3', Chapter 12

$$\prod_{\sigma \in G} \sigma(P) = (P_1 P_2 \cdots P_g)^{ef} = (p)^f = (p^f), \quad \text{where } P_i \cap \mathbb{Z} = p\mathbb{Z}.$$

Since $N(P) = |D/P| = p^f$, this completes the proof. $\square$

Proposition 14.1.3. *Let $K/\mathbb{Q}$ be Galois with group G. Let $\alpha \in D$ and let $A = (\alpha)$ be the principal ideal generated by α. Let $N\alpha$ be the norm of α. Then $N(A) = |N(\alpha)|$.*

PROOF. $(N(A)) = \prod \sigma(A) = \prod \sigma((\alpha)) = \prod (\sigma\alpha) = (\prod \sigma(\alpha)) = (N(\alpha))$. Thus $N(A)$ and $N(\alpha)$ differ by a unit. Since they are both in $\mathbb{Z}$ they can differ only by sign. Since $N(A)$ is, by definition, positive, we have $N(A) = |N(\alpha)|$ as asserted. $\square$

We remark that the above proposition is true even if $K/\mathbb{Q}$ is not a Galois extension. The proof in the general case is somewhat more complicated.

§2 The Power Residue Symbol

Let m be a positive integer, and denote by D_m the ring of integers in $\mathbb{Q}(\zeta_m)$. Let P be a prime ideal in D_m not containing m. Let $q = N(P) = |D_m/P|$. By Proposition 13.2.3 we know that the cosets of $1, \zeta_m, \ldots, \zeta_m^{m-1}$ are distinct and $q \equiv 1 \ (m)$.

Proposition 14.2.1. *Let $\alpha \in D_m$, $\alpha \notin P$. There is an integer i, unique modulo m, such that*

$$\alpha^{(q-1)/m} \equiv \zeta_m^i \ (P).$$

PROOF. Since the multiplicative group of D_m/P has $q - 1$ elements we have $\alpha^{q-1} \equiv 1 \ (P)$. Thus

$$\prod_{i=0}^{m-1} (\alpha^{(q-1)/m} - \zeta_m^i) \equiv 0 \ (P).$$

Since P is a prime ideal there is an integer $i, 0 \leq i < m$ such that $\alpha^{(q-1)/m} \equiv \zeta_m^i \ (P)$. If $i \not\equiv j \ (m)$ then $\zeta_m^i \not\equiv \zeta_m^j \ (P)$, so i is unique modulo m. $\square$

Definition. For $\alpha \in D_m$ and P a prime ideal not containing m, define the mth power residue symbol, $(\alpha/P)_m$, as follows:

(a) $(\alpha/P)_m = 0$ if $\alpha \in P$.
(b) If $\alpha \notin P$, $(\alpha/P)_m$ is the unique mth root of unity such that $\alpha^{(NP-1)/m} \equiv (\alpha/P)_m \ (P)$.

Proposition 14.2.2.

(a) $(\alpha/P)_m = 1$ iff $x^m \equiv \alpha \ (P)$ is solvable in D_m.
(b) For all $\alpha \in D_m$, $\alpha^{(NP-1)/m} \equiv (\alpha/P)_m \ (P)$.
(c) $(\alpha\beta/P)_m = (\alpha/P)_m(\beta/P)_m$.
(d) If $\alpha \equiv \beta \ (P)$ then $(\alpha/P)_m = (\beta/P)_m$.

PROOF. Since the result has been proven earlier for $m = 2, 3$, and 4 we may safely leave the details to the reader. $\square$

Corollary. Suppose P is a prime ideal not containing m. Then

$$\left(\frac{\zeta_m}{P}\right)_m = \zeta_m^{(NP-1)/m}.$$

PROOF. From part (b) of the proposition, both sides of the above equality are congruent modulo P. Since they are both mth roots of unity and $m \notin P$, it follows that they are equal. $\square$

It is important to extend the definition of $(\alpha/P)_m$ in such a way that $(\alpha/\beta)_m$ makes sense when β is prime to m. This is done as follows:

Definition. Suppose $A \subset D_m$ is an ideal prime to m. Let $A = P_1 P_2 \cdots P_n$ be the prime decomposition of A. For $\alpha \in D_m$ define $(\alpha/A)_m = \prod_i (\alpha/P_i)_m$. If $\beta \in D_m$ and β is prime to m define $(\alpha/\beta)_m = (\alpha/(\beta))_m$.

Proposition 14.2.3. *Suppose A and B are ideals prime to (m). Then*

(a) $(\alpha\beta/A)_m = (\alpha/A)_m(\beta/A)_m$.
(b) $(\alpha/AB)_m = (\alpha/A)_m(\alpha/B)_m$.
(c) *If α is prime to A and $x^m \equiv \alpha \ (A)$ is solvable in D_m then $(\alpha/A)_m = 1$.*

PROOF. All three assertions are straightforward to prove using the last proposition and the above definition. We remark that the converse of part (c) is not true. $\square$

We will need to see how the symbol $(\alpha/A)_m$ behaves with respect to automorphisms in the Galois group G of $\mathbb{Q}(\zeta_m)/\mathbb{Q}$.

From now on we will use exponential notation for automorphisms. If $\sigma \in G$ and $\alpha \in \mathbb{Q}(\zeta_m)$ we will write α^σ instead of $\sigma\alpha$. Similarly if A is an ideal, we will write A^σ instead of $\sigma(A)$. This notation is, in fact, more conventional and it has certain advantages.

Proposition 14.2.4. *Let A be an ideal prime to m and $\sigma \in G$. Then*

$$\left(\frac{\alpha}{A}\right)_m^\sigma = \left(\frac{\alpha^\sigma}{A^\sigma}\right)_m.$$

PROOF. Since both sides of the asserted equality are multiplicative in A it will be enough to check the case where $A = P$ is a prime ideal. By definition

$$\alpha^{(NP-1)/m} \equiv \left(\frac{\alpha}{P}\right)_m \ (P).$$

Applying σ to this congruence we find

$$(\alpha^\sigma)^{(NP-1)/m} \equiv \left(\frac{\alpha}{P}\right)_m^\sigma \ (P^\sigma).$$

It follows that $(\alpha^\sigma/P^\sigma)_m \equiv (\alpha/P)_m^\sigma \ (P^\sigma)$ and so $(\alpha^\sigma/P^\sigma)_m = (\alpha/P)_m^\sigma$. Note that we have used $N(P^\sigma) = N(P)$. $\square$

We end this section by stating the Eisenstein reciprocity law. We need an important definition first.

Let l be an odd prime number. Recall that in D_l we have $(l) = (1 - \zeta_l)^{l-1}$ and $(1 - \zeta_l)$ is a prime ideal of degree 1.

Definition. A nonzero element $\alpha \in D_l$ is called primary if it is not a unit and is prime to l and congruent to a rational integer modulo $(1 - \zeta_l)^2$.

In the case $l = 3$ we demanded $\alpha \equiv 2 \ (1 - \zeta_3)^2$ so the above definition is a bit weaker in this case. It is, however, sufficient for our purposes. The following lemma shows that primary elements are plentiful.

Lemma. Suppose $\alpha \in D_l$ and α is prime to l. There is an integer $c \in \mathbb{Z}$, unique modulo l, such that $\zeta_l^c \alpha$ is primary.

PROOF. Let $\lambda = 1 - \zeta_l$. Since the prime ideal (λ) has degree 1 there is an integer $a \in \mathbb{Z}$ such that $\alpha \equiv a \ (\lambda)$. Now, $(\alpha - a)/\lambda \in D_l$ so there is a $b \in \mathbb{Z}$ such that $(\alpha - a)/\lambda \equiv b \ (\lambda)$. Consequently, $\alpha \equiv a + b\lambda \ (\lambda^2)$.

Since $\zeta_l = 1 - \lambda$ we have $\zeta_l^c \equiv 1 - c\lambda \ (\lambda^2)$. It follows that

$$\zeta_l^c \alpha \equiv a + (b - ac)\lambda \ (\lambda^2).$$

The integer a is not divisible by l since otherwise $\lambda \mid \alpha$ and we are assuming α is prime to l. Choose c to be a solution to $ax \equiv b\ (l)$. Then $\zeta_l^c \alpha \equiv a\ (\lambda^2)$ and so $\zeta_l^c \alpha$ is primary.

The uniqueness of c modulo l is clear from the proof. $\qquad\qquad\square$

Theorem 1 (The Eisenstein Reciprocity Law). *Let l be an odd prime, $a \in \mathbb{Z}$ prime to l, and $\alpha \in D_l$ a primary element. Suppose moreover that α and a are prime to each other. Then*

$$\left(\frac{\alpha}{a}\right)_l = \left(\frac{a}{\alpha}\right)_l.$$

The proof of this elegant theorem will be given in Section 5. It is a consequence of the Stickelberger relation which will be stated in the next section and proven in Section 4. Since this process is long, and somewhat involved, the reader may wish to skip to the last part of the chapter, Section 6, where three interesting applications of Eisenstein reciprocity are given.

§3 The Stickelberger Relation

From the very way they are defined Gauss sums are elements of cyclotomic fields. We will investigate the prime ideal decomposition of Gauss sums in these fields.

Let F be a finite field with $p^f = q$ elements, χ a multiplicative character of order m, and ψ a nontrivial additive character. Then the values of χ are mth roots of unity and the values of ψ are pth roots of unity. Consequently, $g(\chi, \psi) = \sum_{t \in F} \chi(t)\psi(t) \in \mathbb{Q}(\zeta_m, \zeta_p)$. The arithmetic of this field was dealt with in the last chapter.

Before beginning it is necessary to normalize matters by specifying the characters χ and ψ. This is done as follows.

Let P be a prime ideal in $D_m \subset \mathbb{Q}(\zeta_m)$ and suppose $m \notin P$. Let $p\mathbb{Z} = P \cap \mathbb{Z}$ and $N(P) = q = p^f$. Finally set $F = D_m/P$. Recall that $p^f \equiv 1\ (m)$.

We define a multiplicative character χ_P on F as follows. Let $0 \neq t \in F$ and let $\gamma \in D_m$ be such that $\bar{\gamma} = t$. Here $\bar{\gamma}$ is the residue class of γ modulo P. Let

$$\chi_P(t) = \left(\frac{\gamma}{P}\right)_m^{-1} = \overline{\left(\frac{\gamma}{P}\right)_m}.$$

By Proposition 14.2.2, $\chi_P(t)$ is well defined and is a multiplicative character. The reason for taking the inverse of the power residue symbol instead of the symbol itself will become apparent later.

For the additive character we choose the character ψ defined in Chapter 10, Section 3. We recall the definition. First one defines tr: $F \to \mathbb{Z}/p\mathbb{Z}$ by tr(t) = $t + t^p + t^{p^2} + \cdots + t^{p^{f-1}}$. Then ψ is defined by $\psi(t) = \zeta_p^{\text{tr}(t)}$.

With these choices we define $g(P) = g(\chi_P, \psi)$. We also define $\Phi(P) = g(P)^m$.

Proposition 14.3.1.

(a) $g(P) \in \mathbb{Q}(\zeta_m, \zeta_p)$.
(b) $|g(P)|^2 = q$.
(c) $\Phi(P) \in \mathbb{Q}(\zeta_m)$.

PROOF. (a) has already been discussed. (b) follows in the same way as when F is the prime field. (c) follows from Proposition 8.3.3 which is stated over $\mathbb{Z}/p\mathbb{Z}$ but generalizes easily to F.

We will give another proof of (c) based on Galois theory. Consider the diagram of fields

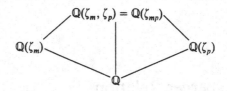

The Galois group of $\mathbb{Q}(\zeta_{mp})/\mathbb{Q}$ is given by the automorphisms σ_c where $(c, pm) = 1$. We remark

(i) σ_c leaves $\mathbb{Q}(\zeta_m)$ element-wise fixed iff $c \equiv 1 \ (m)$.
(ii) σ_c leaves $\mathbb{Q}(\zeta_p)$ element-wise fixed iff $c \equiv 1 \ (p)$.

To show $\Phi(P) \in \mathbb{Q}(\zeta_m)$ it will suffice to show $\Phi(P)^{\sigma_c} = \Phi(P)$ whenever $c \equiv 1 \ (m)$.

Apply σ_c with $c \equiv 1(m)$ to $g(P) = \sum \chi_P(t)\psi(t)$. Since $\chi_P(t)^{\sigma_c} = \chi_P(t)$ and $\psi(t)^{\sigma_c} = \psi(t)^c = \psi(ct)$ we have

$$g(P)^{\sigma_c} = \sum \chi_P(t)\psi(ct) = \chi_P(c)^{-1}g(P).$$

Raising both sides to the mth power shows that $\Phi(P)$ is invariant under σ_c as asserted. $\square$

Before proceeding to discuss the prime decomposition of $g(P)$ and $\Phi(P)$ in the general case it is illuminating to review the situation when $m = 2, 3$, and 4.

When $m = 2$, $\mathbb{Q}(\zeta_2) = \mathbb{Q}$. If p is the positive generator of P we have $g(P)^2 = (-1)^{(p-1)/2}p$.

When $m = 3$, $\mathbb{Q}(\zeta_3) = \mathbb{Q}(\sqrt{-3})$. Suppose P has degree 1 and $P = (\pi)$ where π is primary. From the results of Chapter 9, Section 4, we may deduce $g(P)^3 = \Phi(P) = p\bar{\pi} = \pi\bar{\pi}^2$ (bar denotes complex conjugation).

For $m = 4$, $\mathbb{Q}(\zeta_4) = \mathbb{Q}(\sqrt{-1})$. Suppose P is a prime ideal of degree 1 and $P = (\pi)$ where π is primary. From Chapter 9, Section 7, we may deduce $g(P)^4 = \Phi(P) = p\bar{\pi}^2 = \pi\bar{\pi}^3$ (again, bar denotes complex conjugation).

To see the pattern, and to state the generalization a notational device known as "symbolic powers" is very useful. Suppose $K/\mathbb{Q}$ is a number field, Galois over $\mathbb{Q}$, with group G. The group ring $\mathbb{Z}[G]$ is defined as the set of formal expressions $\sum_{\sigma \in G} a(\sigma)\sigma$ where the coefficients $a(\sigma) \in \mathbb{Z}$. Later, we will show how to make this set into a ring. If $\alpha \in K$ we define

$$\alpha^{\sum a(\sigma)\sigma} = \prod_\sigma \sigma(\alpha)^{a(\sigma)}.$$

If A is an ideal we define its symbolic power by an element of the group ring in the same way.

Let σ be the nontrivial automorphism of $\mathbb{Q}(\sqrt{-3})/\mathbb{Q}$. Our result for $m = 3$ takes the form $\Phi(P) = \pi^{1+2\sigma}$.

Similarly if τ denotes the nontrivial automorphism of $\mathbb{Q}(\sqrt{-1})/\mathbb{Q}$ our result is $\Phi(P) = \pi^{1+3\tau}$.

In general we cannot expect a factorization of $\Phi(P)$ into irreducible elements since D_m is not always a unique factorization domain. However, these special cases generalize beautifully as follows.

Theorem 2 (The Stickelberger Relation). *Let P be a prime ideal in D_m not containing m. Then*

$$(\Phi(P)) = P^{\sum t\sigma_t^{-1}}.$$

The sum is over all $1 \le t < m$ which are relatively prime to m.

The proof of Theorem 2 is long. It will occupy the next section entirely.

§4 The Proof of the Stickelberger Relation

We begin with three elementary results which will be needed later.

Lemma 1. *Let $p > 1$ be a positive integer. Every positive integer can be written uniquely in the form $\sum_{i=0}^n a_i p^i$ where $0 \le a_i < p$.*

PROOF. Let a be a positive integer. There is a unique nonnegative integer n such that $p^n \le a < p^{n+1}$. By the division algorithm we have $a = a_n p^n + r$ where $0 \le r < p^n$. The number a_n is less than p since otherwise $a \ge p^{n+1}$. Apply the same process to r, etc. In finitely many steps we have an expression for a of the required form.

The uniqueness can be shown as follows. Suppose $\sum a_i p^i = \sum b_i p^i$ where $0 \le a_i, b_i < p$. Then p divides $a_0 - b_0$. Since $|a_0 - b_0| < p$ we have $a_0 = b_0$.

Subtract a_0 from both sides, divide by p, and repeat the reasoning. This yields $a_1 = b_1$. In finitely many steps we see $a_i = b_i$ for all i. $\square$

Definition. Let $q = p^f$. If $0 \le a < q - 1$ write $a = \sum_{i=0}^{f-1} a_i p^i$ with $0 \le a_i < p$ and define $S(a) = \sum_{i=0}^{f-1} a_i$. For an arbitrary positive integer a define $S(a) = S(r)$ where $a \equiv r \ (q - 1)$ and $0 \le r < q - 1$.

Definition. For a real number u define $\langle u \rangle$ as $u - [u]$ where $[u]$ is the largest integer less than or equal to u. The number $\langle u \rangle$, which is in the interval $[0, 1)$, is called the fractional part of u.

Lemma 2. $S(a) = (p - 1) \sum_{i=0}^{f-1} \langle p^i a/(q - 1) \rangle$.

PROOF. Both sides are unchanged if a multiple of $q - 1$ is added to a. Thus we may assume $1 \le a < q - 1$.

Write $a = a_0 + a_1 p + \cdots + a_{f-1} p^{f-1}$ where $0 \le a_i < p$. Since $p^f = q \equiv 1 \ (q - 1)$ we have

$$a = a_0 + a_1 p + \cdots + a_{f-1} p^{f-1},$$

$$pa \equiv a_{f-1} + a_0 p + \cdots + a_{f-2} p^{f-1} \ (q - 1),$$

$$p^2 a = a_{f-2} + a_{f-1} p + \cdots + a_{f-3} p^{f-1} \ (q - 1), \text{ etc.}$$

The right-hand sides of these congruences are all less than $q - 1$ so that $\langle p^i a/(q - 1) \rangle$ is equal to the right-hand side of the ith congruence divided by $q - 1$. Thus

$$\sum_{i=0}^{f-1} \left\langle \frac{p^i a}{q - 1} \right\rangle = \frac{1}{q - 1} S(a)(1 + p + \cdots + p^{f-1}).$$

This yields the lemma. $\square$

Lemma 3. $\sum_{a=1}^{q-2} S(a) = (f(p - 1)(q - 2))/2$.

PROOF. Write $a = a_0 + a_1 p + \cdots + a_{f-1} p^{f-1}$ with $0 \le a_i < p$. Notice that $q - 1 = (p - 1) + (p - 1)p + \cdots + (p - 1)p^{f-1}$. It follows that $q - 1 - a = (p - 1 - a_0) + (p - 1 - a_1)p + \cdots + (p - 1 - a_{f-1})p^{f-1}$ and so

$$S(a) + S(q - 1 - a) = f(p - 1).$$

Sum both sides from $a = 1$ to $a = q - 2$. The result is $2 \sum_{a=1}^{q-2} S(a) = f(p - 1)(q - 2)$. $\square$

The Gauss sum $g(P)$ considered in the last section is an element of $\mathbb{Q}(\zeta_m, \zeta_p)$. The proof of Theorem 2 which we will give requires that we work in the bigger field $\mathbb{Q}(\zeta_{q-1}, \zeta_p)$. This has the advantage that all the $(q - 1)$st roots of unity can be used freely. On the other hand, more fields means more confusion. We will try to minimize the confusion by carefully keeping track of which field we are working in.

The following diagram will be useful in following the arguments.

$$
\begin{array}{ccc}
\mathscr{P} \subset D_{(q-1)p} & \to & D_{(q-1)p}/\mathscr{P} \\
| \quad\quad | & & | \\
\mathfrak{P} \subset D_{q-1} & \to & D_{q-1}/\mathfrak{P} \\
| \quad\quad | & & | \\
P \subset D_m & \to & D_m/P \\
| \quad\quad | & & | \\
p \subset \mathbb{Z} & \to & \mathbb{Z}/p\mathbb{Z}
\end{array} \right\} f
$$

In the above diagram P, $\mathfrak{P}$, and $\mathscr{P}$ are prime ideals in the indicated ring of integers. Recall from Section 3 that $p \nmid m$, f is the order of p modulo m, so that $p^f \equiv 1 \ (m)$, and $q = p^f$. For the remainder of this section $\lambda_p = 1 - \zeta_p$.

Lemma 4.

(1) $\mathrm{ord}_{\mathscr{P}}(pD_{(q-1)p}) = p - 1$.
(2) $\mathrm{ord}_{\mathscr{P}}(\lambda_p) = 1$.
(3) $\mathrm{ord}_{\mathscr{P}}(P) = p - 1$.

PROOF. To prove (1) apply Proposition 13.2.9 with m (in the notation of that proposition) replaced by $q - 1$. Since $\mathscr{P}$ lies over p it appears in the decomposition of pD and one has $\mathrm{ord}_{\mathscr{P}} \, pD_{(q-1)p} = p - 1$. Again by the same proposition and Proposition 13.2.7 one has $pD_{p(q-1)} = (pD_p)D_{p(q-1)} = \lambda_p^{p-1} D_{p(q-1)} = (\mathscr{P}_1 \cdots \mathscr{P}_h)^{p-1}$, where, say, $\mathscr{P}_1 = \mathscr{P}$. Hence $\lambda_p D_{p(q-1)} = \mathscr{P}_1 \cdots \mathscr{P}_h$ and (2) follows. To prove (3) one sees easily using Theorem 2 of Chapter 13 and Proposition 13.2.9 that $PP_2 \cdots P_h \cdot D_{(q-1)p} = (\mathscr{P}\mathscr{P}_2 \cdots \mathscr{P}_h)^{p-1}$ where all the primes are distinct and P, P_2, ..., P_h are pairwise relatively prime. Thus $PD_{(q-1)p} = \mathscr{P}^{p-1}$ and the result follows.   □

Lemma 5. $D_m/P \approx D_{q-1}/\mathfrak{P}$.

PROOF. There is a natural monomorphism from D_m/P to $D_{q-1}/\mathfrak{P}$. To show this is an isomorphism it suffices to show both fields have the same number of elements. By Theorem 2 of Chapter 13 we have $|D_{q-1}/\mathfrak{P}| = p^{f'}$ where f' is the smallest positive integer such that $p^{f'} \equiv 1 \ (q - 1)$. Since $q = p^f$ it is clear that $f' = f$ and so $|D_{q-1}/\mathfrak{P}| = p^f = |D_m/P|$.   □

By Proposition 13.2.3 we know that the elements $1, \zeta_{q-1}, \ldots, \zeta_{q-1}^{q-2}$ have distinct images in $D_{q-1}/\mathfrak{P}$. The following definition imitates the definition of the mth power residue symbol.

Definition. For $\alpha \in D_{q-1}$ define

(a) $(\alpha/\mathfrak{P}) = 0$ if $\alpha \in \mathfrak{P}$.
(b) If $\alpha \notin \mathfrak{P}$, $(\alpha/\mathfrak{P})$ is the unique $(q - 1)$st root of unity such that $\alpha \equiv (\alpha/\mathfrak{P}) \ (\mathfrak{P})$.

One easily checks that $(\alpha\beta/\mathfrak{P}) = (\alpha/\mathfrak{P})(\beta/\mathfrak{P})$ and $\alpha \equiv \beta\ (\mathfrak{P})$ implies $(\alpha/\mathfrak{P}) = (\beta/\mathfrak{P})$. The following lemma is also clear from the definitions.

Lemma 6. *If $\alpha \in D_m$, $(\alpha/\mathfrak{P})^{(q-1)/m} = (\alpha/P)_m$.*

We now define a multiplicative character on $\mathbb{F} \approx D_{q-1}/\mathfrak{P}$ as follows

$$\omega(t) = \left(\frac{\gamma}{\mathfrak{P}}\right),$$

where $\gamma \in D_{q-1}$ is such that $\bar{\gamma} = t$. The proof that ω is well defined and is a multiplicative character is immediate from the previous remarks.

Lemma 7. $\omega(\zeta_{q-1}^{p^i}) = \zeta_{q-1}^i$.

PROOF. Immediate from the definition.   □

Consequently, ω has order $q - 1$ and thus generates the group of multiplicative characters on $\mathbb{F}$.

Definition. Let a be a nonnegative integer. Define $g_a = g(\omega^{-a}, \psi)$.

We note that $g(P)$, defined in the last section, is equal to g_a for $a = (q - 1)/m$.

Theorem 2 is a consequence of the following result.

Theorem 3. $\mathrm{ord}_{\mathscr{P}}(g_a) = S(a)$, *where* $1 \le a < q$.

PROOF. To begin with we show that $\mathrm{ord}_{\mathscr{P}}(g_1) = 1$. Recall

$$g_1 = \sum_{t \in \mathbb{F}} \omega(t)^{-1} \zeta_p^{\mathrm{tr}(t)}.$$

Using Lemma 7 we will convert this into a sum over the powers of ζ_{q-1}. Let m_i be a positive integer such that $m_i \equiv \mathrm{tr}(\zeta_{q-1}^i)\ (p)$. Also recall that $\zeta_p = 1 - \lambda_p$. Then

$$g_1 = \sum_{i=0}^{q-2} \zeta_{q-1}^{-i}(1 - \lambda_p)^{m_i}.$$

Using the binomial theorem we see $(1 - \lambda_p)^{m_i} \equiv 1 - m_i\lambda_p\ (\mathscr{P}^2)$ and so

$$g_1 \equiv -\left(\sum_{i=0}^{q-2} m_i\zeta_{q-1}^{-i}\right)\lambda_p\ (\mathscr{P}^2).$$

Now, $m_i\lambda_p \equiv (\zeta_{q-1}^i + \zeta_{q-1}^{pi} + \cdots + \zeta_{q-1}^{p^{f-1}i})\lambda_p\ (\mathscr{P}^2)$. Substituting we find

$$g_1 \equiv -\sum_{i=0}^{q-2} \zeta_{q-1}^{-i}(\zeta_{q-1}^i + \zeta_{q-1}^{pi} + \cdots + \zeta_{q-1}^{p^{f-1}i})\lambda_p\ (\mathscr{P}^2).$$

All the sums $\sum_{i=0}^{q-2} \zeta_{q-1}^{(p^j-1)i}, j = 1, 2, \ldots, f - 1$ are zero while $j = 0$ gives the value $q - 1$. Since $q = p^f \equiv 0 \ (\mathscr{P}^2)$ we have

$$g_1 \equiv \lambda_p \ (\mathscr{P}^2).$$

By Lemma 4, part (2), we see $\lambda_p \in \mathscr{P}$ but $\lambda_p \notin \mathscr{P}^2$. Thus $\operatorname{ord}_{\mathscr{P}} g_1 = 1$.

Let $\tilde{s}(a) = \operatorname{ord}_{\mathscr{P}} g_a$. We will establish a number of properties of the function $\tilde{s}(a)$.

(i) $\tilde{s}(a + b) \leq \tilde{s}(a) + \tilde{s}(b)$ provided $1 \leq a, b, a + b < q - 1$.

By Theorem 1 of Chapter 8 we have $g_a g_b = J(\omega^{-a}, \omega^{-b})g_{a+b}$. Taking $\operatorname{ord}_{\mathscr{P}}$ of both sides yields the result.

(ii) $\tilde{s}(a + b) \equiv \tilde{s}(a) + \tilde{s}(b) \ (p - 1)$.

Notice that the Jacobi sum $J(\omega^{-a}, \omega^{-b})$ is in $\mathbb{Q}(\zeta_{q-1})$. It then follows from the fact that $\mathfrak{P}D_{(q-1)p} = \mathscr{P}^{(p-1)}$ that $p - 1$ divides $\operatorname{ord}_{\mathscr{P}}(J(\omega^{-a}, \omega^{-b}))$. The result is thus again an immediate consequence of the relation $g_a g_b = J(\omega^{-a}, \omega^{-b})g_{a+b}$.

(iii) $\tilde{s}(pa) = \tilde{s}(a)$.

To see this observe $g_{pa} = \sum \omega(t)^{-pa}\psi(t) = \sum \omega(t^p)^{-a}\psi(t^p)$. We have used the fact that $\operatorname{tr}(t) = \operatorname{tr}(t^p)$ which is clear from the definition of trace. Now $t \to t^p$ is an automorphism of $\mathbb{F}$. We conclude that $g_{pa} = g_a$ and so $\tilde{s}(pa) = \tilde{s}(a)$.

In the first part of the proof we found $\tilde{s}(1) = 1$. Using (i) and (ii) we see $\tilde{s}(a) = a$ for $1 \leq a < p$.

For any a between 1 and $q - 1$ write $a = a_0 + a_1 p + \cdots + a_{f-1}p^{f-1}$, $0 \leq a_i < p$. Using (i) and (iii) we find

$$\tilde{s}(a) \leq \sum_{j=0}^{f-1} \tilde{s}(a_j p^j) = \sum_j \tilde{s}(a_j) = \sum_j a_j = S(a).$$

We now have $\tilde{s}(a) \leq S(a)$ for all a in the range under consideration. To prove the theorem it will be enough, in the light of Lemma 3, to show

(iv) $$\sum_{a=1}^{q-2} \tilde{s}(a) = \frac{f(p-1)(q-2)}{2}.$$

In general, for Gauss sums, we have the relation $g(\chi^{-1}) = \chi(-1)\overline{g(\chi)}$ (here "bar" denotes complex conjugation). Thus $g_a g_{q-1-a} = \omega(-1)^a q = \omega(-1)^a p^f$. We know by Lemma 4 that $\operatorname{ord}_{\mathscr{P}}(p) = p - 1$. It follows that

$$\tilde{s}(a) + \tilde{s}(q - 1 - a) = f(p - 1).$$

Sum both sides over a from 1 to $q - 2$. The result is $2\sum_{a=1}^{q-2} \tilde{s}(a) = f(p - 1)(q - 2)$.

This completes the proof of Theorem 3.   □

Corollary. $\mathrm{ord}_P(\Phi(P)) = (m/(p-1))S((q-1)/m)$.

PROOF. Using Lemma 4, part (3), we have $(p-1)\,\mathrm{ord}_P(\Phi(P)) = \mathrm{ord}_{\mathscr{P}}(\Phi(P))$. Now, $\mathrm{ord}_{\mathscr{P}}(\Phi(P)) = m\,\mathrm{ord}_{\mathscr{P}}(g(P)) = mS((q-1)/m)$ where the last equality follows from the theorem because $g(P) = g_a$ with $a = (q-1)/m$. $\qquad\square$

This corollary gives the first step in deriving the full prime decomposition of $\Phi(P)$. To go further we first notice that the only prime ideals in D_m containing $\Phi(P)$ are those containing p. This follows from parts (b) and (c) of Proposition 14.3.1 which show

$$|\Phi(P)|^2 = q^m = p^{fm}.$$

If P' is another prime ideal of D_m containing p then by Proposition 12.3.3 there is an automorphism σ_t of $\mathbb{Q}(\zeta_m)/\mathbb{Q}$ such that $P' = P^{\sigma_t^{-1}}$. For $1 \le t < m$ and $(t, m) = 1$ define $P_t = P^{\sigma_t^{-1}}$.

Lemma 8. $\mathrm{ord}_{P_t}(\Phi(P)) = (m/(p-1))S(t((q-1)/m))$.

PROOF. It follows quickly from the definitions that

$$\mathrm{ord}_{P_t}(\Phi(P)) = \mathrm{ord}_P(\Phi(P)^{\sigma_t}).$$

Choose an integer t' such that $t' \equiv t\ (m)$ and $t' \equiv 1\ (p)$. Then

$$g(P)^{\sigma_{t'}} = \left(\sum_{r\in F}\chi_P(r)\psi(r)\right)^{\sigma_{t'}} = \sum_{r\in F}\chi_P(r)^t\psi(r).$$

Thus, we have

$$\Phi(P)^{\sigma_t} = \left(\sum_{r\in F}\chi_P^t(r)\psi(r)\right)^m.$$

The second term in the above equality is g_a^m where $a = t((q-1)/m)$. The proof of the lemma is now concluded by the same reasoning as in the corollary to Theorem 3. $\qquad\square$

We may now, finally, conclude the proof of Theorem 2.

By the corollary to Theorem 2 of Chapter 13 the group

$$G(P) = \{\sigma \in G(\mathbb{Q}(\zeta_m)/\mathbb{Q})\,|\,P^\sigma = P\}$$

is the cyclic group generated by σ_p.

Let $t_1, t_2, \ldots, t_g$ be a set of integers representing the cosets of $U(\mathbb{Z}/m\mathbb{Z})$ modulo the cyclic subgroup generated by the image of p. In other words, if $1 \le t < m$, $(t, m) = 1$ then $t \equiv t_i p^j\ (m)$ for a unique pair (i, j), $0 \le j < f$, $1 \le i \le g$. By Lemma 8 the prime decomposition of $\Phi(P)$ is given by

$$P^{\gamma'} \quad \text{where } \gamma' = \frac{m}{p-1}\sum_{i=1}^{g} S\left(t_i\frac{q-1}{m}\right)\sigma_{t_i}^{-1}.$$

Using Lemma 2 we can write γ' as follows

$$\gamma' = m \sum_i \left(\sum_j \left\langle \frac{p^j t_i}{m} \right\rangle \right) \sigma_{t_i}^{-1}.$$

The index i goes from 1 to g and the index j goes from 0 to $f - 1$. Since σ_p leaves P fixed, γ' has the same effect on P as

$$\gamma = m \sum_i \sum_j \left\langle \frac{p^j t_i}{m} \right\rangle \sigma_{t_i}^{-1} \sigma_{pj}^{-1}$$

$$= m \sum_{t \bmod m} \left\langle \frac{t}{m} \right\rangle \sigma_t^{-1}$$

$$= \sum t \sigma_t^{-1} \quad \text{where } 1 \le t < m \quad \text{and} \quad (t, m) = 1.$$

This concludes the proof.   □

For future reference we note

$$(\Phi(P)) = P^{m\theta}$$

where $\theta = \sum_{t \bmod m} \langle t/m \rangle \sigma_t^{-1}$, $(t, m) = 1$. The element $\theta \in \mathbb{Q}[G]$ is called the Stickelberger element.

§5 The Proof of the Eisenstein Reciprocity Law

We will need two results on roots of unity.

Lemma 1. *The only roots of unity in* $\mathbb{Q}(\zeta_m)$ *are* $\pm \zeta_m^i$, $i = 1, 2, \ldots, m$.

PROOF. In the proof of Theorem 1 we only need this result when m iş an odd prime. We will leave the proof for general m as an exercise and assume $m = l$, an odd prime.

Suppose $\zeta_n \in \mathbb{Q}(\zeta_l)$. If $4 | n$ then $\sqrt{-1} \in \mathbb{Q}(\zeta_l)$. However, 2 is ramified in $\mathbb{Q}(\sqrt{-1})$ and is not ramified in $\mathbb{Q}(\zeta_l)$. Thus $4 \nmid n$. If $n = 2n_0$, n_0 odd, then $\{\zeta_n^j\} = \{\pm \zeta_{n_0}^j\}$, we may assume that n is odd. If l' is an odd prime dividing n then $\zeta_{l'} \in \mathbb{Q}(\zeta_l)$. However, l' is ramified in $\mathbb{Q}(\zeta_{l'})$ and l is the only prime ramified in $\mathbb{Q}(\zeta_l)$. Thus $l = l'$ and n must be a power of l, l^a say. Since $\phi(l^a) = l^{a-1}(l - 1)$ is the dimension of $\mathbb{Q}(\zeta_{l^a})$ over $\mathbb{Q}$ and $l - 1$ is the dimension of $\mathbb{Q}(\zeta_l)$ over $\mathbb{Q}$ we must have $a = 1$. The result follows from this.   □

Lemma 2. *Let* $K/\mathbb{Q}$ *be an algebraic number field and let* $\sigma_1, \sigma_2, \ldots, \sigma_n$ *be the* $n = [K : \mathbb{Q}]$ *isomorphisms of* K *into* $\mathbb{C}$. *If* $\alpha \in K$ *is such that* $|\alpha^{\sigma_i}| \le 1$ *for all* $i = 1, 2, \ldots, n$ *then* α *is a root of unity.*

PROOF. α is a root of

$$f(x) = \prod_{i=1}^{n} (x - \alpha^{\sigma_i}) \in \mathbb{Z}[x].$$

The hypothesis of the lemma implies that the coefficient of x^m in $f(x)$ is an integer bounded by the binomial coefficient $\binom{n}{m}$. Thus only finitely many polynomials of degree n is $\mathbb{Z}[x]$ can arise in this way.

If α satisfies the hypothesis of the lemma so do all the powers of α. Since finitely many polynomials can have only finitely many roots it follows that two distinct powers of α must be equal. Thus α is a root of unity. $\square$

The next step is to define $\Phi(A)$ for an arbitrary ideal of D_m, A prime to m, and to investigate the properties of this function. In particular, it will be important to determine Φ on principal ideals.

Definition. Let $A \subset D_m$ be an ideal and assume A is prime to m. Let $A = P_1 P_2 \cdots P_n$ be the prime decomposition of A. Define

$$\Phi(A) = \Phi(P_1)\Phi(P_2) \cdots \Phi(P_n).$$

Proposition 14.5.1. *Let $A, B \subset D_m$ be ideals prime to m, $\alpha \in D_m$ an element prime to m, and recall $\gamma = \sum t\sigma_t^{-1} \, 1 \leq t < m$ and $(t, m) = 1$. Then*

(a) $\Phi(A)\Phi(B) = \Phi(AB)$.
(b) $|\Phi(A)|^2 = (NA)^m$.
(c) $(\Phi(A)) = A^\gamma$
(d) $\Phi((\alpha)) = \varepsilon(\alpha)\alpha^\gamma$ *where $\varepsilon(\alpha)$ is a unit in D_m.*

PROOF. (a) is clear from the definition.

Since both sides of (b) are multiplicative in A we can assume A is a prime ideal P. In that case $|\Phi(P)|^2 = |g(P)|^{2m} = (NP)^m$ by Proposition 14.3.1, part (b).

Both sides of (c) are multiplicative in A so again we may suppose A is a prime ideal P. In this case the result is the assertion of Theorem 2.

To do part (d) notice

$$(\Phi((\alpha))) = (\alpha)^\gamma = (\alpha^\gamma)$$

by part (c). Thus $\Phi((\alpha))$ and α^γ generate the same principal ideal. $\square$

From now on we will write $\Phi(\alpha)$ instead of $\Phi((\alpha))$.

It will be important to determine the unit $\varepsilon(\alpha)$ more closely. In fact we will show it is a root of unity.

Lemma 3. *Suppose $A \subset D_m$ is an ideal prime to m and let σ be an automorphism of $\mathbb{Q}(\zeta_m)/\mathbb{Q}$. Then*

$$\Phi(A)^\sigma = \Phi(A^\sigma).$$

PROOF. To see this it is convenient to write $g(P)$ in the following form

$$g(P) = \sum \left(\frac{\alpha}{P}\right)_m^{-1} \zeta_p^{\text{tr}(\bar{\alpha})},$$

where the sum is over a set of representatives for the cosets of D_m/P.

Let $\tilde{\sigma}$ be an automorphism of $\mathbb{Q}(\zeta_m, \zeta_p)/\mathbb{Q}$ which restricts to σ on $\mathbb{Q}(\zeta_m)$ and the identity on $\mathbb{Q}(\zeta_p)$ (see the proof to Lemma 8). By Proposition 14.2.4 we have

$$g(P)^{\tilde{\sigma}} = \sum \left(\frac{\alpha^{\sigma}}{P^{\sigma}}\right)_m^{-1} \zeta_p^{\text{tr}(\bar{\alpha})}.$$

Since $\text{tr}(\bar{\alpha}) \in \mathbb{Z}/p\mathbb{Z}$ we have $\text{tr}(\bar{\alpha}^{\sigma}) = \text{tr}(\bar{\alpha})$. It follows that $g(P)^{\tilde{\sigma}} = g(P^{\sigma})$. Raising both sides to the mth power gives the result when A is a prime ideal. By multiplicativity the result follows in general.   □

Lemma 4. For $\alpha \in D_m$, $|\alpha^{\gamma}|^2 = |N\alpha|^m$.

PROOF. The automorphism σ_{-1} is complex conjugation on $\mathbb{Q}(\zeta_m)$ since it takes ζ_m to $\zeta_m^{-1} = \bar{\zeta}_m$. Thus

$$|\alpha^{\gamma}|^2 = \alpha^{\gamma}\alpha^{\gamma\sigma_{-1}} = \alpha^{\gamma(1+\sigma_{-1})}.$$

Now, $\sigma_{-1}\gamma = \sigma_{-1}\sum t\sigma_t^{-1} = \sum t\sigma_{-t}^{-1}$. Clearly, $\sigma_{m-t} = \sigma_{-t}$, and $\gamma = \sum (m-t)\sigma_{m-t}^{-1}$. Thus, using $t = m - (m-t)$ we find

$$(1 + \sigma_{-1})\gamma = m\sum \sigma_t^{-1}.$$

Since $N\alpha = \prod \alpha^{\sigma_t^{-1}} = \alpha^{\sum \sigma_t^{-1}}$ the result follows.   □

Proposition 14.5.2. Let $\alpha \in D_m$, α prime to m. Then $\Phi(\alpha) = \varepsilon(\alpha)\alpha^{\gamma}$ where $\varepsilon(\alpha) = \pm\zeta_m^i$ for some i.

PROOF. In the light of part (d) of the last proposition it is enough to prove the assertion about $\varepsilon(\alpha)$. We have $|\Phi(\alpha)|^2 = (N(\alpha))^m$ by Proposition 14.5.1 and $|\alpha^{\gamma}|^2 = |N\alpha|^m$ by Lemma 4. By Proposition 14.1.3, $N(\alpha) = |N\alpha|$.

Putting all this together we conclude that $|\varepsilon(\alpha)| = 1$. Using Lemma 3 we find in the same way that $|\varepsilon(\alpha)^{\sigma}| = 1$ for all $\sigma \in G$. It now follows from Lemma 2 that $\varepsilon(\alpha)$ is a root of unity. Finally since $\varepsilon(\alpha) \in \mathbb{Q}(\zeta_m)$ we have $\varepsilon(\alpha) = \pm\zeta_m^i$ by Lemma 1.   □

We are now in a position to begin the proof of the Eisenstein reciprocity law. The pattern of proof of the following proposition should be familiar from our proofs of quadratic, cubic, and biquadratic reciprocity. It is itself a "reciprocity" statement.

Proposition 14.5.3. Suppose P, $P' \subset D_m$ are prime ideals both prime to m. Suppose further that NP and NP' are relatively prime. Then

$$\left(\frac{\Phi(P)}{P'}\right)_m = \left(\frac{NP'}{P}\right)_m.$$

PROOF. Let $q' = p'^{f'} = NP'$. Recall $q' \equiv 1\ (m)$. The following congruences are taken modulo p' in D_m

$$g(P)^{q'} \equiv \sum \chi_P(t)^{q'} \psi(t)^{q'}$$

$$\equiv \sum \chi_P(t)\psi(q't)$$

$$\equiv \left(\frac{q'}{P}\right)_m g(P).$$

On the other hand

$$g(P)^{q'-1} = \Phi(P)^{(q'-1)/m} \equiv \left(\frac{\Phi(P)}{P'}\right)_m (P').$$

It follows that

$$\left(\frac{\Phi(P)}{P'}\right)_m \equiv \left(\frac{NP'}{P}\right)_m (P').$$

Since $m \notin P'$ the two sides of this congruence must be equal.   □

Corollary 1. *Suppose* $A, B \subset D_m$ *are ideals prime to m and that NA and NB are prime to each other. Then*

$$\left(\frac{NB}{A}\right)_m = \left(\frac{\Phi(A)}{B}\right)_m.$$

PROOF. As usual, the corollary follows from the proposition by multiplicativity.   □

Corollary 2. *Suppose A and B are as in Corollary 1 and moreover that $A = (\alpha)$ is principal. Then*

$$\left(\frac{\varepsilon(\alpha)}{B}\right)_m \left(\frac{\alpha}{NB}\right)_m = \left(\frac{NB}{\alpha}\right)_m.$$

PROOF. To begin with

$$\left(\frac{\Phi(\alpha)}{B}\right)_m = \left(\frac{\varepsilon(\alpha)}{B}\right)_m \left(\frac{\alpha^\gamma}{B}\right)_m.$$

Notice that $(\alpha^{t\sigma_t^{-1}}/B)_m = (\alpha^{\sigma_t^{-1}}/B)_m^t = (\alpha^{\sigma_t^{-1}}/B)_m^{\sigma_t} = (\alpha/B^{\sigma_t})_m$ by Proposition 14.2.4. Thus

$$\left(\frac{\alpha^\gamma}{B}\right)_m = \prod_t \left(\frac{\alpha^{t\sigma_t^{-1}}}{B}\right)_m = \prod_t \left(\frac{\alpha}{B^{\sigma_t}}\right)_m = \left(\frac{\alpha}{NB}\right)_m.$$

To obtain the final equality we have used Proposition 14.1.2.   □

From now on we will assume $m = l$, an odd prime number.

Lemma 5. *If $A \subset D_l$ is an ideal prime to l, then $\Phi(A) \equiv \pm 1\ (l)$.*

PROOF. It is enough to show that $\Phi(P) \equiv -1 \ (l)$ where $P \subset D_l$ is a prime ideal prime to l. Well,

$$\Phi(P) = g(P)^l \equiv \sum_t \chi_P(t)^l \psi(t)^l \ (l)$$

$$\equiv \sum_{t \neq 0} \psi(lt) \equiv -1 \ (l).$$

The last congruence follows from the fact that $l \to \psi(lt)$ is a nontrivial additive character on D_l/P and so the sum of its values over all t is zero. Since $\psi(0) = 1$ the result follows. $\square$

Recall that $\alpha \in D$ is called primary if α is prime to l and $\alpha \equiv x \ (1 - \zeta_l)^2$, for some $x \in \mathbb{Z}$.

Lemma 6. *If $\alpha \in D$ is primary, then $\varepsilon(\alpha) = \pm 1$.*

PROOF. Since $(1 - \zeta_l)$ is the unique prime above l in D_l we have $(1 - \zeta_l)^\sigma = (1 - \zeta_l)$ for all $\sigma \in G$. It follows that $(1 - \zeta_l)^\gamma \subset (1 - \zeta_l)$.

Since $\Phi(\alpha) = \varepsilon(\alpha)\alpha^\gamma$ we have by Lemma 5 that $\varepsilon(\alpha)\alpha^\gamma \equiv \pm 1 \ (l)$.

Since $\alpha \equiv x \ (1 - \zeta_l)^2$ with $x \in \mathbb{Z}$ we find

$$\alpha^\gamma \equiv x^\gamma \equiv x^{1 + 2 + \cdots + (l-1)}(1 - \zeta_l)^2.$$

Now, $x^{(l-1)/2} \equiv \pm 1 \ (l)$, so

$$\alpha^\gamma \equiv (\pm 1)^l \equiv \pm 1 \ (1 - \zeta_l)^2.$$

It follows that $\varepsilon(\alpha) \equiv \pm 1 \ (1 - \zeta_l)^2$. From Proposition 14.5.2 we know $\varepsilon(\alpha) = \pm \zeta_l^i$. To conclude the proof we must show that l divides i. This follows from the uniqueness part of the lemma in Section 2, but it may be worthwhile to do it directly.

We have $\zeta_l^i \equiv \pm 1 \ (1 - \zeta_l)^2$. Writing $\zeta_l = 1 - (1 - \zeta_l)$ we find

$$1 - i(1 - \zeta_l) \equiv \pm 1 \ (1 - \zeta_l)^2.$$

The plus sign must hold since otherwise $1 - \zeta_l$ would divide 2. But then, subtracting 1 from both sides, we see $1 - \zeta_l$ divides i which implies $l | i$. $\square$

Proposition 14.5.4. *If $\alpha \in D_l$ is primary, and B is an ideal prime to l, and NB is prime to α, then*

$$\left(\frac{\alpha}{NB}\right)_l = \left(\frac{NB}{\alpha}\right)_l.$$

PROOF. By Corollary 2 to Proposition 14.5.3 we need only show $(\varepsilon(\alpha)/B)_l = 1$.

Since α is primary $\varepsilon(\alpha) = \pm 1$ by the above lemma. Since l is odd, $(\pm 1)^l = \pm 1$ and we are done. $\square$

We can now complete the proof of Theorem 1.

Let $p \in \mathbb{Z}$ be a prime, $p \neq l$, and p prime to α in D_l. Let P be a prime ideal in D_l containing p. Then $NP = p^f$. In the proposition we have just proven we substitute P for B. The result is

$$\left(\frac{\alpha}{p}\right)_l^f = \left(\frac{p}{\alpha}\right)_l^f.$$

Since $f | l - 1 = [\mathbb{Q}(\zeta_l) : \mathbb{Q}]$ we have $(f, l) = 1$. Thus

$$\left(\frac{\alpha}{p}\right)_l = \left(\frac{p}{\alpha}\right)_l.$$

From this and (one last time) multiplicativity, we deduce $(\alpha/a)_l = (a/\alpha)_l$ for all $a \in \mathbb{Z}$ prime to l and α, provided α is primary.   □

§6 Three Applications

In Chapter 5 we proved that if a is an integer such that $x^2 \equiv a \ (p)$ is solvable for all but finitely many primes then a is a square. This has been generalized to nth powers by E. Trost. The result was later rediscovered by N. C. Ankeny and C. A. Rogers. The result states that if $x^n \equiv a \ (p)$ for all but finitely many primes p then $a = b^n$ if $8 \nmid n$ and $a = b^n$ or $a = 2^{n/2}b^n$ if $8 | n$. Using Eisenstein reciprocity we will prove a portion of this when $n = l$ an odd prime. See also [211], [134] and the Notes to Chapter 5.

Theorem 4. *Suppose $a \in \mathbb{Z}$ and that $l \nmid a$ where l is an odd prime. If $x^l \equiv a \ (p)$ is solvable for all but finitely many primes p then $a = b^l$.*

PROOF. We can restate the theorem as follows. If a is not an lth power then there are infinitely many primes p such that $x^l \equiv a \ (p)$ is not solvable.

Assume a is not an lth power in $\mathbb{Z}$. Let $aD_l = P_1^{a_1}P_2^{a_2} \cdots P_n^{a_n}$ be the prime decomposition of a in D_l. We claim that $l \nmid a_i$ for at least one a_i. To see this, let $p_i \mathbb{Z} = P_i \cap \mathbb{Z}$. Since $l \nmid a$ we have $l \neq p_i$ and so p_i is unramified in D_l. Consequently $\text{ord}_{p_i} a = \text{ord}_{P_i} a = a_i$. If $l | a_i$ for all i it would follow that a is an lth power in $\mathbb{Z}$. We may thus assume $l \nmid a_n$.

Let $\{Q_1, Q_2, \ldots, Q_k\}$ be a finite set of primes Q_i different from the P_i and from $(1 - \zeta_l)$.

Using the Chinese Remainder Theorem we can find an element $\tau \in D_l$ such that $\tau \equiv 1 \ (Q_i)$ for $i = 1, 2, \ldots, k$, $\tau \equiv 1 \ (l)$, $\tau \equiv 1 \ (P_j)$ for $j = 1, 2, \ldots, n - 1$, and $\tau \equiv \alpha \ (P_n)$ where α is chosen so that $(\alpha/P_n)_l = \zeta_l$.

Since $\tau \equiv 1 \ (l)$, τ is primary. Thus, on the one hand

$$\left(\frac{a}{\tau}\right)_l = \left(\frac{\tau}{a}\right)_l = \prod \left(\frac{\tau}{P_i}\right)_l^{a_i} = \zeta_l^{a_n} \neq 1.$$

On the other hand, let $(\tau) = R_1 R_2 \cdots R_m$ be the prime decomposition of τ. Then

$$\left(\frac{a}{\tau}\right)_l = \prod_j \left(\frac{a}{R_j}\right)_l.$$

It follows that for some j, $(a/R_j)_l \neq 1$.

From the congruences that τ satisfies it follows immediately that $R_j \notin \{Q_1, Q_2, \ldots, Q_k\} \cup \{(1 - \zeta_l)\} \cup \{P_1, \ldots, P_n\}$.

We have shown that there are infinitely many prime ideals Q such that $x^l \equiv a\ (Q)$ is not solvable. Let $qZ = Q \cap \mathbb{Z}$. Then $x^l \equiv a\ (q)$ is not solvable and there are infinitely many such q since every rational prime is contained in only finitely many prime ideals in D_l.   □

The second application of Eisenstein reciprocity we wish to make is to Fermat's conjecture. This states that if $n > 2$ is an integer there is no solution to $x^n + y^n + z^n = 0$ in non-zero integers. The fascinating history of this conjecture will be sketched in a later chapter.

It is easy to see that if Fermat's conjecture is true for n then it will be true for any multiple of n. Since any integer bigger than 2 is either divisible by 4 or by an odd prime we may restrict our attention to the cases $n = 4$ or $n = l$ an odd prime. The case $n = 4$ was settled, affirmatively, by L. Euler.

When l is an odd prime it is traditional to consider two cases. We say we are in case one if $x^l + y^l + z^l = 0$ and $l \nmid xyz$. Otherwise we are in case two. In 1909 A. Wieferich published the following important result ([166], Vol. 3).

Theorem 5. *If $x^l + y^l + z^l = 0$ is solvable in non-zero integers such that $l \nmid xyz$ then $2^{l-1} \equiv 1\ (l^2)$.*

It has been shown that the only two primes less than 3×10^9 which satisfy $2^{l-1} \equiv 1\ (l^2)$ are 1093 and 3511. It is not known if there are infinitely many primes of this type.

In 1912 Furtwängler proved a theorem which contains Theorem 5 as a corollary. Namely,

Theorem 6. *Let x, y, and z be non-zero integers, relatively prime in pairs, such that $x^l + y^l + z^l = 0$. Assume $l \nmid yz$. Let p be a prime factor of y. Then $p^{l-1} \equiv 1\ (l^2)$.*

It is a simple exercise to see that the condition that x, y, and z be relatively prime in pairs is no loss of generality.

To see how Theorem 5 follows from Theorem 6, assume $l \nmid xyz$. Since $x^l + y^l + z^l = 0$ not all three numbers x, y, and z can be odd. By symmetry we can assume $2|y$. By Theorem 6 we have $2^{l-1} \equiv 1\ (l^2)$.

We proceed to prove Furtwängler's theorem. Let $\zeta = \zeta_l$ be a primitive lth root of unity. We have

$$(x + y)(x + \zeta y) \cdots (x + \zeta^{l-1}y) = (-z)^l. \qquad (*)$$

Lemma 1. *Suppose $i \neq j$ and $0 \leq i, j < l$. Then $x + \zeta^i y$ and $x + \zeta^j y$ are relatively prime in D_l.*

PROOF. Suppose $A \subset D_l$ is an ideal containing $x + \zeta^i y$ and $x + \zeta^j y$. Then $(\zeta^j - \zeta^i)x$ and $(\zeta^j - \zeta^i)y$ are in A. Since x and y are relatively prime it follows that $\zeta^j - \zeta^i$ is in A. It follows that $\lambda = 1 - \zeta \in A$. Since (λ) is a maximal ideal, either $(\lambda) = A$ or $A = D_l$. If $(\lambda) = A$, then from equation $(*)$ we see $(-z) \in (\lambda)$ which implies $z \in (\lambda)$ and $l|z$, contrary to assumption. Thus $A = D_l$ and we are done. $\qquad \square$

Corollary. *The ideals $(x + \zeta^i y)$ are perfect lth powers.*

Consider the element $\alpha = (x + y)^{l-2}(x + \zeta y)$. We claim

(i) The ideal (α) is a perfect lth power.
(ii) $\alpha \equiv 1 - u\lambda \ (\lambda^2)$ where $u = (x + y)^{l-2}y$.

Property (i) follows from the corollary to the lemma.
To prove property (ii) notice $x + \zeta y = x + y - y\lambda$. Thus,

$$\alpha = (x + y)^{l-1} - \lambda u.$$

Now, $x^l + y^l + z^l \equiv x + y + z \ (l)$. If $l|(x + y)$ it would follow that $l|z$ contrary to assumption. Therefore $l \nmid (x + y)$ and $(x + y)^{l-1} \equiv 1 \ (l)$. Property (ii) follows.
Consider $\zeta^{-u}\alpha$. We have

$$\zeta^{-u}\alpha = (1 - \lambda)^{-u}\alpha \equiv (1 + u\lambda)(1 - u\lambda) \equiv 1 \ (\lambda^2).$$

It follows that $\zeta^{-u}\alpha$ is primary. By Eisenstein reciprocity we have

$$\left(\frac{p}{\zeta^{-u}\alpha}\right)_l = \left(\frac{\zeta^{-u}\alpha}{p}\right)_l = \left(\frac{\zeta}{p}\right)_l^{-u}\left(\frac{\alpha}{p}\right)_l. \qquad (**)$$

Since the ideal $(\zeta^{-u}\alpha) = (\alpha)$ is an lth power, the left-hand side of $(**)$ is equal to 1.
Since $p|y$, $\alpha \equiv (x + y)^{l-1} \ (p)$. Thus

$$\left(\frac{\alpha}{p}\right)_l = \left(\frac{(x + y)^{l-1}}{p}\right)_l = \left(\frac{p}{(x + y)^{l-1}}\right)_l = 1,$$

because the ideal $(x + y)$ is an lth power.
It now follows from $(**)$ that $(\zeta/p)_l^u = 1$. To conclude the proof we must evaluate $(\zeta/p)_l$.
Let $pD_l = P_1 P_2 \cdots P_g$ be the prime decomposition of p in D_l. We know $NP_i = p^f$ and, since $p \neq l$, $e = 1$, and so $gf = l - 1$.

By the corollary to Proposition 14.2.2

$$\left(\frac{\zeta}{p}\right)_l = \prod_i \left(\frac{\zeta}{P_i}\right)_l = \prod_i \zeta^{(p^f-1)/l} = \zeta^{g[(p^f-1)/l]}$$

The relation $(\zeta/p)_l^u = 1$ now leads to the congruence

$$ug\frac{p^f-1}{l} \equiv 0\ (l).$$

Since $g\,|\,l-1$, $l \nmid g$. Since $u = (x+y)^{l-2}y$, $l \nmid u$. Thus

$$\frac{p^f-1}{l} \equiv 0\ (l) \quad \text{or} \quad p^f \equiv 1\ (l^2).$$

The theorem is now immediate since $f\,|\,l-1$. $\qquad\square$

We conclude with an application of Theorem 2 which concerns the structure of the ideal class group of $\mathbb{Q}(\sqrt{-l})$ where $l > 3$ is prime $l \equiv 3\ (4)$. Let p be an odd prime $p \equiv 1(l)$. Then since p splits completely in $\mathbb{Q}(\zeta_l)$ it also splits in $\mathbb{Q}(\sqrt{-l})$ (why?). One can also see this by observing $(-l/p) = (-1)^{(p-1)/2}(p/l)(-1)^{((p-1)/2)((l-1)/2)} = (p/l) = 1$ and applying Proposition 13.1.3. In the ring of integers D of $\mathbb{Q}(\sqrt{-l})$ write $p = \mathfrak{P}\bar{\mathfrak{P}}$. If $\tilde{D}$ denotes the ring of integers in $\mathbb{Q}(\zeta_l)$ we have

Lemma 2. $\mathfrak{P}\tilde{D} = \prod P^{\sigma_s}$ where P is a prime ideal of $\tilde{D}$, $P \cap D = \mathfrak{P}$, and s runs over the nonzero squares modulo l.

PROOF. The set of σ_s in the statement of the lemma form the Galois group of $\mathbb{Q}(\zeta_l)$ over $\mathbb{Q}(\sqrt{-l})$. Since $p\tilde{D} = P^{\sum_{i=1}^{l-1}\sigma_i}$ and $\sigma_n(\mathfrak{P}) = \bar{\mathfrak{P}}$ for a nonsquare n modulo l it follows that $\mathfrak{P}\tilde{D}$ is divisible by precisely the $\sigma_s(P)$, each with exponent 1. $\qquad\square$

By Theorem 2 we have $(g(P)^l) = P^{\sum t\sigma_t^{-1}}$, $t = 1, 2, \ldots, l-1$. Applying $\sum \sigma_s$, s a square modulo l gives $(\alpha)\tilde{D} = \mathfrak{P}^{\sum t\sigma_t^{-1}} \cdot \tilde{D} = \mathfrak{P}^{\sum s} \cdot \bar{\mathfrak{P}}^{\sum n} \cdot \tilde{D}$ where $\alpha \in D$ and n runs over the nonsquares modulo l in the interval $[1, l-1]$. Put $R = \sum s$, $N = \sum n$. By Exercise 34 of Chapter 12 it follows that $\alpha D = \mathfrak{P}^R\bar{\mathfrak{P}}^N$. If $[\mathfrak{A}]$ denotes the equivalence class of the ideal $\mathfrak{A}$ and 1 is the unit class then $[\mathfrak{P}]^{-1} = [\bar{\mathfrak{P}}]$. Thus $[\mathfrak{P}]^{N-R} = 1$. On the other hand if $1 \le r \le l-1$ by Exercise 6 (or Lemma 3, Section 3, Chapter 15), one has $g(P)^{\sigma_r - r} = \beta$ for some $\beta \in \tilde{D}$. Raising to the lth power, using Theorem 2, and applying $\sum \sigma_s$ gives, for r a square $\ne 1$, $(\mathfrak{P}^R\bar{\mathfrak{P}}^N)^{1-r}\tilde{D} = (\gamma)^l\tilde{D}$ for some $\gamma \in D$. It follows that $([\mathfrak{P}]^{(N-R)/l})^{r-1} = 1$ (it is easy to show $l\,|\,N$ and $l\,|\,R$). But from the above $([\mathfrak{P}]^{(N-R)/l})^l = 1$. Since $(r-1, l) = 1$ we have proven the following result.

Proposition 14.6.1. Let $\mathfrak{P}$ be a prime ideal of degree 1 in $\mathbb{Q}(\sqrt{-l})$ for $l \ge 3$ a prime such that $l \equiv 3\ (4)$. Then, $[\mathfrak{P}]^{(N-R)/l} = 1$.

While it is elementary that $(N - R)/l$ is an integer it is by no means obvious that it is positive. All known proofs of this fact use analysis. We will give a short proof due to Moser in the Exercises to Chapter 16. For other proofs of the positivity as well as many other interesting results of this type see the paper by B. Berndt [94]. It turns out, as mentioned in Chapter 13 that $(N - R)/l$ is indeed the class number of $\mathbb{Q}(\sqrt{-l})$ but again the proof is analytic. When, by direct calculation $N - R = l$ it follows that $\mathfrak{P}$ is a principal ideal. If one assumes, as can be shown, that each ideal class contains a prime ideal of degree 1 then one can conclude that for such l, $\mathbb{Q}(\sqrt{-l})$ is a unique factorization domain. In this manner one checks that the imaginary quadratic fields with discriminant -7, -11, -19, -43, -67, -163 all have class number 1. Referring again to the proposition $\mathfrak{P}^{(N-R)/l} = (\alpha)$ where $\alpha = (x + \sqrt{-l}y)/2$; $x, y \in \mathbb{Z}$. Taking norms gives the following interesting corollary.

Corollary. If $p \equiv 1(l), l \equiv 3(4), l > 3$, then $4p^{(N-R)/l} = x^2 + ly^2$, with $x, y \in \mathbb{Z}$.

NOTES

In his paper "Über eine Verallgemeinerung der Kreistheilung" (1890) [224], the Swiss mathematician Ludwig Stickelberger (1850–1936) (see [148]) succeeded in determining the prime decomposition of a Gauss sum attached to an arbitrary multiplicative character defined on a finite field (Theorem 2 of this chapter). Actually he proved a more precise result. Namely, using the notation of this chapter

$$g_a \equiv \frac{-(-\lambda)^{S(a)}}{a_0!\, a_1! \cdots a_{f-1}!} (\mathscr{P}^{S(a)+1}).$$

This, of course, implies Theorem 2. The special case of this theorem when m is prime and $p \equiv 1\,(m)$ had already been proven by Kummer in 1847. It is interesting to note that Kummer derived the result by first determining the decomposition in $\mathbb{Q}(\zeta_m)$ of certain Jacobi sums, which in turn was made possible by the congruence $J(\bar{\omega}^m, \bar{\omega}^n) \equiv - [(m + n)!/n!\, m!](P)$, known to Jacobi, Eisenstein and Cauchy. (See Kummer [164], Vol. 1, pp. 361–364, pp. 448–453, and Exercises 1 and 2). An elegant proof of Kummer's result can also be found in Hilbert's "Zahlbericht" [151] (Theorem 135), where the use of Jacobi sums is avoided by using an argument involving ramification. This special case of Stickelberger's theorem was the missing link in the program initiated by Gauss, Eisenstein and Jacobi to establish higher power reciprocity laws. Indeed, in 1850 [132] Eisenstein published his proof of the reciprocity law bearing his name (Theorem 1), making use of the then relatively new language of ideal numbers due to Kummer. A complete proof can be found also in Vol. 3 of Landau [166] as well as in Hilbert's "Zahlbericht" (Theorem 140), where in order to overcome the restriction that $p \equiv 1\ (l)$ he uses the finiteness of the class number for $\mathbb{Q}(\zeta_l)$! Hilbert views the Eisenstein law as an indispensible lemma for the Kummer reciprocity law. The proof of Theorem 2 that we have given follows that found in the important paper by Hasse and Davenport [23] (see also Chapter 7 of Joly [160]), while the derivation of Eisenstein's law from Kummer's Theorem closely follows the treatment in Weil's elegant historical study "La cyclotomie jadis et naguére" [238]. This paper of Weil along with his review of Eisenstein's "Mathematische Werke" [239] and his introduction to the collected papers of Kummer [164] provide a detailed and insightful history of the efforts of Jacobi, Eisenstein, and Kummer to prove higher power reciprocity laws with the use of Gauss sums. In this text we have followed this development up to the work of Eisenstein. The subsequent development leads to the research of Kummer, Hilbert, Furtwangler, and Takagi, and eventually, to the celebrated Artin law of reciprocity. For the history of these developments see Iyanaga [158] and Hasse [110]. For an interesting, and perhaps more elementary, discussion of the nature of reciprocity laws see Wyman's paper "What is a reciprocity law?" [246].

The use of Theorem 2 to show that the ideal class group of $\mathbb{Q}(\sqrt{-l})$, $l \equiv 3\,(4)$ is annihilated by $(1/l)\sum_{x=1}^{l-1} x(x/l)$ goes back to Kummer and appears as Theorem 145 of Hilbert's "Zahlbericht". The corollary to Proposition 14.6.1 was originally observed by Jacobi who, on its basis, conjectured the class number formula for $\mathbb{Q}(\sqrt{-l})$. (See also the comment of Weil [238], pp. 252–253.) Stickelberger, in the above-mentioned paper, returns to this application of cyclotomy to the arithmetic of quadratic forms and obtains similar results for $\mathbb{Q}(\sqrt{-m})$, for general m.

There are other applications of Theorem 1 to Fermat's Last Theorem. For example, a well-known result of Mirimanoff states that if x, y, and z are integers such that $x^p + y^p + z^p = 0$, $p \nmid xyz$ then $3^{p-1} \equiv 1\,(p^2)$ (see Theorem 1041, Landau [166]). Also Vandiver has shown, using similar methods, that if $x^p + y^p + z^p = 0$, $(x,\,y,\,z) = 1$, $p > 3$ then $x^p \equiv x\,(p^3)$, $y^p \equiv y\,(p^3)$, $z^3 \equiv z\,(p^3)$ (Landau [166], Theorem 1046). For further results on Fermat's Last Theorem that utilize Eisenstein reciprocity see Lecture 9 of the beautiful book by P. Ribenboim *13 Lectures on Fermat's Last Theorem* [206].

EXERCISES

Throughout these exercises the notation is as in this chapter.

1. Show that if $1 \le n < q - 1$, $1 \le m < q - 1$ then
   (a) $J(\omega^{-n}, \omega^{-m}) \equiv -[(m+n)!/n!m!]\,(\mathfrak{P})$
   (b) If $1 < a < q - 1$, $a = a_0 + a_1 p + \cdots + a_{f-1}p^{f-1}$ then $J(\omega^{-1},\ \omega^{-(a-1)}) \equiv -a_0(\mathfrak{P})$.

2. In the proof to Theorem 2 we showed that $g_1 \equiv \lambda_p\,(\mathscr{P}^2)$.
   (a) If $1 \le a < p - 1$ show $g_a \equiv (-1)^{a+1}\lambda_p^a/a!\,(\mathscr{P}^{a+1})$ where $\alpha \equiv \beta\,(\mathscr{P}^n)$ means $\operatorname{ord}_{\mathscr{P}}(\alpha - \beta) \ge n$.

(b) If the Stickelberger congruence $g_a \equiv (-1)^{1+S(a)} \lambda_p^{S(a)} / a_0! a_1! \cdots a_{f-1}! (\mathscr{P}^{1+S(a)})$ holds for some $1 \le a < q - 1$ and $pa < q - 1$ then show it also holds for g_{pa}.

(c) Establish the general Stickelberger congruence.

3. Show that if $m > 2$ then $g(P)p^{-1/2}$ is not an algebraic integer (see also Chowla [113]).

4. Let r and s be positive integers $m \nmid r + s$. Show that $(J(\chi_P^r, \chi_P^s)) = P^\alpha$ where $\alpha = \sum (\langle rt/m \rangle + \langle st/m \rangle - \langle [(r+s)t]/m \rangle) \sigma_t^{-1}$ the sum being over t, $1 \le t < m$, $(t, m) = 1$.

5. Check that the argument in Section 4 showing $g_1 \equiv \lambda_p (\mathscr{P}^2)$ is valid for $p = 2$, m odd.

6. If $(r, pm) = 1, 1 \le r < pm$, then $g(P)^{\sigma_r - r} \in \mathbb{Q}(\zeta_m)$.

7. Verify Lemma 1 of Section 5.

8. Let $p \equiv 1 \ (m)$, where m is prime. Without using Exercise 4 show that $J(\chi, \chi^k)$, $1 \le k \le p - 2$ is a product of distinct prime ideals each with exponent 1. Use Exercise 1 to determine the decomposition of $J(\chi, \chi^k)$ and use Proposition 8.3.3 to give a direct proof, in this case, of Theorem 2 (Kummer).

9. Let K/F be a Galois extension with cyclic Galois group of order p and generator σ. Define, for $x \in K, f(x) = 1 + x + x\sigma(x) + \cdots + x\sigma(x) \cdots \sigma^{p-3}(x)$. Let p be prime, $F = \mathbb{Q}(\zeta_{p-1})$, $K = \mathbb{Q}(\zeta_p, \zeta_{p-1})$. Show that $g(\chi) = \sum_{x=1}^{p-1} \chi(x)\zeta_p^{tx} = \zeta_p f(\zeta_{p-1}\zeta_p^{t-1})$, where t is a primitive modulo p, $\chi(t) = \zeta_{p-1}$ and σ is the automorphism of K/F for which $\sigma(\zeta_{p-1}) = \zeta_{p-1}$ and $\sigma(\zeta_p) = \zeta_p^t$. Conclude that the Gauss sum is the great grandfather of cohomology theory (Kummer [164], p. 10).

10. Use Theorem 2 to show that $\mathbb{Q}(g(P)^m)$ is the fixed field of the decomposition group of p, also known as the decomposition field of p.

11. For a prime l and positive integers $r, s,$ and t satisfying $r + s + t = l$ put $H_{r,s,t} = \{h \mid h \in F_l^*, \widetilde{hr} + \widetilde{hs} + \widetilde{ht} = l\}$ where $\tilde{a}$ denotes the smallest nonnegative residue of a modulo l. Show that $H_{r,s,t}$ is a set of coset representatives for the subgroup of order 2 in F_l^*.

12. Consider the curve Γ over F_p defined by $y^l = x^r(1 - x)^s$, the notation being as in Exercise 11.

(a) Show that the zeta function of Γ can be written $z(u) = g(u)/(1 - u)(1 - pu)$, where $g(u) = \prod_p (1 + J(\bar{\chi}_P^r, \bar{\chi}_P^s)u^f)$ where P ranges over the prime ideals in $\mathbb{Q}(\zeta_l)$ over p and the notation is as in the text; i.e., $\bar{\chi}_P$ is the lth power residue symbol.

(b) Show that $(J\bar{\chi}_P^r, \bar{\chi}_P^s)) = P^{\sum \sigma_k^{-1}}$, where $k \in H_{r,s,t}$.

(c) Show that if the order of p modulo l (i.e., f) is even, then complex conjugation is in the decomposition group of P.

(d) If f is even, $(J(\bar{\chi}_P^r, \bar{\chi}_P^s)) = (p^{f/2})$.

(e) $J(\bar{\chi}_P^r, \bar{\chi}_P^s) = up^{f/2}$, where u is an lth root of unity.

(f) Show that $u = 1$.

(Exercises 11 and 12 are from a paper by B. H. Gross and D. E. Rohrlich [142].)

13. Let l be prime, $\chi \ne \varepsilon$ a multiplicative character of F_l. Put $B_\chi = (1/l) \sum_{a=1}^{l-1} a\chi(a)$. Consider the elements of the group ring of the Galois group G of $\mathbb{Q}(\zeta_l)/\mathbb{Q}$ with

coefficients in $\mathbb{Q}(\zeta_l)$ defined by $\varepsilon_\chi = (1/(l-1))\sum_{a=1}^{l-1} \chi(a)^{-1}\sigma_a$, $\theta = (1/l)\sum_{t=1}^{l-1} t\sigma_t^{-1}$, where $\sigma_a(\zeta_l) = \zeta_l^a$. Show

(a) $\varepsilon_\chi(\zeta_l)\varepsilon_{\chi^{-1}}(\zeta_l) = \chi(-1)(l-1)^{-2}l$.

(b) $-l = (1 - \zeta_l)(\zeta_l + 2\zeta_l^2 + \cdots + (l-1)\zeta_l^{l-1})$.

(c) $\varepsilon_\chi(-\zeta_l/(1 - \zeta_l)) = B_\chi \varepsilon_\chi(\zeta_l)$.

(d) $\theta\varepsilon_\chi = -B_{\chi^{-1}}\varepsilon_\chi$, where one defines $(\sum_{t=1}^{l-1} a_t\sigma_t)(\sum_{t=1}^{l-1} b_t\sigma_t) = \sum_{t=1}^{l-1} c_t\sigma_t$ with $c_t = \sum_{uv \equiv t(l)} a_u b_v$, $1 \le u, v < l$.

(This exercise is taken from Iwasawa [157], pp. 115–117.)

14. Let p and l be prime, $l > 3$. If $p \ne l$ and $\alpha \in \mathbb{Z}[\zeta_l]$ is real, $(\alpha, l) = 1$ show that $(\alpha/p)_l = 1$.

15. Let $p \ne l$ be primes, $l > 3$. Show

(a) $(\zeta_l/p)_l = \zeta_l^{((l-1)/f)((p^f-1)/l)}$, where f is the order of p modulo l.

(b) $(\zeta_l/p)_l = 1$ implies $p^{l-1} \equiv 1 \ (l^2)$.

16. Read Satz 1039 and Satz 1041 in Landau [166], Vol. 3.

17. Let $m = l$, an odd prime, and let $\mathcal{L}$ be a prime ideal in $\mathbb{Q}(\zeta_{pl})$ containing $(1 - \zeta_l)$. Show

(a) $g(P) \equiv -1 \ (1 - \zeta_l)$.

(b) $g(P) \equiv -1 + c(1 - \zeta_l) \ (\mathcal{L}^2)$ with $c \in \mathbb{Z}[\zeta_p]$.

(c) $(-g(P))^{\sigma_t} \equiv (-g(P))^t \ (\mathcal{L}^2)$ for $(t, l) = 1$ and σ_t the automorphism of $\mathbb{Q}(\zeta_{pl})$ such that $\sigma_t(\zeta_p) = \zeta_p$ and $\sigma_t(\zeta_l) = \zeta_l^t$.

(d) $g(P)^{\sigma_t - t} \equiv (-1)^{t+1} \ ((1 - \zeta_l)^2)$.

(e) If $1 \le a, b < l$, $l \nmid a + b$ then $J(\chi_P^a, \chi_P^b) \equiv -1 \ (1 - \zeta_l)^2$.

This exercise is taken from Iwasawa [156].

Chapter 15

Bernoulli Numbers

In this chapter we will introduce an important sequence of rational numbers discovered by Jacob Bernoulli (1654–1705) and discussed by him in a posthumous work Ars Conjectandi (1713). *These numbers, now called Bernoulli numbers, appear in many different areas of mathematics. In the first section we give their definition and discuss their connection with three different classical problems. In the next section we discuss various arithmetical properties of Bernoulli numbers including the Claussen–von Staudt theorem and the Kummer congruences. The first of these results determines the denominators of the Bernoulli numbers, and the second gives information about their numerators. In the last section we prove a theorem due to J. Herbrand which relates Bernoulli numbers to the structure of the ideal class group of $\mathbb{Q}(\zeta_p)$. The material in this section is somewhat sophisticated but we have included it anyway because it provides a beautiful and important application of the Stickelberger relation which was proven in the last chapter.*

§1 Bernoulli Numbers; Definitions and Applications

We begin by discussing three problems, each of historic interest.

The first concerns finding formulas for summing the kth powers of the first n integers. Jacob Bernoulli was aware of the following facts

$$1 + 2 + 3 + \cdots + (n - 1) = \frac{n(n - 1)}{2},$$

$$1^2 + 2^2 + 3^2 + \cdots + (n - 1)^2 = \frac{n(n - 1)(2n - 1)}{6},$$

$$1^3 + 2^3 + 3^3 + \cdots + (n - 1)^3 = \frac{n^2(n - 1)^2}{4},$$

as well as corresponding, less well known, formulas for exponents up to 10. For each exponent k the sum $1^k + 2^k + \cdots + (n - 1)^k$ turned out to be a polynomial in n of degree $k + 1$. In his efforts to determine the coefficients of these polynomials for general k, Bernoulli was led to define the numbers which bear his name. He was completely successful in answering the original problem and proudly remarks (in his book *Ars Conjectandi*) that in less than a half of a quarter of an hour he was able to sum the tenth powers of the first thousand integers [220].

Another outstanding problem of that period was to evaluate the sum

$$\zeta(2) = 1 + \frac{1}{4} + \frac{1}{9} + \frac{1}{16} + \frac{1}{25} + \cdots$$

and more generally $\zeta(2m)$ where $\zeta(s) = \sum_{n=1}^{\infty} n^{-s}$ is the Riemann zeta function. After long effort L. Euler showed in 1734 that $\zeta(2) = \pi^2/6$. Subsequently he determined $\zeta(2m)$ for all positive integers m.

The third problem is the celebrated Fermat's Last Theorem. If n is an integer greater than 2, Fermat asserted that $x^n + y^n = z^n$ has no solution in positive integers. This assertion has never been proved in general. It is easily seen that the conjecture is true if it is true whenever $n = p$, an odd prime. In 1847 E. Kummer proved the conjecture is true for a certain set of primes called regular primes. A prime p is called regular if it does not divide the class number of $\mathbb{Q}(\zeta_p)$. Furthermore, Kummer discovered a beautiful and elementary criterion for regularity which involves divisibility properties of the first $(p - 3)/2$ nonvanishing Bernoulli numbers.

We will discuss these three problems in turn.

Define $S_m(n) = 1^m + 2^m + \cdots + (n - 1)^m$. We first give a simple inductive method for evaluating these sums. The binomial theorem implies

$$(k + 1)^{m+1} - k^{m+1} = 1 + \binom{m + 1}{1}k + \binom{m + 1}{2}k^2 + \cdots + \binom{m + 1}{m}k^m.$$

Substitute $k = 0, 1, 2, \ldots, n - 1$ and add. The result is

$$n^{m+1} = n + \binom{m + 1}{1}S_1(n) + \binom{m + 1}{2}S_2(n) + \cdots + \binom{m + 1}{m}S_m(n). \quad (1)$$

If one has formulas for $S_1(n), S_2(n), \ldots, S_{m-1}(n)$ then Equation (1) allows one to find a formula for $S_m(n)$. Bernoulli observed that $S_m(n)$ is a polynomial of degree $m + 1$ in n with leading term $n^{m+1}/m + 1$. This follows easily by induction from Equation (1). Also, the constant term is always zero. The value of the other coefficients is less obvious. By direct computation one finds the coefficient of n to be $-\frac{1}{2}, \frac{1}{6}, 0, -\frac{1}{30}, 0, \frac{1}{42}, 0, -\frac{1}{30}, 0, \frac{5}{66}$ for $m = 1, 2, \ldots, 10$. Further empirical observation of the formulas led Bernoulli to the following definition and theorem.

Definition. The sequence of numbers $B_0, B_1, B_2, \ldots$, the Bernoulli numbers, are defined inductively as follows. $B_0 = 1$ and if $B_1, B_2, \ldots, B_{m-1}$ are already determined then B_m is defined by

$$(m + 1)B_m = -\sum_{k=0}^{m-1} \binom{m + 1}{k}B_k. \quad (2)$$

Written out this becomes the sequence of linear equations

$$1 + 2B_1 = 0$$

$$1 + 3B_1 + 3B_2 = 0$$

$$1 + 4B_1 + 6B_2 + 4B_3 = 0$$

$$1 + 5B_1 + 10B_2 + 10B_3 + 5B_4 = 0.$$

One finds $B_1 = -\frac{1}{2}, B_2 = \frac{1}{6}, B_3 = 0, B_4 = -\frac{1}{30}, B_5 = 0, B_6 = \frac{1}{42}, \ldots$, etc. We shall prove later that the nonzero Bernoulli numbers alternate in sign. Furthermore we shall see that the Bernoulli numbers with odd index bigger than 1 vanish.

Lemma 1. *Expand $t/(e^t - 1)$ in a power series about the origin as follows* $t/(e^t - 1) = \sum_{m=0}^{\infty} b_m(t^m/m!)$. *Then for all m, $b_m = B_m$.*

PROOF. Multiply both sides by $e^t - 1$ to obtain

$$t = \sum_{n=1}^{\infty} \frac{t^n}{n!} \sum_{m=0}^{\infty} b_m \frac{t^m}{m!}.$$

Equating coefficients of t^{m+1} gives $1 = b_0$ for $m = 0$ and

$$\sum_{k=0}^{m} \binom{m+1}{k} b_k = 0$$

in general. This is the same as the system of Equation (2) which defines the Bernoulli numbers. Since $B_0 = b_0 = 1$ it follows that $B_m = b_m$ for all m. □

We now give the answer obtained by Bernoulli to the question of evaluating the sums $S_m(n)$.

Theorem 1. *For $m \geq 1$ the sums $S_m(n)$ satisfy*

$$(m + 1)S_m(n) = \sum_{k=0}^{m} \binom{m+1}{k} B_k n^{m+1-k}.$$

PROOF. In $e^{kt} = \sum_{m=0}^{\infty} k^m(t^m/m!)$ substitute $k = 0, 1, 2, \ldots, n - 1$ and add. This results in

$$1 + e^t + e^{2t} + \cdots + e^{(n-1)t} = \sum_{m=0}^{\infty} S_m(n) \frac{t^m}{m!}. \tag{3}$$

The left-hand side is

$$\frac{e^{nt} - 1}{e^t - 1} = \frac{e^{nt} - 1}{t} \cdot \frac{t}{e^t - 1} = \sum_{k=1}^{\infty} n^k \frac{t^{k-1}}{k!} \sum_{j=0}^{\infty} B_j \frac{t^j}{j!}. \tag{3'}$$

Equating the coefficients of t^m on the right-hand sides of Equations (3) and (3') and multiplying by $(m + 1)!$ gives the result. □

We may reformulate the result of Theorem 1 by introducing an important class of polynomials known as Bernoulli polynomials. Define

$$B_m(x) = \sum_{k=0}^{m} \binom{m}{k} B_k x^{m-k}.$$

Thus $B_1(x) = x - \frac{1}{2}$, $B_2(x) = x^2 - x + \frac{1}{6}$, etc. Then Theorem 1 may be stated as

$$S_m(n) = \frac{1}{m+1}(B_{m+1}(n) - B_{m+1}).$$

We remark in passing that Lemma 1 yields an easy proof that $B_{2k+1} = 0$ for $k \geq 1$. Since $B_1 = -\frac{1}{2}$ we have

$$\frac{t}{e^t - 1} + \frac{t}{2} = 1 + \sum_{k=2}^{\infty} B_k \frac{t^k}{k!}.$$

The left-hand side is the same as $(t/2)((e^t + 1)/(e^t - 1))$ which is unchanged if t is replaced by $-t$, i.e., it is an even function of t. This implies the coefficients of odd powers of t on the right-hand side are zero.

We now turn to the relationship between Bernoulli numbers and the numbers $\zeta(2m)$ for $m = 1, 2, 3, \ldots$. The following result is due to Euler and constitutes one of his most remarkable calculations. For the history of this result and its relation to the functional equation of the Riemann zeta function the reader should consult the article of Raymond Ayoub [88].

Theorem 2. *For m a positive integer*

$$2\zeta(2m) = (-1)^{m+1} \frac{(2\pi)^{2m}}{(2m)!} B_{2m}.$$

PROOF. The proof of this result requires a fact from classical analysis. Namely, we need the partial fraction expansion for $\cot x$.

$$\cot x = \frac{1}{x} - 2 \sum_{n=1}^{\infty} \frac{x}{n^2 \pi^2 - x^2}. \tag{4}$$

There are several ways to derive this expansion. Perhaps the simplest way is to substitute $t = 1$ in the Fourier expansion of $\cos \alpha t$. Alternatively, the result follows from taking the logarithmic derivative of the infinite product expansion of $\sin x$

$$\sin x = x \prod_{n=1}^{\infty} \left(1 - \frac{x^2}{n^2 \pi^2}\right).$$

This is a standard result in texts on complex variables but it is possible to give a completely elementary proof (see Chapter 2 of Koblitz [162]).

Using the formula for the geometric series we can expand the right-hand side of (4) in a power series about 0. This yields

$$x \cot x = 1 - 2 \sum_{m=1}^{\infty} \zeta(2m) \frac{x^{2m}}{\pi^{2m}}. \tag{5}$$

We expand the left-hand side of Equation (5) in another way. Recall

$$\cos x = \frac{e^{ix} + e^{-ix}}{2} \quad \text{and} \quad \sin x = \frac{e^{ix} - e^{-ix}}{2i}.$$

From these expressions we derive

$$x \cot x = ix + \frac{2ix}{e^{2ix} - 1} = 1 + \sum_{n=2}^{\infty} B_n \frac{(2ix)^n}{n!}. \tag{6}$$

Comparing coefficients of x^{2m} on the right-hand sides of Equations (5) and (6) yields

$$-\frac{2}{\pi^{2m}} \zeta(2m) = (-1)^m \frac{2^{2m}}{(2m)!} B_{2m}.$$

This is Euler's result. ☐

As examples, take $m = 1, 2$ and 3. Since $B_2 = \frac{1}{6}$, $B_4 = -\frac{1}{30}$, and $B_6 = \frac{1}{42}$ we find $\zeta(2) = \pi^2/6$, $\zeta(4) = \pi^4/90$, and $\zeta(6) = \pi^6/945$.

A consequence of Theorem 2 is that $(-1)^{m+1}B_{2m} > 0$ for $m \geq 1$. This is because $\zeta(2m)$ is a positive real number for such m. Thus, the even indexed Bernoulli numbers are not zero and alternate in sign.

Theorem 2 also enables one to estimate the growth of B_{2m}. Namely, one sees

$$|B_{2m}| > \frac{2(2m)!}{(2\pi)^{2m}}. \tag{7}$$

Here we have used the simple observation that $\zeta(2m) > 1$. Using the obvious inequality $e^n > n^n/n!$ (look at the series expansion for e^n) we find

$$|B_{2m}| > 2\left(\frac{m}{\pi e}\right)^{2m}. \tag{8}$$

This shows that the even indexed Bernoulli numbers grow at a very rapid rate. A consequence which we will use later is $|B_{2n}/2n| \to \infty$ as $n \to \infty$.

We summarize the above properties of Bernoulli numbers in the following proposition.

Proposition 15.1.1

(a) For $k > 1$ and odd, $B_k = 0$.
(b) $(-1)^{m+1}B_{2m} > 0$ for $m = 1, 2, \dots$.
(c) $|B_{2m}/2m| \to \infty$ as $m \to \infty$.

   The third problem that we discuss in this section deals with the relationship between Bernoulli numbers and the Fermat equation $x^p + y^p = z^p$. This discussion will be purely expository for the result of Kummer is quite deep and requires analytic techniques that we have not developed. However we will introduce the important notion of a regular prime and state the Claussen–von Staudt congruence which we will prove in the following section. First of all we introduce the notion of a p-integer.

   Let p be a prime number. A rational number $r \in \mathbb{Q}$ is said to be a p-integer if $\mathrm{ord}_p(r) \geq 0$. In other words r is a p-integer if $r = a/b$, $a, b \in \mathbb{Z}$ and $p \nmid b$. One also says with slight ambiguity that p does not divide the denominator of r. It is an important observation that the set of p-integers forms a ring. Denote this ring by $\mathbb{Z}_p$. If r and s belong to $\mathbb{Z}_p$ write $r \equiv s \ (p^n)$ if $\mathrm{ord}_p(r - s) \geq n$, or equivalently, if $r - s = a/b$, $p \nmid b$ and $p^n | a$, $a, b \in \mathbb{Z}$. The following theorem proved independently by T. Claussen and C. von Staudt describes the denominator of B_{2m}. No such complete description of the prime divisors of the numerator is known.

Theorem 3. *For $m \geq 1$, $B_{2m} = A_{2m} - \sum_{p-1|2m} 1/p$ where $A_{2m} \in \mathbb{Z}$ and the sum is over all primes p such that $p - 1 | 2m$.*

Corollary. *If $p - 1 \nmid 2m$ then B_{2m} is a p-integer. If $p - 1 | 2m$ then $pB_{2m} + 1$ is a p-integer. More precisely if $p - 1 | 2m$ then*

$$\mathrm{ord}(pB_{2m} + 1) = \mathrm{ord}\, p\left(B_{2m} + \frac{1}{p}\right) = 1 + \mathrm{ord}\left(B_{2m} + \frac{1}{p}\right) \geq 1$$

so that $pB_{2m} \equiv -1 \ (p)$. Finally we notice that 6 always divides the denominator of B_{2m}, $m \geq 1$, since $2 - 1$ and $3 - 1$ divide 2.

   Kummer introduced the notion of a regular prime as follows.

Definition. An odd prime number $p \in \mathbb{Z}$ is said to be regular if p does not divide the numerator of any of the numbers $B_2, B_4, \ldots, B_{p-3}$. If p is not regular it is called irregular. The prime 3 is regular.

   By the corollary to Theorem 3, $B_2, B_4, \ldots, B_{p-3}$ are p-integers. Therefore p is regular if $\mathrm{ord}_p B_{2i} = 0$ for $i = 1, \ldots, (p - 3)/2$. It is easily seen that the units in $\mathbb{Z}_p$ are precisely the elements x with $\mathrm{ord}_p x = 0$. Thus p is regular if $B_2, B_4, \ldots, B_{p-3}$ are units in $\mathbb{Z}_p$. Equivalently p is irregular if some B_{2i}, $1 < i \leq (p - 3)/2$ is a nonunit in $\mathbb{Z}_p$. The first irregular primes are 37 and 59 for it is known that $\mathrm{ord}_{37}(B_{32}) = 1$ and $\mathrm{ord}_{59}(B_{44}) = 1$ [234]. The first few irregular primes are 37, 59, 67, 101, 103, 149 and 157. It was proven by Jensen in 1915 that there are infinitely many irregular primes of the form $4n + 3$. In the next section we give a short proof due to L. Carlitz (1953) that infinitely many irregular primes exist. It has not been proven that infinitely many regular primes exist. This is somewhat unfortunate in view of the following remarkable result of Kummer (1850).

Theorem 4. *Let p be a regular prime. Then $x^p + y^p = z^p$ has no solution in positive integers.*

Actually Kummer proved that Fermat's conjecture is true if p does not divide the class number of $\mathbb{Q}(\zeta_p)$. In other words the criterion is that for any nonprincipal ideal A in $Z[\zeta_p]$, A^p is not principal. This condition is equivalent to the regularity of p. We will not prove this, but the material in the third section of this chapter is closely related.

C. L. Siegel has given a plausible argument to suggest that the density of irregular primes is $1 - e^{-1/2} = 0.3935\ldots$ W. Johnson has checked this for primes less than 30,000 with good results [159]. S. Wagstaff has established the validity of Fermat's conjecture for all primes less than 125,000 [234]. Furthermore the information found by Johnson has now been extended by him to all primes less than 125,000 [234].

If a prime p is irregular one can ask how many nonzero Bernoulli numbers in the set $\{B_2, B_4, \ldots, B_{p-3}\}$ are divisible by p. This number is called the index of irregularity of p. The first prime of index 2 is 157. One of the most remarkable discoveries made with the aid of the computer is the existence of two primes of index 5 [234]. Finally we point out that thus far no pair p, B_{2i}, $1 \leq i \leq (p-3)/2$ has been found for which $\mathrm{ord}_p B_{2i} > 1$. For the above remarks and their relation to the celebrated Iwasawa invariants see the paper by W. Johnson in the bibliography.

§2 Congruences Involving Bernoulli Numbers

We will now prove a number of arithmetic properties of the Bernoulli numbers.

To begin with we direct our efforts toward proving Theorem 3 of the preceding section. Notice that for $m \geq k$ one has

$$\binom{m+1}{k} = \frac{m+1}{m-k+1} \binom{m}{k}$$

as follows immediately from the definition of the binomial coefficients. Thus, Theorem 1 of the last section becomes

$$S_m(n) = \sum_{k=0}^{m} \binom{m}{k} B_k \frac{n^{m+1-k}}{m+1-k}. \tag{9}$$

Now, using $\binom{m}{k} = \binom{m}{m-k}$ we see that

$$S_m(n) = \sum_{k=0}^{m} \binom{m}{k} B_{m-k} \frac{n^{k+1}}{k+1}$$

$$= B_m n + \binom{m}{1} B_{m-1} \frac{n^2}{2} + \cdots + \frac{n^{m+1}}{m+1}. \tag{10}$$

In addition to Equation (10) we need the following simple lemma.

Lemma 1. *Let p be a prime number and $k \geq 1$ an integer. Then*

(a) $p^k/(k + 1)$ *is p-integral.*
(b) $p^k/(k + 1) \equiv 0 \, (p)$ *if* $k \geq 2$.
(c) $p^{k-2}/(k + 1)$ *is p-integral if* $k \geq 3$ *and* $p \geq 5$.

PROOF. To prove (a) we show that $k + 1 \leq p^k$ for $k \geq 1$. If $k = 1$ the result is true. If $k + 1 \leq p^k$ then $k + 2 \leq p^k + 1 < 2p^k \leq p^{k+1}$. Now write $k + 1 = p^a q$ where $(q, p) = 1$. Then $p^k/(k + 1) = p^{k-a}/q$. Since $p^k/(k + 1) \geq 1$ we conclude that $k \geq a$, i.e., we have proven (a). To prove (b) we notice that $k + 1 < p^k$ for $k \geq 2$. The proof is the same as for (a). Therefore $k > a$ which proves (b).

As for part (c) use induction to show that $k + 1 < p^{k-2}$ for $k \geq 3$ and $p \geq 5$. This time one concludes that $k - 2 > a$, so that $p^{k-2}/(k + 1) = p^{k-2-a}/q$ is p-integral (and in fact divisible by p). □

Proposition 15.2.1. *Let p be a prime and $m \geq 1$ an integer. Then pB_m is p-integral. If $m \geq 2$ is even then $pB_m \equiv S_m(p) \, (p)$.*

PROOF. The first assertion states that if p divides the denominator of B_m then p^2 does not. First of all, $pB_1 = -p/2$ which is indeed p-integral for all p. We proceed by induction.

Suppose $m > 1$. Applying Equation (10) with $n = p$ we see that, since $S_m(p) \in \mathbb{Z}$, it suffices to prove that

$$\binom{m}{k} B_{m-k} \frac{n^{k+1}}{k + 1} = \binom{m}{k} pB_{m-k} \frac{p^k}{k + 1} \tag{11}$$

is p-integral for $k = 1, 2, \ldots, m$. By induction pB_{m-k} is p-integral for $k \geq 1$. Also by Lemma 1, part (a), $p^k/(k + 1)$ is p-integral. It follows pB_m is p-integral.

To establish the congruence it is enough to show that

$$\text{ord}_p \binom{m}{k} \left(pB_{m-k} \frac{p^k}{k + 1} \right) \geq 1 \quad \text{for } k \geq 1.$$

By Lemma 1, part (b) this is true for $k \geq 2$. For $k = 1$ we need to show

$$\text{ord}_p \left(\frac{m}{2} (pB_{m-1})p \right) \geq 1,$$

which is also true since m is even. Actually, for m even, $B_{m-1} = 0$ for $m \geq 4$, and so it is only necessary to check it for $m = 2$ where it is obvious. □

Lemma 2. *Let p be a prime. Then if $p - 1 \nmid m$, $S_m(p) \equiv 0 \, (p)$. If $p - 1 | m$ then $S_m(p) \equiv -1 \, (p)$.*

PROOF. Let g be a primitive root modulo p. Then

$$S_m(p) = 1^m + 2^m + \cdots + (p - 1)^m$$

$$\equiv 1^m + g^m + g^{2m} + \cdots + g^{(p-2)m} \, (p).$$

Thus $(g^m - 1)S_m(p) \equiv g^{m(p-1)} - 1 \equiv 0 \, (p)$. If $p - 1 \nmid m$ then $g^m \not\equiv 1 \, (p)$ and $S_m(p) \equiv 0 \, (p)$. On the other hand, if $p - 1 | m$ then $S_m(p) \equiv 1 + 1 + \cdots + 1 \equiv p - 1 \equiv -1 \, (p)$. $\qquad\square$

We are now in a position to prove Theorem 3. Assume m is even. Then by Proposition 15.2.1 we know pB_m is p-integral and $pB_m \equiv S_m(p) \, (p)$. By the lemma just proven it follows that B_m is a p-integer if $p - 1 \nmid m$ and $pB_m \equiv -1 \, (p)$ if $p - 1 | m$. Thus

$$A_m = B_m + \sum_{p-1|m} \frac{1}{p}$$

is a p-integer for all primes p. It follows that $A_m \in \mathbb{Z}$ and the proof is complete.

The reader may suspect by this time that the consequences of Equation (10) have not been exhausted. The following proposition is another important consequence of that equation. Write the mth Bernoulli number $B_m = U_m/V_m$ where $(U_m, V_m) = 1$ and $V_m > 0$. We are assuming m to be even.

Proposition 15.2.2. *If m is even, $m \geq 2$ then for all $n \geq 1$ we have*

$$V_m S_m(n) \equiv U_m n \, (n^2).$$

PROOF. Consider the terms in Equation (10) for $k \geq 1$ and fixed n

$$\binom{m}{k}\left(B_{m-k}\frac{n^{k-1}}{k+1}\right)n^2 = A_k^m n^2. \qquad (12)$$

We will show that for $p|n$ and $p \neq 2, 3$ $\mathrm{ord}_p(A_k^m) \geq 0$. Furthermore if $2|n$ then $\mathrm{ord}_2(A_k^m) \geq -1$ and if $3|n$ then $\mathrm{ord}_3(A_k^m) \geq -1$. This will imply that the greatest common divisor of n and the denominator of A_k^m is a divisor of 6 and thus this will also be true of the sum of the A_k^m. In other words one can write

$$S_m(n) = B_m n + \frac{An^2}{lB},$$

where $(B, n) = 1$ and $l|6$. Multiplying by BV_m and recalling that $6|V_m$ by Corollary to Theorem 3 the result follows immediately.

In order to prove the ord_p estimates we use the Corollary of Theorem 3 which implies that $\mathrm{ord}_p(B_{m-k}) \geq -1$ for all $m - k \geq 0$ and all p. Assume first of all that $p \neq 2, 3$, $p|n$. The cases $k = 1, 2$ are simple by inspection using the fact that $B_t = 0$ for $t > 1$ and odd, and that $B_1 = -\frac{1}{2}$, and that $\mathrm{ord}_p 3 = 0$. If $k \geq 3$, then

$$\mathrm{ord}_p\left(B_{m-k}\frac{n^{k-1}}{k+1}\right) \geq -1 + (k-1)\mathrm{ord}_p n - \mathrm{ord}_p(k+1)$$

$$\geq k - 2 - \mathrm{ord}_p(k+1) \geq 0 \qquad (13)$$

by part (c) of Lemma 1.

Consider now $p = 2$. If $k = 1$ then $B_{m-1} = 0$ for $m > 2$ (m is even) while for $m = 2$, A_k^m becomes $2 \cdot B_1 \cdot \frac{1}{2} = -\frac{1}{2}$ which has ord -1. For $k > 1$

we notice that $B_{m-k} = 0$ unless k is even or $k = m - 1$. But k even implies $\mathrm{ord}_2(k + 1) = 0$ while for $k = m - 1$, $A^m_{m-1} = -\frac{1}{2}n^{m-2}$ which has ord_2 greater than or equal to -1.

Finally consider the case $p = 3$, $3 \mid n$. Then $\mathrm{ord}_3(A^m_2) \geq -1$ and $\mathrm{ord}_3(A^m_3) \geq 1$ as one easily checks. But for $k \geq 4$ one shows exactly as in the lemma that $\mathrm{ord}_3(3^{k-2}/(k + 1)) \geq 0$ so that $\mathrm{ord}_3(A^m_k) \geq 0$. This completes the proof. $\square$

As a simple numerical illustration of this proposition consider $B_2 = \frac{1}{6}$, $U_2 = 1$, $V_2 = 6$ and let $n = 6$. The congruence reads

$$6(1^2 + 2^2 + 3^2 + 4^2 + 5^2) \equiv 6\,(36)$$

and more generally

$$6(1^2 + 2^2 + \cdots + (n - 1)^2) \equiv n\,(n^2).$$

Corollary. *Let m be even and p a prime such that $p - 1 \nmid m$. Then*

$$S_m(p) \equiv B_m p\,(p^2).$$

PROOF. By Theorem 3, $p \nmid V_m$. In the proposition, put $n = p$, and divide both sides of the resulting congruence by V_m which is permissible since $p \nmid V_m$. The result follows. $\square$

We are now in a position to prove the very useful congruences of G. Voronoi. According to the book of Uspensky and Heaslet [230], Voronoi discovered these congruences in 1889 while still a student.

Proposition 15.2.3. *Let $m \geq 2$ be even and define U_m and V_m as in the last proposition. Suppose a and n are positive integers with $(a, n) = 1$. Then*

$$(a^m - 1)U_m \equiv ma^{m-1}V_m \sum_{j=1}^{n-1} j^{m-1}\left[\frac{ja}{n}\right]\,(n), \tag{14}$$

where $[\alpha]$ is the unique integer k such that $k \leq \alpha < k + 1$.

PROOF. For $1 \leq j < n$ write $ja = q_j n + r_j$ where $0 \leq r_j < n$. Then $[ja/n] = q_j$ and since $(a, n) = 1$ the two sets $\{1, 2, 3, \ldots, n - 1\}$ and $\{r_1, r_2, \ldots, r_{n-1}\}$ are identical. By the binomial theorem

$$j^m a^m \equiv r^m_j + mq_j nr^{m-1}_j\,(n^2).$$

Since $r_j \equiv ja\,(n)$ we have

$$j^m a^m \equiv r^m_j + ma^{m-1}n\left[\frac{ja}{n}\right]j^{m-1}\,(n^2).$$

Summing over $j = 1, 2, \ldots, n - 1$ gives

$$S_m(n)a^m \equiv S_m(n) + ma^{m-1}n\sum_{j=1}^{n-1} j^{m-1}\left[\frac{ja}{n}\right]\,(n^2).$$

The result now follows from the congruence of Proposition 15.2.2. $\square$

Corollary. *Let p be a prime, $p \equiv 3\ (4)$. Set $m = (p + 1)/2$. Then if $p > 3$*

$$2\left(2 - \left(\frac{2}{p}\right)\right)B_m \equiv - \sum_{j=1}^{m-1} \left(\frac{j}{p}\right)(p),$$

where (x/p) denotes the Legendre symbol.

PROOF. Notice $m - 1 = (p - 1)/2$ so by Euler's criterion $a^{m-1} \equiv (a/p)\,(p)$ for all integers a.

In Voronoi's congruence set $a = 2$ and $n = p$. Using the above remark we find

$$\left(2\left(\frac{2}{p}\right) - 1\right)U_m \equiv m\left(\frac{2}{p}\right)V_m \sum_{j=1}^{p-1} \left(\frac{j}{p}\right)\left[\frac{2j}{p}\right](p).$$

Now, $[2j/p] = 0$ for $1 \le j \le m - 1$ and $[2j/p] = 1$ for $m \le j < p$. Also, $2m \equiv 1\ (p)$ and $p \nmid V_m$ by Theorem 3. Thus

$$2\left(2 - \left(\frac{2}{p}\right)\right)B_m \equiv \sum_{j=m}^{p-1} \left(\frac{j}{p}\right)(p).$$

Since $\sum_{j=1}^{p-1} (j/p) = 0$, the proof is complete. □

This corollary can be used to prove an interesting result relating class numbers to Bernoulli numbers. Let p be a prime, $p \equiv 3\ (4)$ and consider the imaginary quadratic number field $\mathbb{Q}(\sqrt{-p})$. Let h denote its class number. It can be shown that if $p > 3$

$$\left(2 - \left(\frac{2}{p}\right)\right)h = \sum_{1 \le x < p/2} \left(\frac{x}{p}\right).$$

For a proof, see Chapter 5, Section 4 of the book by Borevich and Shafarevich [9]. Combining the corollary with this formula for h gives the following remarkable congruence.

$$h \equiv -2B_{(p+1)/2}\,(p).$$

The Voronoi congruences lead to many properties of Bernoulli numbers. The following proposition is often attributed to J. C. Adams. It gives some information about the numerator of B_m.

Proposition 15.2.4. *If $p - 1 \nmid m$ then B_m/m is a p-integer.*

PROOF. By Theorem 3, B_m is a p-integer. Write $m = p^t m_0$ where $p \nmid m_0$. In the Voronoi congruence, Equation (14), put $n = p^t$. Then $(a^m - 1)U_m \equiv 0\ (p^t)$. Choose a to be a primitive root modulo p. Since $p - 1 \nmid m$ we have $p \nmid a^m - 1$. Thus, $U_m \equiv 0\ (p^t)$, and $B_m/m = U_m/mV_m$ is a p-integer. □

As a numerical illustration take $m = 22$ and $p = 11$. Then $B_{22} = 11 \cdot 131 \cdot 593/2 \cdot 3 \cdot 23$ so $B_{22}/22$ is integral at 11. Indeed it is a unit at 11. As a further example take $m = 50$ and $p = 5$. One can factor B_{50} as follows

$$B_{50} = \frac{5 \cdot 5 \cdot 417202699 \cdot 47464429777438199}{2 \cdot 3 \cdot 11}$$

Clearly, $B_{50}/50$ is a unit at 5. Less clear is the fact that the 17 digit number in the numerator is a prime!

The following theorem in the case $e = 1$ is due to Kummer. These congruences are now referred to as the Kummer congruences.

Theorem 5. *Suppose $m \geq 2$ is even, p a prime, and $p - 1 \nmid m$. Define $C_m = (1 - p^{m-1})B_m/m$. If $m' \equiv m\,(\phi(p^e))$ we have $C_{m'} \equiv C_m\,(p^e)$.*

PROOF. Write, as usual, $B_m = U_m/V_m$. Let $t = \mathrm{ord}_p\,m$. Proposition 15.2.4 shows $p^t | U_m$. In Equation (14) set $n = p^{e+t}$. Since p^t divides both m and U_m we may divide the resulting congruence throughout by p^t. Since $(m/p^t)V_m$ is prime to p we arrive at the following congruence

$$\frac{(a^m - 1)B_m}{m} \equiv a^{m-1} \sum_{j=1}^{p^{e+t}-1} j^{m-1} \left[\frac{ja}{p^{e+t}}\right] (p^e). \tag{15}$$

This congruence will lead the way to a full proof of the theorem. We will give the proof first in the case $e = 1$. This case reveals the main idea, which is quite simple, and avoids a slightly messy calculation which is necessary when $e > 1$.

In the above congruence assume $e = 1$. On the right-hand side we may omit those j which are divisible by p. If $p \nmid j$, then $j^{p-1} \equiv 1\,(p)$. Also, since $p \nmid a$, $a^{p-1} \equiv 1\,(p)$. Thus modulo p the right-hand side is unchanged if we replace m by m' with $m' \equiv m\,(p - 1)$. It follows that

$$\frac{(a^{m'} - 1)B_{m'}}{m'} \equiv \frac{(a^m - 1)B_m}{m}\,(p).$$

Choose a to be a primitive root modulo p. Since $p - 1 \nmid m$ we have $a^{m'} - 1 \equiv a^m - 1 \not\equiv 0\,(p)$. Consequently,

$$\frac{B_{m'}}{m'} \equiv \frac{B_m}{m}\,(p).$$

When $e > 1$ this procedure must be modified because the terms involving j divisible by p are not so easily disposed of. What we do is to separate them out and rewrite the corresponding sum. More precisely,

$$\sum_{j=1}^{p^{e+t}-1} j^{m-1} \left[\frac{ja}{p^{e+t}}\right] = \sum_{\substack{j=1 \\ (p,\,j)=1}}^{p^{e+t}-1} j^{m-1} \left[\frac{ja}{p^{e+t}}\right] + p^{m-1} \sum_{i=1}^{p^{e+t-1}-1} i^{m-1} \left[\frac{ia}{p^{e+t-1}}\right].$$

Consider the congruence (15) with e replaced by $e - 1$ and recall that $m - 1 \geq 1$. We find

$$\frac{p^{m-1}(a^m - 1)B_m}{m} \equiv p^{m-1}a^{m-1}\sum_{i=1}^{p^{e+t-1}-1} i^{m-1}\left[\frac{ia}{p^{e+t-1}}\right] (p^e).$$

Putting all this together, yields

$$\frac{(1 - p^{m-1})(a^m - 1)B_m}{m} \equiv a^{m-1}\sum_{\substack{j=1 \\ (p,\,j)=1}}^{p^{e+t}-1} j^{m-1}\left[\frac{ja}{p^{e+t}}\right] (p^e). \tag{16}$$

If $p \nmid j$, and $m' \equiv m\,(\phi(p^e))$ then $j^{m'-1} \equiv j^{m-1}\,(p^e)$. Thus the right-hand side of (16) is unchanged modulo p^e if m is replaced by m' with $m' \equiv m\,(\phi(p^e))$. The proof now proceeds exactly as in the case $e = 1$ and yields the full result. $\square$

We make a short detour to indicate a modern interpretation of the Kummer congruences.

Recall the Riemann zeta function $\zeta(s) = \sum_{n=1}^{\infty} n^{-s}$. In Exercise 25 of Chapter 2 we mentioned that $\zeta(s)$ can be extended to a function holomorphic on the entire complex plane except at $s = 1$ where it has a simple pole with residue 1. Moreover, it can be shown that $\zeta(s)$ satisfies the functional equation

$$\zeta(1 - s) = 2(2\pi)^{-s}\cos\left(\frac{\pi s}{2}\right)\Gamma(s)\zeta(s).$$

The Γ-function is defined and discussed in Chapter 16, Section 6. All we require here is the fact that $\Gamma(m) = (m - 1)!$ when m is a positive integer.

Assume $m \geq 2$ is an even integer. Combining the above functional equation with Theorem 2 we find

$$\zeta(1 - m) = \frac{-B_m}{m}.$$

Define $\zeta^*(s) = (1 - p^{-s})\zeta(s)$. Then $\zeta^*(1 - m) = -(1 - p^{m-1})B_m/m$ and Theorem 5 states that if $m' \equiv m\,(\phi(p^e))$ then

$$\zeta^*(1 - m') \equiv \zeta^*(1 - m)\,(p^e). \tag{17}$$

For a fixed prime p, the function $d(n, m) = p^{-\operatorname{ord}_p(n-m)}$ defines a metric on $\mathbb{Z}$, the p-adic metric. In this metric two integers are close if their difference is divisible by a high power of p. The congruence (17) may be stated informally as follows: if m' and m are close p-adically, and $m' \equiv m\,(p - 1)$, then $\zeta^*(1 - m')$ and $\zeta^*(1 - m)$ are close p-adically. This suggests the possibility of extending ζ^* to the metric completion of $\mathbb{Z}$, the ring of p-adic integers. These ideas were made precise by H. Leopoldt and T. Kubota who were the first to construct p-adic zeta functions and investigate their properties. Since then many other approaches have been devised. In the method due to B. Mazur the Bernoulli numbers are expressed as a certain p-adic integral of the functions x^m. The Kummer congruences have a very natural proof in this context. For details the reader is referred to Chapter 2 of [162]. The truly remarkable fact that properties of p-adic zeta functions (and p-adic L functions) are intimately related to the structure of class groups of cyclotomic fields is due to K. Iwasawa. Iwasawa gives a rather condensed and austere account of his theory in his monograph [155]. Another exposition of these matters is found in S. Lang [167].

We conclude this section with an application of Theorem 5. Namely we will prove that there exist infinitely many irregular primes. This proof is due to L. Carlitz [105].

Theorem 6. *The set of irregular primes is infinite.*

PROOF. Let $\{p_1, \ldots, p_s\}$ be a set of irregular primes. We will find an irregular prime not in this set.

Let $k \geq 2$ be even and set $n = k(p_1 - 1) \cdots (p_s - 1)$. If the set is empty choose $n = k$. By Proposition 15.1.1, part (c), choose k so large that $|B_n/n| > 1$. Choose a prime p with $\mathrm{ord}_p(B_n/n) > 0$. By Claussen–von Staudt $p - 1 \nmid n$. Thus $p \neq p_i, i = 1, \ldots, s$. Also $p \neq 2$. We will show that p is irregular.

Let $n \equiv m(p - 1)$ where $0 \leq m < p - 1$. Then m is even and $m \neq 0$. Thus $2 \leq m \leq p - 3$. By the Kummer congruence

$$\frac{B_n}{n} \equiv \frac{B_m}{m}(p).$$

Since $\mathrm{ord}_p(B_n/n) > 0$ and $\mathrm{ord}_p(B_n/n - B_m/m) > 0$ it follows that

$$\mathrm{ord}_p\left(\frac{B_m}{m}\right) = \mathrm{ord}_p B_m > 0$$

which shows that p is irregular.   □

§3 Herbrand's Theorem

Let D_m be the ring of algebraic integers in the cyclotomic number field $\mathbb{Q}(\zeta_m)$ and let P be a prime ideal of D_m not containing m. Thus if p is the rational prime in P then $p \nmid m$. In Section 3 of Chapter 14 we associated to P a Gauss sum $g(P)$ and showed $g(P)^m = \Phi(P) \in D_m$. The Stickelberger relation proved in Theorem 2 of that section gave the prime ideal decomposition of $\Phi(P)$ in D_m, namely

$$(\Phi(P)) = P^{\Sigma t \sigma_t^{-1}}.$$

Here the exponent is an element of the integral group ring $\mathbb{Z}[G]$ of the Galois group G of $\mathbb{Q}(\zeta_m)$ and t ranges over the integers between 1 and m which are relatively prime to m. The automorphism σ_t sends ζ_m to ζ_m^t. We remind the reader that the above exponential notation is a shorthand for

$(\Phi(P)) = \prod_{\substack{(t, m) = 1 \\ 1 \le t < m}} (\sigma_t^{-1}(P))^t$. If A is an ideal relatively prime to m then A is a product of prime ideals not containing m. It follows that $A^{\sum t \sigma_t^{-1}}$ is principal. The following proposition will be needed. We postpone the proof until later.

Proposition 15.3.1. *Let K be an algebraic number field and let M be a fixed ideal in the ring of integers of K. Then every ideal class of K contains an ideal prime to M.*

If α is in the group ring $\mathbb{Z}[G]$ where G is the Galois group of $\mathbb{Q}(\zeta_m)$ then α operates in the obvious way on the ideal class group of $\mathbb{Q}(\zeta_m)$. The above proposition implies that if $\alpha = \sum t \sigma_t^{-1}$ then α sends every ideal class to the identity class. One says that α annihilates the class group. It is natural to ask if there are other such elements of the group ring. Further annihilating elements are given below. First we need a definition.

Definition. The element $\theta = \sum \langle t/m \rangle \sigma_t^{-1}$, where t runs over a set of representatives for the residue classes relatively prime to m, is called the Stickelberger element. Here $\langle t/m \rangle$ denotes the fractional part of t/m, which depends only on the residue of t modulo m. θ is an element of the rational group ring $\mathbb{Q}[G]$. If b is an integer prime to m let $r_b = (\sigma_b - b)\theta$.

The following proposition, whose proof we will postpone, is very important.

Proposition 15.3.2. *The elements r_b are in $\mathbb{Z}[G]$ and annihilate the class group.*

We will see later that this proposition follows without much difficulty from the Stickelberger relation.

With these preliminaries and assumed propositions in mind we proceed to the principal goal of this section, the statement and proof of Herbrand's theorem.

Let $m = l$, an odd prime. Roughly speaking, Herbrand's theorem states that if l does not divide a certain Bernoulli number, then a piece of the class group of $\mathbb{Q}(\zeta_l)$ is missing. To make this statement precise we need a few definitions.

Let $\mathscr{A}$ be the subgroup of the ideal class group of $\mathbb{Q}(\zeta_l)$ consisting of elements whose order divides l. In other words, an ideal class is in $\mathscr{A}$ if it contains an ideal whose lth power is principal.

Definition. Let $1 \le i \le l - 1$. Define

$$\mathscr{A}_i = \{A \in \mathscr{A} \mid A^{\sigma_t} = A^{t^i}, 1 \le t < l\}.$$

It is easily seen that each $\mathscr{A}_i$ is a subgroup of $\mathscr{A}$. Also, since each element of $\mathscr{A}$ has order dividing l, the exponents can be computed modulo l, i.e., $\mathscr{A}$ is acted on by the group ring $\mathbb{Z}/l\mathbb{Z}[G]$. If $t \in \mathbb{Z}$ we denote by $\bar{t}$ its residue class modulo l.

Lemma 1. $\mathscr{A}$ *is the direct product of the* $\mathscr{A}_i$. *In other words,* $\mathscr{A} = \mathscr{A}_1 \mathscr{A}_2 \cdots \mathscr{A}_{l-1}$ *and* $\mathscr{A}_i \cap \prod_{j \neq i} \mathscr{A}_j = e$ (*the identity class*) *for* $i = 1, 2, \cdots, l - 1$.

PROOF. For each i with $1 \leq i \leq l - 1$ we define elements $\varepsilon_i \in \mathbb{Z}/l\mathbb{Z}[G]$ by the formula

$$\varepsilon_i = -\sum_{t=1}^{l-1} \bar{t}^{-i} \sigma_t.$$

Replacing t by ts in the formula leads to the relation $\sigma_s \varepsilon_i = \bar{s}^i \varepsilon_i$ provided that $l \nmid s$. It follows that $\mathscr{A}^{\varepsilon_i} \subseteq \mathscr{A}_i$. On the other hand, if $A \in \mathscr{A}_i$ then

$$A^{\varepsilon_i} = A^{-\sum \bar{t}^{-i} \sigma_t} = A^{-(l-1)} = A.$$

It follows that $\mathscr{A}^{\varepsilon_i} = \mathscr{A}_i$.

By Lemma 2 of Section 2 we see that $\varepsilon_1 + \varepsilon_2 + \cdots + \varepsilon_{l-1} = \sigma_1$, the identity automorphism. Thus

$$\mathscr{A} = \mathscr{A}^{\varepsilon_1 + \cdots + \varepsilon_{l-1}}$$

$$= \mathscr{A}^{\varepsilon_1} \mathscr{A}^{\varepsilon_2} \cdots \mathscr{A}^{\varepsilon_{l-1}} = \mathscr{A}_1 \mathscr{A}_2 \cdots \mathscr{A}_{l-1}.$$

Suppose $i \neq j$. Using the relation $\sigma_s \varepsilon_i = \bar{s}^i \varepsilon_i$, the definition of ε_i, and Lemma 2 of Section 2, we see $\varepsilon_i \varepsilon_j = 0$. It follows that $\mathscr{A}_j^{\varepsilon_i} = e$. Let $A \in \mathscr{A}_i \cap \prod_{j \neq i} \mathscr{A}_j$. On the one hand $A^{\varepsilon_i} = A$ and on the other hand $A^{\varepsilon_i} = e$. Thus, $A = e$ and the proof is complete. $\square$

The following theorem of J. Herbrand [149] gives a Bernoulli criterion for the triviality of $\mathscr{A}_i$. The proof emerges from the interplay of the Stickelberger relation and the Voronoi congruences.

Theorem 7 (J. Herbrand). *Let* i *be an odd integer* $1 \leq i < l$ *and define* j *by* $i + j = l$.
Then $\mathscr{A}_1 = (e)$. *If* $i \geq 3$ *and* $l \nmid B_j$ *then* $\mathscr{A}_i = (e)$.

PROOF. Let $A \in \mathscr{A}_1$. Then, by Stickelberger's relation

$$e = A^{\sum t \sigma_t^{-1}} = A^{\sum \bar{t}^{-1}} = A^{l-1} = A^{-1}.$$

This shows $\mathscr{A}_1 = (e)$ as asserted.

Now suppose i is odd and $3 \leq i \leq l - 2$. Let $A \in \mathscr{A}_i$. By Proposition 15.3.2 $A^{r_b} = e$ where b is any integer prime to l. We analyze this relation more carefully.

By definition $r_b = (\sigma_b - b)\theta$. Now,

$$\sigma_b \theta = \sum \langle t/l \rangle \sigma_b \sigma_t^{-1} = \sum \langle t/l \rangle \sigma_{b^{-1}t}^{-1} = \sum \langle bt/l \rangle \sigma_t^{-1}.$$

Thus

$$r_b = (\sigma_b - b)\theta = \sum_{t=1}^{l-1} \left(\left\langle \frac{bt}{l} \right\rangle - b \left\langle \frac{t}{l} \right\rangle \right) \sigma_t^{-1}.$$

Write $bt = q_t l + s_t$ with $0 \le s_t < l$. Then $\langle bt/l \rangle - b\langle t/l \rangle = s_t/l - bt/l = -q_t = -[bt/l]$. This shows $r_b = -\sum [bt/l]\sigma_t^{-1} \in \mathbb{Z}[G]$.

Suppose $A \in \mathcal{A}_i$. Applying σ_t^{-1} to A has the effect of raising A to the power $t^{l-1-i} = t^{j-1}$. Thus, applying r_b to A has the effect of raising A to the power $-\sum [bt/l] t^{j-1}$.

Write $B_j = U_j/V_j$ with $(U_j, V_j) = 1$. The Voronoi congruence, Proposition 15.2.3, shows after some relabeling

$$(b^j - 1)U_j \equiv jb^{j-1}V_j \sum_{t=1}^{l-1} \left[\frac{bt}{l}\right] t^{j-1} \ (l).$$

By the previous observations the right-hand side of this congruence annihilates any element $A \in \mathcal{A}_i$. Thus, for such an element $A^{(b^j - 1)U_j} = e$. Choosing b to be a primitive root modulo l we see $l \nmid b^j - 1$ and so $A^{U_j} = e$. If $l \nmid B_j$, then $l \nmid U_j$ and so $A = e$. Thus $l \nmid B_j$ implies $\mathcal{A}_i = (e)$ as asserted. $\square$

We remark that the converse of Herbrand's theorem was established by K. Ribet in 1976 [208]. Namely, he showed that if j is even and $2 \le j \le l - 3$ then $l | B_j$ implies $\mathcal{A}_i \ne (e)$ for $i = l - j$. This beautiful existence theorem depends on subtle arithmetic properties of modular forms and is, unfortunately, beyond the scope of this book.

Write $\mathcal{A} = \mathcal{A}^+ \mathcal{A}^-$ where $\mathcal{A}^+ = \mathcal{A}_2 \mathcal{A}_4 \cdots \mathcal{A}_{l-1}$ and $\mathcal{A}^- = \mathcal{A}_3 \mathcal{A}_5 \cdots \mathcal{A}_{l-2}$. Then $\mathcal{A} = \mathcal{A}^+ \mathcal{A}^-$ and $\mathcal{A}^+ \cap \mathcal{A}^- = (e)$ (see Exercise 23). The theorem of Herbrand implies $|\mathcal{A}^-| = 1$ if $l \nmid B_j$ for $j = 2, 4, \ldots, l - 3$. This was already known to Kummer who also showed, in essence, that $|\mathcal{A}^-| = 1$ implies $|\mathcal{A}^+| = 1$. Thus, as we mentioned earlier, Kummer showed that $l \nmid B_j$ for $j = 2, 4, \ldots, l - 3$ implies the class number of $\mathbb{Q}(\zeta_l)$ is not divisible by l.

One of the most famous open problems in algebraic number theory is the conjecture of H. S. Vandiver. This states that the group $\mathcal{A}^+$ of the previous paragraph is always trivial. It is not too hard to show this is equivalent to the assertion that the class number of $\mathbb{Q}(\zeta_l + \zeta_l^{-1}) = \mathbb{Q}(\cos(2\pi/l))$ is not divisible by l. Vandiver made this conjecture around 1920. See his article on Fermat's Last Theorem [231]. If true the conjecture has many important consequences. S. Wagstaff has shown Vandiver's conjecture is true for all primes less than 125,000. This seems to be impressive evidence, but Larry Washington has shown on probabilistic grounds that 125,000 is too small for the evidence to be convincing.

We conclude this chapter by giving proof of Propositions 15.3.1 and 15.3.2.

We begin with Proposition 15.3.1. Let K be an algebraic number field and D its ring of integers. Let $M \subset D$ be a fixed ideal. For any ideal A in D let $\bar{A}$ denote its ideal class. Given A we will construct an ideal C such that $(C, M) = 1$ and $\bar{A}^{-1} = \bar{C}$. This shows the inverse of any class contains an ideal prime to M. Thus every class contains an ideal prime to M. To construct

C we proceed as follows. Let $\{P_1, P_2, \ldots, P_t\}$ be the set of primes dividing M which do not divide A. This set may be empty. If $P \mid A$ let, as usual, $a(P) = \operatorname{ord}_P A$ denote the exponent of P in the prime decomposition of A. Choose

$$\pi(P) \in P^{a(P)} - P^{a(P)+1}.$$

By the Chinese Remainder Theorem we can find an $\alpha \in D$ such that

$$\alpha \equiv \pi(P)\,(P^{a(P)+1}) \quad \text{for } P \mid A$$

$$\alpha \equiv 1\,(P_i) \qquad\qquad \text{for } i = 1, 2, \ldots, t.$$

One checks easily that $(\alpha) = AC$ with $(C, M) = 1$. Thus $\bar{A}^{-1} = \bar{C}$ and the proof is complete. $\qquad\square$

Finally, we turn to the proof of Proposition 15.3.2. We will need the following lemma which is proven in the same way as the special case $m = l$ done during the proof of Theorem 7.

Lemma 2. *Let G denote the Galois group of $\mathbb{Q}(\zeta_m)/\mathbb{Q}$. The element $r_b = (\sigma_b - b)\theta \in \mathbb{Z}[G]$. In fact, $r_b = -\sum [bt/m]\sigma_t^{-1}$ where the sum is over $1 \le t < m$ with $(t, m) = 1$.*

Let P be a prime ideal in D_m the ring of integers in $\mathbb{Q}(\zeta_m)$. Assume $m \notin P$ and let $P \cap \mathbb{Z} = (p)$. As in Section 3 of Chapter 14 we associate a Gauss sum $g(P)$ to P. We know $g(P) \in \mathbb{Q}(\zeta_m, \zeta_p) = \mathbb{Q}(\zeta_{pm})$.

Lemma 3. *Let b be an integer prime to m. Determine b' by the conditions $b' \equiv b\,(m)$ and $b' \equiv 1\,(p)$. Let $\sigma_{b'}$ be the corresponding automorphism of $\mathbb{Q}(\zeta_{pm})$. Then*

$$g(P)^{\sigma_{b'} - b} \in \mathbb{Q}(\zeta_m).$$

PROOF. The automorphisms of $\mathbb{Q}(\zeta_{pm})$ which leave ζ_m fixed are of the form σ_c where $(c, pm) = 1$ and $c \equiv 1\,(m)$. Let

$$\Omega_b(P) = g(P)^{\sigma_{b'} - b}.$$

We will show $\Omega_b(P)^{\sigma_c} = \Omega_b(P)$. This proves, by Galois theory, that $\Omega_b(P) \in \mathbb{Q}(\zeta_m)$.

Recall that $g(P) = \sum \chi_p(t)\psi(t)$ where the sum is over a reduced residue system modulo m. Since $\chi_p(t) \in \mathbb{Q}(\zeta_m)$ and $\psi(t) \in \mathbb{Q}(\zeta_p)$ we have

$$g(P)^{\sigma_{b'}} = \sum \chi_p(t)^b \psi(t)$$

and

$$g(P)^{\sigma_{b'}\sigma_c} = \sum \chi_p(t)^b \psi(t)^c$$
$$= \sum \chi_p(t)^b \psi(ct).$$

Thus

$$g(P)^{\sigma_{b'}\sigma_c} = \chi_p(c)^{-b} g(P)^{\sigma_{b'}}. \tag{1}$$

Similarly, we find

$$g(P)^{\sigma_c} = \chi_p(c)^{-1}g(P). \tag{2}$$

Raising both sides of (2) to the bth power and dividing the result into Equation (1) give $\Omega_b(P)^{\sigma_c} = \Omega_b(P)$ as asserted. $\qquad\Box$

We are now in a position to complete the proof of Proposition 15.3.2. Let $P \subset D_m$ be a prime ideal not containing m. Stickelberger's relation asserts that $g(P)^m \in \mathbb{Q}(\zeta_m)$ and $(g(P)^m) = P^{m\theta}$. Applying $\sigma_{b'} - b$ to both sides shows that $(\Omega_b(P)^m) = P^{mrb}$. By Lemmas 2 and 3 above, this becomes, in D_m, the equation $(\Omega_b(P))^m = (P^{rb})^m$. It follows from unique factorization for ideals that $P^{rb} = (\Omega_b(P))$. Thus P^{rb} is a principal ideal and therefore A^{rb} is principal for any ideal A relatively prime to (m). By Proposition 15.3.1 we conclude that r_b annihilates the class group of D_m. This completes the proof. $\qquad\Box$

NOTES

In 1960 Vandiver published a survey article in which he remarks that some 1500 papers on Bernoulli numbers had been published [232]. Clearly, this sequence of numbers has considerable fascination and importance. The most extensive treatise that has appeared on Bernoulli numbers is the classic by N. Nielsen [199]. A more accessible modern source is the first two chapters of the book on analytic number theory by H. Rademacher [204]. This book has an exposition of the Euler–MacLaurin summation formula, an important application of the Bernoulli numbers which we have not considered.

The evaluation of $\zeta(s)$ at the positive even integers by Euler was a major accomplishment. It is surprising that almost nothing is known about the values of $\zeta(s)$ at positive odd integers. In 1978 the French mathematician R. Apery created a sensation by finding an extraordinarily ingenious proof that $\zeta(3)$ is irrational. See the entertaining article by A. van der Poorten [233].

The relation of Bernoulli numbers to Fermat's Last Theorem and the arithmetic of cyclotomic number fields is very close as is evident from the numerous references to them in the scholarly book by P. Ribenboim [206]; see, in particular, Section 2 of Lecture VI. The short expository article by Vandiver [231] is also worth consulting.

The paper [159] by Johnson has a very readable discussion of regular and irregular primes and mentions a number of interesting open problems. We follow his brief history of the calculation of irregular primes. Kummer himself determined that 8 of the first 37 primes were irregular. In the 1930s Vandiver and others extended the calculation to all primes less than 618. In 1955 Vandiver, D. H. Lehmer, Emma Lehmer, Selfridge, and Nicole worked up to 4001. In 1964 Selfridge and Pollack announced computations up to 25,000. These were not published. In 1970 Kobelev published tables up to 5500 and in 1973 Johnson attained 8000. In 1975 Johnson made it up to 30,000. As stated earlier, the current record is due to Wagstaff, 125,000. The art of computing has come a long way!

The following result was discovered independently by T. Metsänkylä [188] and H. Yokoi [247]. Let $m > 2$ be an integer and H a proper subgroup of $U(\mathbb{Z}/m\mathbb{Z})$. There exist infinitely many irregular primes p such that the congruence class of p modulo m is not in H. By contrast, there is not a single modulus $m > 2$ known for which there exist infinitely many irregular primes $p \equiv 1 \ (m)$.

The main theorem of Section 3 was published by Herbrand in 1932 [149]. A proof relying on p-adic numbers and congruences for generalized Bernoulli numbers can be found in Ribet's paper [208]. See also Chapter 1 of Lang's book [167]. There are a number of important conjectures which concern the p-primary component of the class group of $\mathbb{Q}(\zeta_p)$. The introduction to the paper of A. Wiles [242] describes a conjecture which makes Herbrand's theorem more precise.

EXERCISES

1. Using the definition of the Bernoulli numbers show $B_{10} = 5/66$ and $B_{12} = -691/2730$.

2. If $a \in \mathbb{Z}$, show $a(a^m - 1)B_m \in \mathbb{Z}$ for all $m > 0$.

3. If $a \in \mathbb{Z}$, show $a^m(a^m - 1)B_m/m \in \mathbb{Z}$ for all $m > 0$.

4. If $m \geq 4$ is even, show $2B_m \equiv 1 \ (4)$.

5. If p is an odd prime and $p - 1 | m$ show $(B_m + p^{-1} - 1)/m$ is p-integral. This result is due to L. Carlitz.*

6. For $m \geq 3$, show $|B_{2m+2}| > |B_{2m}|$. (Hint: Use Theorem 2.)

7. Let $m \geq 2$ be an even integer. Show there exist infinitely many $n \geq m$ such that $B_n - B_m \in \mathbb{Z}$. [Hint: Let q be a prime such that $q \equiv 1 \ ((m + 1)!)$ and try $n = qm$. The existence of infinitely many such primes q is shown in Chapter 16. This result is due to R. Rado.]

8. Consider the power series expansion of $\tan(x)$ about the origin;
$$\sum_{k=1}^{\infty} T_k \frac{x^{2k-1}}{(2k-1)!}.$$
Show $T_k = (-1)^{k-1}(B_{2k}/2k)(2^{2k} - 1)2^{2k}$. Note that $T_k \in \mathbb{Z}$ for all k by Exercise 3.

9. Using Lemma 1 in Section 1 show the radius of convergence of $\sum_{n=0}^{\infty} B_n(t^n/n!)$ is 2π. As a consequence show that for any $C, k > 0$ there are infinitely many n such that $|B_n| > Cn^k$. (This result is weaker than the estimate given by Equation (8) of Section 1. On the other hand, it is much easier to obtain.)

---

* L. Carlitz. Some congruences for the Bernoulli numbers. *Amer. J. Math.*, **75** (1953), 163–172.

10. Use the Voronoi congruences to obtain the following result of Kummer.

$$\sum_{k=0}^{r} (-1)^k \binom{r}{k} \frac{B_{2n+k(p-1)}}{2n + k(p - 1)} \equiv 0 \, (p^r)$$

provided $2 \leq r + 1 \leq 2n$ and $p - 1 \nmid 2n$. This is a bit tricky. With minor changes in notation the proof is contained in Section 8 of Chapter IX of the book by Uspensky and Heaslet [230].

11. Those familiar with the approach of B. Mazur to p-adic zeta and L-functions can try the following. Let μ_α be the normalized "Mazur measure" on $\mathbb{Z}_p$. Use the Voronoi congruences to prove $\int_{\mathbb{Z}_p}^* x^{k-1} \, d\mu_\alpha = (\alpha^{-k} - 1)(1 - p^{k-1})(-B_k/k)$. For the notation and the definition of the Mazur measure the reader can consult Koblitz [162].

12. Recall the definition of the Bernoulli polynomials;

$$B_m(x) = \sum_{k=0}^{m} \binom{m}{k} B_k x^{m-k}.$$

Show that $te^{tx}/(e^t - 1) = \sum_{m=0}^{\infty} B_m(x)(t^m/m!)$.

13. Show $B_m(x + 1) - B_m(x) = mx^{m-1}$.

14. Use Exercise 13 to give a new proof of Theorem 1.

15. Suppose $f(x) = \sum_{k=0}^{n} a_k x^k$ is a polynomial with complex coefficients. Use Exercise 13 to find a polynomial $F(x)$ such that $F(x + 1) - F(x) = f(x)$.

16. For $n \geq 1$ show $(d/dx)B_n(x) = nB_{n-1}(x)$.

17. Show $B_n(1 - x) = (-1)^n B_n(x)$.

18. Use Exercises 13 and 17 to give a new proof that $B_n = 0$ for n odd and $n > 1$.

19. Suppose n and F are integers and $n, F > 0$. Show that

$$B_n(Fx) = F^{n-1} \sum_{a=0}^{F-1} B_n\left(x + \frac{a}{F}\right).$$

(*Hint*: Use Exercise 12.)

20. Suppose $H(x)$ is a polynomial of degree n with complex coefficients. Suppose that for all integers n, $F > 0$ we have $H(Fx) = F^{n-1} \sum_{a=0}^{F-1} H(x + (a/F))$. Show that $H(x) = CB_n(x)$ for some constant C. (*Hint*: Use Exercise 16 and induction on n.)

21. Show $B_n(\frac{1}{2}) = (1 - 2^{n-1})B_n$.

22. More generally, show that $(1 - F^{n-1})B_n = \sum_{a=1}^{F-1} B_n(a/F)$.

23. Prove the assertions; $\mathscr{A} = \mathscr{A}^+ \mathscr{A}^-$ and $\mathscr{A}^+ \cap \mathscr{A}^- = (e)$.

Chapter 16

Dirichlet *L*-functions

The theory of analytic functions has many applications in number theory. A particularly spectacular application was discovered by Dirichlet who proved in 1837 that there are infinitely many primes in any arithmetic progression $b, b + m, b + 2m, \ldots$, where $(m, b) = 1$. To do this he introduced the L-functions which bear his name. In this chapter we will define these functions, investigate their properties, and prove the theorem on arithmetic progressions. The use of Dirichlet L-functions extends beyond the proof of this theorem. It turns out that their values at negative integers are especially important. We will derive these values and show how they relate to Bernoulli numbers.

For the most part we will use only basic calculus. However, in Section 6 where we discuss the value of the L-functions at 1 we use complex function theory in an essential way. This can be avoided but to do so involves sacrificing both depth and elegance. All the necessary background can be found in any standard treatise. The book of L. Ahlfors [85] is a convenient reference. In Sections 1–4 the letter s will stand for a real variable, $s > 1$.

§1 The Zeta Function

The Riemann zeta function $\zeta(s)$ is defined by $\zeta(s) = \sum_{n=1}^{\infty} n^{-s}$. It converges for $s > 1$ and converges uniformly for $s \geq 1 + \delta > 1$, for each $\delta > 0$.

Proposition 16.1.1. *For $s > 1$*

$$\zeta(s) = \prod_{p} (1 - p^{-s})^{-1},$$

where the product is over all primes $p > 0$.

PROOF. For $s > 1$, $p^{-s} < 1$, so we have $(1 - p^{-s})^{-1} = \sum_{m=0}^{\infty} p^{-ms}$. By the theorem of unique factorization

$$\prod_{p \leq N} (1 - p^{-s})^{-1} = \sum_{n \leq N} n^{-s} + R_N(s).$$

Clearly, $R_N(s) \le \sum_{n=N+1}^{\infty} n^{-s}$. Since $\zeta(s)$ converges, $R_N(s) \to 0$ as $N \to \infty$. The result follows. □

The behavior of $\zeta(s)$ as $s \to 1$ is very important. Since $\sum_{n=1}^{\infty} n^{-1}$ diverges we, of course, suspect $\zeta(s) \to \infty$ as $s \to 1$. In fact,

Proposition 16.1.2. *Assume $s > 1$. Then*

$$\lim_{s \to 1} (s - 1)\zeta(s) = 1.$$

PROOF. For fixed s, t^{-s} is a monotone decreasing function of t. Thus,

$$(n + 1)^{-s} < \int_{n}^{n+1} t^{-s} \, dt < n^{-s}.$$

Summing from $n = 1$ to ∞,

$$\zeta(s) - 1 < \int_{1}^{\infty} t^{-s} \, dt < \zeta(s).$$

The value of the integral is $(s - 1)^{-1}$. It follows that $1 < (s - 1)\zeta(s) < s$. Taking the limit as $s \to 1$ gives the result. □

Corollary. *As $s \to 1$ we have*

$$\frac{\ln \zeta(s)}{\ln(s - 1)^{-1}} \to 1.$$

PROOF. Let $(s - 1)\zeta(s) = \rho(s)$. Then $\ln(s - 1) + \ln \zeta(s) = \ln \rho(s)$ so we have $\ln \zeta(s)/\ln(s - 1)^{-1} = 1 + (\ln \rho(s)/\ln(s - 1)^{-1})$.

As $s \to 1$, $\rho(s) \to 1$ by the proposition. Therefore, $\ln \rho(s) \to 0$ and the result follows. □

Proposition 16.1.3. $\ln \zeta(s) = \sum_p p^{-s} + R(s)$ *where $R(s)$ remains bounded as $s \to 1$.*

PROOF. We use the formula $-\ln(1 - x) = x + x^2/2 + x^3/3 + \cdots$ which is valid for $-1 < x < 1$.

By Proposition 16.1.1 we have

$$\zeta(s) = \prod_{p \le N} (1 - p^{-s})^{-1} \lambda_N(s),$$

where $\lambda_N(s) \to 1$ as $N \to \infty$. Taking the logarithm of both sides yields $\ln \zeta(s) = \sum_{p \le N} \sum_{m=1}^{\infty} m^{-1} p^{-ms} + \ln \lambda_N(s)$.

Taking the limit as $N \to \infty$

$$\ln \zeta(s) = \sum_p \sum_{m=1}^{\infty} m^{-1} p^{-ms}$$

$$= \sum_p p^{-s} + \sum_p \sum_{m=2}^{\infty} m^{-1} p^{-ms}.$$

The second sum is less than

$$\sum_p \sum_{m=2}^{\infty} p^{-ms} = \sum_p p^{-2s}(1 - p^{-s})^{-1}$$

$$\leq (1 - 2^{-s})^{-1} \sum_p p^{-2s} \leq 2\zeta(2).$$

Throughout we have used the assumption that $s > 1$. $\qquad\qquad\square$

Definition. A set of positive primes $\mathscr{P}$ is said to have Dirichlet density if

$$\lim_{s \to 1} \frac{\sum_{p \in \mathscr{P}} p^{-s}}{\ln(s - 1)^{-1}}$$

exists. If the limit exists we set it equal to $d(\mathscr{P})$ and call $d(\mathscr{P})$ the Dirichlet density of $\mathscr{P}$.

Proposition 16.1.4. *Let $\mathscr{P}$ be a set of positive prime numbers. Then*

(a) *If $\mathscr{P}$ is finite, then $d(\mathscr{P}) = 0$.*
(b) *If $\mathscr{P}$ consists of all but finitely many positive primes, then $d(\mathscr{P}) = 1$.*
(c) *If $\mathscr{P} = \mathscr{P}_1 \cup \mathscr{P}_2$ where $\mathscr{P}_1$ and $\mathscr{P}_2$ are disjoint and $d(\mathscr{P}_1)$ and $d(\mathscr{P}_2)$ both exist, then $d(\mathscr{P}) = d(\mathscr{P}_1) + d(\mathscr{P}_2)$.*

PROOF. Parts (a) and (c) are clear from the definition of Dirichlet density. Part (b) follows quickly from the corollary to Proposition 16.1.2 and Proposition 16.1.3. $\qquad\qquad\square$

We are now in a position to state the main theorem of this chapter. The proof will be spread out over the next three sections.

Theorem 1 (L. Dirichlet). *Suppose $a, m \in \mathbb{Z}$, with $(a, m) = 1$. Let $\mathscr{P}(a; m)$ be the set of positive primes p such that $p \equiv a \ (m)$. Then $d(\mathscr{P}(a; m)) = 1/\phi(m)$.*

Note that Theorem 1 certainly implies $\mathscr{P}(a; m)$ is infinite, since if it were finite its density would be zero.

§2 A Special Case

We will first prove Theorem 1 in the case where $m = 4$. The basic ideas of the proof are all present in this special case but the details are more transparent.

Define a function χ from $\mathbb{Z}$ to $\{0, \pm 1\}$ as follows; $\chi(n) = 0$ if n is even, $\chi(n) = 1$ if $n \equiv 1 \ (4)$, and $\chi(n) = -1$ if $n \equiv 3 \ (4)$. It is easily seen that $\chi(mn) = \chi(m)\chi(n)$ for all $m, n \in \mathbb{Z}$.

Define $L(s, \chi) = \sum_{n=1}^{\infty} \chi(n)n^{-s} = 1 - 3^{-s} + 5^{-s} - 7^{-s} + \cdots$. For all n we have $|\chi(n)n^{-s}| \le n^{-s}$. It follows that the terms of $L(s, \chi)$ are dominated in absolute value by the terms of $\zeta(s)$. Thus $L(s, \chi)$ converges and is continuous for $s > 1$. Since χ is completely multiplicative the proof of Proposition 16.1.1. shows that

$$L(s, \chi) = \prod_{p} (1 - \chi(p)p^{-s})^{-1}.$$

It is useful to modify $\zeta(s)$ so as to suppress the even terms. Define $\zeta^*(s) = \sum_{n \text{ odd}} n^{-s}$. Since

$$\zeta(s) = \sum_{n=1}^{\infty} n^{-s} = \sum_{n \text{ odd}} n^{-s} + \sum_{n \text{ even}} n^{-s} = \zeta^*(s) + 2^{-s}\zeta(s)$$

we have $\zeta^*(s) = (1 - 2^{-s})\zeta(s)$ and so

$$\zeta^*(s) = \prod_{p \text{ odd}} (1 - p^{-s})^{-1}.$$

Using the method of proof of Proposition 16.1.3 we find

$$\ln L(s, \chi) = \sum_{p \text{ odd}} \chi(p)p^{-s} + R_1(s), \qquad \text{(i)}$$

$$\ln \zeta^*(s) = \sum_{p \text{ odd}} p^{-s} + R_2(s), \qquad \text{(ii)}$$

where $R_1(s)$ and $R_2(s)$ remain bounded as $s \to 1$.

We have $1 + \chi(p) = 2$ if $p \equiv 1$ (4) and $1 + \chi(p) = 0$ if $p \equiv 3$ (4). Similarly, $1 - \chi(p) = 2$ if $p \equiv 3$ (4) and $1 - \chi(p) = 0$ if $p \equiv 1$ (4). From (i) and (ii) we deduce

$$\ln \zeta^*(s) + \ln L(s, \chi) = 2 \sum_{p \equiv 1 (4)} p^{-s} + R_3(s), \qquad \text{(iii)}$$

$$\ln \zeta^*(s) - \ln L(s, \chi) = 2 \sum_{p \equiv 3 (4)} p^{-s} + R_4(s), \qquad \text{(iv)}$$

where $R_3(s)$ and $R_4(s)$ remain bounded as $s \to 1$.

The next step is to show that $\ln L(s, \chi)$ remains bounded as $s \to 1$. To see this write $L(s, \chi) = 1 - 3^{-s} + 5^{-s} - \cdots = (1 - 3^{-s}) + (5^{-s} - 7^{-s}) + \cdots = 1 - (3^{-s} - 5^{-s}) - (7^{-s} - 9^{-s}) - \cdots$. It follows that for all $s > 1$ we have $\frac{2}{3} < L(s, \chi) < 1$. Thus, for $s > 1$ we have $\ln \frac{2}{3} < \ln L(s, \chi) < \ln 1 = 0$.

As a final preparatory step we note that $\ln \zeta^*(s) = \ln(1 - 2^{-s}) + \ln \zeta(s)$ so by the corollary to Proposition 16.1.2. we have $\ln \zeta^*(s)/\ln(s - 1)^{-1} \to 1$ as $s \to 1$.

Now divide each term of Equations (iii) and (iv) by $\ln(s - 1)^{-1}$ and take the limit as $s \to 1$. The result is

Proposition 16.2.1. $d(\mathscr{P}(1;4)) = \frac{1}{2}$ *and* $d(\mathscr{P}(3;4)) = \frac{1}{2}$.

To prove Theorem 1 in the general case we need to generalize χ and $L(s, \chi)$. This leads to the introduction of Dirichlet characters and Dirichlet L-functions.

§3 Dirichlet Characters

The function χ considered in the last section can be obtained from the following construction. Consider the group $U(\mathbb{Z}/4\mathbb{Z})$. This group has two elements $1 + 4\mathbb{Z}$ and $3 + 4\mathbb{Z}$. Define $\chi': U(\mathbb{Z}/4\mathbb{Z}) \to \{\pm 1\}$ by $\chi'(1 + 4\mathbb{Z}) = 1$ and $\chi'(3 + 4\mathbb{Z}) = -1$. Then χ' is a homomorphism from $U(\mathbb{Z}/4\mathbb{Z})$ to $\mathbb{C}^*$. For $n \in \mathbb{Z}$ define $\chi(n) = 0$ if $(n, 4) > 1$ and $\chi(n) = \chi'(n + 4\mathbb{Z})$ if $(n, 4) = 1$. This function $\chi: \mathbb{Z} \to \mathbb{C}^*$ coincides with the function χ of the last section.

This construction is easy to generalize. Let m be a fixed positive integer. Let $\chi': U(\mathbb{Z}/m\mathbb{Z}) \to \mathbb{C}^*$ be a homomorphism. Given χ' define $\chi: \mathbb{Z} \to \mathbb{C}^*$ as follows; if $(n, m) > 1$ set $\chi(n) = 0$, if $(n, m) = 1$ set $\chi(n) = \chi'(n + m\mathbb{Z})$. The functions χ defined in this manner are called *Dirichlet characters modulo m*. Another characterization is given by the following three conditions on a function $\chi: \mathbb{Z} \to \mathbb{C}^*$

(a) $\chi(n + m) = \chi(n)$ for all $n \in \mathbb{Z}$
(b) $\chi(kn) = \chi(k)\chi(n)$ for all $k, n \in \mathbb{Z}$
(c) $\chi(n) \neq 0$ if and only if $(n, m) = 1$.

It is an easy exercise to see that these three conditions specify the set of Dirichlet characters modulo m.

To investigate the properties of Dirichlet characters we begin by studying a more general problem.

Let A be a finite abelian group (written multiplicatively). A character on A is a homomorphism from A to $\mathbb{C}^*$. The set of such characters will be denoted by $\hat{A}$. If $\chi, \psi \in \hat{A}$ define $\chi\psi$ to be the function which takes $a \in A$ to $\chi(a)\psi(a)$. Then $\chi\psi$ is also a character. We show that this product makes $\hat{A}$ into a group. Define χ_0, the trivial character, by $\chi_0(a) = 1$ for all $a \in A$. If $\chi \in \hat{A}$ define χ^{-1} by $\chi^{-1}(a) = \chi(a)^{-1}$ for all $a \in A$. It is easily seen that $\chi^{-1} \in \hat{A}$ and $\chi\chi^{-1} = \chi_0$. With these definitions $\hat{A}$ becomes an abelian group with χ_0 as the identity element. We omit the more or less obvious details.

Let n be the order of A. If $a \in A$, then $a^n = e$, the identity element of A. So, if $\chi \in \hat{A}$ we have $\chi(a)^n = 1$, i.e., the values of χ are nth roots of unity. It follows that $\overline{\chi(a)} = \chi(a)^{-1} = \chi^{-1}(a)$, where bar denotes complex conjugation. Thus χ^{-1} is sometimes written $\bar{\chi}$ and called the conjugate character of χ.

Two questions present themselves immediately. How big is $\hat{A}$? What is its structure? The questions are easy to answer when A is cyclic. In the general case we will use a theorem from group theory which asserts that a finite abelian group is a direct product of cyclic groups (see I. Herstein [150]). When $A = U(\mathbb{Z}/m\mathbb{Z})$, the case of interest to us, this result follows from Theorem 3 of Chapter 4.

Suppose that A is cyclic and generated by an element g of order n. Let $\zeta_n = e^{2\pi i/n}$. If $\chi \in \hat{A}$ we have $\chi(g) = \zeta_n^e$ for some uniquely determined integer e such that $0 \leq e < n$. Since $\chi(g^m) = \chi(g)^m$, χ is determined by its value at g. Conversely, if $0 \leq e < n$ define $\chi(g^m) = \zeta_n^{me}$. It is easy to see χ is well defined and is a character. Thus there are exactly n characters on A. Let $\chi_1 \in \hat{A}$ be such that $\chi_1(g) = \zeta_n$. If $\chi \in \hat{A}$ and $\chi(g) = \zeta_n^e$, then $\chi(g) = \chi_1^e(g)$ which implies $\chi = \chi_1^e$. This shows that $\hat{A}$ is cyclic and generated by χ_1. Thus $A \simeq \hat{A}$.

In general A is a direct product of cyclic groups. This means that there are elements $g_1, g_2, \ldots, g_t \in A$ such that

(i) The order of g_i is n_i.
(ii) Every element $a \in A$ can be uniquely written in the form $a = g_1^{m_1} g_2^{m_2} \cdots g_t^{m_t}$ where $0 \leq m_i < n_i$ for all i.

If the order of A is n, then clearly $n = n_1 n_2 \cdots n_t$.

Suppose $\chi \in \hat{A}$. Then χ is determined by the values $\chi(g_i) = \zeta_{n_i}^{e_i}$ where $0 \leq e_i < n_i$. Conversely, given a t-tuple $(e_1, e_2, \ldots, e_t)$ with $0 \leq e_i < n_i$ for all i we can define a character χ as follows. For $a \in A$ write $a = g_1^{m_1} g_2^{m_2} \cdots g_t^{m_t}$ as in (ii) and set $\chi(a) = \zeta_{n_1}^{m_1 e_1} \zeta_{n_2}^{m_2 e_2} \zeta_{n_t}^{m_t e_t}$. It can easily be checked that χ is a character. There are thus $n_1 n_2 \cdots n_t = n$ characters on A. Moreover, let χ_i be specified by the conditions $\chi_i(g_i) = \zeta_{n_i}$ and $\chi_i(g_j) = 1$ for $i \neq j$. Then χ_i has order n_i and $\hat{A}$ is the direct product of the cyclic subgroups generated by the χ_i. This shows $A \simeq \hat{A}$.

The next two results will be of importance in the next section.

Proposition 16.3.1. *Let A be a finite abelian group. If $\chi, \psi \in \hat{A}$ and $a, b \in A$, then*

(i) $\sum_{a \in A} \chi(a)\overline{\psi(a)} = n\delta(\chi, \psi)$ *where* $\delta(\chi, \chi) = 1$ *and* $\delta(\chi, \psi) = 0$ *if* $\chi \neq \psi$.
(ii) $\sum_{\chi \in \hat{A}} \chi(a)\overline{\chi(b)} = n\delta(a, b)$ *where* $\delta(a, a) = 1$ *and* $\delta(a, b) = 0$ *if* $a \neq b$.

PROOF. Since $\sum_{a \in A} \chi(a)\overline{\psi(a)} = \sum_{a \in A} \chi\psi^{-1}(a)$ it will be enough to show (i) that we can prove $\sum_{a \in A} \chi(a) = n$ if $\chi = \chi_0$ and $\sum_{a \in A} \chi(a) = 0$ if $\chi \neq \chi_0$. The first assertion is clear by definition. Assume $\chi \neq \chi_0$. Then there is a $b \in A$ such that $\chi(b) \neq 1$. We have $\sum_a \chi(a) = \sum_a \chi(ba) = \chi(b) \sum_a \chi(a)$ and so $(\chi(b) - 1) \sum_a \chi(a) = 0$. Since $\chi(b) - 1 \neq 0$ this implies $\sum_a \chi(a) = 0$.

To prove (ii) we first note that $\sum_\chi \chi(a)\overline{\chi(b)} = \sum_\chi \chi(ab^{-1})$. It suffices to show $\sum_\chi \chi(a) = n$ if $a = e$ and $\sum_\chi \chi(a) = 0$ if $a \neq e$. The first assertion is clear. Assume $a \neq e$. We claim there is a character ψ such that $\psi(a) \neq 1$. To see this write $a = g_1^{m_1} g_2^{m_2} \cdots g_t^{m_t}$ with $0 \leq m_i < n_i$ for all i. Since $a \neq e$ at least one $m_i \neq 0$. Then $\chi_i(a) = \chi_i(g_i)^{m_i} = \zeta_{n_i}^{m_i} \neq 1$. Take $\psi = \chi_i$. Then,

$\sum_{\chi} \chi(a) = \sum_{\chi} \psi\chi(a) = \psi(a)\sum_{\chi} \chi(a)$ and so $(\psi(a) - 1)\sum_{\chi} \chi(a) = 0$. Since $\psi(a) - 1 \neq 0$ we have $\sum_{\chi} \chi(a) = 0$.   □

The relations given by (i) and (ii) are called the orthogonality relations. We now interpret these for Dirichlet characters modulo m. Here we take $A = U(\mathbb{Z}/m\mathbb{Z})$. Dirichlet characters are defined on $\mathbb{Z}$ but induce and are induced by elements in the character group of $U(\mathbb{Z}/m\mathbb{Z})$. Hence there are exactly $\phi(m)$ Dirichlet characters modulo m. From the definition and the last proposition we deduce

Proposition 16.3.2. *Let χ and ψ be Dirichlet characters modulo m, and $a, b \in \mathbb{Z}$. Then*

(i) $\sum_{a=0}^{m-1} \chi(a)\overline{\psi(a)} = \phi(m)\delta(\chi, \psi)$,
(ii) $\sum_{\chi} \chi(a)\overline{\chi(b)} = \phi(m)\delta(a, b)$.

In part (ii) *the sum is over all Dirichlet characters modulo m, and $\delta(a, b) = 1$ if $a \equiv b \ (m)$ and $\delta(a, b) = 0$ if $a \not\equiv b \ (m)$.*

§4 Dirichlet L-functions

Let χ be a Dirichlet character modulo m. We define the *Dirichlet L-function associated to χ* by the formula

$$L(s, \chi) = \sum_{n=1}^{\infty} \chi(n)n^{-s}.$$

Since $|\chi(n)n^{-s}| \leq n^{-s}$ we see that the terms of $L(s, \chi)$ are dominated in absolute value by the corresponding terms of $\zeta(s)$. Thus $L(s, \chi)$ converges and is continuous for $s > 1$. Moreover, since χ is completely multiplicative we have a product formula for $L(s, \chi)$ in exactly the same way as for $\zeta(s)$. Namely,

$$L(s, \chi) = \prod_{p} (1 - \chi(p)p^{-s})^{-1}.$$

Since $\chi(p) = 0$ for $p|m$ the above product is over positive primes not dividing m. The formula is valid for $s > 1$.

There is a close connection between $L(s, \chi_0)$ and $\zeta(s)$. In fact,

$$L(s, \chi_0) = \prod_{p \nmid m} (1 - p^{-s})^{-1}$$

$$= \prod_{p|m} (1 - p^{-s}) \prod_{p} (1 - p^{-s})^{-1}$$

$$= \prod_{p|m} (1 - p^{-s}) \zeta(s).$$

From Proposition 16.1.2 we see $\lim_{s \to 1} (s - 1)L(s, \chi_0) = \prod_{p|m} (1 - p^{-1}) = \phi(m)/m$. In particular $L(s, \chi_0) \to \infty$ as $s \to 1$.

To generalize the proof of Proposition 16.2.1 we will need to consider $\ln L(s, \chi)$. Even if we restrict s to be real, the values of $L(s, \chi)$ are in general complex so it is necessary to worry about the fact that $\ln z$ is multivalued as a function of a complex variable z. One way around this is to define $\ln L(s, \chi)$ by an infinite series.

Let χ be a Dirichlet character and define $G(s, \chi) = \sum_p \sum_{k=1} (1/k)\chi(p^k)p^{-ks}$. Since $|(1/k)\chi(p^k)p^{-ks}| \leq p^{-ks}$ and since $\zeta(s)$ converges for $s > 1$ and converges uniformly for $s \geq 1 + \delta > 1$ we can conclude the same assertions are true for $G(s, \chi)$. Consequently $G(s, \chi)$ is continuous for $s > 1$. Moreover, for z a complex number with $|z| < 1$ we have $\exp(\sum_{k=1}^{\infty} (1/k)z^k) = (1 - z)^{-1}$, where exp denotes the usual exponential function. Substituting $z = \chi(p)p^{-s}$ we find $\exp(\sum_{k=1}^{\infty} (1/k)\chi(p^k)p^{-ks}) = (1 - \chi(p)p^{-s})^{-1}$ and a simple argument then shows $\exp G(s, \chi) = L(s, \chi)$ for all $s > 1$. Thus the infinite series $G(s, \chi)$ provides an unambiguous definition for $\ln L(s, \chi)$. To avoid confusion we work directly with $G(s, \chi)$.

From the definition and the argument used in the proof of Proposition 16.1.3 we find

$$G(s, \chi) = \sum_{p \nmid m} \chi(p)p^{-s} + R_\chi(s), \tag{i}$$

where $R_\chi(s)$ remains bounded as $s \to 1$. Multiply both sides of (i) by $\overline{\chi(a)}$ where $a \in \mathbb{Z}$, $(a, m) = 1$. Then sum over all Dirichlet characters modulo m. The result is

$$\sum_\chi \overline{\chi(a)}G(s, \chi) = \sum_{p \nmid m} p^{-s} \sum_\chi \overline{\chi(a)}\chi(p) + \sum_\chi \overline{\chi(a)}R_\chi(s).$$

Using Proposition 16.3.2, part (ii), we see

$$\sum_\chi \overline{\chi(a)}G(s, \chi) = \phi(m) \sum_{p \equiv a(m)} p^{-s} + R_{\chi, a}(s), \tag{ii}$$

where $R_{\chi, a}(s)$ remains bounded as $s \to 1$.

To conclude the proof of Theorem 1 we need the following proposition.

Proposition 16.4.1. *If χ_0 denotes the trivial character modulo m, then $\lim_{s \to 1}$ $G(s, \chi_0)/\ln(s - 1)^{-1} = 1$. If χ is a nontrivial Dirichlet character modulo m, then $G(s, \chi)$ remains bounded as $s \to 1$.*

PROOF. The first assertion is easy. $L(s, \chi_0)$ is a real valued function of positive real numbers. We have seen $L(s, \chi_0) = \prod_{p|m} (1 - p^{-s})\zeta(s)$. It follows that $G(s, \chi_0) = \sum_{p|m} \ln(1 - p^{-s}) + \ln \zeta(s)$. The assertion now follows from the corollary to Proposition 16.1.2.

The second assertion is quite deep. It is the most difficult part of the proof of Dirichlet's theorem on arithmetic progressions. We postpone the proof to the next section.

Now, assuming the above proposition, the proof of Dirichlet's theorem follows quickly from Equation (ii). We simply divide all the terms on both sides by $\ln(s - 1)^{-1}$ and take the limit as $s \to 1$. By the above proposition, the limit on the left-hand side is 1 whereas the limit on the right-hand side is $\phi(m)d(\mathscr{P}(a; m))$. Thus $d(\mathscr{P}(a; m)) = 1/\phi(m)$ and we are done. $\qquad\square$

§5 The Key Step

Up to now all our functions have been defined for $s > 1$. We will show how to extend the domain of definition to $s > 0$. In particular, if χ is nontrivial we will see that $L(1, \chi)$ is a well defined complex number and prove that $L(1, \chi) \neq 0$. This is the key step. Once we know this it is a relatively simple matter to show $G(s, \chi)$ remains bounded as $s \to 1$. This was what was left unproved in Section 4.

In what follows we will consider s as a complex variable. Write $s = \sigma + it$ where σ and t are real. The symbol σ will be used throughout to denote the real part of s.

If $a > 0$ is real then $|a^s| = a^\sigma$. From this observation we see that the series defining $\zeta(s)$ and $L(s, \chi)$ converge and define an analytic function of the complex variable s in the half plane $\{s \in \mathbb{C} \,|\, \sigma > 1\}$.

Lemma 1. *Suppose $\{a_n\}$ and $\{b_n\}$ for $n = 1, 2, 3, \ldots$ are sequences of complex numbers such that $\sum_{n=1}^{\infty} a_n b_n$ converges. Let $A_n = a_1 + a_2 + \cdots + a_n$ and suppose $A_n b_n \to 0$ as $n \to \infty$. Then*

$$\sum_{n=1}^{\infty} a_n b_n = \sum_{n=1}^{\infty} A_n(b_n - b_{n+1}).$$

PROOF. Let $S_N = \sum_{n=1}^{N} a_n b_n$. Set $A_0 = 0$. Then

$$S_N = \sum_{n=1}^{N} (A_n - A_{n-1})b_n = \sum_{n=1}^{N} A_n b_n - \sum_{n=1}^{N} A_{n-1} b_n$$

$$= \sum_{n=1}^{N} A_n b_n - \sum_{n=1}^{N-1} A_n b_{n+1}$$

$$= A_N b_N + \sum_{n=1}^{N-1} A_n(b_n - b_{n+1}).$$

Taking the limit as $N \to \infty$ yields the result. $\qquad\square$

Proposition 16.5.1. $\zeta(s) - (s - 1)^{-1}$ *can be continued to an analytic function on the region $\{s \in \mathbb{C} \,|\, \sigma > 0\}$.*

PROOF. Assume $\sigma > 1$. Then, by the lemma

$$\zeta(s) = \sum_{n=1}^{\infty} n^{-s} = \sum_{n=1}^{\infty} n(n^{-s} - (n + 1)^{-s}).$$

For a real number x recall that $[x]$ is the greatest integer less than or equal to x and $\langle x \rangle = x - [x]$. From the above expression for $\zeta(s)$ we find

$$\zeta(s) = s \sum_{n=1}^{\infty} n \int_{n}^{n+1} x^{-s-1} \, dx$$

$$= s \sum_{n=1}^{\infty} \int_{n}^{n+1} [x] x^{-s-1} \, dx$$

$$= s \int_{1}^{\infty} [x] x^{-s-1} \, dx$$

$$= s \int_{1}^{\infty} x^{-s} \, dx - s \int_{1}^{\infty} \langle x \rangle x^{-s-1} \, dx$$

$$= \frac{s}{s-1} - s \int_{1}^{\infty} \langle x \rangle x^{-s-1} \, dx.$$

Since $|\langle x \rangle| \leq 1$ for all x the last integral converges and defines an analytic function for $\sigma > 0$. The result follows. $\qquad\square$

We will use the same technique to extend $L(s, \chi)$ but first we need another lemma.

Lemma 2. *Let χ be a nontrivial character modulo m. For all $N > 0$ we have $|\sum_{n=0}^{N} \chi(n)| \leq \phi(m)$.*

PROOF. Write $N = qm + r$ where $0 \leq r < m$. Since $\chi(n + m) = \chi(n)$ for all n we see

$$\sum_{n=1}^{N} \chi(n) = q \left(\sum_{n=0}^{m-1} \chi(n) \right) + \sum_{n=0}^{r} \chi(n).$$

By the Proposition 16.3.2, (part i), we have $\sum_{n=0}^{m-1} \chi(n) = 0$. Thus,

$$\left| \sum_{n=0}^{N} \chi(n) \right| = \left| \sum_{n=0}^{r} \chi(n) \right| \leq \sum_{n=0}^{m-1} |\chi(n)| = \phi(m). \qquad\square$$

Proposition 16.5.2. *Let χ be a nontrivial Dirichlet character modulo m. Then, $L(s, \chi)$ can be continued to an analytic function in the region $\{s \in \mathbb{C} \,|\, \sigma > 0\}$.*

PROOF. Define $S(x) = \sum_{n \leq x} \chi(n)$.

By Lemma 1 we have for $\sigma > 1$,

$$L(s, \chi) = \sum_{n=1}^{\infty} S(n)(n^{-s} - (n+1)^{-s})$$

$$= s \sum_{n=1}^{\infty} S(n) \int_{n}^{n+1} x^{-s-1} \, dx$$

$$= s \int_{1}^{\infty} S(x) x^{-s-1} \, dx.$$

By Lemma 2, $|S(x)| \leq \phi(m)$ for all x. It follows that the above integral converges and defines an analytic function for all s such that $\sigma > 0$. $\square$

Our goal is to show that for χ nontrivial $L(1, \chi) \neq 0$. The next proposition will enable us to give a simple proof in the case where χ is a complex character, i.e., a character which takes on nonreal values.

Proposition 16.5.3. *Let $F(s) = \prod_\chi L(s, \chi)$ where the product is over all Dirichlet characters modulo m. Then, for s real and $s > 1$ we have $F(s) \geq 1$.*

PROOF. Assume s is real and $s > 1$. Recall that

$$G(s, \chi) = \sum_p \sum_{k=1}^{\infty} \frac{1}{k} \chi(p^k) p^{-ks}.$$

Summing over χ and using Proposition 16.3.2, part (ii), we find

$$\sum_\chi G(s, \chi) = \phi(m) \sum \frac{1}{k} p^{-ks}$$

where the sum is over all primes p and integers k such that $p^k \equiv 1 \ (m)$.

The right-hand side of the above equation is nonnegative (in fact, it is positive). Taking the exponential of both sides shows $\prod_\chi L(s, \chi) \geq 1$ as asserted. $\square$

Proposition 16.5.4. *If χ is a nontrivial complex character modulo m, then $L(1, \chi) \neq 0$.*

PROOF. From the series defining $L(s, \chi)$ we see that for s real, $s > 1$, $\overline{L(s, \chi)} = L(s, \bar{\chi})$. Letting s tend towards 1 it follows that $L(1, \chi) = 0$ implies $L(1, \bar{\chi}) = 0$.

Assume $L(1, \chi) = 0$ where χ is a complex character. The functions $L(s, \chi)$ and $L(s, \bar{\chi})$ are distinct and both have a zero at $s = 1$. In the product $F(s) = \prod_\chi L(s, \chi)$ we know $L(s, \chi_0)$ has a simple pole at $s = 1$ and all the other factors are analytic about $s = 1$. It follows that $F(1) = 0$. However, Proposition 16.5.3 shows $F(s) \geq 1$ for all real $s > 1$. This is a contradiction. Therefore, $L(1, \chi) \neq 0$. $\square$

It remains to consider the case where χ is a nontrivial real character, i.e., $\chi(n) = 0, 1$, or -1 for all $n \in \mathbb{Z}$. Dirichlet was able to prove $L(1, \chi) \neq 0$ by using his class number formula for quadratic number fields (to be more accurate, for equivalence classes of binary quadratic forms of fixed discriminant). We will use an elegant proof due to de la Vallée Poussin (1896), following the exposition of Davenport [119].

Lemma 3. *Suppose f is a nonnegative, multiplicative function on $\mathbb{Z}^+$, i.e., for all m, $n > 0$ with $(m, n) = 1$, $f(mn) = f(m)f(n)$. Assume there is a constant*

c such that $f(p^k) < c$ for all prime powers p^k. Then $\sum_{n=1}^{\infty} f(n)n^{-s}$ converges for all real $s > 1$. Moreover

$$\sum_{n=1}^{\infty} f(n)n^{-s} = \prod_p \left(1 + \sum_{k=1}^{\infty} f(p^k)p^{-ks}\right).$$

PROOF. Fix $s > 1$. Let $a(p) = \sum_{k=1}^{\infty} f(p^k)p^{-ks}$. Then $a(p) < cp^{-s}\sum_{k=0}^{\infty} p^{-ks} = cp^{-s}(1 - p^{-s})^{-1}$, and so $a(p) < 2cp^{-s}$. For positive x one has $1 + x < \exp x$. Thus

$$\prod_{p \leq N} (1 + a(p)) < \prod_{p \leq N} \exp a(p) = \exp \sum_{p \leq N} a(p).$$

Now, $\sum_{p \leq N} a(p) < 2c\sum_p p^{-s} = M$. From the definition of $a(p)$ and the multiplicativity of f we see $\sum_{n=1}^{N} f(n)n^{-s} < \prod_{p \leq N} (1 + a(p))$. It follows that $\sum_{n=1}^{N} f(n)n^{-s} < \exp M$ for all N. Since f is, by assumption, nonnegative we have $\sum_{n=1}^{\infty} f(n)n^{-s}$ converges.

The last assertion of the lemma follows from the same reasoning used in the proof of Proposition 16.1.1. $\square$

Theorem 2. *Let χ be a nontrivial Dirichlet character modulo m. Then $L(1, \chi) \neq 0$.*

PROOF. Having already proved that $L(1, \chi) \neq 0$ if χ is complex we assume χ is real.

Assume $L(1, \chi) = 0$ and consider the function

$$\psi(s) = \frac{L(s, \chi)L(s, \chi_0)}{L(2s, \chi_0)}.$$

The zero of $L(s, \chi)$ at $s = 1$ cancels the simple pole of $L(s, \chi_0)$ so the numerator is analytic on $\sigma > 0$. The denominator is nonzero and analytic for $\sigma > \frac{1}{2}$. Thus $\psi(s)$ is analytic on $\sigma > \frac{1}{2}$. Moreover, since $L(2s, \chi_0)$ has a pole at $s = \frac{1}{2}$ we have $\psi(s) \to 0$ as $s \to \frac{1}{2}$.

We assume temporarily that s is real and $s > 1$. Then $\psi(s)$ has an infinite product expansion

$$\psi(s) = \prod_p (1 - \chi(p)p^{-s})^{-1}(1 - \chi_0(p)p^{-s})^{-1}(1 - \chi_0(p)p^{-2s})$$

$$= \prod_{p \nmid m} \frac{(1 - p^{-2s})}{(1 - p^{-s})(1 - \chi(p)p^{-s})}.$$

If $\chi(p) = -1$ the p-factor is equal to 1. Thus

$$\psi(s) = \prod_{\chi(p) = 1} \frac{1 + p^{-s}}{1 - p^{-s}},$$

where the product is over all p such that $\chi(p) = 1$. Now,

$$\frac{1 + p^{-s}}{1 - p^{-s}} = (1 + p^{-s})\left(\sum_{k=0}^{\infty} p^{-ks}\right) = 1 + 2p^{-s} + 2p^{-2s} + \cdots +.$$

Applying Lemma 3 we find that $\psi(s) = \sum_{n=1}^{\infty} a_n n^{-s}$ where $a_n \geq 0$ and the series converges for $s > 1$. Note that $a_1 = 1$. (It is possible, but unnecessary to give an explicit formula for a_n).

We once again consider $\psi(s)$ as a function of a complex variable and expand it in a power series about $s = 2$, $\psi(s) = \sum_{m=0}^{\infty} b_m(s - 2)^m$. Since $\psi(s)$ is analytic for $\sigma > \frac{1}{2}$ the radius of convergence of this power series is at least $\frac{3}{2}$. To compute the b_m we use Taylor's theorem, i.e., $b_m = \psi^{(m)}(2)/m!$ where $\psi^{(m)}(s)$ is the mth derivative of $\psi(s)$. Since $\psi(s) = \sum_{n=1}^{\infty} a_n n^{-s}$ we find $\psi^{(m)}(2) = \sum_{n=1}^{\infty} a_n(- \ln n)^m n^{-2} = (-1)^m c_m$ with $c_m \geq 0$. Thus $\psi(s) = \sum_{m=0}^{\infty} c_m(2 - s)^m$ with c_m nonnegative and $c_0 = \psi(2) = \sum_{n=1}^{\infty} a_n n^{-2} \geq a_1 = 1$. It follows that for real s in the interval $(\frac{1}{2}, 2)$ we have $\psi(s) \geq 1$. This contradicts $\psi(s) \to 0$ as $s \to \frac{1}{2}$, and so $L(1, \chi) \neq 0$. $\qquad\square$

We are now in a position to prove Proposition 16.4.1. Suppose χ is a nontrivial Dirichlet character. We want to show $G(s, \chi)$ remains bounded as $s \to 1$ through real values $s > 1$.

Since $L(1, \chi) \neq 0$ there is a disc D about $L(1, \chi)$ such $0 \notin D$. Let $\ln z$ be a single-valued branch of the logarithm defined on D. There is a $\delta > 0$ such that $L(s, \chi) \in D$ for $s \in (1, 1 + \delta)$. Consider $\ln L(s, \chi)$ and $G(s, \chi)$ for s in this interval. The exponential of both functions is $L(s, \chi)$. Thus there is an integer N such that $G(s, \chi) = 2\pi i N + \ln L(s, \chi)$ for $s \in (1, 1 + \delta)$. This implies $\lim_{s \to 1} G(s, \chi)$ exists and is equal to $2\pi i N + \ln L(1, \chi)$. Since $G(s, \chi)$ has a limit as $s \to 1$ it clearly remains bounded.

§6 Evaluating $L(s, \chi)$ at Negative Integers

In the last section we showed how to analytically continue $L(s, \chi)$ into the region $\{s \in \mathbb{C} \mid \sigma > 0\}$. Riemann showed how to analytically continue these functions to the whole complex plane. As noted earlier this fact has important consequences for number theory. For example, the values $L(1 - k, \chi)$, where k is a positive integer, are closely related to the Bernoulli numbers. A knowledge of these numbers has deep connections with the theory of cyclotomic fields. We will analytically continue $L(s, \chi)$ and evaluate the numbers $L(1 - k, \chi)$ following a method due to D. Goss [141].

Before beginning we need to discuss some properties of the Γ-function. This is defined by

$$\Gamma(s) = \int_0^{\infty} e^{-t} t^{s-1} \, dt. \tag{i}$$

It is not hard to see that the integral converges and defines an analytic function on the region $\{s \in \mathbb{C} \mid \sigma > 0\}$. For $\sigma > 1$ we integrate by parts and find

$$\Gamma(s) = -e^{-t} t^{s-1} \Big|_0^{\infty} + (s - 1) \int_0^{\infty} e^{-t} t^{s-2} \, dt$$

It follows that $\Gamma(s) = (s - 1)\Gamma(s - 1)$ for $\sigma > 1$. Since $\Gamma(1) = \int_0^\infty e^{-t} \, dt$ $= 1$ we see $\Gamma(n + 1) = n!$ for positive integers n.

The functional equation $\Gamma(s) = (s - 1)\Gamma(s - 1)$ enables us to analytically continue $\Gamma(s)$ by a step by step process.

If $\sigma > -1$ we define $\Gamma_1(s)$ by

$$\Gamma_1(s) = \frac{1}{s}\Gamma(s + 1). \tag{ii}$$

For $\sigma > 0, \Gamma_1(s) = \Gamma(s)$. Moreover, $\Gamma_1(s)$ is analytic on $\sigma > -1$ except for a simple pole at $s = 0$.

Similarly, if k is a positive integer we define

$$\Gamma_k(s) = \frac{1}{s(s + 1)\cdots(s + k - 1)}\Gamma(s + k).$$

$\Gamma_k(s)$ is analytic on $\{s \in \mathbb{C} | \sigma > -k\}$ except for simple poles at $s = 0, -1, \ldots,$ $1 - k$ and $\Gamma_k(s) = \Gamma(s)$ for $\sigma > 0$. These functions fit together to give an analytic continuation of $\Gamma(s)$ to the whole complex plane with poles at the nonpositive integers and nowhere else. From now on $\Gamma(s)$ will denote this extended function. We remark, without proof, that $\Gamma(s)^{-1}$ is entire.

We will now show how to analytically continue $\zeta(s)$ by the same process. It is necessary to express $\zeta(s)$ as an integral. In Equation (i) substitute nt for t. We find, for $\sigma > 1$

$$n^{-s}\Gamma(s) = \int_0^\infty e^{-nt}t^{s-1} \, dt. \tag{iii}$$

Sum both sides of (iii) for $n = 1, 2, 3, \ldots$. It is not hard to justify interchanging the sum and the integral. The result is

$$\Gamma(s)\zeta(s) = \int_0^\infty \frac{e^{-t}}{1 - e^{-t}} t^{s-1} \, dt. \tag{iv}$$

If we tried to integrate by parts at this stage we would be blocked by the fact that $1 - e^{-t}$ is zero when $t = 0$. To get around this we use a trick. In (iv) substitute $2t$ for t. We find

$$2^{1-s}\Gamma(s)\zeta(s) = 2\int_0^\infty \frac{e^{-2t}}{1 - e^{-2t}} t^{s-1} \, dt. \tag{v}$$

Define $\zeta^*(s) = (1 - 2^{1-s})\zeta(s)$ and $R(x) = x/(1 - x) - 2(x^2/(1 - x^2))$. Subtracting (v) from (iv) yields

$$\Gamma(s)\zeta^*(s) = \int_0^\infty R(e^{-t})t^{s-1} \, dt. \tag{vi}$$

What has been gained? A simple algebraic manipulation shows $R(x) = x/(1 + x)$. Thus $R(e^{-t}) = e^{-t}/(1 + e^{-t})$ has a denominator that does not vanish at $t = 0$. The integral in Equation (vi) thus converges for $\sigma > 0$ and this equation provides a continuation for $\zeta(s)$ to the region $\{s \in \mathbb{C} \mid \sigma > 0\}$.

Let $R_0(t) = R(e^{-t})$ and for $m \geq 1$, $R_m(t) = (d^m/dt^m)R(e^{-t})$. It is easy to see that $R_m(t) = e^{-t}P_m(e^{-t})(1 + e^{-t})^{-2m}$ where P_m is a polynomial. It follows that $R_m(0)$ is finite and $R_m(t)/e^{-t}$ is bounded as $t \to \infty$. These facts enable us to repeatedly integrate by parts in Equation (vi).

Take $u = R(e^{-t})$ and $dv = t^{s-1}\, dt$. Then $du = R_1(t)dt$ and $v = t^s/s$. Thus

$$\Gamma(s)\zeta^*(s) = \frac{1}{s} t^s R_0(t)\Big|_0^\infty - \frac{1}{s}\int_0^\infty R_1(t)t^s\, dt$$

and so

$$\Gamma(s + 1)\zeta^*(s) = -\int_0^\infty R_1(t)t^s\, dt. \tag{vii}$$

The integral in (vii) converges to an analytic function in $\{s \in \mathbb{C} \mid \sigma > -1\}$, and provides an analytic continuation of $\zeta(s)$ to this region. Continuing this process we find for k a positive integer

$$\Gamma(s + k)\zeta^*(s) = (-1)^k \int_0^\infty R_k(t)t^{s+k-1}\, dt, \tag{viii}$$

where the integral converges to an analytic function of s for $\sigma > -k$. This procedure provides an analytic continuation of $\zeta(s)$ to the whole complex plane. We continue to use the notation $\zeta(s)$ for the extended function.

Proposition 16.6.1. *Let k be a positive integer. Then, $\zeta(0) = -\frac{1}{2}$ and for $k > 1$, $\zeta(1 - k) = -B_k/k$ where B_k is the kth Bernoulli number.*

PROOF. In Equation (viii) substitute $s = 1 - k$. The result is $\zeta^*(1 - k) = (-1)^k \int_0^\infty R_k(t)dt$. Since $R_k(t) = (d/dt)R_{k-1}(t)$ we deduce $(1 - 2^k)\zeta(1 - k) = (-1)^{k-1}R_{k-1}(0)$. By definition $R_{k-1}(t)$ is the $(k - 1)$st derivative of

$$\frac{e^{-t}}{1 - e^{-t}} - 2\frac{e^{-2t}}{1 - e^{-2t}} = \frac{1}{t}\left(\frac{t}{e^t - 1} - \frac{2t}{e^{2t} - 1}\right).$$

By Taylor's theorem, $R_{k-1}(0)$ is $(k - 1)!$ times the coefficient of t^{k-1} in the power series expansion of this function about $t = 0$. Since $t/(e^t - 1) = \sum_{k=0}^\infty (B_k/k!)t^k$ we find $\zeta(1 - k) = (-1)^{k-1}B_k/k$. If $k = 1$, then $\zeta(0) = B_1 = -\frac{1}{2}$. If $k > 1$ and odd, then $B_k = 0$. Thus for $k > 1$, $\zeta(1 - k) = -B_k/k$. $\square$

Assume now that χ is a nontrivial character modulo m. To handle $L(s, \chi)$ we proceed in exactly the same way as for $\zeta(s)$. In Equation (iii) multiply both sides by $\chi(n)$ and sum over n. The result is $\Gamma(s)L(s, \chi) = \int_0^\infty F_\chi(e^{-t})t^{s-1}\, dt$, where

$$F_\chi(e^{-t}) = \sum_{n=1}^\infty \chi(n)e^{-nt} = \sum_{a=1}^m \chi(a)\sum_{k=0}^\infty e^{-(a+km)t} = \sum_{a=1}^m \chi(a)\frac{e^{-at}}{1 - e^{-mt}}.$$

If we define $L^*(s, \chi) = (1 - 2^{1-s})L(s, \chi)$, then in the same way as we derived Equation (vi) we find

$$\Gamma(s)L^*(s, \chi) = \int_0^\infty R_\chi(e^{-t})t^{s-1}\, dt, \qquad \text{(ix)}$$

where

$$R_\chi(x) = F_\chi(x) - 2F_\chi(x^2)$$

$$= \sum_{a=1}^m \chi(a)\left(\frac{x^a}{1 - x^m} - 2\frac{x^{2a}}{1 - x^{2m}}\right)$$

$$= \sum_{a=1}^m \chi(a)x^a\left(\frac{1 + x^m - 2x^a}{(1 - x)(1 + x + \cdots + x^{2m-1})}\right).$$

For each value of a we see $x = 1$ is a root of $1 + x^m - 2x^a$, and it follows that $R_\chi(x)$ has the form

$$R_\chi(x) = \frac{xf(x)}{1 + x + \cdots + x^{2m-1}},$$

where $f(x)$ is a polynomial. Let $R_{\chi,0}(t) = R_\chi(e^{-t})$ and $R_{\chi,n} = (d^n/dt^n)R_\chi(e^{-t})$. By repeated integration by parts we find in the same way that we derived Equation (viii) that

$$\Gamma(s + k)L^*(s, \chi) = (-1)^k \int_0^\infty R_{\chi,k}(t)t^{s+k-1}\, dt. \qquad \text{(x)}$$

The integral in (x) converges to an analytic function in $\{s \in \mathbb{C} \mid \sigma > -k\}$. These formulas provide an analytic continuation of $L^*(s, \chi)$ and thus $L(s, \chi)$ to the whole complex plane.

Before attempting to evaluate $L(s, \chi)$ at the negative integers we need a definition.

Definition. Let χ be a nontrivial Dirichlet character modulo m. The generalized Bernoulli number $B_{n,\chi}$ is defined by the following formula

$$\sum_{a=1}^m \chi(a)\frac{te^{at}}{e^{mt} - 1} = \sum_{n=0}^\infty \frac{B_{n,\chi}}{n!} t^n. \qquad \text{(xi)}$$

In the literature it is usual to define $B_{n,\chi}$ in this manner only if χ is a primitive character modulo m. We will discuss this point later.

Lemma 1. $tF_\chi(e^{-t}) = \sum_{n=0}^\infty (-1)^n(B_{n,\chi}/n!)t^n$.

PROOF. Simply substitute $-t$ for t in Equation (xi). $\square$

Proposition 16.6.2. *Let k be a positive integer. Then $L(1 - k, \chi) = -B_{k,\chi}/k$.*

PROOF. In Equation (x) substitute $s = 1 - k$. The result is $(1 - 2^k)L(1 - k, \chi)$ $= (-1)^k \int_0^\infty R_{\chi, k}(t)dt$. Since $R_{\chi, k}(t) = (d/dt)R_{\chi, k-1}(t)$ it follows that $(1 - 2^k)L(1 - k, \chi) = (-1)^{k-1}R_{\chi, k-1}(0)$. Since

$$R_{\chi, k-1}(t) = \frac{d^{k-1}}{dt^{k-1}} R_\chi(e^{-t})$$

and $R_\chi(e^{-t}) = F_\chi(e^{-t}) - 2F_\chi(e^{-2t}) = (1/t) \sum_{k=1}^\infty (-1)^k (1 - 2^k)(B_{k, \chi}/k!)t^k$ (by Lemma 1) we see that $(-1)^{k-1}R_{\chi, k-1}(0) = -(1 - 2^k)(B_{k, \chi}/k)$. Thus, $L(1 - k, \chi) = -B_{k, \chi}/k$ as asserted.   □

It follows from Equation (xi) that the numbers $B_{k, \chi}$ are in the field generated over $\mathbb{Q}$ by the values of χ. Thus, in particular, they are algebraic numbers.

As mentioned earlier it is usual to define $B_{n, \chi}$ by Equation (xi) only when χ is a primitive character modulo m. This means that χ when restricted to $\{n \in \mathbb{Z} \,|\, (n, m) = 1\}$ does not have a smaller period than m. The trivial character is primitive only for the modulus 1. From Equation (xi) we then have

$$\sum_{n=0}^\infty \frac{B_{n, \chi_0}}{n!} t^n = \frac{te^t}{e^t - 1} = t + \frac{t}{e^t - 1} = 1 + \tfrac{1}{2}t + \sum_{n=2}^\infty \frac{B_n}{n!} t^n.$$

Thus $-B_{1, \chi_0} = B_1$ and $B_{n, \chi_0} = B_n$ for $n \neq 2$. It is in this sense that the $B_{n, \chi}$ are "generalized Bernoulli numbers."

The $B_{n, \chi}$ have many interesting arithmetic properties. The interested reader should consult Chapter 2 of Iwasawa's monograph [155]. This monograph is devoted to showing how the equation $L(1 - k, \chi) = -B_{k, \chi}/k$ leads to p-adic L-functions and to the remarkable connection between these functions and the theory of cyclotomic fields. Another approach to these topics are the books of S. Lang [167] and [171]. More accessible to the novice than these works is the book of N. Koblitz [162].

NOTES

Legendre attempted, without success, to prove the existence of infinitely many primes in an arithmetic progression $a + bn$, $(a, b) = 1$. Dirichlet states that, unable to overcome the difficulties in completing Legendre's argument, he was subsequently led to study a class of infinite series and products analogous to those considered by Euler (see [124]). The results of Dirichlet's investigation are far reaching for the development of algebraic and analytic number theory. In addition to proving the existence of primes in an arithmetic progression Dirichlet was able, using the analytic techniques he introduced, to derive explicit formulas, conjectured in part by Jacobi (see the Notes to Chapter 14), for the class numbers of quadratic number fields. For example, if p is prime, $p > 3$ then the class number of $\mathbb{Q}(\sqrt{-p})$ is $(\pi/\sqrt{p})L(1, \chi)$ where χ is the Dirichlet character associated to the Legendre symbol. The well-known expression $(-\sum x\chi(x))/p$ for the class number is then obtained by deriving a closed form for $L(1, \chi)$ (see [9], p. 343). This in turn is obtained using the value of the classical Gauss sum. Since class numbers are positive we see that this approach shows $L(1, \chi) \neq 0$.

If F is a Galois extension of $\mathbb{Q}$ of degree n then one may show by an extension of the methods of this chapter, that the set of prime numbers p that split completely in F, i.e., that are the product of n distinct prime ideals in F, has Dirichlet density $1/n$. As a corollary it can be shown that if $f(x)$ is an irreducible polynomial with integer coefficients then the set of primes p for which $f(x)$ is the product of linear factors modulo p has density $1/n$ where n is the degree of the splitting field of $f(x)$.

The generalized Bernoulli numbers for quadratic characters appear in A. Hurwitz [153]. In this paper Hurwitz derives the functional equation for $L(s, \chi)$, χ quadratic, through consideration of the partial zeta functions $\sum_{t=0}^{\infty} 1/(mt + a)^s$. The values at negative integers of these latter functions may be found by either the classical method or that of Goss, as done in this chapter. A suitable linear combination of these values then yields the expression for $L(1 - k, \chi)$ (Proposition 16.6.2). N. C. Ankeny, E. Artin, and S. Chowla also introduced generalized Bernoulli numbers for quadratic characters in connection with certain remarkable congruences relating the class number of a real quadratic field and the components of the fundamental unit [86]. The definition and basic properties of generalized Bernoulli numbers are given in H. Leopoldt [178] who employs them elsewhere to obtain a generalization, to arbitrary abelian extensions of $\mathbb{Q}$, of Kummer's criterion for the divisibility of the class number of $\mathbb{Q}(\zeta_p)$ (see the comment following Theorem 4, Chapter 15). Leopoldt proves in this paper a theorem of the von Staudt–Claussen type of $B_{n,\chi}$. See also Carlitz [104] and the monograph on p-adic L-functions by K. Iwasawa [155].

EXERCISES

1. Using the method of Section 2 compute the density of the set of primes congruent to 1 modulo 3.

2. Let $p_1, \ldots, p_n$ be primes congruent to 1 modulo 4. If p is a prime dividing $(2 \prod_{i=1}^{n} p_i)^2 + 1$ show that $p \equiv 1 \ (4)$ and $p \neq p_i$, $i = 1, \ldots, n$.

3. Compute the set of Dirichlet characters modulo 8 and modulo 12.

4. Let χ be the nontrivial Dirichlet character modulo 3. Show that
$$L(1, \chi) = \sum_{n=0}^{\infty} \frac{1}{(3n + 1)(3n + 2)}.$$
Can you find the exact value of $L(1, \chi)$? (See Exercise 8.)

5. Use Theorem 2, Chapter 13 to determine the Dirichlet density of the set of primes p which factor into the product of 4 distinct prime ideals in the ring of integers in $\mathbb{Q}(\zeta)$, $\zeta = e^{2\pi i/5}$.

6. Generalize Exercise 5 to $\mathbb{Q}(\zeta_m)$ for general m.

7. By considering $\Phi_m(x)$ modulo p give an algebraic proof that there are an infinite number of primes in the progression $mk + 1, k = 1, 2, 3, \ldots$.

8. Let $g(\chi)$ be the classical Gauss sum $\sum_{x=1}^{p-1} \chi(x)\zeta^x$, χ the Legendre symbol, $\zeta = e^{2\pi i/p}$, p prime. Define $P = \prod (1 - \zeta^n) \prod (1 - \zeta^r)^{-1}$ where n, r run over respectively the nonsquares and squares modulo p. Show that

$$P = \exp(g(\chi)L(1, \chi)).$$

9. Using Exercise 8 compute $L(1, \chi)$ where χ is the nontrivial quadratic character modulo 5.

10. (Chowla) The notation being as in Exercise 8 show that $P \neq 1$ (and thus $L(1, \chi) \neq 0!!$) as follows. Choose C a nonsquare modulo p. Prove that $P = 1$ implies

$$1 + x + \cdots + x^{p-1} | \prod \left(\frac{1 - x^{Cr}}{1 - x^r} \right) - 1.$$

Obtain a contradiction by specializing x!

11. Use Dirichlet's theorem to show that Galois extensions of $\mathbb{Q}$ exist with any pre-scribed finite cyclic group as group of automorphisms.

12. Derive the irreducibility over $\mathbb{Q}$ of the cyclotomic polynomial $\Phi_n(x)$ from Dirichlet's theorem (Landau [166], Vol. 2).

13. Let χ be a Dirichlet character modulo m, $\chi(2) \neq 0$. Show

$$L(s, \chi) = (1 - 2^{-s}\chi(2))^{-1} \sum_{n=0}^{\infty} \frac{\chi(2n + 1)}{(2n + 1)^s}.$$

The following exercises adapted from Moser [193] give a short proof that there are more squares than nonsquares on the interval $[1, (p - 1)/2]$ for $p \equiv 3$ (4), p prime. In Exercises 14, 15, 16, 17, $p \equiv 3$ (4).

14. Let $p \equiv 3$ (4). Show that

$$\sum_{x=1}^{p-1} \left(\frac{x}{p} \right) \sin \frac{2\pi x}{p} = \sqrt{p}.$$

15. Show that, using Exercise 14,

$$\sum_{n \text{ odd}} \left(\frac{n}{p} \right) \frac{1}{n} = \frac{1}{\sqrt{p}} \sum_{t=1}^{p-1} \left(\frac{t}{p} \right) \sum_{m \text{ odd}} \frac{\sin(2\pi t m/p)}{m}.$$

[*Hint*: replace x by nt and sum.]

16. Using the elementary fact from Fourier series

$$\sum_{n=1}^{\infty} \frac{\sin(2n - 1)x}{2n - 1} = \begin{cases} \pi/4, & \text{if } 0 < x < \pi, \\ -\pi/4, & \text{if } \pi < x < 2\pi, \end{cases}$$

show that

$$\sum_{n \text{ odd}} \left(\frac{n}{p} \right) \frac{1}{n} = \frac{\pi}{4\sqrt{p}} \left[\sum_{t=1}^{(p-1)/2} \left(\frac{t}{p} \right) - \sum_{t=(p+1)/2}^{p-1} \left(\frac{t}{p} \right) \right]$$

$$= \frac{\pi}{2\sqrt{p}} \sum_{t=1}^{(p-1)/2} \left(\frac{t}{p} \right).$$

17. Since $\sum_{t=1}^{(p-1)/2} (t/p) \neq 0$ (why?) conclude that $\sum_{n \, odd} (n/p)(1/n) > 0$ and thus $\sum_{t=1}^{(p-1)/2} (t/p) > 0$. Recall $p \equiv 3 \ (4)$.

18. Let $m \geq 2, (a, m) = 1$. If a has order f in the group of units modulo m show that there are infinitely many primes p such that $(p) = P_1 \cdots P_t$, $t = \phi(m)/f$, P_i distinct prime ideals in $\mathbb{Q}(\zeta_m)$. What is the density of this set of primes?

Chapter 17

Diophantine Equations

In Chapter 10 we discussed Diophantine equations over finite fields. In this chapter we consider special Diophantine equations with integral coefficients and seek integral or rational solutions. The techniques used vary from elementary congruence considerations to the use of more sophisticated results in algebraic number theory. In addition to establishing the existence or nonexistence of solutions we also obtain results of a quantitative nature, as in the determination of the number of representations of an integer as the sum of four squares. All of the equations considered in this chapter are classical, each playing an important role in the historical development of the subject.

§1 Generalities and First Examples

By a Diophantine equation will be understood a polynomial equation

$$f(x_1, x_2, x_3, \ldots, x_n) = 0, \tag{1}$$

whose coefficients are rational integers. If this equation has a solution in integers $x_1, \ldots, x_n$ then we shall say that $(x_1, \ldots, x_n)$ is an integral solution. If (1) is homogeneous then a solution distinct from $(0, \ldots, 0)$ is called nontrivial. A solution to (1) with rational $x_1, \ldots, x_n$ is called a rational solution. Clearly, in the homogeneous case the problem of finding a rational solution is equivalent to that of finding an integral solution.

While the degree of $f(x_1, \ldots, x_n)$ controls to some extent the difficulty of the problem, the existence or nonexistence of a solution is often related to subtle invariants and even perhaps the complex differential geometry of (1) over the complex numbers.

We begin by considering the linear Diophantine equation

$$a_1 x_1 + a_2 x_2 + \cdots + a_n x_n = m. \tag{2}$$

Here $a_1, \ldots, a_n, m$ are rational integers. Then by Chapter 1 (see Exercises 6, 13, 14 of that chapter) it follows that a solution in integers exists iff the greatest common divisor of $a_1, \ldots, a_n$ divides m.

If $n = 2$ and $d = (a_1, a_2)$ the Euclidean algorithm gives an explicit procedure for constructing a solution to $a_1 x_1 + a_2 x_2 = d$ (Exercises 2 and 4, Chapter 1). Multiplying the solution by m/d gives a solution to (2). For

269

$n > 2$ one may proceed by induction using the simple observation that $((a_1, \ldots, a_{n-1}), a_n) = (a_1, \ldots, a_n)$.

If (1) has an integral solution then for each prime p the congruence

$$f(x) \equiv 0 \ (p) \tag{3}$$

has a solution. If therefore one can find a prime p for which (3) has no solution then (1) also has no solution. This method can be applied in many special cases to obtain nonexistence theorems. We will consider several examples of this technique.

For example, consider the equation

$$y^2 = x^3 + 7. \tag{4}$$

If (4) has a solution then x is odd. For otherwise reduction modulo 4 would imply that 3 is a square modulo 4 which is not the case. Write (4) as

$$y^2 + 1 = (x + 2)(x^2 - 2x + 4) \tag{5}$$
$$= (x + 2)((x - 1)^2 + 3).$$

Now since $(x - 1)^2 + 3$ is of the form $4n + 3$ there is a prime p of the form $4n + 3$ dividing it and reduction of (5) modulo p implies that -1 is a square modulo p. But this contradicts Proposition 5.1.2, Corollary 3. Of course this ingenious argument works only because one chose $x^3 + 7$. There are many results concerning the rational and integral solutions of the equation

$$y^2 = x^3 + k \tag{6}$$

for special values of k (see Section 10). The interested reader should consult Mordell [189] for an indication of the vast array of techniques used to discuss (6). We mention in passing that it follows from deep theorems of Mordell and Siegel that (6) has only a finite number of integral solutions. The question of rational solutions leads to the famous conjectures of Birch and Swinnerton-Dyer. A statement of these conjectures will be given in the next chapter.

Consider next the equation

$$y^3 = px + 2. \tag{7}$$

Here p is a prime $p \equiv 1 \ (3)$. We note that this Diophantine equation is equivalent to the congruence

$$y^3 \equiv 2 \ (p). \tag{8}$$

By Proposition 9.6.2, Equation (7) has a solution iff $p = C^2 + 27D^2$ for suitable integers C and D. Thus the Diophantine problem (7) is related to the question of the representability of p by the quadratic form $x^2 + 27y^2$.

In a similar manner quadratic reciprocity can be used to show that

$$y^2 = 41x + 3 \tag{9}$$

has no solution. For reduction modulo 41 shows that 3 is a square modulo 41. But since $41 \equiv 1 \ (4)$ quadratic reciprocity implies 41 is a square modulo 3 which is not the case.

A well-known Diophantine equation is given by

$$x^2 + y^2 = z^2. \tag{10}$$

The solutions in integers are known as Pythagorean triples. We solve this problem using Proposition 1.4.1 which states that $\mathbb{Z}[i]$ is a unique factorization domain. A proof that does not use complex numbers can be found, for example, in Hardy and Wright [40], p. 190. Assume that (10) has a solution and that $(x, y) = 1$. Thus x and y are not both even and reduction of (10) modulo 4 shows that z is odd. Factor (10) in $\mathbb{Z}[i]$ to obtain

$$(x + iy)(x - iy) = z^2. \tag{11}$$

If π is an irreducible in $\mathbb{Z}[i]$ that divides $x + iy$ and $x - iy$ then π divides $2x$ and $2y$. Since z is odd $(\pi) \neq (1 + i)$ for otherwise $\pi\bar{\pi} = 2|z^2$. Thus $\pi|x$ and $\pi|y$. Taking norms shows that $N(\pi) = p|x$ and $p|y$ which contradicts the fact that $(x, y) = 1$. Thus $x + iy$ and $x - iy$ are relatively prime. If $z = u\pi_1^{a_1} \cdots \pi_s^{a_s}$, u a unit, is a factorization of z in $\mathbb{Z}[i]$ then, by unique factorization,

$$x + iy = u\beta^2. \tag{12}$$

Writing $\beta = a + bi$ and taking $u = 1$ gives the solutions

$$x = a^2 - b^2,$$

$$y = 2ab,$$

$$z = a^2 + b^2.$$

The other choices of the unit give essentially (i.e., up to sign) the same solution. The identity $(a^2 - b^2)^2 + (2ab)^2 = (a^2 + b^2)^2$ shows that (10) has infinitely many solutions. The above argument shows that there are no others.

We conclude this section by giving a simple example of a homogeneous cubic equation with no nontrivial solution. For any prime p consider

$$x^3 + py^3 + p^2z^3 = 0. \tag{13}$$

Assume that (13) has an integer solution (x, y, z), x, y, z not all divisible by p. Then $p|x^3$ so $p|x$. Putting $x = px'$ and cancelling shows that $p|y^3$ so that $p|y$. Substituting $y = py'$ and cancelling shows that $p|z^3$ or $p|z$ which is a contradiction. This elegant example is due to Euler (see Hurwitz [154], p. 455).

§2 The Method of Descent

This method, first enunciated by P. Fermat may be used to handle several important Diophantine equations. The technique is best illustrated by examples. Consider therefore the Diophantine equation

$$x^4 + y^4 = z^2. \tag{14}$$

We show that (14) has no integral solution with $xyz \neq 0$, $z > 0$. Assuming that (14) has such an integral solution we construct another solution with smaller *positive z*. This is clearly impossible as it leads to an infinite sequence of decreasing positive integers. The details are as follows.

We may assume that $(x, y, z) = 1$, $z > 0$. Next x and y cannot both be odd since otherwise reduction modulo 4 would give $z^2 \equiv 2$ (4) which is impossible. Let then x be odd, y even so that z is odd. Write $y^4 = (z - x^2)(z + x^2)$ and observe that, since any prime p dividing the two factors on the right must also divide $2z$ and $2x^2$, one must have $(z - x^2, z + x^2) = 2$. But the product of the two factors is a fourth power. The possibilities are therefore

$$z - x^2 = 2a^4, \qquad a > 0,$$

$$z + x^2 = 8b^4, \tag{15}$$

$$a \text{ odd}, \quad (a, b) = 1,$$

or

$$z - x^2 = 8b^4,$$

$$z + x^2 = 2a^4, \qquad a > 0 \tag{16}$$

$$a \text{ odd}, \quad (a, b) = 1.$$

The first case implies $x^2 = -a^4 + 4b^4$ which is impossible since otherwise $1 \equiv -1$ (4). Thus (16) holds and $z = a^4 + 4b^4$. Note that $0 < a < z$. Also eliminating z in (16) shows that $4b^4 = (a^2 - x)(a^2 + x)$. Since $(a, b) = 1$ it follows that $(a, x) = 1$ and arguing as earlier one sees that $(a^2 - x, a^2 + x) = 2$. Writing $a^2 - x = 2c^4$ and $a^2 + x = 2d^4$ one obtains

$$a^2 = c^4 + d^4.$$

Thus we have found a solution to (14) with smaller positive value for z and the proof is complete. $\qquad\qquad\square$

In particular $x^4 + y^4 = z^4$ has no solution, $xyz \neq 0$. This is a special case of Fermat's Last Theorem.

§3 Legendre's Theorem

In this section we consider the Diophantine equation

$$ax^2 + by^2 + cz^2 = 0, \tag{17}$$

where a, b, c are square free, pairwise relatively prime integers. We would like to have necessary and sufficient conditions in order that (17) have a nontrivial integral solution. In order that a solution exist it is of course necessary to assume that a, b and c are neither all positive nor all negative.

If m and n are nonzero integers let $m \, R \, n$ denote the fact that m is a square modulo n. In other words there is an integer x with $x^2 \equiv m \, (n)$. Legendre discovered the following beautiful theorem.

Proposition 17.3.1. *Let a, b, c be nonzero integers, square free, pairwise relatively prime and not all positive nor all negative. Then (17) has a nontrivial integral solution iff the following conditions are satisfied*

(i) $-ab \, R \, c$.
(ii) $-ac \, R \, b$.
(iii) $-bc \, R \, a$.

It is convenient to prove this result in the following equivalent form.

Proposition 17.3.2. *Let a and b be positive square free integers. Then*

$$ax^2 + by^2 = z^2 \tag{18}$$

has a nontrivial solution iff the following three conditions are satisfied

(i) $a \, R \, b$.
(ii) $b \, R \, a$.
(iii) $-(ab/d^2) \, R \, d$, where $d = (a, b)$.

In order to see that Proposition 17.3.2 implies Proposition 17.3.1 consider $ax^2 + by^2 + cz^2 = 0$ as in Proposition 17.3.1 and assume that a and b are positive while c is negative. Then $-acx^2 - bcy^2 - z^2 = 0$ is easily seen to satisfy the conditions of Proposition 17.3.2. If (x, y, z) is a solution then since c is square free $c \mid z$. Putting $z = cz'$ and cancelling we arrive at a solution to (17). That Proposition 17.3.1 implies Proposition 17.3.2 is left as an exercise.

We now proceed to the proof of Proposition 17.3.2. If $a = 1$ the proposition is obvious. Furthermore we may assume $a > b$. For if $b > a$ just interchange x and y. If $a = b$ then by (iii) -1 is a square modulo b. By Exercise 25 at the end of this chapter one can find integers r and s such that $b = r^2 + s^2$. A solution is then given by $x = r$, $y = s$, $z = r^2 + s^2$.

With these preliminaries we proceed to construct a new form $Ax^2 + by^2 = z^2$ satisfying the same hypotheses as (18), $0 < A < a$, and such that if it has a nontrivial solution then so does (18). After a finite number of steps, interchanging A and b in case A is less than b we arrive at one of the cases $A = 1$ or $A = b$, each of which has been settled. Now for the details.

By (ii) there exist, T and c such that

$$c^2 - b = aT = aAm^2; \qquad A, m \in \mathbb{Z} \tag{19}$$

where A is square-free, and $|c| \le a/2$. First of all we show that $0 < A < a$. This follows from (19) since first of all one has $0 \le c^2 = aAm^2 + b < a(Am^2 + 1)$. Thus $A \ge 0$. But since b is square-free $A > 0$ by (19). Furthermore by (19) $aAm^2 < c^2 \le a^2/4$ so that $A \le Am^2 < a/4 < a$.

Next we verify that $A R b$. Put $b = b_1 d$, $a = a_1 d$ with $(a_1, b_1) = 1$ and note that $(a_1, d) = (b_1, d) = 1$ since a and b are square-free. Then (19) becomes

$$c^2 - b_1 d = a_1 d A m^2 \tag{20}$$

and since d is square-free $d \mid c$. Put $c = c_1 d$ and cancel to obtain

$$d c_1^2 - b_1 = a_1 A m^2. \tag{21}$$

Thus $A a_1 m^2 \equiv -b_1 \ (d)$ or $A a_1^2 m^2 \equiv -a_1 b_1 \ (d)$. But $(m, d) = 1$ since by (21) a common factor would divide b_1 and d and thus b would not be square-free. Using (iii) and the fact that m is a unit modulo d we conclude that $A R d$. Furthermore $c^2 \equiv a A m^2 \ (b_1)$. Since $a R b$ one has $a R b_1$. Also $(a, b_1) = 1$ since a common divisor would divide d and b_1 contradicting the fact that $b = b_1 d$ is square-free. Similarly $(m, b_1) = 1$ which shows that $A R b_1$. By Exercise 26, $A R d b_1$ or $A R b$.

Next write $A = r A_1, b = r b_2, (A_1, b_2) = 1$. We must verify that $-A_1 b_2 R r$. From (19) we conclude that

$$c^2 - r b_2 = a r A_1 m^2. \tag{22}$$

But r is square-free so $r \mid c$. If $c = r c_1$ then

$$a A_1 m^2 \equiv -b_2 \ (r).$$

Since $a R b$ we have $a R r$. Finally writing

$$-a A_1 b_2 m^2 \equiv b_2^2 \ (r)$$

and observing that $(a, r) = (m, r) = 1$ we conclude $-A_1 b_2 R r$.

Assume now that $A X^2 + b Y^2 = Z^2$ has a nontrivial solution. Then

$$A X^2 = Z^2 - b Y^2. \tag{23}$$

Multiplying (23) by (19) one has

$$\begin{aligned}
a(A X m)^2 &= (Z^2 - b Y^2)(c^2 - b) \\
&= (Z c + b Y)^2 - b(c Y + Z)^2.
\end{aligned}$$

(Note the use of the multiplicativity of the norm map on $\mathbb{Q}(\sqrt{b})$!). Thus (18) has a solution with

$$x = A X m,$$

$$y = c Y + Z,$$

$$z = Z c + b Y.$$

This completes the proof since $X \neq 0$; and $m \neq 0$ as follows from the fact that b is square-free. $\square$

An important corollary of Proposition 17.3.1 is a special case of the so called "Hasse Principle." This principle states roughly that local solvability implies global solvability. Here local solvability means that the equation under consideration has a nontrivial solution modulo p^m for all primes p and all positive integers m, as well as a real solution while global solvability refers to a solution in integers. For quadratic forms this principle is true but it fails for equations of higher degree. For example, the equation $x^4 - 17y^4 = 2z^4$ has a nontrivial solution modulo p^m for all p and m, and a real solution, but it has no nontrivial solution in integers [205].

Corollary. *Let a, b, c be square-free, pairwise relatively prime integers not all of the same sign. If for each prime power p^m the congruence*

$$ax^2 + by^2 + cz^2 \equiv 0 \ (p^m)$$

has a solution in integers (x, y, z) not all divisible by p then $ax^2 + by^2 + cz^2 = 0$ has a nontrivial integral solution.

PROOF. Let $m = 2$ and suppose $p \,|\, a$. Then if (x, y, z) is a solution as in the corollary we show that $p \nmid yz$. For if $p \,|\, y$, say, then $p \,|\, cz^2$ which implies, since $(a, c) = 1$, that $p \,|\, z$. Thus $p^2 \,|\, ax^2$ and since $p \nmid x$ we obtain the contradiction $p^2 \,|\, a$. Similarly $p \nmid z$. Thus $by^2 + cz^2 \equiv 0 \ (p)$ and division (mod p) shows that $-bc \ R \ p$. This being the case for every $p \,|\, a$ it follows that $-bc \ R \ a$ (Exercise 26). Similarly $-ab \ R \ c$ and $-ac \ R \ b$ and the corollary now follows by Proposition 17.3.1. □

§4 Sophie Germain's Theorem

In Chapter 14 we proved that if Fermat's equation for an odd prime p

$$x^p + y^p + z^p = 0 \tag{24}$$

had a solution with $p \nmid xyz$ then a very strong congruence held, namely

$$2^{p-1} \equiv 1 \ (p^2).$$

In 1823 Sophie Germain proved the following remarkable result by completely elementary considerations.

Proposition 17.4.1. *If p is an odd prime such that $2p + 1 = q$ is also prime then (24) has no integral solution with $p \nmid xyz$,*

PROOF. Assume on the contrary that such a solution exists and suppose that $(x, y, z) = 1$. Write

$$-x^p = (y + z)(z^{p-1} - z^{p-2}y + \cdots + y^{p-1}). \tag{25}$$

The two factors on the right are relatively prime. For clearly $p \nmid y + z$ and if $r \neq p$ is a prime dividing both factors then since $y \equiv -z \ (r)$ one has

$$0 \equiv z^{p-1} - z^{p-2}y + \cdots + y^{p-1} \equiv py^{p-1} \ (r),$$

which implies that $r\,|\,y$. This in turn implies that $r\,|\,z$ (by (24)) contradicting the assumption that $(x, y, z) = 1$. By unique factorization in $\mathbb{Z}$ we conclude that

$$y + z = A^p \tag{26}$$

$$z^{p-1} - z^{p-2}y + \cdots + y^{p-1} = T^p \tag{27}$$

for suitable integers A and T. Similarly

$$x + y = B^p \tag{28}$$

$$x + z = C^p. \tag{29}$$

Since $p = (q - 1)/2$ reducing (24) modulo q gives

$$x^{(q-1)/2} + y^{(q-1)/2} + z^{(q-1)/2} \equiv 0 \ (q).$$

If $q \nmid xyz$ then each of the terms on the left-hand side is ± 1 modulo q. This is impossible since $q > 5$. Thus, by symmetry, we may assume that $q\,|\,x$. From (26), (28) and (29) we conclude that

$$B^p + C^p - A^p = 2x$$

so that

$$B^{(q-1)/2} + C^{(q-1)/2} - A^{(q-1)/2} \equiv 0 \ (q). \tag{30}$$

Once again it follows that $q\,|\,ABC$. However, since $q\,|\,x$, (28) and (29) imply that $q\,|\,BC$ is impossible. Thus $q\,|\,A$. By (26) and (27) we see that

$$T^p \equiv py^{p-1} \ (q)$$

By (28), $y \equiv B^p \ (q)$; and since $(A, T) = 1$, $q \nmid T$. Thus, since $p = (q - 1)/2$ we have $\pm 1 \equiv p \ (q)$ which is impossible. Thus the proof is complete. $\qquad \square$

Unfortunately it is not known whether there are infinitely many "Germain" primes, i.e., primes p such that $2p + 1$ is prime. The interested reader should consult Lecture IV in the book by Ribenboim [206].

§5 Pell's Equation

Let d be a positive square-free integer. The Diophantine equation to be considered is

$$x^2 - dy^2 = 1. \tag{31}$$

That this equation has an infinite number of solutions was conjectured by Fermat in 1657 and eventually solved by Lagrange. It seems that Pell had nothing to do with it, the error in attaching his name to it being due to Euler. For the whole story, and much more, the interested reader should consult the book by Edwards [128]. See also Davenport [22], and A. Weil [240].

The solution to (30) depends upon the following proposition of Dirichlet and is an application of the pigeon hole principle.

Proposition 17.5.1. *If ξ is irrational then there are infinitely many rational numbers x/y, $(x, y) = 1$ such that $|x/y - \xi| < 1/y^2$.*

PROOF. Partition the half-open interval $[0, 1)$ by

$$[0, 1) = \left[0, \frac{1}{n}\right) \cup \left[\frac{1}{n}, \frac{2}{n}\right) \cup \cdots \cup \left[\frac{n-1}{n}, 1\right).$$

If $[\alpha]$ denotes, as usual, the largest integer less than or equal to α then the fractional part of α is defined by $\alpha - [\alpha]$. It lies in a unique member of the partition. Consider the fractional parts of $0, \xi, 2\xi, \ldots, n\xi$. At least two of these must lie in the same subinterval. In other words there exist j, k with $j > k$, $0 \le j, k \le n$ such that

$$|j\xi - [j\xi] - (k\xi - [k\xi])| < \frac{1}{n}. \tag{32}$$

Put $y = j - k$, $x = [k\xi] - [j\xi]$ so that (32) becomes $|x - y\xi| < 1/n$. Here we may assume that $(x, y) = 1$ since division by (x, y) only strengthens the inequality. But $0 < y < n$ implies that $|x/y - \xi| < 1/ny < 1/y^2$. To obtain infinitely many solutions note that $|x/y - \xi| \ne 0$ and choose an integer $m > 1/|x/y - \xi|$. The above procedure gives the existence of integers x_1, y_1 such that $|x_1/y_1 - \xi| < 1/my_1 < |x/y - \xi|$ and $0 < y_1 < m$. This procedure leads to an infinite number of solutions.   □

This proposition will be applied to show that $|x^2 - dy^2|$ assumes the same value infinitely often.

Lemma 1. *If d is a positive square-free integer then there is a constant M such that $|x^2 - dy^2| < M$ has infinitely many integral solutions.*

PROOF. Write $x^2 - dy^2 = (x + \sqrt{d}y)(x - \sqrt{d}y)$. By Proposition 17.5.1 there exist infinitely many pairs of relatively prime integers (x, y), $y > 0$ satisfying $|x - \sqrt{d}y| < 1/y$. It follows that

$$|x + \sqrt{d}y| < |x - \sqrt{d}y| + 2\sqrt{d}|y| < \frac{1}{y} + 2\sqrt{d}y.$$

Hence $|x^2 - dy^2| < |1/y + 2\sqrt{d}y|1/y \le 2\sqrt{d} + 1$ and the proof is complete.   □

The main result of this section is as follows.

Proposition 17.5.2. *If d is a positive square-free integer then $x^2 - dy^2 = 1$ has infinitely many integral solutions. Furthermore there is a solution (x_1, y_1) such that every solution has the form $\pm(x_n, y_n)$ where $x_n + \sqrt{d}y_n = (x_1 + \sqrt{d}y_1)^n$, $n \in \mathbb{Z}$.*

PROOF. By Lemma 1 there is an $m \in \mathbb{Z}$ such that $x^2 - dy^2 = m$ for infinitely many integral pairs (x, y), $x > 0$, $y > 0$. We may assume that the x components are distinct. Furthermore since there are only finitely many residue classes modulo $|m|$ one can find (x_1, y_1), (x_2, y_2), $x_1 \neq x_2$ such that $x_1 \equiv x_2 \,(|m|)$, $y_1 \equiv y_2 \,(|m|)$. Put $\alpha = x_1 - y_1\sqrt{d}$, $\beta = x_2 - y_2\sqrt{d}$. If $\gamma = x - y\sqrt{d}$ let $\gamma' = x + y\sqrt{d}$ denote the conjugate of γ and $N(\gamma) = x^2 - dy^2$ denote the norm of γ. Recall that $N(\alpha\beta) = N(\alpha)N(\beta)$. A short calculation shows that $\alpha\beta' = A + B\sqrt{d}$ where $m|A$, $m|B$. Thus $\alpha\beta' = m(u + v\sqrt{d})$ for integers u and v. Taking norms of both sides given $m^2 = m^2(u^2 - v^2 d)$. Thus

$$u^2 - v^2 d = 1. \tag{33}$$

It remains to see that $v \neq 0$. However if $v = 0$ then $u = \pm 1$ and $\alpha\beta' = \pm m$. Multiplying by β gives $\alpha m = \pm m\beta$ or $\alpha = \pm \beta$. But this implies that $x_1 = x_2$. Thus Pell's equation has a solution with $xy \neq 0$.

To prove the second assertion let us say that a solution (x, y) is greater than a solution (u, v) if $x + y\sqrt{d} > u + v\sqrt{d}$. Now consider the smallest solution α with $x > 0$, $y > 0$. Such a solution clearly exists (why?) and is unique. It is called the fundamental solution.

Consider any solution $\beta = u + v\sqrt{d}, u > 0, v > 0$. We show that there is a positive integer n such that $\beta = \alpha^n$. For otherwise chose $n > 0$ so that $\alpha^n < \beta < \alpha^{n+1}$. Then since $\alpha' = \alpha^{-1}$, $1 < (\alpha')^n\beta < \alpha$. But if $(\alpha')^n\beta = A + B\sqrt{d}$, (A, B) is a solution to Pell's equation and $1 < A + B\sqrt{d} < \alpha$. Now $A + B\sqrt{d} > 0$ so $A - B\sqrt{d} = (A + B\sqrt{d})^{-1} > 0$. Thus $A > 0$. Also $A - B\sqrt{d} = (A + B\sqrt{d})^{-1} < 1$ so $B\sqrt{d} > A - 1 \geq 0$. Thus $B > 0$. This contradicts the choice of α. If $\beta = a + b\sqrt{d}$ is a solution $a > 0$, $b < 0$ then $\beta^{-1} = a - b\sqrt{d} = \alpha^n$ by the above so $\beta = \alpha^{-n}$. The cases $a < 0, b > 0$ and $a < 0$, $b < 0$ lead obviously to $-\alpha^n$ for $n \in \mathbb{Z}$. The proof is now complete. $\qquad\square$

For a solution to special cases of Pell's equation using cyclotomy see Dirichlet [126] and Hartung [145].

§6 Sums of Two Squares

If p is prime, $p \equiv 1 \,(4)$ then by Proposition 8.3.1 the Diophantine equation $x^2 + y^2 = p$ has an integral solution which is essentially unique. There are many proofs of this result. It will be recalled that the proof in Chapter 8 made use of the ring of Gaussian integers. By further exploiting the arithmetic of this ring we will determine the number of representations of an arbitrary positive integer as the sum of two squares. The result is conveniently stated and in fact proved using the nontrivial Dirichlet character modulo 4 introduced in

Section 2 of Chapter 16. Recall that this character χ is defined on $\mathbb{Z}$ by $\chi(d) = 1$ if $d \equiv 1$ (4), $\chi(d) = -1$ if $d \equiv 3$ (4) and $\chi(2k) = 0$.

Proposition 17.6.1. *The number of integral solutions* (x, y), $x > 0$, $y \geq 0$ *to the equation* $x^2 + y^2 = n$ *is* $\sum_{d|n} \chi(d)$.

In other words the number of representations of n as the sum of two nonnegative squares the first of which is positive is the excess of the number of divisors of the form $4n + 1$ over the number of divisors of the form $4n + 3$. The total number of solutions (x, y), $x, y \in \mathbb{Z}$ is then easily seen to be $4 \sum_{d|n} \chi(d)$.

Before proceeding to the proof we derive two corollaries.

Corollary 1. *The equation* $x^2 + y^2 = n$, $n > 0$ *has an integral solution iff* $\text{ord}_p n$ *is even for every prime* $p \equiv 3$ (4). *When that is the case the number of solutions is* $\prod_{p \equiv 1 \,(4)} (1 + \text{ord}_p n)$.

PROOF. Since $\chi(n)$ is multiplicative it follows by Exercise 10, Chapter 2 that $\sum_{d|n} \chi(d)$ is multiplicative. If $p \equiv 1$ (4) then $\sum_{d|p^n} \chi(d) = n + 1$ while if $p \equiv 3$ (4) then $\sum_{d|p^n} \chi(d)$ is 0 or 1 according as n is odd or even. The result follows.   □

Corollary 2. *Let* m *be a positive odd integer. The number of integral solutions* (x, y), $x > 0$, $y > 0$ *to* $x^2 + y^2 = 2m$ *is* $\sum_{d|m} \chi(d)$.

PROOF. Since $2m \equiv 2$ (4), y is positive. On the other hand $\chi(2d) = 0$ for any divisor $2d$ of $2m$.   □

We now proceed to the proof of the proposition. Consider the ring $\mathbb{Z}[i]$ of Gaussian integers. By Exercise 33, Chapter 1 the units are ± 1, $\pm i$. Thus each nonzero $\alpha \in \mathbb{Z}[i]$ has a unique associate $x + iy$, $x > 0$, $y \geq 0$. If $N(x + iy) = x^2 + y^2$ is the norm mapping then clearly the number of solutions to $x^2 + y^2 = n$, $x > 0$, $y \geq 0$ is the number of ideals (α) with $N(\alpha) = n$. Denote this number by a_n. Recall further that every ideal $(\alpha) \neq 0$ may be uniquely written (up to order) as $(\pi_1)^{f_1} \cdots (\pi_s)^{f_s}$ where π_i is irreducible. Finally according to Section 7 of Chapter 9 the irreducibles are given, up to a unit, by $1 + i$, π with $\pi\bar{\pi} = p \equiv 1$ (4), and q, a rational prime, $q \equiv 3$ (4). Also π and $\bar{\pi}$ are not associates.

We now introduce the formal Dirichlet series $\sum_{n=1}^{\infty} a_n/n^s$. This series is known as the zeta function of the ring $\mathbb{Z}[i]$. We view this expression formally and shall not need any analytic properties of the associated function of a complex variable. Using the unique factorization of ideals in $\mathbb{Z}[i]$ proved in Section 4 of Chapter 1 one sees, using the same argument as in Exercise 25, Chapter 2, that

$$\sum_{n=1}^{\infty} \frac{a_n}{n^s} = \prod_{(\pi)} \left(\frac{1}{1 - 1/N(\pi)^s} \right), \tag{34}$$

the product being over the set of (unassociated) irreducibles in $\mathbb{Z}[i]$. The right-hand side of (34) becomes, by the above classification of irreducibles

$$\left(\frac{1}{1 - 1/2^s}\right) \prod_{p \equiv 1 \,(4)} \left(\frac{1}{1 - 1/p^s}\right)^2 \prod_{q \equiv 3 \,(4)} \left(\frac{1}{1 - 1/q^{2s}}\right). \tag{35}$$

Next recall that

$$\zeta(s) = \prod_p \frac{1}{1 - 1/p^s} = \sum_{n=1}^{\infty} \frac{1}{n^s}.$$

Noting that $1/(1 - q^{-2s}) = (1/(1 - q^{-s}))(1/(1 + q^{-s}))$ we see by rearrangement of terms that (35) becomes

$$\zeta(s) \prod_{p \equiv 1 \,(4)} \frac{1}{1 - 1/p^s} \prod_{q \equiv 3 \,(4)} \frac{1}{1 + 1/q^s}. \tag{36}$$

This may be written as

$$\zeta(s) \prod_p \frac{1}{1 - \chi(p)/p^s}. \tag{37}$$

Finally, using the fact that χ is multiplicative we see that (37) may be written as

$$\zeta(s) \sum_{n=1}^{\infty} \frac{\chi(n)}{n^s}. \tag{38}$$

Recall that the second factor in (38) is the Dirichlet L-series introduced in Chapter 16, Section 2 in order to compute the density of primes $p \equiv 1$ (4). We have shown

$$\sum_{n=1}^{\infty} \frac{a_n}{n^s} = \left(\sum_{n=1}^{\infty} \frac{1}{n^s}\right)\left(\sum_{n=1}^{\infty} \frac{\chi(n)}{n^s}\right). \tag{39}$$

Proposition 17.6.1 follows immediately from (39) for the coefficient in the right-hand side of (39) is, by the very definition of Dirichlet multiplication $\sum_{d|n} \chi(d)$. This completes the proof. □

It should be noted that the rearrangement step in the above proof is purely formal and does not require any analytic properties of the infinite products.

§7 Sums of Four Squares

In 1621 Bachet stated without proof that every positive integer is the sum of four squares. This assertion was proved in 1770 by Lagrange. In 1834 Jacobi was able to give a remarkably simple formula for the total number of repre-

sentations of an integer as the sum of four squares from which the result of
Lagrange follows immediately.

We begin this section by giving the standard proof of Lagrange's theorem.
The technique is that of descent. Having established the result for primes the
general result follows from a formal identity due to Euler expressing the
fact that the norm of a quaternion is a multiplicative function. In the last, and
somewhat lengthier part of this section we prove Jacobi's theorem. The proof
is based upon a letter (1856) from Dirichlet to Liouville ([122], pp. 201–208)
simplifying Jacobi's proof. See also Weil [237].

We begin with a diophantine problem modulo p.

Lemma 1. *If p is prime the congruence $x^2 + y^2 + 1 \equiv 0 \ (p)$ has a solution in
integers x, y.*

PROOF. Denote by S the set of squares modulo p. Then S and $\{-1 - x | x \in S\}$
$= S'$ each have $(p + 1)/2$ elements. Thus S and S' are not disjoint and the
result follows.

By the above lemma there is an integer m such that $mp = 1 + x^2 + y^2$ has
an integral solution and furthermore by adjusting the residues one may assume
$|x| < p/2, |y| < p/2$. Thus $mp < 1 + p^2/4 + p^2/4$ so that $m < p$.

Lemma 2. *Suppose for a prime p there is an integer m, $1 < m < p$ such that mp
is the sum of four squares. Then there is an n, $0 < n < m$ such that np is the sum
of four squares.*

PROOF. Write

$$mp = x_1^2 + x_2^2 + x_3^2 + x_4^2. \tag{40}$$

Let $x_i \equiv y_i \ (m)$ with $-m/2 < y_i \leq m/2$. Then $y_1^2 + y_2^2 + y_3^2 + y_4^2 \equiv 0 \ (m)$ so
that there is an integer $r \geq 0$ such that

$$rm = y_1^2 + y_2^2 + y_3^2 + y_4^2. \tag{41}$$

Now $rm \leq m^2/4 + m^2/4 + m^2/4 + m^2/4 = m^2$ so that $r \leq m$. First of all
$r \neq 0$ for otherwise $y_i = 0, i = 1, \ldots, 4$ which would imply by (40) that $m | p$,
a contradiction. Also $r \neq m$, since otherwise $y_i = m/2$; then $x_i^2 \equiv m^2/4 \ (m^2)$
and (40) implies that $mp \equiv m^2 \ (m^2)$ or $m | p$. Multiplying (40) and (41) gives,
by Exercise 28, the identity

$$m^2 rp = (x_1 y_1 + x_2 y_2 + x_3 y_3 + x_4 y_4)^2 + (x_1 y_2 - x_2 y_1 + x_3 y_4 - x_4 y_3)^2$$
$$+ (x_1 y_3 - x_3 y_1 - x_2 y_4 + x_4 y_2)^2 + (x_1 y_4 - x_4 y_1 + x_2 y_3 - x_3 y_2)^2 \tag{42}$$

Using $x_i \equiv y_i \ (m)$ one sees that each term on the right-hand side of (42) is
divisible by m^2. Cancelling m^2 shows that rp is the sum of four squares and the
proof is complete.   □

Proposition 17.7.1. *Any positive integer is the sum of four squares.*

PROOF. This follows immediately from Lemmas 1 and 2 and Exercise 28. □

Let us now turn to the statement and proof of Jacobi's theorem. The result that we will establish is the following.

Proposition 17.7.2. *Let n be a positive integer n ≡ 4 (8). The number of integral solutions (x, y, z, w), x, y, z, w positive and odd to the equation*

$$x^2 + y^2 + z^2 + w^2 = n \qquad (43)$$

is the sum of the positive odd divisors of n.

We leave to the Exercises the following corollary.

Corollary. *Let n be a positive integer. The number of integral solutions (x, y, z, w) to $x^2 + y^2 + z^2 + w^2 = n$ is $8 \sum_{d|n} d$, if n odd and $24 \sum_{d|n} d$, d odd, if n is even.*

The proof of the proposition is divided into several lemmas. Let N denote the number of integral solutions (x, y, z, w) to (43) with x, y, z, w positive and odd. Since $n \equiv 4$ (8) we may write $n = 2m$, $m \equiv 2$ (4).

Lemma 3. *N is the number of solutions (x, y, z, u, v) to the system of Diophantine equations*

$$x^2 + y^2 = 2u,$$
$$z^2 + w^2 = 2v, \qquad (44)$$
$$u + v = m,$$

with x, y, z, u, v odd and positive.

PROOF. This is left as a simple exercise.   □

As in Section 5 let χ denote the nontrivial Dirichlet character modulo 4.

Lemma 4. *$N = \sum \chi(de) = \sum (-1)^{(de-1)/2} = \sum (-1)^{(d-e)/2}$ the sum over all solutions (d, e, t, s) in positive odd integers to $ds + et = m$.*

PROOF. By Lemma 3 and Corollary 2 of Proposition 17.6.1 we see easily that

$$N = \sum_{\substack{u, v \\ u+v = m}} \left(\sum_{\substack{d|u \\ e|v}} \chi(d)\chi(e) \right). \qquad (45)$$

Write $u = ds$, $v = et$ so that the terms in (45) are in one-to-one correspondence with solutions (d, e, t, s), d, e, t, s positive, odd and satisfying $ds + et = m$. This proves the first equality in the lemma. The second follows from the definition of χ and the fact that $(d - 1)/2 + (e - 1)/2 \equiv (de - 1)/2$ (2) when d and e are odd. $\qquad\qquad\qquad\qquad\qquad\qquad\qquad\qquad\qquad\qquad\square$

Consider now the terms in $\sum \chi(de)$, the sum being as in Lemma 3, for which $d = e$. For each odd $d \mid m$, $s + t = m/d$ has $m/2d$ solutions in positive odd, s, t. The total number of solutions is therefore $\sum_{d\mid m} m/2d = \sum_{d\mid m} d$. Each solution of $ds + et = m$, $d = e$ contributes $\chi(d^2) = 1$ to N by Lemma 4. The proof of Proposition 17.7.2 will follow if one shows $\sum \chi(de) = 0$ the sum as in Lemma 4 and $d \neq e$. Pairing (d, e, t, s) with (e, d, s, t) shows that is enough to prove $\sum \chi(de) = 0$, $d > e$.

Denote by S the set of all (d, e, t, s), $d > e$, $ds + et = m$, d, e, t, s positive and odd. The idea behind the remainder of the proof is to construct a bijection of S that sends $\sum_S \chi(de)$ to its negative. This, of course, will imply that $\sum_S \chi(de) = 0$.

For a positive integer n put

$$A_n = \begin{pmatrix} n + 1 & n + 2 \\ n & n + 1 \end{pmatrix}$$

and define (d', e', t', s') by

$$A_n \begin{pmatrix} t \\ s \end{pmatrix} = \begin{pmatrix} d' \\ e' \end{pmatrix},$$

$$A_n^{-1} \begin{pmatrix} d \\ e \end{pmatrix} = \begin{pmatrix} t' \\ s' \end{pmatrix}.$$

$$(46)$$

Since

$$A_n^{-1} = \begin{pmatrix} n + 1 & -n - 2 \\ -n & n + 1 \end{pmatrix}$$

one checks quickly that

$$A_n \begin{pmatrix} t & d \\ s & -e \end{pmatrix} = \begin{pmatrix} d' & t' \\ e' & -s' \end{pmatrix}.$$

Taking determinants one sees that

$$ds + et = d's' + e't'.$$

$$(47)$$

Thus for each n we have a mapping from $\mathbb{Z}^4$ to $\mathbb{Z}^4$, which we denote by ψ_n.

Lemma 5. *Given* $(d, e, t, s) \in S$ *there is a unique* $n \in \mathbb{Z}^+$ *such that* $\psi_n(d, e, t, s) \in S$.

PROOF. One sees immediately using (46) that d', e', t', s' are odd, $d' > e', d' > 0$, $e' > 0$. Furthermore the conditions $s' > 0$, $t' > 0$ are equivalent to, by (46), $e/(d - e) - 1 < n < e/(d - e)$. But $d - e$ is positive and even and e is odd from which it follows that this inequality is satisfied for a unique $n \geq 0$. This concludes the proof. $\qquad\qquad\qquad\qquad\qquad\qquad\qquad\qquad\qquad\qquad\square$

Denote the mapping from S to S defined by Lemma 5 by Φ.

Lemma 6. Φ *is a bijection.*

PROOF. We will show that Φ^2 is the identity map. For if $(d, e, t, s) \in S$ then

$$\Phi^2(d, e, t, s) = \Phi\left(\left(A_n\binom{t}{s}\right)^*, \left(A_n^{-1}\binom{d}{e}\right)^*\right)$$

$$= \left(\left(A_k A_n^{-1}\binom{d}{e}\right)^*, \left(A_k^{-1} A_n\binom{t}{s}\right)^*\right), \qquad (48)$$

where the asterisk denotes transpose. Here k and n are defined by Lemma 5. But the integer k is uniquely defined by the condition that the right-hand side of (48) is in S and that is true if $k = n$. Thus $\Phi^2(d, e, t, s) = (d, e, t, s)$ and the proof of the lemma is complete.   □

In order to complete the proof observe from (46) that $d' - e' = s + t$. But $\chi(de) = (-1)^{(d-e)/2}$. Since $ds + et \equiv 2$ (4) one sees that $(d - e)/2$ is even iff $(s + t)/2$ is odd. Thus $\chi(de) = -\chi(d'e')$. Finally $M = \sum_S \chi(de) = -\sum_S \chi(d'e') = -M$ from which it follows that $M = 0$ and the proof is complete.   □

§8 The Fermat Equation: Exponent 3

The Fermat equation

$$x^p + y^p = z^p \qquad (49)$$

has been discussed in special cases in Sections 2 and 4 and in Chapter 14 (Theorem 5). In this section using the arithmetic of $\mathbb{Z}[\omega]$ where $\omega^3 = 1$, $\omega \neq 1$ we give a complete solution to the equation

$$x^3 + y^3 = z^3. \qquad (50)$$

That this equation has no integral solution, $xyz \neq 0$, was first proved essentially by Euler. See, however, G. Bergmann [91].

Instead of (50) we shall study the more general equation

$$x^3 + y^3 = uz^3, \qquad (51)$$

where u is a fixed unit in $\mathbb{Z}[\omega]$ and prove the following result.

Proposition 17.8.1. *The equation* $x^3 + y^3 = uz^3$, *where u a fixed unit in $\mathbb{Z}[\omega]$ has no integral solution (x, y, z), $xyz \neq 0$ where $x, y, z \in \mathbb{Z}[\omega]$.*

This implies, of course, that a nonzero cube in $\mathbb{Z}$ is not the sum of two nonzero cubes in $\mathbb{Z}$.

Proposition 17.8.1 will be proved in a sequence of lemmas. First we recall the basic facts concerning the arithmetic of $\mathbb{Z}[\omega]$, proved in Chapter 9. The ring $\mathbb{Z}[\omega]$ is a principal ideal ring with units $\pm 1, \pm \omega, \pm \omega^2$. Write $\lambda = 1 - \omega$ and recall that $(\lambda)^2 = (3)$, and that λ is irreducible. Each element $\alpha \in \mathbb{Z}[\omega]$ is congruent modulo λ to $+1, -1$ or 0. This fact will be used repeatedly in the following. If $\alpha = u\lambda^n\beta$ where u is a unit and $\lambda \backslash \beta$ then we write $n = \mathrm{ord}_\lambda \alpha$.

First of all we establish the weaker result, the so-called first case, that (51) has no solution with $\lambda \backslash xyz$.

Lemma 1. *The equation* $x^3 + y^3 = uz^3$, u *a unit in* $\mathbb{Z}[\omega]$ *has no solution with* $x, y, z \in \mathbb{Z}[\omega]$, $\lambda \backslash xyz$.

PROOF. Note that since λ is irreducible the condition $\lambda \backslash xyz$ is equivalent to $\lambda \backslash x, \lambda \backslash y, \lambda \backslash z$. If $x \in \mathbb{Z}[\omega]$, $x \equiv 1 \ (\lambda)$ then $x^3 \equiv 1 \ (\lambda^4)$. For if $x = 1 + \lambda t$ then

$$
\begin{aligned}
x^3 - 1 &= (x - 1)(x - \omega)(x - \omega^2) \\
&= \lambda t(1 - \omega + \lambda t)(1 - \omega^2 + \lambda t) \\
&= \lambda t(\lambda + \lambda t)((1 + \omega)\lambda + \lambda t) \\
&= \lambda^3 t(1 + t)(t - \omega^2).
\end{aligned}
$$

Since $\omega^2 \equiv 1 \ (\lambda)$ and t is congruent modulo λ to $+1, -1$ or 0 the congruence follows.

Now assume a solution to (51) exists with $\lambda \nmid xyz$ and reduce modulo λ. Then

$$\pm 1 \pm 1 \equiv \pm u \ (\lambda^4). \tag{52}$$

But it is easy to check that (52) is impossible for any choice of signs and unit. This completes the proof. $\qquad\square$

We pass now to the more difficult situation in which we assume a solution exists with $\lambda | z$ and $(x, y) = 1$. Thus $\lambda \backslash xy$. Under these conditions the following lemma shows that in fact $\lambda^2 | z$.

Lemma 2. *If* $x^3 + y^3 = uz^3$ *for* $x, y, z \in \mathbb{Z}[\omega]$, $\lambda \backslash xy$, $\lambda | z$ *then* $\lambda^2 | z$.

PROOF. Reduction of (51) modulo λ^4 gives

$$\pm 1 \pm 1 \equiv uz^3 \ (\lambda^4).$$

If $0 \equiv uz^3 \ (\lambda^4)$ then $3 \ \mathrm{ord}_\lambda z \geq 4$ so that $\mathrm{ord}_\lambda z \geq 2$. If $\pm 2 \equiv uz^3 \ (\lambda^4)$ then $\lambda | 2$ which is not true. $\qquad\square$

The following lemma constitutes the "descent" step.

Lemma 3. *If* $x^3 + y^3 = uz^3$, $(x, y) = 1$, $\lambda \backslash xy$, $\mathrm{ord}_\lambda z \geq 2$ *then there exist* $u_1, x_1, y_1, z_1 \in \mathbb{Z}[\omega]$, u_1 *a unit*, $\lambda \backslash x_1 y_1$, $\mathrm{ord}_\lambda z_1 = \mathrm{ord}_\lambda z - 1$ *and such that*

$$x_1^3 + y_1^3 = u_1 z_1^3.$$

PROOF. Recall that if $\text{ord}_\lambda \alpha \neq \text{ord}_\lambda \beta$ then $\text{ord}_\lambda(\alpha \pm \beta) = \min(\text{ord}_\lambda \alpha, \text{ord}_\lambda \beta)$. Next

$$(x + y)(x + \omega y)(x + \omega^2 y) = uz^3. \tag{53}$$

Since $\text{ord}_\lambda(uz^3) \geq 6$ at least one factor on the left-hand side of (53) is divisible by λ^2. Replacing if necessary y by ωy or $\omega^2 y$ we may assume that $\text{ord}_\lambda(x + y) \geq 2$. Since $\text{ord}_\lambda(1 - \omega)y = \text{ord}_\lambda \lambda y = 1$ we see that

$$\text{ord}_\lambda(x + \omega y) = \text{ord}_\lambda(x + y - (1 - \omega)y)$$
$$= 1.$$

Similarly $\text{ord}_\lambda(x + \omega^2 y) = 1$. Thus

$$\text{ord}_\lambda(x + y) = 3\,\text{ord}_\lambda z - 2.$$

If π is an irreducible $(\pi) \neq (\lambda)$ then π cannot divide $x + y$ and $x + \omega y$. For otherwise $\pi | (1 - \omega)y = \lambda y$, so that $\pi | y$, $\pi | x$. It follows that $(x + y, x + \omega y) = (\lambda)$. Similarly the other pairs of factors of (53) have greatest common divisor λ. Since unique factorization in $\mathbb{Z}[\omega]$ holds one can write

$$x + y = u_1 \alpha^3 \lambda^t, \qquad t = 3\,\text{ord}_\lambda z - 2, \lambda \nmid \alpha,$$
$$x + \omega y = u_2 \beta^3 \lambda, \qquad \lambda \nmid \beta, \tag{54}$$
$$x + \omega^2 y = u_3 \gamma^3 \lambda, \qquad \lambda \nmid \gamma.$$

In (54) u_1, u_2, u_3 are units and $(\alpha, \beta) = (\alpha, \gamma) = (\beta, \gamma) = 1$. Multiplying the second equation in (54) by ω, the third by ω^2 and adding one obtains

$$0 = u_1 \alpha^3 \lambda^t + \omega u_2 \beta^3 \lambda + \omega^2 u_3 \gamma^3 \lambda. \tag{55}$$

Cancelling $\lambda(!!)$ gives

$$0 = u_1 \alpha^3 \lambda^{3(\text{ord } z - 1)} + \omega u_2 \beta^3 + \omega^2 u_3 \gamma^3. \tag{56}$$

Finally putting $\alpha \lambda^{\text{ord } z - 1} = z_1$, $\beta = x_1$, $\gamma = y_1$, (56) becomes, with units $\varepsilon_1, \varepsilon_2$

$$x_1^3 + \varepsilon_1 y_1^3 = \varepsilon_2 z_1^3. \tag{57}$$

Reducing (57) modulo λ^2 and noting that $\text{ord}_\lambda(z_1^3) > 2$ we find

$$\pm 1 \pm \varepsilon_1 \equiv 0 \,(\lambda^2). \tag{58}$$

An examination of cases leads immediately to $\varepsilon_1 = \pm 1$. Thus, replacing if necessary y_1 by $-y_1$ we arrive at a new relation

$$x_1^3 + y_1^3 = \varepsilon z_1^3,$$

with $\lambda \nmid x_1 y_1$, $\text{ord}_\lambda z_1 = \text{ord}_\lambda z - 1$, ε a unit. This completes the proof. $\square$

To prove Proposition 17.8.1 we proceed as follows. If $\lambda \nmid xyz$ we invoke Lemma 1. If $\lambda \nmid xy$ but $\lambda | z$ then Lemmas 2 and 3 lead to a contradiction. Finally, if $\lambda | x$ but $\lambda \nmid yz$, then $\pm 1 \equiv u\,(\lambda^3)$ which implies $\pm 1 = u$. But then $(\pm z)^3 + (-y)^3 = x^3$ and we are in a situation already disposed of.

§9 Cubic Curves with Infinitely Many Rational Points

In the previous section it was shown that the equation $x^3 + y^3 = z^3$ has no solution in integers x, y, z with $xyz \neq 0$. Division by z^3 shows that the cubic curve $x^3 + y^3 = 1$ has no rational points (x, y), $xy \neq 0$. Similarly from the fact established in Section 2 that $x^4 + y^4 = z^2$ has no integral solution with $xyz \neq 0$ one concludes that the curve defined by $y^2 = x^4 + 1$ has $(0, \pm 1)$ as its only rational points (see Exercise 31).

In this section we give examples of cubic curves with an infinite number of rational points. The proof is based upon the simple observation that the tangent line to a cubic curve at a rational point intersects the curve in a unique, not necessarily new, point which is again rational. We say that an integer a is cube-free if $\mathrm{ord}_p\, a \leq 2$ for all primes p that is, no cube $\neq 1$, -1 divides a.

Proposition 17.9.1. *If $a > 2$ is a cube-free integer such that the cubic curve with equation*

$$x^3 + y^3 = a \tag{59}$$

has a rational point then it has infinitely many rational points.

PROOF. Let (α, β) be a rational point on (59). If $\alpha = x_1/z_1$, $\beta = y_1/z_1'$, $(x_1, z_1) = (y_1, z_1') = 1$ with x_1, y_1, z_1, z_1' integers then it is easy to see that $z_1 = z_1'$. Since $a > 2$ is cube-free $x_1 y_1 \neq 0$ and $x_1 \neq y_1$. The tangent line to (59) at (α, β) is $\alpha^2 x + \beta^2 y = a$. Solving for y and substituting in (59) gives

$$x^3 + \left(\frac{a - \alpha^2 x}{\beta^2}\right)^3 - a = 0. \tag{60}$$

The left-hand side of (60) is a cubic polynomial with α as a double root (at least). If the third root is γ then since the sum of the roots is the negative of the coefficient of x^2, we obtain after a simple calculation,

$$2\alpha + \gamma = \frac{3\alpha^4}{\alpha^3 - \beta^3}. \tag{61}$$

Thus

$$\gamma = \frac{\alpha(\alpha^3 + 2\beta^3)}{\alpha^3 - \beta^3}$$

$$= \frac{x_1}{z_1} \frac{(x_1^3 + 2y_1^3)}{(x_1^3 - y_1^3)}. \tag{62}$$

The corresponding value for $y = (a - \alpha^2 x)/\beta^2$ is

$$\rho = \frac{-y_1}{z_1} \frac{(2x_1^3 + y_1^3)}{(x_1^3 - y_1^3)} \tag{63}$$

and by (60) (γ, ρ) is a rational point on the cubic. The reader may verify directly, of course, that (γ, ρ) satisfies $\gamma^3 + \rho^3 = a$. It remains to show that (γ, ρ) is distinct from (α, β) and moreover that one obtains by this process an infinite number of points on the curve. Define the integer A by $A > 0$ and

$$
\begin{aligned}
Ax_2 &= x_1(x_1^3 + 2y_1^3), \\
Ay_2 &= -y_1(2x_1^3 + y_1^3), \\
Az_2 &= z_1(x_1^3 - y_1^3),
\end{aligned}
\tag{64}
$$

with $(x_2, y_2, z_2) = 1$. Thus A is the greatest common divisor of the integers on the right-hand side of (64). Clearly one has

$$
x_2^3 + y_2^3 = az_2^3, \qquad z_2 \neq 0.
\tag{65}
$$

Since a is cube-free and $(x_2, y_2, z_2) = 1$ we see that $(x_2, y_2) = (x_2, z_2) = (y_2, z_2) = 1$. We claim that A is equal to 1 or 3. For if p is prime and $p|A$ then it follows without difficulty from (64) that $p \nmid x_1 y_1 z_1$. Thus p divides each of the second factors on the right-hand side of (64) and consequently $p|3y_1^3$. Thus p is 1 or 3. Notice, also, that $(A, z_1) = 1$ implies $A|x_1^3 - y_1^3$.

The proof will be completed by showing that $|z_2| > |z_1|$. To this end one has

$$
|z_2| = \frac{|z_1|}{A} |x_1^3 - y_1^3|
$$

$$
= \frac{|z_1|}{A} |x_1 - y_1||x_1^2 + x_1 y_1 + y_1^2|.
\tag{66}
$$

One sees, $4|x_1^2 + x_1 y_1 + y_1^2| = |(2x_1 + y_1)^2 + 3y_1^2| > 4$ and consequently one has the inequality $|z_2| > |z_1||x_1 - y_1|/A$. If $A = 1$ then (66) shows that $|z_2| > |z_1|$. On the other hand, if $A = 3$, then since $A|x_1^3 - y_1^3$ one has $x_1^3 \equiv y_1^3$ (3) which implies that $x_1 \equiv y_1$ (3) and once again (66) implies that $|z_2| > |z_1|$. Continuing in this manner one obtains a succession of points $(x_n/z_n, y_n/z_n)$, $x_n y_n \neq 0$, $(x_n, z_n) = (y_n, z_n) = 1$ and $|z_n| > |z_{n-1}|$, and the proof is complete.   □

§10 The Equation $y^2 = x^3 + k$

The Diophantine equation

$$
y^2 = x^3 + k
\tag{67}
$$

has been studied extensively since its consideration in the seventeenth century by Fermat and Bachet in the special case $k = -2$. The integral values of k for which (67) has a rational solution have not been determined thus far. It was asserted, though not demonstrated, by Bachet and others that given a rational solution (x, y), $xy \neq 0$ the tangent method, used in Section 9, produces an infinite number of solutions. Thus, in modern language, the elliptic curve (67) then has positive rank (see Chapter 18). This result was established with several exceptional cases by Fueter in 1930.

In 1966 Mordell gave a remarkably short proof of Fueter's result [191]. More precisely he proved

Proposition 17.10.1. *If $y^2 = x^3 + k$, k a sixth power-free integer, has a rational solution (x, y), $xy \neq 0$ then there are an infinite number of rational solutions provided $k \neq 1, -432$.*

It is shown in the Exercises that the case $k = -432$ is equivalent to Fermat's equation $x^3 + y^3 = 1$, which by the main result of Section 8 can easily be shown to have only the rational solutions $(1, 0)$, $(0, 1)$. We will not give the details to Proposition 17.10.1, but rather refer the interested reader to Mordell's paper. The proof consists in showing that the tangent method used in the preceding section leads to an infinite number of solutions.

Thus $y^2 = x^3 - 2$ has an infinite number of rational points since it has one, namely $(3, 5)$. However, we point out that there are only a finite number of integral solutions. This is a difficult theorem for general k but in the case $k = -2$ a very short proof can be given using Exercise 36 of Chapter 1. For

$$(y + \sqrt{-2})(y - \sqrt{-2}) = x^3. \tag{68}$$

If π is an irreducible in $\mathbb{Z}[\sqrt{-2}]$ dividing both factors on the left-hand side of (68) then $\pi \mid 2\sqrt{-2}$. Thus $(\pi) = (\sqrt{-2})$, and $\sqrt{-2} \mid x$ which implies, taking norms, that $2 \mid x$. But this implies that $y^2 \equiv 2$ (4) which is impossible. Since $\mathbb{Z}[\sqrt{-2}]$ is a unique factorization ring with units ± 1, (68) shows that

$$y + \sqrt{-2} = (a + b\sqrt{-2})^3.$$

Thus

$$y = a^3 - 6ab^2, \tag{69}$$
$$1 = 3a^2b - 2b^3$$
$$= b(3a^2 - 2b^2), \tag{70}$$

Hence $b = 1$ and one obtains as the only solutions $(3, \pm 5)$.

If d is a positive square free integer then one can find the integer solutions to $y^2 = x^3 - d$ in certain cases using the arithmetic of the imaginary quadratic field $\mathbb{Q}(\sqrt{-d})$. As in the case of Fermat's Last Theorem (see Section 11) it is necessary in this approach to impose a divisibility condition on the class number h of $\mathbb{Q}(\sqrt{-d})$, namely we require that $3 \nmid h$. If, furthermore, we restrict d by assuming $d \neq +1, +3$ and $-d \equiv 2$ or 3 (4) then by Chapter 13 the ring of integers of $\mathbb{Q}(\sqrt{-d})$ is $\mathbb{Z}[\sqrt{-d}]$ and ± 1 are the only units. Under these conditions assume that (x, y) is an integral solution to $y^2 = x^3 - d$. Then (Exercise 32) x is odd and $(x, d) = 1$. Now

$$x^3 = (y + \sqrt{-d})(y - \sqrt{-d}).$$

If $P \subset \mathbb{Z}[\sqrt{-d}]$ is a prime ideal containing $y \pm \sqrt{-d}$ then $2\sqrt{-d} \in P$ and $x \in P$. Thus $N(P)|4d$ and $N(P)|x^2$ which is impossible. It follows that $(y + \sqrt{-d})$ and $(y - \sqrt{-d})$ have no common ideal factors. Since $\mathbb{Z}[\sqrt{-d}]$ is a Dedekind ring we have

$$(y + \sqrt{-d}) = \mathfrak{A}^3$$

for some ideal $\mathfrak{A}$. Since $3 \nmid h$ the ideal class group of $\mathbb{Z}[\sqrt{-d}]$ has no element of order 3 and therefore $\mathfrak{A}$ is principal. Thus, since ± 1 are the only units one has

$$y + \sqrt{-d} = \pm(a + b\sqrt{-d})^3. \tag{71}$$

This implies

$$\begin{aligned} 1 &= \pm b(3a^2 - db^2), \\ y &= \pm a(a^2 - 3db^2), \end{aligned} \tag{72}$$

from which one derives easily $b = \pm 1$ and

$$d = 3a^2 \pm 1. \tag{73}$$

Thus $y^2 = x^3 - d$ has a solution precisely when d lies in one of the quadratic progressions $3a^2 \pm 1$. When this is so one finds easily the value of x to be $a^2 + d$. Thus we have the following proposition.

Proposition 17.10.2. *Let $d > 1$, square-free and $d \equiv 2$ or 1 (4). Assume that the class number of $\mathbb{Q}(\sqrt{-d})$ is not divisible by 3. Then $y^2 = x^3 - d$ has an integral solution iff d is of the form $3t^2 \pm 1$. The solutions are then $(t^2 + d, \pm t(t^2 - 3d))$.*

For a discussion of the real quadratic case, see W. Adams and L. Goldstein [84], Chapter 10, and Mordell [189], Chapter 26.

§11 The First Case of Fermat's Conjecture for Regular Exponent

In this last section we use results from Chapters 12 and 13 on the arithmetic of cyclotomic number fields to prove a special case of Fermat's conjecture. If ζ denotes an lth root of unity different from 1, where l is an odd prime then $\mathbb{Q}(\zeta)$ is an algebraic number field of degree $l - 1$ whose ring of integers is, by Proposition 13.2.10, $\mathbb{Z}[\zeta]$. Thus by Theorem 2, Chapter 12 every nonzero ideal in $\mathbb{Z}[\zeta]$ can be factored uniquely as a product of powers of distinct prime ideals. Recall that l is called regular if $l \nmid h$ where h denotes the class number of $\mathbb{Q}(\zeta)$. Thus if $\mathfrak{A}$ is an ideal such that $\mathfrak{A}^l$ is principal then $\mathfrak{A}$ itself is principal, a fact of central importance in the following.

We need one additional result concerning the arithmetic of $\mathbb{Z}[\zeta]$.

Lemma 1. *If u is a unit in $\mathbb{Z}[\zeta]$ then $\zeta^s u$ is real for some rational integer s.*

PROOF. Observe first of all that complex conjugation is an automorphism of $\mathbb{Q}(\zeta)$ since $\bar{\zeta} = \zeta^{l-1}$. Thus if u is a unit then $\bar{u}$ is a unit and $\tau = u/\bar{u} \in \mathbb{Z}[\zeta]$. Furthermore if ρ is any automorphism of $\mathbb{Q}(\zeta)$ then $\rho(\tau) = \rho(u)/\rho(\bar{u}) = \rho(u)/\overline{\rho(u)}$ so that $|\rho(\tau)| = 1$. By Lemmas 1 and 2, Section 5, Chapter 14, $\tau = \pm\zeta^t$ for some integer t. If $\lambda = 1 - \zeta$ then $\zeta^j \equiv 1 \; (\lambda)$ for all j, so that writing $u = a_0 + a_1\zeta + \cdots + a_{l-2}\zeta^{l-2}$ and using the fact that $\rho(\zeta) = \zeta^k$ for some k we see that $u \equiv \rho(u) \; (\lambda)$. In particular $u \equiv \bar{u}(\lambda)$. If $\tau = -\zeta^t$ then $u = -\zeta^t\bar{u}$ so that $u \equiv -\bar{u}(\lambda)$. Thus $2u \equiv 0 \; (\lambda)$ which is impossible. Therefore $u = \zeta^t\bar{u} = \zeta^{-2s}\bar{u}$ where $-2s \equiv t \; (l)$. Finally $\zeta^s u = \overline{\zeta^s u}$ showing that $\zeta^s u$ is real. $\qquad\square$

The main result of this section is the following.

Proposition 17.11.1. *If l is a regular prime then the diophantine equation*

$$x^l + y^l = z^l \qquad (74)$$

has no solution in rational integers x, y, z with $l \nmid xyz$.

The proof of this proposition will be presented in several lemmas. We begin by factoring the left-hand side of (74)

$$x^l + y^l = (x + y)(x + \zeta y) \cdots (x + \zeta^{l-1}y). \qquad (75)$$

Recall that two ideals $\mathfrak{A}$ and $\mathfrak{B}$ are relatively prime in $\mathbb{Z}[\zeta]$ if $\mathfrak{A} + \mathfrak{B} = \mathbb{Z}[\zeta]$. When this is the case $\mathfrak{A}$ and $\mathfrak{B}$ have no common prime ideal divisors. Assume for the remainder of this section that (74) has a solution in integers x, y, z, $l \nmid xyz$ and that $l \nmid h$. Suppose, as we may, that x, y, z are pairwise relatively prime.

Lemma 2. *The ideals $(x + \zeta^i y)$ and $(x + \zeta^j y)$ are relatively prime if $i \not\equiv j \; (l)$.*

This lemma has already been proven in Section 6, Chapter 14.

Lemma 3. *There exist $u, \beta \in \mathbb{Z}[\zeta]$, u is a real unit such that $x + \zeta y = \zeta^s u\beta$, where $s \in \mathbb{Z}$ and $\beta \equiv n \; (l)$ for some $n \in \mathbb{Z}$.*

PROOF. Using Lemma 2, Corollary in Section 6, Chapter 14, and the fact that the right-hand side of (74) is an lth power we see that $(x + \zeta y) = \mathfrak{A}^l$ for some ideal $\mathfrak{A}$. Since $l \nmid h$ it follows that $\mathfrak{A}$ is principal. Thus $x + \zeta y = \varepsilon\alpha^l$ where $\alpha \in \mathbb{Z}[\zeta]$ and ε is a unit. The result follows from Lemma 1 and the observation that if $\alpha = \sum_{i=0}^{l-2} a_i\zeta^i$ then $\alpha^l \equiv \sum_{i=0}^{l-2} a_i \; (l)$. $\qquad\square$

Taking conjugates one has $x + \zeta^{-1}y = \zeta^{-s}u\bar{\beta}$ so that $\zeta^{-s}(x + \zeta y) - \zeta^s(x + \zeta^{-1}y) = u(\beta - \bar{\beta})$. However $\bar{\beta} \equiv \beta \equiv n \; (l)$ and so we have shown that $\zeta^{-s}(x + \zeta y) - \zeta^s(x + \zeta^{-1}y) \in l\mathbb{Z}[\zeta]$. We state this as

Lemma 4. $x + \zeta y - \zeta^{2s}x - \zeta^{2s-1}y \in l\mathbb{Z}[\zeta]$.

By Proposition 6.4.1. 1, ζ, ζ^2, ..., ζ^{l-2} are linearly independent over $\mathbb{Q}$. Furthermore we may assume $l > 3$ (by Section 8) and $0 \le s \le l - 1$. The proof of Proposition 17.11.1 will be completed by deriving a contradiction from the relation of Lemma 4. By the above comment we need only to examine the cases when two of the powers of ζ are the same. Thus we must examine the cases

(a) $\zeta^{2s} = 1$.
(b) $\zeta^{2s-1} = 1$,
(c) $\zeta^{2s-1} = \zeta$.

In case (a), Lemma 4 implies $-y + \zeta^2 y \in l\mathbb{Z}[\zeta]$ so that $l | y$. In case (c), we find $x - \zeta^2 x \in l\mathbb{Z}[\zeta]$ so that $l | x$, a contradiction. Finally in case (b) we find $(x - y) + (y - x)\zeta \in l\mathbb{Z}[\zeta]$. Thus $x \equiv y\ (l)$. Write Fermat's equation as $x^l + (-z)^l = (-y)^l$. Then, arguing as earlier, we obtain Lemma 4 with a possibly different s. However cases (a) and (c) lead to contradictions and case (b) gives, as above, $x \equiv -z\ (l)$. But $0 = x^l + y^l - z^l \equiv x + y - z\ (l)$. Thus $3x \equiv 0\ (l)$ which implies $l | x$ a contradiction! This completes the proof of the first case of Fermat's Last Theorem for regular exponent. $\square$

The above proof is essentially that given in Borevich and Shafarevich [9].

§12 Diophantine Equations and Diophantine Approximation

In this final section we give a brief discussion of the relationship between diophantine equations and the approximation of algebraic numbers by rational numbers. The technqiues required to prove the results mentioned below are different from those developed in the preceding chapters. Here we can only give an indication of the results and refer the interested reader to the literature.

If α is an irrational number then by Proposition 17.5.1 there are infinitely many rational numbers p/q such that

$$\left| \alpha - \frac{p}{q} \right| < \frac{1}{q^2}.$$

It is natural to ask whether the exponent 2 in this inequality can be increased. A deep result of Roth in 1955 [118], for which he was awarded the Fields Medal in 1958, asserts that if α is algebraic of degree ≥ 2 then for each fixed $\varepsilon > 0$ there are at most finitely many rational numbers p/q, $q > 0$ with

$$\left| \alpha - \frac{p}{q} \right| < \frac{1}{q^{2+\varepsilon}}. \tag{76}$$

It follows that there is a constant $c > 0$ such that for all rationals p/q one has

$$\left| \alpha - \frac{p}{q} \right| > \frac{c}{q^{2+\varepsilon}}.$$

The theorem of Roth was preceded by deep results of A. Thue (1909) and C. L. Siegel (1921) each of which improved an elementary estimate of J. Liouville (1844). This simple result is the following.

Proposition 17.12.1. *If α is a real algebraic number of degree n, $n \geq 2$ then there is a constant $c > 0$ such that for any rational number p/q, $q > 0$*

$$\left| \alpha - \frac{p}{q} \right| > \frac{c}{q^n}.$$

PROOF. It is clearly enough to assume $|\alpha - p/q| \leq 1$. By the mean value theorem $|f(p/q)| = |f(\alpha) - f(p/q)| \leq |\alpha - p/q| A$ where $f(x) \in \mathbb{Z}[x]$ is irreducible, $f(\alpha) = 0$, and $A = \sup |f'(x)|$, $|x - \alpha| \leq 1$. But since α is not rational $f(p/q) \neq 0$ and $|f(p/q)| \geq 1/q^n$. This completes the proof. □

The Thue and Siegel results replaced n by $n/2 + 1$ and $2\sqrt{n}$ respectively. Roth's result is, in a certain sense, the best possible, by Dirichlet's theorem (Proposition 17.5.1). However we shall see that any improvement in the Liouville estimate, i.e., any lowering of the exponent n (but greater than 2!) has profound consequences in the study of certain diophantine equations. In fact, let $a_n x^n + a_{n-1} x^{n-1} + \cdots + a_0$ be a polynomial with integral coefficients, irreducible over $\mathbb{Q}$ and of degree at least 3. For a nonzero integer m consider the diophantine equation

$$a_n x^n + a_{n-1} x^{n-1} y + \cdots + a_0 y^n = m. \tag{77}$$

We will show that if one has an inequality of the form

$$\left| \alpha - \frac{p}{q} \right| > \frac{c}{q^{n-\varepsilon}}, \qquad n - \varepsilon > 2, \tag{78}$$

valid for some $0 < \varepsilon < n$, and all rational numbers p/q then (77) has at most a finite number of integral solutions. This remarkable result follows quite easily from (78). For write (77) in the form

$$\left(\frac{x}{y} - \alpha^{(1)} \right) \left(\frac{x}{y} - \alpha^{(2)} \right) \cdots \left(\frac{x}{y} - \alpha^{(n)} \right) = \frac{m}{a_n y^n}.$$

Put $A = \min |\alpha^{(i)} - \alpha^{(j)}|$, $i \neq j$. Then if (x, y) is an integral solution $y \neq 0$ clearly at most one $\alpha^{(j)}$ satisfies $|x/y - \alpha^{(j)}| < A/2$. For such an $\alpha^{(j)}$ apply (78) and for the remaining terms use $|x/y - \alpha^{(i)}| \geq A/2$. Then

$$\frac{m}{|y|^n} > \frac{T}{|y|^{n-\varepsilon}}$$

for a suitable T depending only on $\alpha^{(1)}, \ldots, \alpha^{(n)}$. Thus

$$m > T|y|^{\varepsilon}, \qquad \varepsilon > 0,$$

from which it follows that $|y|$ is bounded. But for any y the number of x satisfying (77) is bounded and we are through. Thus while $x^2 - 2y^2 = 1$ has infinitely many integral solutions, $x^3 - 2y^3 = 1$ has only finitely many integral solutions.

Among the texts treating in detail this vast area of number theory we recommend K. B. Stolarsky [225], A. Baker [89], and W. M. Schmidt [217].

NOTES

The literature on diophantine equations is vast. We will cite only a few articles and essays that have a relationship with the equations discussed in this chapter. For a good general survey article we recommend W. J. LeVeque, "A Brief Survey of Diophantine Equations" [180], as well as the early essay by G. H. Hardy [39]. The supplement of Heath's edition of *Diophantus* [146], provides a technical study of the equations considered by Fermat and Euler in the seventeenth and eighteenth centuries. See also the scholarly work by J. E. Hoffman [152], where a detailed analysis is made of the results of Fermat and Euler and their relationship to the tangent method for finding rational points on cubic curves described in Sections 9 and 10. Relationships between this process and the corresponding diophantine equations modulo p will be indicated in the following chapter.

Excellent chapters on diophantine problems can be found in various introductory texts on number theory. We mention in particular Adams and Goldstein [84], Hardy and Wright [40], Uspensky and Heaslet [230] Davenport [22], and Niven and Zuckerman [61].

For a broad perspective on the formative period of this branch of mathematics and number theory in general, see the informal lecture by A. Weil [235].

An extensive coverage of diophantine equations by a modern master is given in the text by L. J. Mordell, [189]. A much more sophisticated and abstract approach is taken by S. Lang in his book Diophantine Geometry [170]. For a spirited discussion of the relative merits of these books the interested reader should consult the reviews of Lang's book by Mordell, [190] and the subsequent review of Mordell's book by Lang [172]. See also the advanced surveys by S. Lang [53], [173].

EXERCISES

1. Show that $165x^2 - 21y^2 = 19$ has no integral solution.

2. Find the integral solutions to $y^2 + 31 = x^3$.

3. Show that $x^3 + y^3 = 3z^3$ has no solution $x, y, z \in \mathbb{Z}[\omega]$, $z \neq 0$.

4. (In memoriam Ramanujan) Show that 1729 is the smallest positive integer expressible as the sum of two different integral cubes in two ways.

5. Which of the following have nontrivial solutions?
   (a) $3x^2 - 5y^2 + 7z^2 = 0$.
   (b) $7x^2 + 11y^2 - 19z^2 = 0$.
   (c) $8x^2 - 5y^2 - 3z^2 = 0$.
   (d) $11x^2 - 3y^2 - 41z^2 = 0$.

6. Find the fundamental solutions to $x^2 - 3y^2 = 1$, $x^2 - 6y^2 = 1$, $x^2 - 624y^2 = 1$.

7. Reduce the problem of the integral solutions of $3x^2 + 1 = 4y^3$ to Proposition 17.8.1 as follows:
   (a) Put $t = (3x - 1)/2$; $t \neq 1, -2$, so that $t^2 + t + 1 = 3y^3$, $y \neq 0$.
   (b) $(t + 2)^3 + (1 - t)^3 = (3y)^3$.

8. Find the integral solutions to $y^2 = x^3 - 4$.

9. Find four rational points on $x^3 + y^3 = 9$ using the method of Proposition 17.9.1.

10. Find the integral solutions to $y^2 = x^3 - 1$.

11. Show that if $x^2 - dy^2 = -1$ has an integral solution then so does $x^2 - dy^2 = 1$.

12. List the integral solutions of $x^2 + y^2 + z^2 + w^2 = 15$ and check with Proposition 17.7.2.

13. Let t be an integral cube. Show that $y^2 = x^2 - t$ has an integral solution.

14. Show that if $x^2 - dy^2 = n$, $d > 0$ square-free has an integral solution $xy \neq 0$ it has infinitely many.

15. Let $a + b\sqrt{p}$ be the fundamental solution to $x^2 - py^2 = 1$, where p is prime $p \equiv 1$ (4). The following steps show that $x^2 - py^2 = -1$ has an integral solution x, y, $x \cdot y \neq 0$.
    (a) a is odd.
    (b) $a \pm 1 = 2u^2$, $a \mp 1 = 2pv^2$, $2uv = b$.
    (c) $u^2 - pv^2 = \pm 1$.
    (d) In (c) the negative sign holds.

The following seven exercises establish the corollary to Proposition 17.7.2. Let $A(n)$ denote the number of integral solutions to $x_1^2 + x_2^2 + x_3^2 + x_4^2 = n$. See [52].

16. Show that $A(4n) = A(2n)$.

17. If n is odd show that $16 \sum_{d|n} d + A(n) = A(4n)$.

18. If n is odd let S be the number of solutions to $x_1^2 + x_2^2 + x_3^2 + x_4^2 = 2n$ with $x_1 \equiv x_2 \equiv 1$ (2) and $x_3 \equiv x_4 \equiv 0$ (2). Show that the number of elements of S is $\frac{1}{6}A(2n)$.

19. If $n \equiv 1$ (4) and S is as in Exercise 18 show that the number of elements in S is $\frac{1}{2}A(n)$. Conclude that $A(2n) = 3A(n)$.

20. If $n \equiv 3$ (4) then $A(2n) = 3A(n)$.

21. If n is odd show that $A(n) = 8 \sum_{d|n} d$, $A(2n) = 24 \sum_{d|n} d$.

22. If n is even $n = 2^s m$, $s \geq 1$, m odd show that $A(n) = 24 \sum_{d|m} d$.

23. The discriminant of $t^3 + pt + q$ is $-(4p^3 + 27q^2)$. Reduce the problem of determining the cubics with discriminant 1 and p, q rational to Fermat's equation $x^3 + y^3 = 1$ by putting $x = (3q + 1)/(3q - 1)$, $y = 2p/(3q - 1)$, $q \neq \frac{1}{3}$. Show that the resulting cubics are $t^3 - t \pm \frac{1}{3}$.

24. Show that Proposition 17.3.1 implies Proposition 17.3.2.

25. Show that if b is a positive integer and -1 is a square modulo b then $x^2 + y^2 = b$ has an integral solution.

26. If $(n, m) = 1$ show that $a \, R \, m$, $a \, R \, n$ implies $a \, R \, mn$.

27. Justify the rearrangement steps in Proposition 17.6.1.

28. Let A be the set of complex matrices of the form

$$\begin{pmatrix} \alpha & \beta \\ -\bar{\beta} & \bar{\alpha} \end{pmatrix}.$$

Show that Euler's identity, which states that $(x_1^2 + x_2^2 + x_3^2 + x_4^2)(y_1^2 + y_2^2 + y_3^2 + y_4^2)$ equals the right-hand side of Equation (42), is equivalent to $\det(MN) = (\det M)(\det N)$ for $M, N \in A$.

29. The following argument shows that Proposition 17.8.1 implies that $y^2 = x^3 - 432$ has $(12, \pm 36)$ as its only rational solutions. Fill in the details. Assume a solution (x, y) exists distinct from $(\pm 36, 12)$, $x > 0$.
   (a) Write $y/36 = a/c$, $x/12 = b/c$, with $a \equiv c \equiv 0$ (2).
   (b) Put $r = (a + c)/2$, $s = (c - a)/2$, $t = b > 0$.
   (c) Show that $r^3 + s^3 = t^3$, $rst \neq 0$.

30. The converse to Exercise 29 is also true; Show that if $x^3 + y^3 = z^3$, $xyz \neq 0$, $x, y, z \in \mathbb{Z}$ then putting $r = 36(x - y)/(x + y)$, $s = 12z \cdot (x + y)$ leads to $r^2 = s^3 - 432$.

31. Using the fact that $x^4 + y^4 = z^2$ has no integral solution $xyz \neq 0$ show that $(0, \pm 1)$ are the only rational solutions to $y^2 = x^4 + 1$.

32. Let d be a square-free integer $d \equiv 1$ or 2 modulo 4. Show that if x and y are integers such that $y^2 = x^3 - d$ then $(x, 2d) = 1$.

Chapter 18

Elliptic Curves

Many of the themes studied throughout this book come together in the arithmetic theory of elliptic curves. This is a branch of number theory whose roots go back a long way, but which is, nevertheless, the subject of intense investigation at the present time.

In this chapter we will give a brief overview of some of the relevant definitions, problems, and conjectures about elliptic curves. In particular, it is our purpose to describe a subtle and influential conjecture due to B. J. Birch and H. P. F. Swinnerton-Dyer. For the most part we will omit proofs and be content to give a rough guide to the ideas involved. For curves of the form $y^2 = x^3 + D$ and $y^2 = x^3 - Dx$ we will give a more detailed analysis and show how the global zeta functions of these curves are related to Hecke L-functions. This will yield a special case of an important theorem due to M. Deuring. Our exposition is based on the seminal papers of H. Davenport and H. Hasse [23] and A. Weil [81].

The techniques that are currently being used to study elliptic curves are among the most sophisticated in all of mathematics. We hope that the elementary approach of this chapter will inspire the reader to further study in this fascinating and lively branch of number theory. There is much to be learned and much work yet to be done.

§1 Generalities

We begin with some general observations about curves in projective space. For the terminology the reader may wish to review Chapter 10, Section 1.

Let K be a field and $F(x_0, x_1, x_2) \in K[x_0, x_1, x_2]$ a homogeneous polynomial of degree d. A very general problem is to determine whether $F(x_0, x_1, x_2) = 0$ has a solution in $P^2(K)$.

It is useful to introduce geometric terminology. The equation

$$F(x_0, x_1, x_2) = 0$$

is said to define a curve of degree d over K. The field K is called a field of definition. If L is a field containing K one can consider the zeros of F in

$P^2(L)$. In our previous terminology this is the hypersurface $\bar{H}_F(L)$. A hypersurface in projective 2-space is appropriately called a curve. Notice F sets up a map from fields containing K to sets; $L \to \bar{H}_F(L)$.

A point $a \in \bar{H}_F(L)$ is said to be a nonsingular point if it is not a simultaneous solution to the equations

$$\frac{\partial F}{\partial x_0} = 0, \qquad \frac{\partial F}{\partial x_1} = 0, \qquad \frac{\partial F}{\partial x_2} = 0.$$

In this case, the line

$$0 = \frac{\partial F}{\partial x_0}(a)x_0 + \frac{\partial F}{\partial x_1}(a)x_1 + \frac{\partial F}{\partial x_2}(a)x_2$$

is called the tangent line to F at a. The curve $F(x_0, x_1, x_2) = 0$ is said to be nonsingular if all the points in $\bar{H}_F(L)$ are nonsingular for all extensions L of K. It can be shown that it is enough to check this for algebraic extensions of K. (In Chapter 11 we called this notion absolutely nonsingular).

If two curves intersect at a point, one can define an integer called the intersection multiplicity of the two curves at the point. This is a somewhat delicate notion and we will not go into detail about it (see W. Fulton [135], Chapter 3). In general, if L is algebraically closed, a line in $P^2(L)$ intersects a curve of degree d in d points if multiplicity is taken into account. To get an idea of why this is true, write $x = x_1/x_0$, $y = x_2/x_0$, and $f(x, y) = F(1, x, y)$. We work for the moment in affine 2-space $A^2(L)$. To find the intersection points of $f(x, y) = 0$ with the line $y = mx + b$ one simply substitutes for y and finds the roots of $f(x, mx + b) = 0$. If F has degree d this latter equation will, in general, have degree d, and since L is algebraically closed there will be d roots if multiplicity is taken into account. The only exceptions will be intersections at infinity, in which case $f(x, mx + b)$ will have degree less than d.

As an example, consider $F(x_0, x_1, x_2) = -x_0^3 - x_1^3 + x_0 x_2^2$. Then $f(x, y) = -1 - x^3 + y^2$ so the affine part of the curve is given by $y^2 = x^3 + 1$. The intersection with the line $y = x + 1$ is determined by $(x + 1)^2 = x^3 + 1$ leading to the three points $(-1, 0)$, $(0, 1)$, and $(2, 3)$. On the other hand the line $y = 1$ leads to the equation $x^3 = 0$. This is interpreted as saying that $y = 1$ intersects $y^2 = x^3 + 1$ at the point $(0, 1)$ with multiplicity 3.

The intersections with vertical lines $x = c$ are determined by $f(c, y) = 0$. In our example, $y^2 = c^3 + 1$ so there are two finite points of intersection $(c, \sqrt{c^3 + 1})$ and $(c, -\sqrt{c^3 + 1})$ provided $c^3 + 1 \neq 0$. The third point of intersection is at infinity. If $c^3 + 1 = 0$, then $(c, 0)$ is an intersection point of multiplicity 2.

Finally, the intersections with the line at infinity $x_0 = 0$ can be obtained from the equation $F(0, x_1, x_2) = -x_1^3$, so the point $(0, 0, 1) \in P^2(L)$ is an intersection point of multiplicity 3.

If $a \in \bar{H}_F(L)$ then the tangent line to F at a can be shown to be an intersection point of multiplicity two or greater. If the multiplicity is greater than 2 then a is said to be a flex point.

If F is defined over K then a zero of F in $P^2(K)$ is said to be a rational point over K.

We will say that a nonsingular homogeneous cubic polynomial

$$F(x_0, x_1, x_2) \in K[x_0, x_1, x_2]$$

defines an elliptic curve over K provided there is at least one rational point. The problem of determining all rational points on an elliptic curve has given rise to a vast body of theory.

One of the things which make elliptic curves so interesting is the fact that the set of rational points can be made into an abelian group in a natural way.

Let $F(x_0, x_1, x_2) = 0$ define an elliptic curve over K. If L is a field extension of K we will write $E(L)$ instead of $\bar{H}_F(L)$.

Let 0 be an element of $E(K)$. If P_1, $P_2 \in E(L)$ then the line connecting P_1 and P_2 intersects the curve in a uniquely determined third point P_3 which is easily seen to be in $E(L)$. If $P_1 = P_2$ then the tangent line at P_1 gives rise to a third point P_3. It is tempting to take P_3 as the "sum" of P_1 and P_2. However, this would not define a group structure since there would be no identity. What we do instead is to find the third point of intersection with E of the line connecting 0 with P_3 and call this new point $P_1 + P_2$. With this definition $E(L)$ becomes an abelian group having 0 as the identity element. The proof is not hard except for showing associativity, i.e.,

$$P_1 + (P_2 + P_3) = (P_1 + P_2) + P_3.$$

For a rigorous treatment of this construction see [135], Chapter 5, especially pp. 124 and 125.

If the characteristic of K is not 2 or 3 it can be shown that every elliptic curve over K can be transformed into one of the form

$$x_0 x_2^2 = x_1^3 - A x_0^2 x_1 - B x_0^3, \qquad A, B \in K.$$

This curve has exactly one point at infinity, namely $(0, 0, 1) \in \mathbb{P}^2(K)$. We call this point ∞ and take it as the zero element of our group.

The line at infinity $x_0 = 0$ intersects the curve at the point ∞ with multiplicity 3. If $x_0 \neq 0$ set $x = x_1/x_0$ and $y = x_2/x_0$. Then, in affine coordinates the defining equation of the curve is

$$y^2 = x^3 - Ax - B.$$

The point at infinity is thought of as lying infinitely far off in the direction of the y axis.

A calculation shows that the nonsingularity of

$$F(x_0, x_1, x_2) = x_0 x_2^2 - x_1^3 + A x_0^2 x_1 + B x_0^3$$

is equivalent to the nonvanishing of

$$\Delta = 16(4A^3 - 27B^2).$$

This number is -16 times the discriminant of the polynomial

$$x^3 - Ax - B.$$

Conversely if $\Delta \neq 0$ then F defines an elliptic curve.

The fact that ∞ is a flex point can be used to show that $P_1 + P_2 + P_3 = \infty$ iff P_1, P_2, and P_3 lie on a straight line. In particular, $-P$ is the third point of intersection of the line connecting P and ∞. In affine coordinates this shows $-(a, b) = (a, -b)$ since the line connecting (a, b) and ∞ is the vertical line $x = a$. The points of order 2 are those for which $b = 0$. If $x^3 - Ax - B = (x - a_1)(x - a_2)(x - a_3) \in L[x]$ then the points of order dividing 2 on $E(L)$ are $\infty, (a_1, 0), (a_2, 0), (a_3, 0)$.

As an example of how to add points consider $P_1 = (2, 3)$ and $P_2 = (-1, 0)$ on $y^2 = x^3 + 1$. The line connecting P_1 and P_2 is given by $y = x + 1$. The equation $(x + 1)^2 = x^3 + 1$ has three roots 2, -1, and 0 corresponding to P_1, P_2 and $(0, 1)$. Thus $P_1 + P_2 = (0, -1)$.

Now suppose $K = \mathbb{Q}$, the rational numbers. In 1922 L. J. Mordell proved the following remarkable theorem, conjectured by H. Poincaré in 1901 [203].

Theorem 1. *Let E be an elliptic curve defined over $\mathbb{Q}$. Then $E(\mathbb{Q})$ is a finitely generated abelian group.*

In 1928 A. Weil extended this result to the case where $\mathbb{Q}$ is replaced by an arbitrary algebraic number field. The resulting theorem is referred to as the Mordell–Weil theorem.

The subgroup $E(\mathbb{Q})_t \subseteq E(\mathbb{Q})$, consisting of points of finite order, is finite. It turns out that there is an effective method for computing $E(\mathbb{Q})_t$ in any given case.

It was conjectured for some time that there is a uniform upper bound for $|E(\mathbb{Q})_t|$ as E varies over all elliptic curves defined over $\mathbb{Q}$. It was noticed by G. Shimura and others that the theory of elliptic modular curves could be used to attack this problem. This point of view was extensively developed by A. Ogg who proved a number of partial results and made some rather precise conjectures. Finally, in 1976 B. Mazur proved the following very deep result which had been conjectured by Ogg.

Theorem 2. *Let E be an elliptic curve defined over $\mathbb{Q}$. Then $E(\mathbb{Q})_t$ is isomorphic to one of the following groups: $\mathbb{Z}/m\mathbb{Z}$ for $m \leq 10$ or $m = 12$, or $\mathbb{Z}/2\mathbb{Z} \oplus \mathbb{Z}/2m\mathbb{Z}$ for $m \leq 4$.*

It is also believed that there is a uniform upper bound for $|E(K)_t|$ where E varies over elliptic curves defined over a fixed algebraic number field K. This is not known to be true for a single such $K \neq \mathbb{Q}$, but partial results have been obtained by V. A. Demjanenko, D. Kubert, and Y. Manin, among others.

Another important integer associated to $E(\mathbb{Q})$ has proved to be even more intractable, namely the rank. The rank of an abelian group is the maximal number of independent elements. If A is an abelian group we say a set of elements $a_1, a_2, \ldots, a_t \in A$ is independent if $m_1 a_1 + m_2 a_2 + \cdots + m_t a_t = 0$ with $m_1, m_2, \ldots, m_t \in \mathbb{Z}$ implies $m_1 = m_2 = \cdots = m_t = 0$. We denote the rank of $E(\mathbb{Q})$ by r_E.

The rank r_E has been computed for a large number of elliptic curves over $\mathbb{Q}$. In most examples it is quite small; 0, 1, or 2. A. Néron has shown the existence of an elliptic curve over $\mathbb{Q}$ with rank 11. His method is not constructive. In 1977 A. Brumer and K. Kramer produced an explicit example with $r_E \geq 9$. Here it is

$$y^2 + 525xy = x^3 + 228x^2 - 14972955x + (856475)^2.$$

It is not known if there is an upper bound on the numbers r_E, where E is defined over $\mathbb{Q}$. Cassels considers this to be unlikely ([109], Section 20).

One of the most celebrated conjectures in modern number theory connects the number r_E with the order at $s = 1$ of an analytic function associated with E. This conjecture was formulated by the English mathematicians B. J. Birch and H. P. F. Swinnerton-Dyer. The formulation of their conjecture will be the task of the next section.

§2 Local and Global Zeta Functions of an Elliptic Curve

Let E be the elliptic curve defined over $\mathbb{Q}$ by the equation

$$x_0 x_2^2 = x_1^3 - A x_0^2 x_1 - B x_0^3, \qquad A, B \in \mathbb{Q} \tag{i}$$

The affine equation is obtained by setting $x = x_1/x_0$ and $y = x_2/x_0$.

$$y^2 = x^3 - Ax - B. \tag{ii}$$

The transformation $(x, y) \to (c^2 x, c^3 y)$ transforms this equation into

$$y^2 = x^3 - c^4 Ax - c^6 B. \tag{iii}$$

Thus, we may assume to begin with that $A, B \in \mathbb{Z}$ and we make this assumption from now on. The number $\Delta = 16(4A^3 - 27B^2)$ is called the discriminant of E. As we have seen $\Delta \neq 0$.

Let $p \in \mathbb{Z}$ be a prime and consider the congruence

$$y^2 \equiv x^3 - Ax - B\,(p),$$

or equivalently the equation,

$$y^2 = x^3 - \bar{A}x - \bar{B}, \qquad \bar{A}, \bar{B} \in \mathbb{Z}/p\mathbb{Z} = \mathbb{F}_p. \qquad \text{(iv)}$$

This equation defines an elliptic curve E_p over $\mathbb{F}_p$ provided that $p \nmid \Delta$. In what follows only such primes will be considered unless explicitly stated otherwise. The curve E_p is called the reduction of E modulo p.

Let N_{p^m} be the number of points in $E_p(\mathbb{F}_{p^m})$. Then, as in Chapter 11, we may consider the zeta function

$$Z(E_p, u) = \exp\left(\sum_{m=1}^{\infty} N_{p^m} \frac{u^m}{m}\right). \qquad \text{(v)}$$

By use of the Riemann–Roch theorem it can be shown that

$$Z(E_p, u) = \frac{1 - a_p u + p u^2}{(1 - u)(1 - pu)}, \qquad a_p \in \mathbb{Z}. \qquad \text{(vi)}$$

In special cases this can be proved using the methods of Chapter 11. H. Hasse was able to prove that $a_p^2 \le 4p$. It follows that

$$1 - a_p u + p u^2 = (1 - \pi u)(1 - \bar{\pi} u), \qquad \text{(vii)}$$

where $\bar{\pi}$ is the complex conjugate of π. Clearly, $\pi\bar{\pi} = p$, $a_p = \pi + \bar{\pi}$. Also, $|\pi| = |\bar{\pi}| = \sqrt{p}$. This is the "Riemann Hypothesis" for elliptic curves over $\mathbb{F}_p$.

By logarithmically differentiating (v), (vi), and (vii) and comparing coefficients one finds

$$N_{p^m} = p^m + 1 - \pi^m - \bar{\pi}^m. \qquad \text{(viii)}$$

In particular, $N_p = p + 1 - a_p$. Thus, if one calculates N_p this determines a_p. Since π and $\bar{\pi}$ are the roots of $T^2 - a_p T + p = 0$, Equation (viii) yields N_{p^m} for all $m \ge 1$.

A very special case which will be useful later is the following. If $N_p = p + 1$ then

$$Z(E_p, u) = \frac{1 + pu^2}{(1 - u)(1 - pu)}.$$

It is useful to change the variable from u to p^{-s}. We define

$$\zeta(E_p, s) = \frac{1 - a_p p^{-s} + p^{1-2s}}{(1 - p^{-s})(1 - p^{1-s})}. \qquad \text{(ix)}$$

The function $\zeta(E_p, s)$ is called the local zeta function of E at p.

It is illuminating to see that $\zeta(E_p, s)$ can be obtained from another point of view which makes the connection with the Riemann zeta function much clearer.

The ring $\mathbb{F}_p[x]$ and its quotient field $\mathbb{F}_p(x)$ is analogous to $\mathbb{Z}$ and its quotient field $\mathbb{Q}$. Let $K = \mathbb{F}_p(x)(\sqrt{x^3 - Ax - B})$ and let D be the integral closure of $\mathbb{F}_p[x]$ in K, i.e., D consists of all the elements in K which satisfy monic polynomials with coefficients in $\mathbb{F}_p[x]$. D is a Dedekind domain and every nonzero ideal is of finite index in D. If $I \subset D$ is a nonzero ideal let $NI = |D/I|$, and define $\zeta_D(s) = \sum NI^{-s}$, where the sum is over all nonzero ideals in D. It is not hard to show that $\zeta_D(s)$ converges for Re $s > 1$. Moreover, one can prove that $\zeta_D(s) = (1 - p^{-s})\zeta(E_p, s)$. See also Section 1 of Chapter 11.

The point of view outlined here is that taken by E. Artin in his thesis [2].

We have defined $\zeta(E_p, s)$ for those primes p such that $p \nmid \Delta$. If $p \mid \Delta$ we define

$$\zeta(E_p, s) = \frac{1}{(1 - p^{-s})(1 - p^{1-s})}.$$

This is not the best definition but it will suffice for our purposes.

Now that we have defined a local zeta function for all primes p, we define a global zeta function by simply taking the product of the local zeta functions.

$$\zeta(E, s) = \prod_p \zeta(E_p, s). \tag{x}$$

From the definitions we see that $\zeta(E, s) = \zeta(s)\zeta(s - 1)L(E, s)^{-1}$ where

$$L(E, s) = \prod_{p \nmid \Delta} (1 - a_p p^{-s} + p^{1-2s})^{-1}. \tag{xi}$$

The function $L(E, s)$ is called the L-function of E. Recalling Hasse's result that $(1 - a_p p^{-s} + p^{1-2s}) = (1 - \pi p^{-s})(1 - \bar{\pi} p^{-s})$ with $|\pi| = |\bar{\pi}| = \sqrt{p}$ one can show fairly easily that the product for $L(E, s)$ converges for Re $s > \frac{3}{2}$.

It was conjectured by Hasse that $\zeta(E, s)$ can be analytically continued to all of $\mathbb{C}$. This was first shown to be true in special cases by Weil [81]. After that M. Deuring proved the result for an important class of elliptic curves which are said to possess "complex multiplication."

Lang [169], Chapter 10, has an exposition of Deuring's results. Y. Taniyama, and later A. Weil, conjectured that every elliptic curve over $\mathbb{Q}$ can be parameterized by elliptic modular forms. See the article by Swinnerton-Dyer in [226] for a precise statement of this conjecture. For such curves Hasse's conjecture is true. Thus, the evidence for the truth of Hasse's conjecture seems overwhelming.

Assuming $L(E, s)$ can be continued to all of $\mathbb{C}$ it makes sense to speak of the analytic behavior of $L(E, s)$ about $s = 1$.

On the basis of extensive empirical work on curves of the form $y^2 = x^3 - Dx$, Birch and Swinnerton-Dyer were led to the following remarkable conjecture.

Conjecture. *Suppose E is an elliptic curve defined over $\mathbb{Q}$. Then the rank of E, r_E, is equal to the order of the zero of $L(E, s)$ at $s = 1$.*

This conjecture can be supplemented. Assuming the conjecture we can define a nonzero constant $B_E = \lim_{s \to 1} (s - 1)^{-r_E} L(E, s)$. Birch and Swinnerton-Dyer give an expression for B_E which depends on subtle arithmetic invariants of E. It would take us too far afield to discuss these here. See Cassels [109] or J. Tate [227].

In an important paper [114] published in 1977, J. Coates and A. Wiles made significant progress on the above conjecture. Their main result was subsequently generalized by N. Arthaud [87]. We would need to enter into the theory of complex multiplication to even state this result in full generality so we will be content with a special case.

Theorem 3. *Let E be an elliptic curve defined over $\mathbb{Q}$ and suppose that E has complex multiplication. If $L(E, 1) \neq 0$, then $E(\mathbb{Q})$ is finite.*

Most of the work we have been discussing is of a very advanced nature and is beyond the scope of this book. In the following sections we will discuss elliptic curves of two types; $y^2 = x^3 + D$ and $y^2 = x^3 - Dx$. For these curves we will analyze the local and global zeta functions and show on the basis of a fundamental result of E. Hecke that the global zeta function of these curves can be analytically continued to all of $\mathbb{C}$. This will give the reader a sample, at least, of the extensive arithmetic theory of elliptic curves.

§3 $y^2 = x^3 + D$, the Local Case

Let D be a nonzero integer. We will consider the elliptic curve E defined by $x_0 x_2^2 - x_1^3 - D x_0^3 = 0$, or in affine coordinates $y^2 = x^3 + D$. The discriminant Δ of E is $-2^4 3^3 D^2$ so we will only consider primes $p \neq 2$ or 3 and $p \nmid D$.

The curve $y^2 = x^3 + \bar{D}$ over $\mathbb{F}_p$ has one point at infinity. Thus, $N_p = 1 + N(y^2 = x^3 + \bar{D})$ where we use the notation introduced in Chapter 8. By means of Jacobi sums we will derive an explicit formula for N_p. From now on we will write D instead of $\bar{D}$ so by "abuse of notation" D will represent the coset of D modulo p.

If $p \equiv 2\,(3)$ then $x \to x^3$ is an automorphism of $\mathbb{F}_p^*$. It follows easily (see Exercise 1) that $N_p = p + 1$ in this case.

If $p \equiv 1\,(3)$ let χ be a character of order 3 and ρ a character of order 2 of $\mathbb{F}_p^*$. Then

$$N(y^2 = x^3 + D) = \sum_{u+v=D} N(y^2 = u)N(x^3 = -v)$$

$$= \sum_{u+v=D} (1 + \rho(u))(1 + \chi(-v) + \chi^2(-v))$$

$$= p + \sum_{u+v=D} \rho(u)\chi(v) + \sum_{u+v=D} \rho(u)\chi^2(v).$$

We have used the fact that $\chi(-1) = 1$. Making the substitutions $u = Du'$ and $v = Dv'$ we find

$$N_p = p + 1 + \rho\chi(D)J(\rho, \chi) + \overline{\rho\chi(D)}\,\overline{J(\rho, \chi)}, \tag{i}$$

where bar denotes complex conjugation.

In order to analyze Equation (i) still further the following lemma will be useful.

Lemma. *Let p be an odd prime, ρ a character of order 2 and ξ any nontrivial character of $\mathbb{F}_p^*$. Then $J(\rho, \xi) = \xi(4)J(\xi, \xi)$.*

PROOF.

$$J(\rho, \xi) = \sum_{u+v=1} \rho(u)\xi(v)$$

$$= \sum_{u+v=1} (1 + \rho(u))\xi(v) = \sum_{u+v=1} N(t^2 = u)\xi(v)$$

$$= \sum_t \xi(1 - t^2) = \xi(4) \sum_t \xi\left(\frac{1-t}{2}\right)\xi\left(\frac{1+t}{2}\right) = \xi(4)J(\xi, \xi). \quad \square$$

Using the lemma, Equation (i) can be transformed into

$$N_p = p + 1 + \rho\chi(4D)J(\chi, \chi) + \overline{\rho\chi(4D)J(\chi,\chi)}. \tag{ii}$$

We want to specify ρ and χ. Since $p \equiv 1\,(3)$, $p = \pi\bar\pi$ in $\mathbb{Z}[\omega]$ (recall that $\omega = e^{2\pi i/3}$) where we can take π and $\bar\pi$ to be primary, i.e., $\pi \equiv \bar\pi \equiv 2\,(3)$. Let $(a/\pi)_6$ be the sixth power residue symbol and take $\rho(a) = (a/\pi)_6^3$ and $\chi(a) = (a/\pi)_6^2 = (a/\pi)_3$. Then $\rho\chi(a) = \rho(a)\chi(a) = (a/\pi)_6^5 = \overline{(a/\pi)_6}$. Finally, if we set $\chi_\pi(a) = (a/\pi)_3$ then Lemma 1 of Section 4, Chapter 9 shows $J(\chi_\pi, \chi_\pi) = \pi$. Substituting this information into Equation (ii) we find

Theorem 4. *Suppose $p \neq 2$ or 3 and $p \nmid D$. Consider the elliptic curve $y^2 = x^3 + D$ over $\mathbb{F}_p$. If $p \equiv 2\,(3)$ then $N_p = p + 1$. If $p \equiv 1\,(3)$ let $p = \pi\bar\pi$ with $\pi \in \mathbb{Z}[\omega]$ and $\pi \equiv 2\,(3)$. Then*

$$N_p = p + 1 + \left(\frac{4D}{\pi}\right)_6 \pi + \left(\frac{4D}{\pi}\right)_6 \bar\pi.$$

Theorem 4 completely determines the local zeta function of $y^2 = x^3 + D$.

As an example consider the curve $y^2 = x^3 + 1$ over $\mathbb{F}_{13}$. We find $13 = (-1 + 3\omega)(-1 + 3\omega^2)$ and $-1 + 3\omega \equiv 2\,(3)$. To apply the formula in the theorem we must know $(4/-1 + 3\omega)_6 = (2/-1 + 3\omega)_3$. Since $2^{(13-1)/3} = 2^4 \equiv 3 \equiv \omega^2(-1 + 3\omega)$ it follows that $(2/-1 + 3\omega)_3 = \omega^2$. The formula in the theorem gives

$$N_{13} = 13 + 1 + \omega(-1 + 3\omega) + \omega^2(-1 + 3\omega^2)$$

$$= 14 + 2(\omega^2 + \omega) = 14 - 2 = 12.$$

One checks that the points on $y^2 = x^3 + 1$ with coefficients in $\mathbb{F}_{13}$ are $\infty, (4, 0), (10, 0), (12, 0), (0, \pm 1), (2, \pm 3), (5, \pm 3),$ and $(6, \pm 3)$.

§4 $y^2 = x^3 - Dx$, the Local Case

Let D be a nonzero integer and consider the elliptic curve E defined by $x_0 x_2^2 - x_1^3 + Dx_1 x_0^2 = 0$ or, in affine coordinates, $y^2 = x^3 - Dx$. The discriminant of E is $\Delta = 2^6 D^3$. We will only consider primes p such that $p \neq 2$ and $p \nmid D$.

The curve $y^2 = x^3 - Dx$ over $\mathbb{F}_p$ (we continue to write D instead of $\bar{D}$) has one point at infinity so that $N_p = 1 + N(y^2 = x^3 - Dx)$. The methods of Chapter 8 are not immediately applicable in this case. We will first transform the curve $y^2 = x^3 - Dx$ into the curve $u^2 = v^4 + 4D$. The number of solutions to $u^2 = v^4 + 4D$ can then be handled by our previous methods.

For the moment let C denote the curve $y^2 = x^3 - Dx$ and C' denote the curve $u^2 = v^4 + 4D$. Define a transformation T as follows

$$T(u, v) = (\tfrac{1}{2}(u + v^2), \tfrac{1}{2}v(u + v^2)).$$

A simple calculation shows that T maps C' to C. The point $(0, 0)$ on C is not in the image since $4D = u^2 - v^4 = (u - v^2)(u + v^2)$ shows $u + v^2 \neq 0$.
Define a transformation S by

$$S(x, y) = \left(2x - \frac{y^2}{x^2}, \frac{y}{x}\right).$$

It is easily shown that S maps $C - \{(0, 0)\}$ to C' and moreover TS is the identity on $C - \{(0, 0)\}$ and ST is the identity on C'. Let $N' = N(u^2 = v^4 + 4D)$ and $N = N(y^2 = x^3 - Dx)$. We have shown that $N - 1 = N'$.

If $p \equiv 3\,(4)$ then -1 is a quadratic nonresidue so every element of $\mathbb{F}_p$ is of the form $\pm w^2$. Thus every square is automatically a fourth power. Consequently,

$$N' = N(u^2 = v^4 + 4D) = N(u^2 = v^2 + 4D) = p - 1.$$

Thus we find that if $p \equiv 3\,(4)$, $N_p = 1 + N = 2 + N' = 2 + p - 1 = p + 1$.

Suppose now that $p \equiv 1$ (4). Let λ be a character of order 4 of $\mathbb{F}_p$ and set $\rho = \lambda^2$. Then, by the now familiar process, we find

$$N(u^2 = v^4 + 4D) = \sum_{r+s=4D} N(u^2 = r)N(v^4 = -s)$$

$$= p - 1 + \overline{\lambda(-4D)J(\rho, \lambda)} + \lambda(-4D)\overline{J(\rho, \lambda)}. \quad \text{(i)}$$

We have used the fact that for $p \equiv 1$ (4), $J(\rho, \rho) = -1$ (see Chapter 8, Section 3, Theorem 1). By the lemma of the previous section we have $J(\rho, \lambda) = \lambda(4)J(\lambda, \lambda)$. Thus, $\overline{\lambda(-4D)J(\rho, \lambda)} = \overline{\lambda(D)\lambda(-1)J(\lambda, \lambda)}$.

We now specify λ. Since $p \equiv 1$ (4), $p = \pi\bar{\pi}$ in $\mathbb{Z}[i]$ with π primary, i.e., $\pi \equiv 1 (2 + 2i)$. Identify $\mathbb{F}_p$ with $\mathbb{Z}[i]/\pi\mathbb{Z}[i]$ and chose λ to be the biquadratic residue symbol, $\lambda(a) = (a/\pi)_4$. Then, by Proposition 9.9.4 we have

$$-\lambda(-1)J(\lambda, \lambda) = \pi.$$

Starting from Equation (i) and substituting all this information we arrive at

Theorem 5. *Suppose $p \neq 2$ and $p \nmid D$. Consider the elliptic curve $y^2 = x^3 - Dx$ over $\mathbb{F}_p$. If $p \equiv 3$ (4) then $N_p = p + 1$. If $p \equiv 1$ (4) let $p = \pi\bar{\pi}$ with $\pi \in \mathbb{Z}[i]$ and $\pi \equiv 1 (2 + 2i)$. Then*

$$N_p = p + 1 - \left(\frac{\overline{D}}{\pi}\right)_4 \pi - \left(\frac{D}{\pi}\right)_4 \bar{\pi}.$$

As an example, consider $y^2 = x^3 - x$ over $\mathbb{F}_{13}$. One sees

$$13 = (3 + 2i)(3 - 2i)$$

and $3 + 2i \equiv 1 (2 + 2i)$. The formula of the theorem tells us that $N_{13} = 13 + 1 - (3 + 2i) - (3 - 2i) = 14 - 6 = 8$. In fact, a short calculation shows the points on $y^2 = x^3 - x$ with coefficients in $\mathbb{F}_{13}$ are ∞, $(0, 0)$, $(1, 0)$, $(-1, 0)$, $(5, \pm 4)$, and $(-5, \pm 6)$.

§5 Hecke L-functions

In two important papers published in 1918 and 1920 the German mathematician E. Hecke introduced a new class of characters and L-functions. These can be defined over arbitrary algebraic number fields. We shall confine our attention to algebraic Hecke characters over CM fields of a certain type (the terminology will be explained below). For the applications we have in mind this will suffice.

Let $K/\mathbb{Q}$ be an algebraic number field. An isomorphism σ of K into $\mathbb{C}$ is called real if $\sigma(K) \subset \mathbb{R}$, otherwise it is called complex. K is said to be totally real if every isomorphism of K into $\mathbb{C}$ is real. K is said to be totally complex if every isomorphism of K into $\mathbb{C}$ is complex. K is called a CM field if it is a totally complex quadratic extension of a totally real subfield K_0. For example, if $d \in \mathbb{Q}$ with $d > 0$ then $\mathbb{Q}(\sqrt{-d})$ is a *CM* field. Other examples are provided by cyclotomic fields $\mathbb{Q}(\zeta_m)$. The totally real subfield of $\mathbb{Q}(\zeta_m)$ is $\mathbb{Q}(\zeta_m + \zeta_m^{-1})$.

Let $K \subset \mathbb{C}$ be a *CM* field such that $K/\mathbb{Q}$ is a Galois extension. Let j be the restriction of complex conjugation to K. Then it is easily seen (Exercise 2) that j is in the center of G, the Galois group of $K/\mathbb{Q}$. Moreover, K_0 is the fixed field of j. From now on we assume K satisfies these conditions.

Let $\mathcal{O} \subset K$ be the ring of integers and $M \subseteq \mathcal{O}$ an ideal. An algebraic Hecke character modulo M is a function χ from the ideals of $\mathcal{O}$ to $\mathbb{C}$ subject to the following conditions.

(i) $\chi(\mathcal{O}) = 1$.
(ii) $\chi(A) \neq 0$ if and only if A is relatively prime to M.
(iii) $\chi(AB) = \chi(A)\chi(B)$.
(iv) There is an element $\theta = \sum n(\sigma)\sigma \in \mathbb{Z}[G]$ such that if $\alpha \in \mathcal{O}$, $\alpha \equiv 1 \ (M)$, then $\chi((\alpha)) = \alpha^\theta$.
(v) There is an integer $m > 0$ such that $n(\sigma) + n(j\sigma) = m$ for all $\sigma \in G$.

The last condition is easily seen to be equivalent to $(1 + j)\theta = mN$, where $N = \sum \sigma$ is the norm element in $\mathbb{Z}[G]$.

The number m in condition (v) is called the weight of χ.

Another thing to note is that by condition (iii) χ is completely determined by its values on prime ideals not dividing M.

Proposition 18.5.1. *Let χ be an algebraic Hecke character of weight m. Then if $(A, M) = (1)$, $|\chi(A)| = NA^{m/2}$.*

PROOF. Let I_M be the set of ideals in $\mathcal{O}$ which are relatively prime to M. We put an equivalence relation on I_M as follows; if $A, B \in I_M$ we say $A \sim B$ if there exist $\alpha, \beta \in \mathcal{O}$ such that $\alpha, \beta \equiv 1 \ (M)$ and $(\alpha)A = (\beta)B$. It can be shown that the equivalence classes are finite in number and form a group C_M. The product in this group takes the equivalence class of A and the equivalence class of B to the equivalence class of AB. If $M = \mathcal{O}$, this construction yields the ideal class group of $\mathcal{O}$ (see Chapter 12, Section 1). Let h be the number of elements in C_M.

If $A \in I_M$ there exist $\alpha, \beta \in \mathcal{O}$, $\alpha, \beta \equiv 1 \ (M)$, such that $(\alpha)A^h = (\beta)$. Thus,

$$\alpha^\theta \chi(A)^h = \beta^\theta.$$

Take complex conjugates of both sides and multiply. This yields

$$(\alpha^\theta)^{1+j}|\chi(A)|^{2h} = (\beta^\theta)^{1+j},$$

or, by (v)

$$(N\alpha)^m|\chi(A)|^{2h} = (N\beta)^m.$$

Since $(\alpha)A^h = (\beta)$ we also have

$$N\alpha NA^h = N\beta.$$

Comparing these last two equations we find $|\chi(A)|^{2h} = NA^{mh}$ and $|\chi(A)| = NA^{m/2}$. □

It should be noted that the proof shows that the values $\chi(A)$ are algebraic numbers (in fact, hth roots of elements of K). This is a partial explanation of why χ is called an "algebraic" Hecke character.

We now proceed to attach an *L*-function to an algebraic Hecke character χ. Namely, define

$$L(s, \chi) = \prod_P (1 - \chi(P)NP^{-s})^{-1}$$

$$= \sum_A \chi(A)NA^{-s}.$$

The product is over all prime ideals in $\mathcal{O}$, and the sum over all ideals in $\mathcal{O}$.

Simple estimates show that the product converges absolutely for Re $s >$ $1 + m/2$ and uniformly for Re $s \geq 1 + m/2 + \delta$ for any $\delta > 0$. Indeed, the product converges absolutely if and only if $\sum_P |\chi(P)NP^{-s}|$ converges. By Proposition 18.5.1, if s is real

$$|\chi(P)NP^{-s}| = NP^{(m/2)-s} \leq p^{-(s-(m/2))}$$

where p is the rational prime below P. Since every rational prime has at most $[K : \mathbb{Q}]$ primes above it in $\mathcal{O}$ we see

$$\sum |\chi(P)NP^{-s}| \leq [K : \mathbb{Q}] \sum_p p^{-(s-(m/2))},$$

which converges for $s > 1 + m/2$.

Using the fact that the product for $L(s, \chi)$ converges absolutely for Re $s > 1 + m/2$ it can be shown the sum also converges in this region and that the two are equal.

The crucial fact which we need about Hecke *L*-functions is given by the following theorem. We will not give the proof which is long and difficult.

Theorem 6. *Let χ be an algebraic Hecke character and $L(s, \chi)$ the corresponding L-function. If $\chi(A)$ is not equal to 0 or 1 for some A, then $L(s, \chi)$ can be analytically continued to an entire function on all of $\mathbb{C}$.*

It should be pointed out that this theorem is true for all number fields and all Hecke *L*-functions, not only those which come from algebraic Hecke characters. Moreover, Hecke established a very important functional equation for his *L*-functions. When χ is an algebraic Hecke character of weight m the functional equation relates $L(s, \chi)$ with $L(m + 1 - s, \bar{\chi})$.

Some authors normalize by defining $\tilde{\chi}(A) = \chi(A)/NA^{m/2}$. Then $L(s, \tilde{\chi}) = \prod_P (1 - \tilde{\chi}(P)NP^{-s})^{-1}$ converges for Re $s > 1$ using the same reasoning as for $L(s, \chi)$ together with the fact that for $(A, M) = (1)$ one has $|\tilde{\chi}(A)| = 1$. We will work directly with the Hecke character χ.

In the next two sections we will show the L-function, $L(E, s)$, for elliptic curves of the form $y^2 = x^3 + D$ and $y^2 = x^3 - Dx$ are Hecke L-functions. In the first case we will construct an algebraic Hecke character on $\mathbb{Q}(\omega)$ and in the second case on $\mathbb{Q}(i)$.

One final comment. In Chapter 14 we defined by means of Gauss sums a function $\Phi(A)$ on the ideals of $\mathbb{Q}(\zeta_m)$ which are prime to m. It can be shown that $\Phi(A)$ extends to an algebraic Hecke character for the modulus (m^2) of weight m. This was first shown by A. Weil in [81]. In a later paper [236] he points out that the case where m is an odd prime goes back to Eisenstein.

§6 $y^2 = x^3 - Dx$, the Global Case

We will now analyze the global zeta function of the elliptic curve E defined by $y^2 = x^3 - Dx$, $D \in \mathbb{Z}$. It is enough to consider the associated L-function $L(E, s)$. Since $\Delta = 2^6 D^3$ in this case we have (see Equation (xi) of Section 2)

$$L(E, s) = \prod_{p \nmid 2D} (1 - a_p p^{-s} + p^{1-2s})^{-1}.$$

The numbers a_p are determined by $N_p = p + 1 - a_p$ and N_p has been determined in Theorem 5, Section 4.

We are going to construct an algebraic Hecke character χ on $\mathbb{Z}[i]$ with respect to the modulus $(8D)$ such that $L(E, s) = L(s, \chi)$.

To construct χ it is enough to specify $\chi(P)$ for prime ideals P in $\mathbb{Z}[i]$. If P divides $2D$ define $\chi(P) = 0$. Suppose P does not divide $2D$. If $NP = p$, then $p \equiv 1$ (4) and $P = (\pi)$ with $\pi \equiv 1$ $(2 + 2i)$. Define $\chi(P) = \overline{(D/\pi)_4} \pi$. If $NP = p^2$, then $p \equiv 3$ (4) and $P = (p)$. Define $\chi(P) = -p$.

Lemma. *Suppose $p \equiv 3$ (4). Then $(D/p)_4 = 1$.*

PROOF. Let P be the prime ideal in $\mathbb{Z}[i]$ generated by p. Then $(D/p)_4 = (D/P)_4 \equiv D^{(NP-1)/4}$ (P). Since $NP = p^2$ we have $(NP - 1)/4 = (p^2 - 1)/4 = (p - 1)(p + 1)/4$. By Fermat's Little Theorem $D^{p-1} \equiv 1$ (p) which implies $(D/p)_4 \equiv 1$ (P) and so $(D/p)_4 = 1$. $\square$

As a consequence of the lemma we can define $\chi(P)$ uniformly for prime ideals P not dividing $2D$. If $P = (\pi)$ where $\pi \equiv 1 (2 + 2i)$ then $\chi(P) = \overline{(D/\pi)_4} \pi$.

Theorem 7. *Let E be the elliptic curve defined by $y^2 = x^3 - Dx$ with $D \in \mathbb{Z}$. The character χ defined above is an algebraic Hecke character of weight 1 for the modulus $(8D)$. Moreover, $L(E, s) = L(s, \chi)$.*

PROOF. Assume to begin with that $p \equiv 3$ (4) and $p \nmid 2D$. By Theorem 5, $N_p = p + 1$ so that $a_p = 0$. Let $P = (p)$. Then $NP = p^2$ and $\chi(P) = -p$. Thus

$$1 - a_p p^{-s} + p^{1-2s} = 1 + p^{1-2s} = 1 - \chi(P)NP^{-s}.$$

Now suppose $p \equiv 1$ (4) and $p \nmid 2D$. Write $p\mathbb{Z}[i] = P\bar{P}$, $P = (\pi)$ and $\pi \equiv 1 (2 + 2i)$. Then, $NP = p$ and, by Theorem 5, $a_p = (D/\pi)_4 \pi + (D/\pi)_4 \bar{\pi}$. Thus

$$1 - a_p p^{-s} + p^{1-2s} = \left(1 - \left(\frac{\overline{D}}{\pi}\right)_4 \pi p^{-s}\right)\left(1 - \left(\frac{D}{\pi}\right)_4 \bar{\pi} p^{-s}\right)$$

$$= (1 - \chi(P)NP^{-s})(1 - \chi(\bar{P})N\bar{P}^{-s}).$$

We have used $\overline{(D/\bar{\pi})}_4 = (D/\pi)_4$. Putting these facts together yields

$$L(E, s) = \prod_P (1 - \chi(P)NP^{-s})^{-1} = \sum_A \chi(A)NA^{-s} = L(s, \chi).$$

It remains to show that χ is an algebraic Hecke character of weight 1 for the modulus $(8D)$.

It is clear for A relatively prime to $2D$ that $\chi(A) = (\overline{D/\alpha})_4 \alpha$ where α is the unique generator of A such that $\alpha \equiv 1 (2 + 2i)$. The theorem will be proved if we can show $\alpha \equiv 1 (8D)$ implies $(D/\alpha)_4 = 1$. To do this we will have to separate the cases $D \equiv 1$ (4), $D \equiv 3$ (4), and D even.

If $D \equiv 1$ (4), then by Proposition 9.9.8 we have $(D/\alpha)_4 = (\alpha/D)_4$. Since $\alpha \equiv 1 (D)$, $(\alpha/D)_4 = 1$ and we are done in this case.

Before going further we need a remark about $(i/\alpha)_4$. If $\alpha \equiv 1$ (8) we claim $(i/\alpha)_4 = 1$. To see this note first that $(i/\alpha)_4 = i^{(N\alpha - 1)/4}$. If $\alpha \equiv a + bi \equiv 1$ (8), then $a - 1 \equiv 0$ (8) and $b \equiv 0$ (8). Thus, $N\alpha - 1 = a^2 + b^2 - 1 = (a^2 - 1) + b^2 \equiv 0 (16)$. This proves the assertion.

Now suppose $D \equiv 3$ (4). Assume $\alpha \equiv 1 (8D)$. Using Proposition 9.9.8 and the above remark we have $(D/\alpha)_4 = (i^2 D/\alpha)_4 = (-D/\alpha)_4 = (\alpha/D)_4 = 1$.

It remains to treat the case where D is even. Write $D = 2^t D_0$ where D_0 is odd. Assume $\alpha \equiv 1 (8D)$. By what has been proved to this point $(D_0/\alpha)_4 = 1$. It thus suffices to show $(2/\alpha)_4 = 1$. For this we need a supplement to the law of biquadratic reciprocity. Namely, assume $\alpha = a + bi$ is primary. Then

$$\left(\frac{1 + i}{\alpha}\right)_4 = i^{(a-b-b^2-1)/4}.$$

A proof of this in the case when α is a prime element has been outlined in the Exercises to Chapter 9. It is not difficult to go from the case of α a prime to that of α primary.

If $\alpha \equiv 1 (8D)$ and D is even then $\alpha \equiv 1 (16)$. It follows that $a - 1 \equiv 0 (16)$ and $b \equiv 0 (16)$ and so $(1 + i/\alpha)_4 = 1$. Thus

$$1 = \left(\frac{1 + i}{\alpha}\right)_4^2 = \left(\frac{2i}{\alpha}\right)_4 = \left(\frac{2}{\alpha}\right)_4.$$

The proof is now complete.   □

§7 $y^2 = x^3 + D$, the Global Case

In order to analyze the L-function of the elliptic curve defined by $y^2 = x^3 + D$, $D \in \mathbb{Z}$, we proceed as in the last section. Since the discriminant in this case is $\Delta = -2^4 3^3 D^2$ we have

$$L(E, s) = \prod_{p \nmid 6D} (1 - a_p p^{-s} + p^{1-2s})^{-1}.$$

The numbers a_p are determined by $N_p = p + 1 - a_p$ and N_p has been determined by Theorem 4, Section 3.

We will construct an algebraic Hecke character χ on $\mathbb{Z}[\omega]$ of weight 1 with respect to the modulus $(12D)$, and show that $L(E, s) = L(s, \chi)$.

Let $P \subset \mathbb{Z}[\omega]$ be a prime ideal. If P divides $6D$ define $\chi(P) = 0$. Assume now that $P \nmid 6D$. If $NP = p$, then $p \equiv 1\,(3)$ and $P = (\pi)$ with π primary, i.e., $\pi \equiv 2\,(3)$. Define $\chi(P) = -\overline{(4D/\pi)}_6 \pi$. If $NP = p^2$, then $p \equiv 2\,(3)$ and $P = (p)$. Define $\chi(P) = -p$.

Lemma 1. *Suppose p is an odd prime and $p \equiv 2\,(3)$. Then $(4D/p)_6 = 1$.*

PROOF. It follows from the hypotheses that $p + 1$ is divisible by 6. We know $(4D)^{p-1} \equiv 1\,(p)$. Raising both sides of this congruence to the $((p + 1)/6)$th power gives the result. $\qquad\square$

Lemma 1 permits us to give a uniform definition of $\chi(P)$. If $P \nmid 6D$ write $P = (\pi)$ with $\pi \equiv 2\,(3)$. Then $\chi(P) = -\overline{(4D/\pi)}_6 \pi$.

Lemma 2. *Suppose $\alpha \in \mathbb{Z}[\omega]$ and $(\alpha, 2D) = (1)$. Define $(D/\alpha)_2$ to be $(D/\alpha)_6^3$. Then $(D/\alpha)_2 = (D/N\alpha)$, where this last symbol is the Jacobi symbol (see Chapter 5, Section 2).*

PROOF. Both $(D/\alpha)_2$ and $(D/N\alpha)$ are multiplicative in α. Thus it is enough to check that they are equal when $\alpha = \pi$, a prime element.

Suppose $\pi = p \neq 2$, a rational prime with $p \equiv 2\,(3)$. Then $Np = p^2$ and so $(D/Np) = (D/p)^2 = 1$. On the other hand

$$\left(\frac{D}{p}\right)_2 = \left(\frac{D}{p}\right)_6^3 \equiv D^{(p^2-1)/2} \equiv (D^{p-1})^{(p+1)/2} \equiv 1\,(p).$$

Thus, $(D/Np) = 1 = (D/p)_2$.

Assume now that π is a complex prime and so $N\pi = p \equiv 1\,(3)$. Then

$$\left(\frac{D}{\pi}\right)_2 = \left(\frac{D}{\pi}\right)_6^3 \equiv D^{(p-1)/2} \equiv \left(\frac{D}{p}\right)(\pi).$$

Since $p = N\pi$, it follows that $(D/\pi)_2 = (D/N\pi)$ and the proof is complete. $\quad\square$

Theorem 8. *Let E be the elliptic curve over $\mathbb{Q}$ defined by $y^2 = x^3 + D$, $D \in \mathbb{Z}$. The character χ defined above is an algebraic Hecke character of weight 1 for the modulus $(12D)$. Moreover, $L(E, s) = L(s, \chi)$.*

PROOF. Assume first that $p \equiv 2\ (3)$ and $p \nmid 6D$. By Theorem 4, $N_p = p + 1$ so that $a_p = 0$. Let $P = (p)$. P is a prime ideal in $\mathbb{Z}[\omega]$ and $\chi(P) = -p$. Thus

$$1 - a_p p^{-s} + p^{1-2s} = 1 + p^{1-2s} = 1 - \chi(P)NP^{-s}.$$

Now suppose $p \equiv 1\ (3)$ and $p \nmid 6D$. Write $p\mathbb{Z}[\omega] = P\bar{P}$ where $P = (\pi)$ with $\pi \equiv 2\ (3)$. Then $NP = p$ and by Theorem 4, $a_p = -\overline{(4D/\pi)_6}\,\pi - (4D/\pi)_6\,\bar{\pi}$. Thus

$$1 - a_p p^{-s} + p^{1-2s} = \left(1 + \overline{\left(\frac{4D}{\pi}\right)}_6 \pi p^{-s}\right)\left(1 + \left(\frac{4D}{\pi}\right)_6 \bar{\pi} p^{-s}\right)$$

$$= (1 - \chi(P)NP^{-s})(1 - \chi(\bar{P})N\bar{P}^{-s}).$$

We have used the fact that $\overline{(4D/\bar{\pi})}_6 = (4D/\pi)_6$. Putting these facts together,

$$L(E, s) = \prod_P (1 - \chi(P)NP^{-s})^{-1} = \sum_A \chi(A)NA^{-s} = L(s, \chi).$$

It remains to show that χ is an algebraic Hecke character of weight 1 for the modulus $12D$.

It is clear that for A relatively prime to $12D$ we have $\chi(A) = \overline{(4D/\alpha)}_6 \alpha$, where α is the unique generator of A such that $\alpha \equiv 1\ (3)$. We will be done if we can show $\alpha \equiv 1\ (12D)$ implies $(4D/\alpha)_6 = 1$.

Since $1 = (4D/\alpha)_6 (4D/\alpha)_6^2 (4D/\alpha)_6^3$ it is enough to show $\alpha \equiv 1\ (12D)$ implies $(4D/\alpha)_3 = 1$ and, by Lemma 2, that $(4D/N\alpha) = 1$. We do both implications in turn.

Assume $3 \nmid D$. Since $\alpha \equiv 1\ (3)$ and α is relatively prime to $4D$, we have by cubic reciprocity (Theorem 1, Chapter 9), $(4D/\alpha)_3 = (-\alpha/4D)_3 = (\alpha/4D)_3 = 1$. The last equality follows from $\alpha \equiv 1\ (4D)$.

If $3 | D$, write $D = 3^t D_0$ with $3 \nmid D_0$. Then $(4D/\alpha)_3 = (3/\alpha)_3^t (4D_0/\alpha)_3 = (3/\alpha)_3^t$. We must show $\alpha \equiv 1\ (12D)$ and $3 | D$ implies $(3/\alpha)_3 = 1$. The hypotheses imply $\alpha \equiv 1\ (9)$. We need the supplements to the law of cubic reciprocity. These can be stated as follows. If $\gamma \in \mathbb{Z}[\omega]$ is primary, then $\gamma = a + b\omega \equiv 2\ (3)$. Write $a = 3m - 1$ and $b = 3n$. Then

$$\left(\frac{\omega}{\gamma}\right)_3 = \omega^{m+n} \quad \text{and} \quad \left(\frac{1 - \omega}{\gamma}\right)_3 = \omega^{2m}.$$

A proof is outlined in the Exercises to Chapter 9. Now, $3 = -\omega^2(1 - \omega)^2$ so $\alpha \equiv 1\ (9)$ implies $(3/\alpha)_3 = 1$ as desired.

It remains to show $\alpha \equiv 1\ (12D)$ implies $(4D/N\alpha) = 1$. Now, $\alpha \equiv 1\ (12D)$ implies $N\alpha \equiv 1\ (4)$ and $N\alpha \equiv 1\ (D)$. If D is odd we have

$$\left(\frac{4D}{N\alpha}\right) = \left(\frac{D}{N\alpha}\right) = \left(\frac{N\alpha}{D}\right) = 1.$$

We have used the law of quadratic reciprocity. If D is even, write $D = 2^t D_0$
with D_0 odd. Then

$$\left(\frac{4D}{N\alpha}\right) = \left(\frac{2}{N\alpha}\right)^t \left(\frac{D_0}{N\alpha}\right) = \left(\frac{2}{N\alpha}\right)^t.$$

The final thing to prove is that D even and $\alpha \equiv 1$ $(12D)$ implies $(2/N\alpha) = 1$.
The hypotheses imply $\alpha \equiv 1$ (8) so that $N\alpha \equiv 1$ (8) and so $(2/N\alpha) = 1$. $\square$

We conclude by observing that Theorems 6, 7, and 8 show that for
elliptic curves E of the form $y^2 = x^3 - Dx$ or $y^2 = x^3 + D$, the L-function,
$L(E, s)$, can be analytically continued to all of $\mathbb{C}$. This proves Hasse's conjecture for these curves!

§8 Final Remarks

In this chapter we have considered special types of elliptic curves defined
over $\mathbb{Q}$ and investigated their local and global zeta functions. It is possible
to generalize these considerations to algebraic varieties defined over algebraic
number fields. We will go a short way along this path by considering curves
defined by a single polynomial with coefficients in an algebraic number
field. After giving the relevant definitions we will investigate the Fermat
curves $x_0^l + x_1^l + x_2^l = 0$, l an odd prime. In this connection we will en-
counter a class of algebraic Hecke characters defined by Jacobi sums.

Let K be an algebraic number field and $\mathcal{O} \subset K$ its ring of integers. Let
$f(x_0, x_1, x_2) \in \mathcal{O}[x_0, x_1, x_2]$ be a nonsingular homogeneous polynomial
of positive degree, and let C denote the algebraic curve defined by the
equation $f(x_0, x_1, x_2) = 0$. If P is a prime ideal of $\mathcal{O}$ we may reduce the
coefficients of f modulo P to obtain a polynomial $\bar{f} \in \mathcal{O}/P[x_0, x_1, x_2]$. It
may be shown that there is a finite set of primes $\mathcal{S}$ such that for $P \notin \mathcal{S}$ the
reduced polynomial $\bar{f}$ is nonsingular. Let C_P be the curve defined over
$\mathcal{O}/P$ by the equation $\bar{f}(x_0, x_1, x_2) = 0$. In Section 1, Chapter 11, we showed
how to attach a zeta function to C_P. Namely,

$$Z(C_P, u) = \exp \sum_{m=1}^{\infty} \frac{N_m(P)u^m}{m},$$

where $N_m(P)$ is the number of (projective) solutions to $\bar{f}(x_0, x_1, x_2) = 0$
in the extension of $\mathcal{O}/P$ of degree m. Recall that this extension is unique up
to isomorphism so that $N_m(P)$ is well defined.

Using the Riemann–Roch theorem one may show there is a polynomial
$H(C_P, u) \in \mathbb{Z}[u]$ with constant term equal to one such that

$$Z(C_P, u) = \frac{H(C_P, u)}{(1 - u)(1 - NPu)}. \tag{i}$$

If $P \in \mathscr{S}$ it is not easy to decide on the appropriate definition. For our purposes we simply define $H(C_P, u) = 1$ if $P \in \mathscr{S}$.

The local zeta function of C at P is obtained by setting $u = NP^{-s}$ in Equation (i). Namely,

$$\zeta(C_P, s) = \frac{H(C_P, NP^{-s})}{(1 - NP^{-s})(1 - NP^{1-s})}. \tag{ii}$$

This generalizes Equation (ix) of Section 2.

The global zeta function of C is defined by

$$\zeta(C, s) = \prod_P \zeta(C_P, s) \tag{iii}$$

The product is over all nonzero prime ideals in $\mathcal{O}$.

The product $\prod_P (1 - NP^{-s})^{-1}$ is called the zeta function of K and is denoted by $\zeta_K(s)$. This function was first investigated by Dedekind. It converges for $\operatorname{Re}(s) > 1$ and it was shown by Hecke that it can be continued to a meromorphic function on all of $\mathbb{C}$ and satisfies a functional equation. The only pole is a simple pole at $s = 1$.

Define $L(C_P, s) = H(C_P, NP^{-s})^{-1}$ and $L(C, s) = \prod_P L(C_P, s)$. Then from Equations (ii) and (iii)

$$\zeta(C, s) = \frac{\zeta_K(s)\zeta_K(s - 1)}{L(C, s)}. \tag{iv}$$

It follows that if we wish to investigate whether $\zeta(C, s)$ can be analytically continued to all of $\mathbb{C}$ it is enough to concentrate on the function $L(C, s)$.

Fix an odd prime l. From now on we will consider the curve C defined by $x_0^l + x_1^l + x_2^l = 0$. It will be convenient to consider C as being defined over $K = \mathbb{Q}(\zeta_l)$ rather than over $\mathbb{Q}$. We set $\mathcal{O} = \mathbb{Z}[\zeta_l]$, the ring of integers in K.

It is easy to see that the exceptional set $\mathscr{S}$ consists, in this case, of the single prime ideal $\mathscr{L} = (1 - \zeta_l)$. If $P \neq \mathscr{L}$ we know l divides $NP - 1$. It is this fact which makes K a more convenient field of definition.

Assume $P \neq \mathscr{L}$ and apply Theorem 2 of Section 3, Chapter 11 to the curve C_P over $\mathcal{O}/P$. We find

$$H(C_P, u) = \prod_{\chi_0, \chi_1, \chi_2} (1 + NP^{-1} g(\chi_0) g(\chi_1) g(\chi_2) u), \tag{v}$$

where the product is over 3-tuples of characters of $(\mathcal{O}/P)^*$ of order l such that $\chi_0 \chi_1 \chi_2 = \varepsilon$, the trivial character.

Since $g(\chi_1) g(\chi_2) = J(\chi_1, \chi_2) g(\chi_1 \chi_2)$ and $\chi_0 \chi_1 \chi_2 = \varepsilon$ we find that $g(\chi_0) g(\chi_1) g(\chi_2) = \chi_1 \chi_2(-1) NP \, J(\chi_1, \chi_2)$. Since $-1 = (-1)^l$, $\chi_1 \chi_2(-1) = 1$. Substituting this information into Equation (v) we find

$$H(C_P, u) = \prod_{\chi_1, \chi_2} (1 + J(\chi_1, \chi_2) u), \tag{vi}$$

where the product is over pairs of characters of order l such that $\chi_1 \chi_2 \neq \varepsilon$.

Let $\chi_P(\alpha) = (\alpha/P)_l^{-1}$ for $\alpha \in \mathcal{O}$. This is the inverse of the lth power residue symbol (see Chapter 14, Section 2). If $1 \leq a, b \leq l-1$ and $a+b \neq l$ define $\lambda_{a,b}(P) = -J(\chi_P^a, \chi_P^b)$. With this notation we have

$$H(C_P, u) = \prod_{\substack{a,b=1 \\ a+b \neq l}}^{l-1} (1 - \lambda_{a,b}(P)u) \tag{vii}$$

and so

$$L(C_P, s) = \prod_{\substack{a,b=1 \\ a+b \neq l}}^{l-1} (1 - \lambda_{a,b}(P)NP^{-s})^{-1}. \tag{viii}$$

Let us define $\lambda_{a,b}(\mathcal{L}) = 0$ and $L(s, \lambda_{a,b}) = \prod_P (1 - \lambda_{a,b}(P)NP^{-s})^{-1}$. We have shown

$$L(C, s) = \prod_{\substack{a,b=1 \\ a+b \neq l}}^{l-1} L(s, \lambda_{a,b}).$$

At this point it is certainly reasonable to hope that $\lambda_{a,b}$ extends to an algebraic Hecke character. This is indeed the case! $\lambda_{a,b}$ is an algebraic Hecke character of weight 1 for the modulus (l^2). The corresponding group ring element is

$$\sum_{t=1}^{l-1} \left(\left\langle \frac{at}{l} \right\rangle + \left\langle \frac{bt}{l} \right\rangle - \left\langle \frac{(a+b)t}{l} \right\rangle \right) \sigma_t^{-1}.$$

The proof of these facts will be outlined in the Exercises. Here we simply remark that since $L(C, s)$ is a product of Hecke L-functions, the fundamental result of Hecke, Theorem 6, shows that $L(C, s)$ can be analytically continued to an entire function on all of $\mathbb{C}$ and, moreover, satisfies a functional equation connecting $L(C, s)$ with $L(C, 2-s)$.

Notes

The notion of local and global zeta functions attached to an algebraic curve defined over an algebraic number field goes back to Hasse. In the late 1930's, Hasse proposed to one of his students the problem of showing that the global zeta function can be analytically continued to all of $\mathbb{C}$ and satisfies a functional equation. Weil was asked by G. DeRham for his opinion of this problem. At the time Weil could see no reason why the global zeta function should have the properties ascribed to it by Hasse. Moreover, he thought the problem too difficult for a beginner ("... trop difficile pour un débutant..."). For this and other enlightening comments see Weil's *Complete Works* [241], Vol. II, pp. 529–530.

In spite of his initial pessimism Weil later gained confidence in Hasse's conjecture through working out special cases, initially $y^2 = x^4 + 1$ (this is equivalent to the curve $y^2 = x^3 - \frac{1}{4}x$). His work along these lines culminated in his famous paper "Jacobi Sums as Grössencharaktere" [81]. In this paper Weil treats curves of the form $y^e = \gamma x^f + \delta$ where $2 \le e \le f$ and $\gamma\delta \ne 0$. At the end of the paper he notes the cases $e = 2$ and $f = 3$ or 4 correspond to elliptic curves with complex multiplication. These are, in essence, the curves we have treated in this chapter. He goes on to say "... it would be of considerable interest to investigate more general elliptic curves with complex multiplication from the same point of view." This suggestion was taken up by M. Deuring with complete success.

In passing it is worth noting that what we have called Hecke characters were called by Hecke "Grössencharktere." In the older literature algebraic Hecke characters are referred to as characters of type A_0.

In his 1954 paper "Abstract versus Classical Algebraic Geometry" [241] (Vol. II, pp. 550–558) Weil defines local and global zeta functions for a nonsingular algebraic variety defined over an algebraic number field. He raises the question of whether these functions can be analytically continued to all of $\mathbb{C}$ and satisfy a functional equation of an appropriate type. Having verified that these properties hold in many examples, he writes, "It is tempting to surmise that this is always so, but I have little hope that a general proof may soon be found." This conjecture is now known as the Hasse-Weil conjecture. Although there has been much progress due to Weil himself, Taniyama, Shimura, and others, the Hasse–Weil conjecture remains very much an open problem.

For a comprehensive survey of the various zeta and L-functions that have been defined and studied since the nineteenth century see the article on zeta functions in the *Encyclopedic Dictionary of Mathematics*, Vol. II, Section 436 (M.I.T. Press, 1977).

EXERCISES

1. Let p be prime $p \equiv 2$ (3) and consider the curve E_p defined over F_p by $y^2 = x^3 + a$, $a \in F_p$. Show that $N(y^2 = x^3 + a) = p + 1$ (projective points).

2. Let $K \subset \mathbb{C}$ be a *CM* field which is Galois and let j be the restriction to K of complex conjugation. Show that the fixed field of j is the unique totally real subfield of K of degree $\frac{1}{2}[K : \mathbb{Q}]$ and that $j\sigma = \sigma j$ for all σ in the Galois group of K over $\mathbb{Q}$.

3. Let A, $B \in \mathbb{Z}$, $\Delta = 16(4A^3 - 27B^2) \ne 0$, and E be the elliptic curve defined by $y^2 = x^3 - Ax - B$. If p is prime, $p \nmid \Delta$ let N_p denote the number of projective points on the reduced curve E_p over F_p. The prime p is said to be anomolous for E if $\sum_{x=0}^{p-1}((x^3 - Ax - B)/p) \equiv -1$ (p). Put $f_p = -\sum_{x=0}^{p-1}((x^3 - Ax - B)/p)$. Show
   (a) p is anomolous for E iff $p|N_p$.
   (b) Assume the Riemann hypothesis for E_p (see Chapter 11, Section 3). If $p > 5$ then $f_p = 1 \Leftrightarrow p$ is anomolous for E.
   (c) Let $B = 0$, $p \equiv 1$ (4), $p \nmid \Delta$. Then f_p is even. If $p > 5$ p is not anomolous.
   (d) If $B = 0$ then 5 is anomolous $\Leftrightarrow A \equiv 2$ (5).
   This exercise is taken from Olson [202].

4. Consider the underlying abelian group of rational points on the elliptic curve E, defined by $y^2 = x^3 + c$. If $p \nmid 6c$ then it is known that the torsion subgroup (i.e., the points of finite order) of E is isomorphic to a subgroup of the torsion subgroup of the reduced curve modulo p. Use Exercise 1 and Dirichlet's theorem on the density of primes in an arithmetic progression to show that the torsion subgroup of the above curve can have only 1, 2, 3, 4 or 6 elements. This exercise is taken from Olson [201].

In the following exercises the notation is as in Chapter 14, Section 3. Furthermore $G = \mathbb{Z}/m\mathbb{Z} \oplus \mathbb{Z}/m\mathbb{Z}$, and T denotes the subset of G consisting of (a, b) with $a \neq 0$, $b \neq 0$, $a + b \neq 0$.

5. Generalize Exercise 13, Chapter 6, as follows. If $x = (x_1, x_2)$, $y = (y_1, y_2) \in G$ define $\langle x, y \rangle = x_1 y_1 + x_2 y_2$. For a $\mathbb{C}$ valued map f defined on G define $\hat{f}(x) = (1/m^2) \sum_y f(y) \zeta_m^{-\langle x, y \rangle}$, $y \in G$. Show
   (a) $f(x) = \sum_y \hat{f}(y) \zeta_m^{\langle x, y \rangle}$.
   (b) $\sum_x |\hat{f}(x)|^2 = (1/m^2) \sum_x |f(x)|^2$.
   (c) Assume f maps G to the unit circle and $\hat{f}$ is integer valued. Show that $f(x, y) = f(0, 0) \zeta_m^{ax+by}$ for a suitable (a, b). Conclude that if $f(0, 0) = f(1, 0) = f(0, 1) = 1$ then f is identically 1.

6. For $(a, b) \in G$ define, for $P \subset D_m$, P a prime ideal, $m \notin P$, $\lambda_{a, b}(P)$ as follows:
   (i) If $(a, b) \in T$, $\lambda_{a, b}(P) = -J(\chi_P^a, \chi_P^b)$.
   (ii) If $(a, b) \neq (0, 0)$, $a + b = 0$ put $\lambda_{a, b}(P) = +\chi_P^a(-1)$.
   (iii) If $a + b \neq 0$ and a or b is 0 put $\lambda_{a, b}(P) = 1$.
   (iv) $\lambda_{0, 0}(P) = -(N(P) - 2)$.
   Show that if one modifies the convention in Chapter 8 concerning the trivial character by putting $\varepsilon(0) = 0$ then $\lambda_{a, b}(P) = -J(\chi_P^a, \chi_P^b)$ for all $(a, b) \in G$.

7. For $(c, d) \in G$ define $N_{c, d}$ as the number of solution (x, y), $x, y \in \mathbb{F}_q$ $(q = N(P))$ to the equations $x + y = 1$, $\chi_P(y) = \zeta_m^d$, and $\chi_P(x) = \zeta_m^c$. Show that $J(\chi_P^a, \chi_P^b) = \sum_{c, d} N_{c, d} \zeta_m^{ac + bd}$. Conclude that $-N_{c, d} = \hat{\lambda}_{c, d}(P)$.

8. Extend $\lambda_{a, b}(P)$ to all ideals $\mathfrak{A} \subset D_m$, $m \notin \mathfrak{A}$, by multiplicativity. Show
   (a) $\lambda_{0, 0}(\mathfrak{A}) N(\mathfrak{A}) \equiv 1$ (m^2).
   (b) If $\alpha \in D_m$, $\alpha \neq 0$, $(a, b) \in T$ then $\lambda_{a, b}((\alpha)) = u(a, b) \alpha^{\gamma(a, b)}$, where $u(a, b) \in D_m$, $|u(a, b)| = 1$ and

$$\gamma(a, b) = \sum_{(t, m) = 1} \left(\left\langle \frac{at}{m} \right\rangle + \left\langle \frac{bt}{m} \right\rangle - \left\langle \frac{(a + b)t}{m} \right\rangle \right) \sigma_t^{-1}.$$

   (c) $\hat{\lambda}_{a, b}(\mathfrak{A}) \in \mathbb{Z}$.

9. Assume $\alpha \equiv 1$ (m^2). Define $u(a, b)$, for fixed α, by Exercise 8 if $(a, b) \in T$. If $(a, b) \notin T$, $(a, b) \neq (0, 0)$ put $u(a, b) = \lambda_{a, b}((\alpha))$, and $u(0, 0) = 1$. Show
   (a) $u(a, b) \equiv \lambda_{a, b}((\alpha))$ (m^2) for all $(a, b) \in G$.
   (b) $\hat{u}(a, b) \in D_m$, all $(a, b) \in G$.
   (c) $\hat{u}(a, b) \in \mathbb{Z}$, all $(a, b) \in G$.
   (d) Apply (c) of Exercise 5 to show that $u(a, b) = 1$ for all $(a, b) \in G$, and conclude that $\lambda_{a, b}$ is an algebraic Hecke character for D_m with a defining modulus m^2.

Exercises 5-9 are adapted from Lang [171], Chapter 1, Section 4.

10. Give an example of a nonabelian CM field.

Chapter 19

The Mordell–Weil Theorem

In this chapter we prove the celebrated theorem of Mordell–Weil for elliptic curves defined over the field of rational numbers. Our treatment is elementary in the sense that no sophisticated results from algebraic geometry are assumed. It is our desire to present a self-contained treatment of this important result. The significance and implications of this theorem for contemporary research in diophantine geometry are far-reaching. In the following chapter a summary without proofs of these developments to the present time is sketched. We hope that these two chapters will inspire the interested student to continue this study by consulting the more comprehensive texts on the arithmetic of elliptic curves listed in the bibliography to this chapter.

Our proof of the Mordell–Weil theorem is based on Weil's 1929 paper [W4] "Sur un théorème de Mordell" and an interesting simplification of the "weak Mordell–Weil" theorem appearing in J.W.S. Cassels's paper entitled "The Mordell–Weil Group of Curves of Genus 2" [Ca2].

§1 The Addition Law and Several Identities

Let k be an arbitrary field of characteristic zero with a fixed algebraic closure $\bar{k}$. Consider an elliptic curve E defined over k with an affine equation in Weierstrass form

$$y^2 = x^3 + ax + b = f(x). \tag{1}$$

Here a and b are constants in k subject only to the condition that the curve E is nonsingular. It is a simple exercise to see that this is equivalent to the condition that $f(x)$ have three distinct roots in $\bar{k}$. Denote these roots by θ_1, θ_2, θ_3 so that we have

$$y^2 = f(x) = (x - \theta_1)(x - \theta_2)(x - \theta_3). \tag{2}$$

For completeness we include the proof of the following well-known result from classical algebra.

Lemma 1. $[(\theta_1 - \theta_2)(\theta_2 - \theta_3)(\theta_1 - \theta_3)]^2 = -(4a^3 + 27b^2)$.

PROOF. By substituting $x = \theta_i$ in the formal derivative of $f(x)$ one obtains $3\theta_i^2 + a = (\theta_i - \theta_j)(\theta_i - \theta_k)$, i, j, k distinct. Multiplication now shows that the negative of the left-hand side of the statement in the lemma is

$$27(\theta_1\theta_2\theta_3)^2 + 9a(\theta_1^2\theta_2^2 + \theta_1^2\theta_3^2 + \theta_2^2\theta_3^2) + 3a^2(\theta_1^2 + \theta_2^2 + \theta_3^2) + a^3.$$

But $\theta_1 + \theta_2 + \theta_3 = 0$, $\theta_1\theta_2 + \theta_1\theta_3 + \theta_2\theta_3 = a$, $\theta_1\theta_2\theta_3 = -b$, and several applications of the identity $(x + y + z)^2 = x^2 + y^2 + z^2 + 2(xy + xz + yz)$ completes the proof. $\qquad\square$

Recall from Chapter 18 that the nonzero quantity $-16(4a^3 + 27b^2)$ is called the discriminant Δ of the curve E. We will see in Chapter 20 that the prime divisors of Δ enter into the precise formulation of the conjectures of Birch and Swinnerton-Dyer.

We view E as a projective curve whose points are the affine points (x, y) satisfying (1) along with a single point on the line at infinity denoted by ∞. As mentioned in Chapter 18 the "chord and tangent" process defines a group structure on E. We now make this definition precise and derive several identities that are needed in what follows.

The identity element is taken to be the point at infinity ∞. The group structure is defined by the requirement that three points P, Q, R on E are collinear if and only if $P + Q + R = \infty$. If $P = (\alpha, \beta)$ is a point of E, then $Q = (\alpha, -\beta)$ is also on E and P, Q, ∞ are collinear. Thus $P + Q = \infty$ and we see that $-P = (\alpha, -\beta)$. The points of order 2 are therefore $(\theta_i, 0)$, $i = 1, 2, 3$. It is important to realize that these points need not be rational over k. Now let $P = (x_1, y_1)$, $Q = (x_2, y_2)$ be two affine points on E with $x_1 \neq x_2$. Intersecting E with the line through P and Q shows that the polynomial in x

$$x^3 + ax + b - \left[y_1 + \left(\frac{y_2 - y_1}{x_2 - x_1}\right)(x - x_1)\right]^2$$

has roots x_1 and x_2. Hence, if x_3 is the third root,

$$x_1 + x_2 + x_3 = \left(\frac{y_2 - y_1}{x_2 - x_1}\right)^2$$

so that

$$x_3 = -x_1 - x_2 + \left(\frac{y_2 - y_1}{x_2 - x_1}\right)^2. \tag{3}$$

If (x_3, y_3) is the third point of intersection of the line between P and Q with E then

$$y_3 = y_1 + \left(\frac{y_2 - y_1}{x_2 - x_1}\right)(x_3 - x_1),$$

and it follows that if P and Q are rational over k, then so is (x_3, y_3). Now by definition of the group law one has

$$P + Q + (x_3, y_3) = \infty$$

or

$$P + Q = -(x_3, y_3) = (x_3, -y_3). \tag{4}$$

Finally, if $P = (x_1, y_1)$, $y_1 \neq 0$, we must derive a formula for $2P$. The tangent line to E at (x_1, y_1) has equation

$$y = y_1 + \frac{3x_1^2 + a}{2y_1}(x - x_1).$$

In other words, one calculates easily that the polynomial

$$f(x) - \left[y_1 + \frac{3x_1^2 + a}{2y_1}(x - x_1)\right]^2$$

has x_1 as a double root. Again, as the coefficient of x^2 in $f(x)$ is 0, one has

$$2x_1 + x_3 = \left(\frac{3x_1^2 + a}{2y_1}\right)^2,$$

where x_3 is the third root. Hence,

$$x_3 = -2x_1 + \left(\frac{3x_1^2 + a}{2y_1}\right)^2.$$

If (x_3, y_3) is the third point of intersection with E, one has

$$y_3 = y_1 + \frac{3x_1^2 + a}{2y_1}(x_3 - x_1).$$

Thus,

$$2P + (x_3, y_3) = \infty,$$

and we see that

$$2P = -(x_3, y_3) = (x_3, -y_3). \tag{5}$$

As mentioned in Chapter 18, the proof of the associative law is not obvious. There we referred the reader to Fulton [135] for a geometric approach. Since the first edition of this text was published, several new texts have appeared. We recommend, in particular, J.H. Silverman [Si] and D. Husemöller [Hus] for a thorough treatment of this matter. We do mention, however, that if one uses the parameterization of E by the Weierstrass $\wp$-function and its derivative (at least when $k \subset \mathbb{C}$), then one sees that the group law is precisely the "addition formula" and "duplication formula" from the classical theory of elliptic functions. Thus, the group law on E is the "transport" to E of the natural additive structure on the complex torus whose lattice defines the Weierstrass functions. Thus, associativity is "obvious." It is, of course, a nontrivial fact that every elliptic curve arises in such a fashion. With our purely algebraic definition of the group law, the proof of associativity becomes a straightforward, if somewhat tedious, exercise in algebra.

We see that the set of points on E that are rational over k form a group, denoted by $E(k)$. This group is clearly abelian, as follows from the geometric definition of addition law and is visible again in (3). We may now state the main theorem of this chapter. Let $k = \mathbb{Q}$, the field of rational numbers.

Theorem. *$E(\mathbb{Q})$ is a finitely generated group.*

The addition formulas were obtained by using the fact that the sum of the roots of a polynomial is the negative of the trace term. Beginning with (2) and using the corresponding observation for the product of the roots we obtain relations that will be needed later. Replace x by $t + \theta$ in (2), where θ is one of the roots $\theta_1, \theta_2, \theta_3$. Let θ', θ'' be the other two roots.

If $P = (x_1, y_1)$, $Q = (x_2, y_2)$ are points on E, with $x_1 \neq x_2$, then, as before, the polynomial

$$\left[y_1 + \left(\frac{y_2 - y_1}{x_2 - x_1} \right)(t + \theta - x_1) \right]^2 - t(t + \theta - \theta')(t + \theta - \theta'')$$

has roots $x_1 - \theta$, $x_2 - \theta$, $x_3 - \theta$. Thus,

$$(x_1 - \theta)(x_2 - \theta)(x_3 - \theta) = \left[y_1 + (\theta - x_1)\left(\frac{y_2 - y_1}{x_2 - x_1} \right) \right]^2$$

$$= \left[\frac{y_1(x_2 - \theta) - y_2(x_1 - \theta)}{x_2 - x_1} \right]^2. \tag{6}$$

Similarly, if $P = (x_1, y_1)$, $y_1 \neq 0$ is on E, then

$$\left[y_1 + \frac{f'(x_1)}{2y_1}(x - x_1) \right]^2 - (x - \theta_1)(x - \theta_2)(x - \theta_3)$$

has x_1 as a double root. Let, as usual, x_3 be the third root. By putting $x - \theta = t$ one sees immediately that

$$(x_1 - \theta)^2 (x_3 - \theta) = \left[y_1 + \frac{f'(x_1)(\theta - x_1)}{2y_1} \right]^2$$

or

$$x_3 - \theta = \frac{1}{(x_1 - \theta)^2} \left[\frac{2y_1^2 + (3x_1^2 + a)(\theta - x_1)}{2y_1} \right]^2 \tag{7}$$

Now

$$y_1^2 = x_1^3 + ax_1 + b - \theta^3 - a\theta - b$$
$$= (x_1 - \theta)(x_1^2 + x_1\theta + \theta^2 + a).$$

Substituting into (7) gives the relation

$$x_3 - \theta = \left[\frac{-x_1^2 + 2\theta^2 + 2\theta x_1 + a}{2y_1} \right]^2. \tag{8}$$

We require one final relation. From (3) one sees that

$$x_3 = \frac{-(x_1^2 - x_2^2)(x_1 - x_2) + y_2^2 - 2y_1y_2 + y_1^2}{(x_2 - x_1)^2}.$$

Using $y_1^2 = x_1^3 + ax_1 + b$ and $y_2^2 = x_2^3 + ax_2 + b$ we obtain, after a simple calculation,

$$x_3 = \frac{(x_1x_2 + a)(x_1 + x_2) + 2b - 2y_1y_2}{(x_2 - x_1)^2}. \tag{9}$$

In formula (9) we are assuming, of course, that $x_1 \neq x_2$.

This completes the list of identities that will be needed in the proof of the Mordell–Weil theorem.

§2 The Group $E/2E$

In this section k remains an arbitrary field of characteristic zero. Using the notation of §1 consider the residue class ring $k[x]/(f(x)) = k[\xi]$, ξ being the class of $x \bmod f(x)$. This ring is a k-algebra of dimension 3 over k. If $f(x) = f_1(x)f_2(x) \ldots f_n(x)$ is the decomposition of $f(x)$ as a product of distinct irreducibles, then

$$k[x]/(f(x)) \cong \bigoplus_{i=1}^{n} k[x]/(f_i(x)) \tag{10}$$

by the polynomial version of the Chinese remainder theorem. Here n may take the values 1, 2, or 3 according as $f(x)$ is, respectively, irreducible, the product of a linear and an irreducible quadratic, or the product of three distinct linear factors. If one of the factors is linear, say $f_1(x) = x - \alpha$, then, of course, $k[x]/(f_1(x))$ is naturally isomorphic to k by the map which sends the class of $g(x)$ to $g(\alpha)$. The linear factors $x - \alpha$ correspond to the k-rational points $(\alpha, 0)$ of order 2 on the elliptic curve E defined by $y^2 =$

$f(x)$. Denote the group of units of the ring $k[\xi]$ by U. These elements are the residues modulo $f(x)$ of polynomials $h(x)$ that are coprime to $f(x)$. Furthermore, this group of units is isomorphic to the direct product of the unit groups of the factors in the preceding decomposition.

The purpose of this section is to construct a homomorphism ϕ from E to the group U/U^2 with kernel precisely $2E$. In the context of the Weierstrass parameterization of E by elliptic functions this map was defined by Weil in his 1929 paper [W4]. We follow an algebraic adaptation and simplification of Weil's construction due to Cassels [Ca2].

The mapping ϕ is defined as follows: First, $\phi(\infty) = 1$ where 1 denotes the identity of the group U/U^2. Next, if $P = (\alpha, \beta)$ is a point on E distinct from the points of order 2, i.e., $\beta \neq 0$, then since $\alpha - x$ is prime to $f(x)$, $\alpha - \xi$ is in U and $\phi(P)$ is defined to be the image of $\alpha - \xi$ in U/U^2. It remains to define $\phi(P)$ when $P = (\alpha, 0)$ is a point of order 2 on E. Write $f(x) = (x - \alpha)g(x)$. Then

$$k[\xi] \cong k[x]/(x - \alpha) \oplus k[x]/(g(x)) \tag{11}$$

where $g(x)$ is a polynomial, not necessarily irreducible, of degree 2. Identifying the first factor in the preceding decomposition with k, as mentioned earlier, we see that the element $(f'(\alpha), (\alpha - x)\bmod g(x))$ is a unit corresponding in $k[\xi]$ to a unique element, say $h(\xi) \bmod U^2$, in U/U^2. The reason for the choice $f'(\alpha)$ in the component where $\alpha - \xi$ ceases to be a unit is made partially clear by the proof the map ϕ so defined is indeed a homomorphism. For an explanation using a little more algebraic geometry see Cassels's original paper ([Ca2], §1.3). See also §2 of Brumer and Kramer [Br-Kr].

Lemma 2. ϕ *is a homomorphism.*

PROOF. If $P = (\alpha, \beta)$, since the definition of ϕ is independent of β and $-P = (\alpha, -\beta)$, then $\phi(P) = \phi(-P)$. Now if ρ is in U/U^2, then $\rho^2 = 1$, so that $\phi(P + Q) = \phi(P)\phi(Q)$ is equivalent to $\phi(P + Q)\phi(P)\phi(Q) = \phi(P + Q)\phi(-P)\phi(-Q) = 1$. Thus, to establish that ϕ is a homomorphism we must show that if $A + B + C = \infty$ on E, then

$$\phi(A)\phi(B)\phi(C) = 1 \tag{12}$$

in U/U^2. The condition $A + B + C = \infty$ is, by §1, simply the condition that A, B, C, are colinear. Put $A = (x_1, y_1)$, $B = (x_2, y_2)$, $C = (x_3, y_3)$, and assume that A, B, C are distinct points. If $x_1 = x_2$, then the points are A, $-A$ and infinity. The result follows noting that $\phi(-A) = \phi(A)$. Let $x_1 \neq x_2$ and assume none of the points has order 2. The collinearity of A, B, C simply amounts to the existence of a linear form $cx + d$ such that

$$f(x) - (cx + d)^2 = (x - x_1)(x - x_2)(x - x_3) \tag{13}$$

and the result follows by reduction mod $f(x)$. Next suppose that precisely one of the points, say, $A = (\alpha, 0)$, has order 2. We check (12) in each of the two summands of (11). The result holds in the second factor by reducing (13) mod $g(x)$. Furthermore, differentiating (13) and putting $x = \alpha$ shows $f'(\alpha) = (\alpha - x_2)(\alpha - x_3)$ so that, by definition, the first component of (12) is $(f'(\alpha))^2$. The final case to check is $A = (\theta_1, 0)$, $B = (\theta_2, 0)$, $C = (\theta_3, 0)$, but again differentiating (13) and putting $x = \theta_i$ one sees that the three components of (12) in the decomposition of $k[\xi]$ as the direct sum of three copies of k corresponding to the roots of $f(x)$ are the squares $f'(\theta_1)^2, f'(\theta_2)^2, f'(\theta_3)^2$. □

We mention that one can also use the explicit formulas (6) and (7) of §1 applied to the various factors in the decomposition (10). Once again we refer to Cassels [Ca2] for a proof of this statement that avoids the examination of special cases. The last result of this section is the proof that the kernel of ϕ is $2E$.

Lemma 3. ker $\phi = 2E$.

PROOF. Since $\phi(2P) = \phi(P)^2 = 1$ we see that $2E \subseteq \ker \phi$. Thus, consider a point P, which we may assume different from ∞, such that $\phi(P) = 1$. Write $P = (\alpha, \beta)$, $\alpha, \beta \in k$. Then $\alpha - \xi$ is a square in $k[\xi]$. Note that this holds even when $2P = \infty$, for then $\alpha - \xi$ is 0 in one of the components of (10). Thus, we may write

$$\alpha - \xi = (\alpha_1 \xi^2 + \alpha_2 \xi + \alpha_3)^2, \tag{14}$$

where $\alpha_1, \alpha_2, \alpha_3 \in k$. It is easy to see, using $\xi^3 = -a\xi - b$, that one can write

$$e_1 \xi + f_1 = (\alpha_1 \xi^2 + \alpha_2 \xi + \alpha_3)(-\alpha_1 \xi + \alpha_2), \tag{15}$$

where $e_1, f_1 \in k$. Now $\alpha_1 \neq 0$, for otherwise, by linear independence of 1, ξ, ξ^2, (14) would give a contradiction. Thus, squaring (15), substituting (14), and dividing by α_1^2 gives the relation

$$(e\xi + e')^2 = (\alpha - \xi)(h - \xi)^2 \tag{16}$$

for $\alpha, e, e', h \in k$. This implies that $(ex + e')^2 - (\alpha - x)(h - x)^2$ is a multiple of $f(x)$, and, since the latter polynomial is a monic cubic, we see that

$$f(x) = (ex + e')^2 - (\alpha - x)(h - x)^2. \tag{17}$$

But geometrically this says that the line $y = ex + e'$ intersects E at (α, β) or $(\alpha, -\beta)$ and (h, t) for suitable t with (h, t) counted twice. Thus, by definition of the group structure on E we have

$$(\alpha, \pm\beta) + 2(h, t) = 0$$

for a suitable choice of the sign of β. This implies that

$$P = (\alpha, \beta) = 2Q$$

for $Q = (h, \pm t)$, again adjusting the sign of t. We have thus shown ker ϕ $\subseteq 2E$.   □

§3 The Weak Dirichlet Unit Theorem

If the ground field k is an algebraic number field, then the existence of an injection of $E/2E$ into U/U^2 can be used to show that $E/2E$ is a finite group. This result is often referred to as the Weak Mordell–Weil Theorem. To derive this result one needs, in addition to the finiteness of the class number of an algebraic number field, the Dirichlet unit theorem. The fact that the group of ideal classes is finite is proved in Theorem 1 of Chapter 12. The structure of the group of units in the ring of integers of an algebraic number field is stated without proof on page 192, Chapter 13. However, the full statement of the unit theorem is unnecessary if one is interested only in the finite generation of $E(\mathbb{Q})$. What is needed is only the fact that the group of units of the ring of integers of an algebraic number field is finitely generated, and this follows without difficulty, via the standard "logarithmic embedding," from the fact that a discrete subgroup of the additive group $\mathbb{R}^n$ is a lattice. In view of our desire to keep the proof of our main result self-contained, we include a proof of this weaker form of the Dirichlet unit theorem. Those who are willing to accept this fact can proceed directly to the following section where the proof of Mordell–Weil is concluded.

Let K be an algebraic number field of degree n. We consider as in Chapter 12 the n distinct isomorphisms from K to $\mathbb{C}$, but we order them in the following way: Let $\sigma_1, \ldots, \sigma_s$ be the isomorphisms such that $\sigma_i(K) \subset \mathbb{R}$. The remaining isomorphisms occur in distinct conjugate pairs. There are t such pairs, and we choose one from each pair, denoting these elements by $\sigma_{s+1}, \ldots, \sigma_{s+t}$. The set all n isomorphisms is then $\{\sigma_1, \ldots, \sigma_s, \sigma_{s+1}, \overline{\sigma}_{s+1}, \ldots, \sigma_{s+t}, \overline{\sigma}_{s+t}\}$, which we also list as $\{\tau_1, \ldots, \tau_n\}$ when a uniform notation is convenient. Let $V = \mathbb{R}^s \times \mathbb{C}^t$, and define a mapping ϕ from K to V by $\phi(\alpha) = (\sigma_1(\alpha), \ldots, \sigma_{s+t}(\alpha))$. Fix an integral basis $\alpha_1 \cdots \alpha_n$ of $K/\mathbb{Q}$. Then by Proposition 12.1.3, $(\det(\tau_j(\alpha_i)))^2$ is not zero, being the discriminant of K. It is then a simple exercise to show that the vectors $\phi(\alpha_1), \ldots, \phi(\alpha_n)$ are $\mathbb{R}$-linearly independent in V. Now a lattice in V is, by definition, an additive subgroup of V, which may be written in the form $\mathbb{Z}v_1 + \cdots + \mathbb{Z}v_l$, where $v_1, \ldots, v_l$ are $\mathbb{R}$-linearly independent elements of V. If $\mathbb{D}$ is the ring of integers in K, we have shown the following:

Lemma 1. $\phi(\mathbb{D})$ *is a lattice.*

By a discrete subset of $\mathbb{R}^n$ is meant any subset A for which $A \cap T$ is finite whenever T is compact. It is, of course, sufficient to take T a closed ball of finite radius. It is a simple exercise to show that a lattice is discrete. We need the following converse. Let W be any finite dimensional vector space over $\mathbb{R}$.

Lemma 2. *A discrete additive subgroup A of W is a lattice.*

PROOF. Let $v_1, \ldots, v_m$ be a maximal set of $\mathbb{R}$-independent elements in A. Then any element a of A may be written in the form $a = r_1 v_1 + \cdots + r_m v_m$ where $r_i \in \mathbb{R}$. Now A contains the lattice $\Gamma = \mathbb{Z} v_1 + \cdots + \mathbb{Z} v_m$. If $T = \{c_1 v_1 + \cdots + c_m v_m \mid 0 \le c_i \le 1\}$, then T is compact and clearly any element $a \in A$ can be written as $\gamma + t$ for $\gamma \in \Gamma$ and $t \in T \cap A$. But $T \cap A = \{a_1, \ldots, a_r\}$ is a finite set. It follows that Γ is a subgroup of finite index in A and so $dA \subset \Gamma$ for some positive integer d. Then $A \subset 1/d \cdot \Gamma$ where $1/d \cdot \Gamma$ is a free $\mathbb{Z}$-module of finite rank m. By a standard result in algebra A is then a free $\mathbb{Z}$-module of rank $l \le m$, generated by, say, $w_1, \ldots, w_l$. Now $v_1, \ldots, v_m \in A$ and the $\mathbb{R}$-module generated by them has dimension m and is contained in the $\mathbb{R}$-module generated by $w_1, \ldots, w_l$. It follows that $m = l$ and $w_1, \ldots, w_m$ are $\mathbb{R}$-linearly independent. Thus, we have shown that A is a lattice. $\square$

In order to discuss the structure of the group of units of $\mathbb{D}$ in the context of lattices, we define a map λ from the open subset of $\mathbb{R}^s \times \mathbb{C}^t$ consisting of all points no coordinate of which is zero to $\mathbb{R}^{s+t}$ by $\lambda(\alpha_1, \ldots, \alpha_{s+t}) = (\ln|\alpha_1|, \ldots, \ln|\alpha_s|, 2 \ln|\alpha_{s+1}|, \ldots, 2 \ln|\alpha_{s+t}|)$. Then $\mu = \lambda\phi$ maps K to $\mathbb{R}^{s+t}$. It is simple to see that $\lambda^{-1}(T)$ is compact when T is a compact subset of $\mathbb{R}^{s+t}$. The map μ is clearly a homomorphism from the multiplicative group K^* to the additive group $\mathbb{R}^{s+t}$, which, by lemma 2, §5, Chapter 14, has the group of roots of unity in K as kernel. Denote by $\mathscr{E}$ the group of units of $\mathbb{D}$.

Lemma 3. $\mu(\mathscr{E})$ *is a lattice.*

PROOF. If T is a compact set in $\mathbb{R}^{s+t}$, then $S = \lambda^{-1}(T \cap \mu(\mathscr{E})) \subset \phi(\mathbb{D})$, and the comment preceding this lemma together with lemma 1 shows that S is finite. Hence, $T \cap \mu(\mathscr{E})$ is finite, and thus, $\mu(\mathscr{E})$ is discrete. But clearly μ is a homomorphism from the multiplicative group K^* to the additive group $\mathbb{R}^{s+t}$, and lemma 2 now shows that $\mu(\mathscr{E})$ is a lattice. $\square$

Lemma 4. $\mathscr{E}$ *is finitely generated.*

PROOF. Choose a lattice basis for $\mu(\mathscr{E})$, say, $v_1, \ldots, v_l$, and let $u_1, \ldots, u_l$ be units with $\mu(u_i) = v_i$. If $u \in \mathscr{E}$, put $\mu(u) = c_1 v_1 + \cdots + c_l v_l$, $c_i \in \mathbb{Z}$. If $\zeta = u^{-1} u_1^{c_1} u_2^{c_2} \cdots u_l^{c_l}$, then clearly $\mu(\zeta) = 0$. This implies, by the comment preceding lemma 3, that ζ is a root of unity. But the set of roots of unity in K is finite and $u = \zeta^{-1} u_1^{c_1} u_2^{c_2} \cdots u_l^{c_l}$. $\square$

§4 The Weak Mordell–Weil Theorem

Assume now that the ground field is the field $\mathbb{Q}$ of rational numbers. In §2 we established the existence of a homomorphism ϕ from the group $E(\mathbb{Q})$, of $\mathbb{Q}$-rational points on E to the multiplicative group U/U^2, where U is the group of units of the ring $R = \mathbb{Q}[x]/(f(x))$. As in that section, R may be identified with the direct sum of the fields $\mathbb{Q}(\theta_i) = K_i$, and the image $\phi(E(\mathbb{Q}))$ may be viewed as a subgroup of the direct product of the groups $K_i^*/(K_i^*)^2 = G_i$. With these identifications we will show that if P is a point on $E(\mathbb{Q})$, then the ith component of $\phi(P)$ lies in a finite subgroup of G_i. We may assume that $P \neq \infty$ and write $P = (\alpha/\beta, w)$, where α and β are coprime rational integers. Let $\theta = \theta_i$ be a fixed root of $f(x)$, and write $f(x) = (x - \theta)g(x)$. Then $\alpha - \beta\theta$ and $h_{\alpha,\beta} = g(\alpha/\beta)\beta^2$ are algebraic integers in $K = K_i$, and we put $I(P) = (\alpha - \beta\theta, h_{\alpha,\beta})$, the ideal generated by them. In the remainder of this section, all algebraic integers, ideals, and units are in the ring of integers of K.

Lemma 1. *The set of ideals* $I(P)$ *is finite.*

PROOF. $g(x) - g(\theta) = (x - \theta)t(x)$, where $t(x)$ is a linear polynomial with coefficients in $\mathbb{Z}[\theta]$. Substituting $x = \alpha/\beta$ gives $g(\theta)\beta^2 = h_{\alpha,\beta} - (\alpha - \beta\theta)t(\alpha/\beta)\beta$. Hence, $g(\theta)\beta^2 \in I(P)$. Similarly, one calculates

$$g(\theta)x^2 - g(x)\theta^2 = g(\theta)(x^2 - \theta^2) + \theta^2(g(\theta) - g(x))$$
$$= (x - \theta)l(x),$$

which shows, putting $x = \alpha/\beta$, that

$$g(\theta)\alpha^2 = (\alpha - \beta\theta)l(\alpha/\beta)\beta + \theta^2 g(\alpha/\beta)\beta^2.$$

Hence, $g(\theta)\alpha^2 \in I(P)$. It follows that $I(P)$ divides the ideal $(g(\theta)\alpha^2, g(\theta)\beta^2)$. But α and β are relatively prime, and we conclude that $I(P)$ divides the principal ideal $(g(\theta))$. But $g(\theta) \neq 0$, and therefore, $(g(\theta))$ has only a finite number of ideal divisors. $\qquad\square$

The denominator β of the first coordinate of P is the square of an integer. This elementary fact is shown, in a homogeneous context, at the beginning of the next section, but it can be seen, without fear of redundancy, quickly as follows. If $w = c/d$, $(c, d) = 1$, then $\beta^3 c^2 = d^2(\alpha^3 + a\alpha\beta^2 + b\beta^3)$. Then $\beta^3 | d^2$ and using $(c, d) = 1$ we conclude $\beta^3 = d^2$, from which it follows that β is a square.

Lemma 2. $(\alpha - \beta\theta) = I(P)C^2$ *for some ideal* C.

PROOF. Since $I(P)$ is the greatest common divisor of $\alpha - \beta\theta$ and $h_{\alpha,\beta}$, we may write $(\alpha - \beta\theta) = I(P)A$, $(h_{\alpha,\beta}) = I(P)B$, where A and B are coprime ideals. But $P \in E(\mathbb{Q})$, so there is a rational number r/s so that $(r/s)^2 = f(\alpha/\beta)$. Thus, $\beta^3 r^2 = s^2(\alpha - \beta\theta)h_{\alpha,\beta}$. It follows that $(s)^2 I(P)^2 AB$ is a square and, since A and B are coprime, we conclude that A is the square of an ideal.   □

Recall from Chapter 12 that the group of ideal classes of the ring of integers in K is finite. Let $C_1, \ldots, C_h$ be representatives for the ideal classes. Then, by definition, if J is an ideal, there is an index i and algebraic integers μ, ν such that $\mu J = \nu C_i$.

Lemma 3. *There is a finite set of algebraic integers S such that for any $P = (\alpha, \beta) \in E(\mathbb{Q})$ one can write*

$$\alpha - \beta\theta = u\gamma\tau^2$$

for a suitable unit u, an algebraic number τ, and $\gamma \in S$.

PROOF. If C is as in the preceding lemma, then C is equivalent to C_s for some s. Therefore, $I(P)C_s^2$ is eqivalent to the principal ideal $I(P)C^2 = (\alpha - \beta\theta)$ and is thus a principal ideal, say, (γ). By Lemma 1 the set $\{(\gamma)\}$ is finite depending only on $E(\mathbb{Q})$ and not on the particular P. Now there exist algebraic integers ρ, τ_1 such that $\rho C = \tau_1 C_s$. Hence, $(\rho^2(\alpha - \beta\theta)) = I(P)\tau_1^2 C_s^2 = (\gamma\tau_1^2)$. It follows that $\rho^2(\alpha - \beta\theta) = u\gamma\tau_1^2$ for some unit u. The lemma follows by putting $\tau = \tau_1/\rho$.   □

We may now prove the finiteness of $E/2E$.

Theorem 19.4.1. *$E/2E$ is a finite group.*

PROOF. It is enough to show that $\phi(E)$ is finite. We may assume that $P \neq \infty$ and that P does not have order 2. Then $\phi(P)$ is defined as the coset modulo U^2 of $\alpha/\beta - x$, where $P = (\alpha/\beta, w)$, in the group U. If we consider the ith component $K = K_i = \mathbb{Q}(\theta_i)$ of $\mathbb{Q}[x]/f(x)$, the preceding lemma shows that in $K_i^*/(K_i^*)^2$ the image of $\alpha/\beta - x$ is the coset of $(1/\beta)\,u\gamma$. As we have seen, β is the square of a rational integer, and by the weak Dirichlet unit theorem, the group $\mathscr{E}$ of units in the ring of algebraic integers of K is finitely generated with basis, say, $u_1, u_2, \ldots, u_t$. It follows that the coset of $(1/\beta)u\gamma$ mod $(K^*)^2$ has a representative of the form $u_1^{\varepsilon_1}u_2^{\varepsilon_2} \cdots u_t^{\varepsilon_t}\gamma$, where each ε_i is 0 or 1. Since γ varies over a finite set, the ith component of the image $\phi(E)$ is finite and the result follows.   □

§5 The Descent Argument

In this section we take the ground field k to be the field of rational numbers $\mathbb{Q}$. The algebraic closure $\bar{k}$ is then the field of algebraic numbers, which we assume to be a subfield of the complex numbers $\mathbb{C}$. If $\alpha \in \mathbb{C}$, denote by $|\alpha|$ its ordinary absolute value. The coefficients a and b of the elliptic curve E are assumed to be rational integers. We write the equation defining E in homogeneous form

$$y^2z = x^3 + axz^2 + bz^3 \tag{1}$$

and use homogeneous coordinates (x_0, y_0, z_0) for a point P on E. Thus, x_0, y_0, z_0 are determined up to a nonzero proportionality factor, and since P is assumed to be rational over $\mathbb{Q}$, we may assume that x_0, y_0, z_0 are integers with greatest common divisor 1. Suppose $P \neq \infty$ and put $Z_0 = \gcd(x_0, z_0)$, $X_0 = x_0/Z_0$ so that $x_0 = X_0Z_0$, $Z_0|x_0$, $Z_0|z_0$. Finally, for uniformity of notation put $Y_0 = y_0$.

From (1) one sees immediately that

$$X_0^3 Z_0^3 = z_0(Y_0^2 - ax_0z_0 - bz_0^2). \tag{2}$$

Now $\gcd(Z_0, Y_0) = \gcd((x_0, z_0), y_0) = 1$ and $Z_0|z_0$ so that z_0 is coprime to the second factor on the right-hand side of (2). Hence, $Z_0^3|z_0$ and we may define t by $Z_0^3 t = z_0$. Substituting this value of z_0 for the first term on the right side of (2) and canceling gives the relation

$$X_0^3 = t(Y_0^2 - ax_0z_0 - bz_0^2). \tag{3}$$

Now $1 = \gcd(x_0/Z_0, z_0/Z_0) = \gcd(X_0, Z_0^2 t)$. Since for any prime p such that $p|t$ one has $p|X_0$ it follows that, after a sign adjustment, $t = 1$. Thus, $z_0 = Z_0^3$ and $\gcd(X_0, Z_0) = 1$. We can therefore write $(x_0, y_0, z_0) = (X_0Z_0, Y_0, Z_0^3)$ with $(X_0, Z_0) = (Y_0, Z_0) = 1$. The corresponding affine coordinates for P now become $(X_0/Z_0^2, Y_0/Z_0^3)$, where we observe that both terms are written in lowest terms. In what follows an affine point on E is always written in this form.

Substituting the values X_0Z_0, Z_0^3, respectively, for x_0, z_0 in (2) gives

$$Y_0^2 = X_0^3 + aX_0Z_0^4 + bZ_0^6. \tag{4}$$

We now introduce the important concept of the height $H(P)$ of a point P on E that is rational over $\mathbb{Q}$. First, $H(\infty) = 1$. For $P = (X_0Z_0, Y_0, Z_0^3)$ we define

$$H(P) = \max(|X_0|, Z_0^2). \tag{5}$$

Thus, it is the maximum, in absolute value, of the numerator and denominator of the first coordinate of P in affine form. It may be thought of as a measure of the "size" of the point P. This function and its associated logarithmic function are discussed further in Chapter 20, where it will enter in an important way in the precise statement of the Birch and Swinnerton-Dyer conjectures.

The descent argument in the proof of the Mordell–Weil theorem requires an estimate of the rate of growth of the height of P when P is doubled.

From relation (4) we see that

$$|Y_0| \leq \sqrt{1 + |a| + |b|}\,(H(P))^{3/2} \leq CH(P)^{3/2}, \tag{6}$$

where C is a positive constant depending only on E and not on the particular point P. The following basic lemma follows immediately from the definition and (4):

Lemma 1. *If C is a fixed constant, then there are at most a finite number of points P on E, rational over $\mathbb{Q}$ with $H(P) \leq C$.*

Now fix a point $Q \neq \infty$ on E, and let P be any $\mathbb{Q}$-rational point on E such that $P - Q$ is not of order 2, i.e., $2P \neq 2Q$. Write, using homogeneous coordinates as above,

$$P = (x_1 z_1, y_1, z_1^3)$$
$$Q = (ce, d, e^3)$$
$$2P - Q = (x_2 z_2, y_2, z_2^3). \tag{7}$$

Write $2P = (2P - Q) + Q$, where, by assumption, the two points on the right side are distinct. Now apply (9) of §1 to conclude that, writing $2P = (x_3 z_3, y_3, z_3^3)$,

$$\frac{x_3}{z_3^2} = \frac{(x_2/z_2^2 \; c/e^2 + a)(x_2/z_2^2 + c/e^2) + 2b - 2y_2 d/z_2^3 e^3}{(c/e^2 - x_2/z_2^2)^2}$$

$$= \frac{(x_2 c + az_2^2 e^2)(x_2 e^2 + cz_2^2) + 2bz_2^4 e^4 - 2z_2 y_2 de}{(cz_2^2 - e^2 x_2)^2}.$$

Denote the numerator and denominator, respectively, of this last expression by A and B. Since $\gcd(x_3, z_3) = 1$ we see that $x_3 | A$, $z_3^2 | B$. In particular, $|x_3| \leq |A|$, $z_3^2 \leq |B|$, and by definition of height, we see that $H(2P) \leq \max(|A|, |B|)$. From (6) we know that $|y_2| \leq C_1(H(2P - Q))^{3/2}$ for a constant C_1 and trivially $|z_2| \leq H(2P - Q)^{1/2}$. We conclude, examining the expressions for A and B, that

$$H(2P) \leq CH(2P - Q)^2, \tag{8}$$

where C is a constant depending only on Q.

Recall that $f(x) = (x - \theta_1)(x - \theta_2)(x - \theta_3)$, where now $\theta_1, \theta_2, \theta_3$ are distinct algebraic integers. The discriminant of $f(x)$, $-(4a^3 + 27b^2)$, is a nonzero rational integer which we denote by δ. From relation (8) of §1, we conclude, after simplification, that for the point P and its double $2P$ one has the relation

$$x_3 - \theta_i z_3^2 = z_3^2 \left[\frac{x_1^2 + 2\theta_i^2 z_1^4 + 2\theta_i x_1 z_1^2 + az_1^4}{2z_1 y_1} \right]^2$$
$$= \alpha_i^2.$$

Since the left-hand side is an algebraic integer, it follows that the α_i are algebraic integers. Write, furthermore,

$$\alpha_i = A + B\theta_i + C\theta_i^2 = z_3 \left[\frac{x_1^2 + 2\theta_i^2 z_1^4 + 2\theta_i x_1 z_1^2 + az_1^4}{2z_1 y_1} \right], \tag{9}$$

where A, B, C are rational numbers. Cramer's rule shows that δA, δB, δC are elements of $\mathbb{Z}[\alpha_1, \alpha_2, \alpha_3, \theta_1, \theta_2, \theta_3]$ that are, in fact, linear in α_1, α_2, α_3. Thus, δA, δB, δC are algebraic integers that are rational, and we conclude by Proposition 6.1.1 that they are, in fact, integers. From (9) one sees easily that

$$2A - aC = \frac{z_3}{z_1 y_1} \cdot x_1^2$$

$$C = \frac{z_3}{z_1 y_1} \cdot z_1^4. \tag{10}$$

Therefore, $\delta(2A - C) = \delta z_3/z_1 y_1 \cdot x_1^2$ and $\delta C = \delta z_3/z_1 y_1 \cdot z_1^4$ are rational integers. If we write $\delta z_3/z_1 y_1 = m/n$, $\gcd(m, n) = 1$, then $mx_1^2 = nR$, $mz_1^4 = nS$ for integers R and S. But $\gcd(x_1, z_1) = 1$. Hence, $n = 1$ and we have consequently established the fact that $\delta z_3/z_1 y_1$ is an integer. It now follows from (10) that

$$x_1^2 \leq |\delta(2A - C)|$$

$$z_1^4 \leq |\delta C|. \tag{11}$$

Since $\alpha_i^2 = x_3 - \theta_i z_3^2$ we see that $|\alpha_i| \leq C_1 \sqrt{H(2P)}$ for a suitable constant C_1. Furthermore, as noted, $\delta(2A - C)$ and δC are linear combinations of α_1, α_2, α_2 with coefficients in $\mathbb{Q}(\theta_1, \theta_2, \theta_3)$. It follows now from (11) that, for a suitable constant C_2, one has

$$H(P) \leq C_2 H(2P)^{1/4}. \tag{12}$$

Combining this relation with (8), we arrive at the important result that there exists a constant C_3 depending only on Q such that for any point P we have

$$H(P) \leq C_3(H(2P - Q))^{1/2}. \tag{13}$$

Here C_3 has been adjusted to handle the finite number of exceptional P for which $2P = 2Q$. Now allowing Q to vary in a fixed finite set $Q_1, \ldots, Q_{n_0}$ we have shown the following lemma. All points are assumed rational over $\mathbb{Q}$.

Lemma 2. Let $\{Q_1, \ldots, Q_{n_0}\}$ be a fixed set of points on E. Then there is a constant C depending only on E and this set such that for any point P one has

$$H(P) \leq C(H(2P - Q_i))^{1/2}, \qquad i = 1, \ldots, n_0.$$

We are now ready to use a descent argument to complete the proof of the Mordell–Weil theorem.

Recall from §3 that $E/2E$ is a finite group. Let $Q_1, \ldots, Q_{n_0}$ be a set of representatives in E for this group. Thus, for any point P there is a j, $1 \leq j \leq n_0$, such that $P + Q_j = 2P'$ for some point $P' \in E$.

Theorem 19.5.1. *The group $E(\mathbb{Q})$ of $\mathbb{Q}$-rational points on E is a finitely generated abelian group.*

PROOF. Let P be an arbitrary point on E, rational over $\mathbb{Q}$. Then $P + Q_{a_1} = 2P_1$ for some a_1 and P_1. We have by the preceding lemma

$$H(P_1) \leqq C(H(2P_1 - Q_{a_1}))^{1/2} = C(H(P))^{1/2}.$$

Similarly, write $P_1 + Q_{a_2} = 2P_2$ so that $P = 2P_1 - Q_{a_1} = 4P_2 - 2Q_{a_2} - Q_{a_1}$, and $H(P_2) \leqq C(H(2P_2 - Q_{a_2}))^{1/2} = CH(P_1)^{1/2} \leqq C^{1+1/2} H(P)^{1/4}$. Continuing in this manner we arrive at a sequence of points P_r with

$$H(P_r) \leqq C^{1+1/2+\cdots+1/2^r} H(P)^{1/(2^{r+1})} \tag{14}$$

and

$$P = 2^r P_r - 2^{r-1} Q_{a_r} - \cdots - Q_{a_1}. \tag{15}$$

Now the right-hand side of (14) approaches C^2 as r approaches infinity, and therefore, there is an integer r_0 satisfying the condition that if $r \geqq r_0$ then $H(P_r) \leqq C^2 + 1$. But by lemma 1 this last inequality is satisfied by only a finite set of points, say, $P'_1, \ldots, P'_{s_0}$. Finally (15) shows that P may be written as a finite linear combination, with integer coefficients, of the points $Q_1, \ldots, Q_{n_0}, P'_1, \ldots, P'_{s_0}$. $\qquad\square$

NOTES

The Mordell–Weil theorem and the arithmetic of elliptic curves in general have a long and rich history. In these notes we mention only a few salient points and refer the interested reader to the references at the end of this chapter. The literature in this subject is vast, and the references we have listed represent only a starting point for further study.

Diophantus (circa 250) in Book 4 of his Arithmetica ([He], problem 24, p. 124) asks that a given integer be divided into two parts so that the product is the volume of a cube less its side. In geometric language this amounts to finding the rational points on the cubic $y^3 - y = x(n - x)$. He illustrates the method by choosing $n = 6$ and, after an informed guess, puts $y = 3x - 1$. Substitution leads to a cubic with zero as a double root, and he computes the third root to be 26/27. In geometric language one observes that the above line is tangent to the cubic at $(0, -1)$, and $(26/27, 136/27)$ is the third intersection point. Similarly, in problem 26 of Book 2 two numbers are sought such that their product added to either is a cube. The propitious choice of $8x$ and $x^2 - 1$ leads to the problem of finding rational points on the cubic $y^3 = (x^2 - 1)(8x + 1)$, and in modern terms, Diophantus intersects the curve with the line $y = 2x - 1$. This line intersects the cubic at $(0, -1)$ and infinity. He computes the third point to be $(112/13, 27/109)$. All this is accomplished without the aid of present algebraic notation and, of course, there is no indication of a geometric interpretation of the process, since he lived well over a thousand years before the advent of analytic geometry. The use of the chord method to locate a third point given two points is less easy to find in the ancient literature, and as Weil points out in his historical study of number theory ([W 3], p. 108), it is none other than Newton, who, in a paper written in the 1670s ([N], vol. 4, pp. 112–115), states that, beginning with three noncollinear points, iteration of the chord process leads, in general, to infinitely many rational points. However, no examples are given.

The method of descent is invariably associated with Fermat, who used it to show, among other things, that a positive square is not the difference of two fourth powers. This new point of view was to be contrasted with the generation of new points through the process of doubling, that is, by iterating the tangent process. The actual iteration of the tangent method seems to have been initiated by Fermat, who developed the techniques of Diophantus, Viete, Bachet, and others, yet he does not appear to have used the chord process or to have interpreted these methods geometrically (see [W 3], p. 110).

The efforts of Fermat were continued a century later by Euler, who gave rigorous proofs of many, but not all of Fermat's assertions. In this connection the scholarly studies of Hoffman [Hof] and Bashmakova [B] as well as Weil [W3] are particularly useful. Lagrange, whose interest in number theory was stimulated and encouraged by Euler, also utilized the method of descent. In his memoir of 1777, concerning the equation $2x^4 - y^4 = z^2$, Lagrange praises the method of Fermat, stating, "Le principe de la démonstration de Fermat est un des plus féconds dans toute la Théorie des nombres, et surtout dans celle des nombre entiers." (The principle of Fermat's proof is one the most fruitful in number theory, particularly over the integers.)

In a long memoir, "Sur les propriétés arithmetiques des courbes algébriques," published in 1901 [P], Poincaré initiated a program (". . . plutôt un programme d'étude qu'une véritable théorie") to study the arithmetic of algebraic curves over the rationals of any genus, emphasizing the birational point of view. The major portion of the paper deals with elliptic curves. Using a Weierstrass parameterization ("argument elliptique"), he shows how to generate subgroups of rational points on the curve ("formule 1," p. 492) and states, "On peut se proposer de choisir les arguments . . . de tel façon que la formule (1) comprenne tous les points rationels de la cubique." (One may propose to choose the arguments in such a way that all the rational points on the cubic are contained in the equation 1.) He defines the rank as the minimum number of "fundamental points" necessary to generate the group and asks, "Quelles valeurs peut-on attribuer au nombre entier que nous avons appelé le rang d'une cubique rationelle?" (Which values are assumed by the integer we have called the rank of the cubic?)

This is, of course, still an open question. Only curves of relatively low rank have been found. In 1982 Mestre [Mes] showed that there exists a curve of rank at least 12 and that, assuming a variety of unsolved conjectures, it has exact rank 12. In an important survey of Zagier [Z], it is stated that Mestre also found examples of curves of rank as large as 14. Whether or not the rank is bounded for curves defined over the rational numbers is unsolved, although A. Néron in his annotations to Poincaré's paper states, "L'existence de cette borne est cependent considerée comme probable." (The existence of such a bound seems, however, likely.) However, Zagier mentions in the survey that it is conjectured that all values can occur. Indeed Cassels ([Ca 3], p. 257), in his now classical survey of the arithmetic of elliptic curves, argues that the rank may well be unbounded but that examples of curves with large rank may be difficult to find since "an abelian variety can only have high rank if it is defined by equations with very large coefficients." For example, in 1986 Kretschmer [Kr] proved that the the curve $y^2 = x^3 + ax^2 + bx$ where $a = 12273038545$ and $b = 2^{10}.3^6.17.19.23.29.31.37.41.43.53$ has rank 10. We mention that A. Néron, in 1954, was able to prove that infinitely many elliptic curves of rank at least 11 exist. Kretschmer [Kr] gives a summary of the various results bounding the rank from below. The plausibility of the hypothesis that the rank is unbounded is also strengthened by the fact that in 1967 Tate and Shafarevich [Sh-T] proved that the analogous conjecture for curves defined over a field of rational functions of one variable with coefficients in a finite field is true.

Now the finiteness of the rank must be considered as part of Poincaré's program. Indeed 16 years later Hurwitz [Hur], in a paper in which certain elliptic curves are constructed with rank 0 or 1, emphasized the conjectural status of Poincaré's statement by stating, at the conclusion of his paper, "Wenn aber die Anzahl der rationale Punkte auf der Kurve unendlich ist, so spricht a priori nichts dafür, dass auch dann immer endlich viele fundamentale Punkte vorhanden sind. Bis also dieses nicht bewiesen ist, sind die auf diese Annahme gegründeten Bemerkungen von Poincaré in seiner mehrfach zitierten Arbeit entsprechend zu modifizieren." (If, however, the number of rational points is infinite, then it isn't clear, a priori, that a finite basis exists. Until this is shown the remarks of Poincaré in his often cited article that are based on this assumption should be modified.) Five years later Poincaré's intuition (or oversight) was vindicated with the 1922 publication Mordell [M1] of the first proof that the rank is finite. Cassels has written an interesting analysis of Mordell's paper ([Ca 2]) which should be studied by anyone interested in the history of this fundamental result.

This proof and the subsequent proofs of this and its various generalizations follow the same general strategy. First one shows that E/nE is finite and then a descent argument using properties of an appropriately defined height function completes the proof. The weak theorem can be proved by constructing a nondegenerate pairing between E/nE and the galois group of the extension obtained by adjoining to k the coordinates of all points P, algebraic over k, such that mP is rational. This can be shown to be a finite extension and the result follows. (See [L2] and [Si].) It should be mentioned that the weak (or as Weil calls it, the "petit") Mordell–Weil theorem, namely, the finiteness of E/nE, is discussed (for $n = 3$) in section 8 of Poincaré's memoir, using his "cubiques derivées." As A. Châtelet mentions in his annotations to Poincaré's memoir ([P], p. 546), this section is the basis of the proofs of Mordell and Weil.

The chronology of Weil's research into these matters is engagingly recorded by Weil himself in the annotations to his collected papers ([W2], vol. 1, pp. 524–526). After Mordell's proof was brought to his attention by chance, he saw the possibility of using his own work to generalize the descent argument in Mordell to curves of arbitrary genus defined over an algebraic number field, the elliptic curve being replaced by the group of rational points on the Jacobian of the curve. This is accomplished in Weil's thesis of 1928 ([W1], pp. 11–45). With the development of the general theory of abelian varieties, due also to Weil, it became possible to extend the theorem to abelian varieties defined over a number field. (See Lang [L2], chapter 5, and the historical notes on pages 88–90.)

In 1929 Weil published the short note that is presented, without the use of elliptic functions and with an interesting simplification due to Cassels ([Ca2], pp. 31–34), in this chapter. Weil mentions that since his thesis would be difficult for some to read, it would perhaps be useful to publish a simplified proof for the case of genus 1. This proof avoids the use of his decomposition theorem. At the conclusion of the introduction to this paper Weil states "Je ne pretends pas que la demonstration qu'on va lire soit essentiellement differente de celle de Mordell: et je serai satisfait si j'ai contribué a mieux mettre en valeur les idées du mathematicien anglais." (I am not suggesting that the proof below differs essentially from Mordell's and I would be satisfied if I have contributed to a better understanding of the ideas of this English mathematician.) This proof also appears in Lang ([L2], pp. 101–105) and Mordell ([M2], chapter 16).

In 1961 and 1970 J. Tate lectured on the arithmetic of elliptic curves at Haverford College. These excellent lectures, available for years in barely visible mimeograph form, became the basis for Husemöller's book on elliptic curves [Hus]. In a forthcoming book, Silverman and Tate ([Si-T]) have revised and expanded the Haverford Lectures, maintaining the elementary nature of the original presentation. We also mention the delightful little book of Chowla [Cho], as well as one by Chalal [Cha], for elementary treatments of the Mordell–Weil theorem. In Chowla, by assuming that the curve has three rational points of order 2 the proof simplifies and becomes, according to him, "nothing beyond the capacity of a ten year old." In a more sophisticated direction we strongly recommend the excellent text of Silverman [Si], especially Chapter 8. With this well-written text the interested reader can continue the study of arithmetic geometry at a more advanced level. Finally, we recommend the text on elliptic curves and modular forms by Koblitz [Ko]. In this approach to the arithmetic of elliptic curves Koblitz focuses on the essential solution of a classical problem in number theory: the determination of those positive square free integers that can be the area of a right triangle with rational sides. The solution depends on the arithmetic of the elliptic curve $y^2 = x^3 - n^2x$. Further applications of elliptic curves are discussed in the survey of current results in Chapter 20, which also presents additional references.

Bibliography

B. I. Bashmakova. Diophante et Fermat. *Revue d'histoire des sciences,* **19** (1966), 289–306.

Br-Kr. A. Brumer and K. Kramer. The rank of elliptic curves. *Duke Math. J.,* **44** (1977), 715–743.

Ca1. J.W.S. Cassels. Mordell's finite basis theorem revisited. *Math. Proc. Camb. Phil. Soc.,* **100** (1986), 31–41.

Ca2. J.W.S. Cassels. The Mordell–Weil group of curves of genus 2. *Arithmetic and Geometry. Papers Dedicated to I. R. Shaferevich on the Occasion of His Sixtieth Birthday.* Cambridge, Mass: Birkhäuser, 1983, p. 1.

Ca3. J.W.S. Cassels. Diophantine equations with special reference to elliptic curves. *J. London Math. Soc.* **41** (1966), 193–291.

Cha. J.S. Chalal. *Topics in Number Theory.* New York: Plenum, 1988.

Cho. S. Chowla. *The Riemann Hypothesis and Hilbert's Tenth Problem.* New York: Gordon & Breach, 1965.

He. T.L. Heath. *Diophantus of Alexandria: A Study in the History of Greek Algebra.* New York: Dover, 1964.

Ho. J.E. Hoffman. Über Zahlentheoretische Methoden Fermats und Eulers, ihre Zusammenhänge und ihre Bedeutung. *Arch. Hist. Exact Sci.,* **1** (1960–62), 122–159.

Hur. A. Hurwitz. Über die diophantische Gleichungen dritten Grades. *Mathematische Werke,* Vol. 2. Basel: Birkhäuser, 1963, pp. 446–468.

Hus. D. Husemöller. *Elliptic Curves.* New York: Springer-Verlag, 1987. Graduate Texts in Mathematics, Vol. 111, (1987)

Ko. N. Koblitz. *Introduction to Elliptic Curves and Modular Forms.* New York: Springer-Verlag, 1984. Graduate Texts in Mathematics, Vol. 97.

Kr. T.J. Kretschmer. Construction of elliptic curves of large rank. *Math. Comp.,* **46**(174), (1986), 627–635.

L1. S. Lang. *Elliptic Curves: Diophantine Analysis.* New York: Springer-Verlag, 1978.

L2. S. Lang. *Fundamentals of Diophantine Geometry.* New York: Springer-Verlag, 1983.

Mes. J.F. Mestre. Construction of an elliptic curve of rank 12. *Comptes rendus,* **295** (1982), 643–644.

M1. L.J. Mordell. On the rational solutions of the indeterminate equations of the third and fourth degrees. *Proc. London Math. Soc.,* **21** (1922), 179–182.

M2. L.J. Mordell. *Diophantine Equations.* New York: Academic Press, 1969.

N. I. Newton. *Mathematical Papers,* 8 vols. Cambridge: Cambridge University Press, 1967–1981.

P. H. Poincaré. Sur les propriétés arithmetiques des courbes algébriques. *J. math. pures appl.,* **7** (5) (1901), 161–233. Also in: H. Poincaré, *Oeuvres,* Vol. 5. Paris: Gauthier-Villars, 1916, pp. 483–550.

Se. J.P. Serre. *Lectures on the Mordell–Weil Theorem*. Transl. and ed. by Martin Brown. From notes by Michel Waldschmidt. Wiesbaden; Braunschweig: Vieweg, 1989.

Sh-T. I.R. Shafarevich and J. Tate. The rank of elliptic curves. *Am. Math. Soc. Transl.* **8** (1967), 917–920.

Si. J.H. Silverman. *The Arithmetic of Elliptic Curves*. New York: Springer-Verlag, 1986. Graduate Texts in Mathematics, Vol. 106.

Si-T. J. Silverman and J. Tate. *Rational Points on Elliptic Curves*. New York: Springer-Verlag (to appear).

W1. A. Weil. L'arithmetique sur les courbes algébriques. *Acta Math.*, **52** (1928), 281–315. Also in: A. Weil, *Oeuvres Scientifiques*, Vol. 1. New York: Springer-Verlag, pp. 11–45.

W2. A. Weil. *Oeuvres Scientifiques, Collected Papers*, 3 vols. Corrected second printing. New York: Springer-Verlag, 1980.

W3. A. Weil. *Number Theory: An Approach Through History*. Cambridge, Mass.: Birkhäuser, 1983.

W4. A. Weil. Sur un théorème de Mordell. *Bull. Sci. Math.*, **54** (2) (1929), 281–291. Also in: A. Weil, *Oeuvres Scientifiques*, Vol. 1. New York: Springer-Verlag, pp. 47–56.

Z. D. Zagier. *L*-Series of elliptic curves, the Birch-Swinnerton–Dyer conjecture, and the class number problem of Gauss. *Notices Am. Math. Soc.*, **31**(7), (1984), 739–743.

Chapter 20

New Progress in Arithmetic Geometry

The decade of the eighties saw dramatic progress in the field of arithmetic geometry. Problems that were previously thought to be inaccessible by contemporary methods were in fact resolved. It is the purpose of this chapter to survey a portion of these dramatic developments.

The material covered falls into two parts. The first part discusses the resolution of the Mordell conjecture by Gerd Faltings in 1983. The second part summarizes new results by B. Gross, V. Kolyvagin, K. Rubin, and D. Zagier, which deal with the conjecture of Birch and Swinnerton-Dyer that was discussed in Chapter 18.

The resolution of the Mordell conjecture has an immediate application to Fermat's last theorem. In a less transparent manner, the progress on elliptic curves also has a surprising application to Fermat's last theorem. Work of G. Frey, J.P. Serre, and K. Ribet can be combined to show that Fermat's last theorem follows from a standard conjecture, the Taniyama–Weil conjecture, about elliptic curves.

Another surprising application of the progress in the theory of elliptic curves is the resolution of an old conjecture of C.F. Gauss on the class numbers of imaginary quadratic number fields. This comes about by combining work of D. Goldfeld with a theorem of Gross–Zagier, as we shall see.

The material discussed in this chapter is mathematically sophisticated. We give few proofs, and some of the definitions are not precise. Our goals are to sketch these new results and to inspire the reader to learn more by pursuing some of the references listed at the end of the chapter.

§1 The Mordell Conjecture

In 1922 L.J. Mordell published a paper entitled "On the Rational Solutions of the Indeterminate Equation of Third and Fourth Degrees." In the first part of the paper, he states and proves what is now referred to as the Mordell–Weil theorem for elliptic curves over $\mathbb{Q}$. At the end of the paper, he discusses the situation for curves other than elliptic curves and conjectures that curves defined over $\mathbb{Q}$ that have genus greater than 1 can have only finitely many rational points. He further states that this is only a guess and that he has no real evidence or argument for its truth. This conjecture became known as the Mordell conjecture. Many papers were written proving that this or that curve had only finitely many rational points, but no very general result was forthcoming except for a famous theorem of C.L. Siegel (1929) on integral points on affine curves. This states that a curve of positive genus defined by $F(x, y) = 0$, where $F(x, y) \in \mathbb{Z}[x, y]$, has only finitely many solutions in $\mathbb{Z} \times \mathbb{Z}$.

The Mordell conjecture was generalized a bit as the years went by to state that a curve C defined over a number field K and having genus greater than 1 has only finitely many points rational over K, i.e., that $C(K)$ must be finite. Note that if this were true, then $C(L)$ would be finite for every number field L containing K. It is remarkable that until 1983 there was not a single example of a curve known to have this property. In that year G. Faltings created a sensation in the mathematical world by writing a relatively short paper that proved the generalized Mordell conjecture and several other important number-theoretical conjectures all at once. His accomplishment was built on the work of many others. We do not intend to give a history here, but merely mention some of the names of people who did important work that was used by Faltings in his proof: S. Arekelov, H. Grauert, Yu.I. Manin, A.N. Parshin, I.N. Shafarevich, L. Szpiro, J. Tate, and J.G. Zarhin.

In our preceding discussion, the notion of the genus of a curve occurred several times. This is an important concept that arose originally in topology. It is now possible to give several definitions of the genus of a curve, all of which are equivalent. Let C/K be a curve defined over a field K.

(a) Suppose $K \subseteq \mathbb{C}$ and that C is nonsingular. Then $C(\mathbb{C})$ can be given the structure of a compact Riemann surface. Topologically, this is a torus with g holes. The number of holes is the genus of C.
(b) Let $H_1(C(\mathbb{C}), \mathbb{Z})$ be the first homology group of $C(\mathbb{C})$ with coefficients in $\mathbb{Z}$. This is a free abelian group with $2g$ generators. The number g is the genus of C. (This definition is just a precise version of part a.)
(c) The holomorphic differentials on C, $\Omega^1(C(\mathbb{C}))$, form a vector space over $\mathbb{C}$ of dimension g. The number g is the genus.

Although (a) and (b) are hard to adapt to a curve defined over an arbitrary field, (c) can be modified to apply in the general case. One defines algebraic differentials on a curve, and a holomorphic differential is one that has no pole in a purely algebraic sense.

One more definition will enable us to compute the genus of a few concretely given curves. Let C be given by a homogeneous equation $F(x, y, z) = 0$, where $F(x, y, z) \in K[x, y, z]$. We need the notion of an ordinary double point. This is a singularity of a mild type. Recall that $P = (a, b, c)$ is a singular point of C if it is a zero of all three partial derivatives $\partial F/\partial x$, $\partial F/\partial y$, and $\partial F/\partial z$. P is said to be an ordinary double point if it is a singular point and the matrix

$$\begin{bmatrix} \partial^2 F/\partial x^2 & \partial^2 F/\partial x \partial y & \partial^2 F/\partial x \partial z \\ \partial^2 F/\partial y \partial x & \partial^2 F/\partial y^2 & \partial^2 F/\partial y \partial z \\ \partial^2 F/\partial z \partial x & \partial^2 F/\partial z \partial y & \partial^2 F/\partial z^2 \end{bmatrix}$$

has rank 2. A standard example is the point $P = (0, 0, 1)$ on the curve $y^2 z = x^3 + z x^2$.

(d) Let C/K be defined by $F(x, y, z) = 0$ as above. Suppose that F has degree n and that the only singularities in $C(\overline{K})$ are ordinary double points (here $\overline{K}$ is the algebraic closure of K). Then the genus of C is given by $(n - 1)(n - 2)/2 - r$, where r is the number of double points.

A nonsingular conic has genus zero. Here $n = 2$ and $r = 0$. Recall that the problem of Pythagorean triples was equivalent to finding all rational solutions of $x^2 + y^2 = z^2$, a nonsingular conic. Another example, perhaps less obvious, is the lemniscate which was studied by a succession of mathematicians—Fagnano, Bernoulli, Abel, and Gauss, among others. This curve, whose graph resembles a figure eight, is defined by $(x^2 + y^2)^2 = (x^2 - y^2)z^2$. It has degree 4 but there are three ordinary double points, $(0, 0, 1)$, $(1, \sqrt{-1}, 0)$, and $(1, -\sqrt{-1}, 0)$. By (d) above we calculate the genus to be zero.

A nonsingular cubic must have genus 1, again by using (d). If a nonsingular cubic has a rational point over the field of definition, it is an elliptic curve. A singular cubic must have genus 0.

Consider the Fermat curve defined by $x^n + y^n = z^n$. It is easily seen to be nonsingular. Thus, its genus is equal to $(n - 1)(n - 2)/2$. If $n = 2$, the genus is 0; if $n = 3$, the genus is 1; and if $n > 3$, the genus is greater than 2. When $n = 2$ there are infinitely many solutions, as we have seen (Chapter 17, §1). When $n = 3$ Euler showed there were no solutions in positive rational numbers (Chapter 17, §8). Fermat's last theorem asserts there are no solutions in positive rational numbers for any $n > 2$. The Mordell conjecture implies that for all $n > 3$ there are at most finitely many solutions in rational numbers. This is, of course, much weaker than Fermat's assertion, but it is remarkable nevertheless. (As of 1980 Fermat's last theorem had been proved for all prime exponents less than 125,000. That bound has undoubtedly been pushed much further by now.)

As a final example of an interesting family of curves, let us define a curve to be hyperelliptic if it is defined by an equation of the form $y^2 z^{n-2} = a_0 x^n + a_1 x^{n-1} z + \cdots + a_n z^n$, where $a_0 \neq 0$, and the polynomial on the right-hand side of the equation is assumed not to have repeated roots. If $n = 3$, we are again in the situation of a nonsingular cubic, so the genus is 1. If $n > 3$, the only singular point is the point at infinity $(0, 1, 0)$. The singularity is worse than a double point, so that (d) no longer tells us the genus. We simply record the answer. If n is odd, the genus is $(n - 1)/2$; if n is even, the genus is $(n - 2)/2$. If the reader is familiar with Riemann surface theory, the easiest way to see this is to use the Riemann–Hurwitz formula as it applies to a branched covering of the Riemann sphere.

The interesting feature of all this is that the genus, which is essentially a topological invariant, controls the diophantine properties of a curve. We have a threefold division. Let C be a curve defined over a number field K.

If the genus is zero, then either $C(K)$ is empty or $C(K)$ is infinite. This result is due to Hurwitz and Hilbert. We have already seen that there are infinitely many Pythagorean triples. As for the lemniscate, it is possible to give a rational parameterization

$$x = 1 - m^4, \qquad y = 2m - 2m^5, \qquad z = 1 + 6m^2 + m^4.$$

Every $m \in K$ gives rise to a rational point (x, y, z) on the lemniscate.

If the genus is 1, then either $C(K)$ is empty or C is an elliptic curve and consequently by the Mordell–Weil theorem $C(K)$ is a finitely generated abelian group (which may be finite or infinite depending on C and K).

If the genus is greater than 1, then we have Theorem 20.1.1.

Theorem 20.1.1 (Faltings). *Let C/K be a curve of genus greater than 1, defined over a number field K. Then $C(K)$ is finite.* (See [Co-Sil], [B], [Fa-Wu], and [Maz].)

We end our short survey of this topic by mentioning that Paul Vojta found a new proof of the Mordell conjecture in 1989. He was led to his proof by means of a beautiful analogy, which he uncovered between the theory of meromorphic functions in complex analysis (Nevanlinna theory) and the theory of heights in number theory. The proof is in the tradition of diophantine approximation, a topic we touched on briefly in §12 of Chapter 17. These new ideas are very powerful and point the way to generalizations of the Mordell conjecture to higher dimensional algebraic varieties (see Vojta's article in [Co-Sil] and [La1]). Faltings [Fa] has built on Vojta's ideas to prove a conjecture of Serge Lang that deals with subvarieties of abelian varieties. This is a significant advance since the Mordell conjecture is a corollary of Lang's conjecture.

§2 Elliptic Curves

In this section we review some facts about elliptic curves, which we have already discussed, and add some new material as well.

An elliptic curve E, over any field K, may be defined by a Weierstrass equation of the form

$$y^2z + a_1xyz + a_3yz^2 = x^3 + a_2x^2z + a_4xz^2 + a_6z^3,$$

where the coefficients are in K. There is one point at infinity, i.e., when $z = 0$, namely $(0, 1, 0)$. There is also a polynomial condition on the coefficients that ensures that E is nonsingular. When K is of characteristic different from 2 and 3, things are easier. In affine form E can be given by $y^2 = x^3 + ax + b$, where we require that $\Delta_E = -16(4a^3 + 27b^2) \neq 0$. Δ_E is called the discriminant of E.

As we saw in Chapter 19, the rational points on E, namely, $E(K)$, can be made into an abelian group for which the point at infinity is the zero element. We denote this point as O. For any field L containing K, $E(L)$ is also a group and one can inquire about its structure. We will review some of what is known about this.

To begin, suppose $K = \mathbb{F}$ is a finite field with q elements. Then $E(\mathbb{F})$ is contained in $\mathbb{P}^2(\mathbb{F})$, which is a set with $q^2 + q + 1$ elements. Thus, $E(\mathbb{F})$ is a finite group. Let N be the number of elements in $E(\mathbb{F})$. The congruence Riemann hypothesis implies that $|N - q - 1| \leq 2\sqrt{q}$. See Chapter 18, §2, for a discussion in the case $\mathbb{F}$ is a prime field.

If K is a number field, the Mordell–Weil theorem tells us that $E(K)$ is a finitely generated group. There is another class of fields that behaves a lot like number fields. Let $\mathbb{F}(T)$ be a rational function field with coefficients in a finite field, and suppose K is a finite extension of $\mathbb{F}(T)$. K is called an algebraic function field in one variable over a finite field. For such a field, one can show that once again $E(K)$ is finitely generated. Later we will discuss a very recent application of this result to a problem in the geometry of numbers.

If $K = \mathbb{R}$, the real numbers, then $E(\mathbb{R})$ is topologically either a circle or a disjoint union of two circles, the second case occurring when $x^3 + ax + b$ has three real roots and the first when it doesn't. Algebraically, either $E(\mathbb{R}) \cong T^1$ or $T^1 \times \mathbb{Z}/2\mathbb{Z}$, where $T^1 = \{z \in \mathbb{C} \mid |z| = 1\}$ is the unit circle in the complex plane. This fact has an immediate application to the structure of the torsion subgroup of $E(\mathbb{Q})$. Since $E(\mathbb{Q}) \subset E(\mathbb{R})$, it follows that $E(\mathbb{Q})_{\text{tors}}$ is either cyclic or the direct sum of a cyclic group and $\mathbb{Z}/2\mathbb{Z}$.

If $K = \mathbb{C}$, the complex numbers, then $E(\mathbb{C})$ is topologically a torus, i.e., a compact surface with genus 1. Algebraically, $E(\mathbb{C})$ is isomorphic to $T^1 \times T^1$. There is a better way to state this. On E there is a distinguished holomorphic differential dx/y. Remember that the space of such differentials is one dimensional over $\mathbb{C}$, so there is not much choice. If one integrates dx/y over all closed paths on $E(\mathbb{C})$, the resulting set of complex numbers Λ forms a lattice in $\mathbb{C}$, called the period lattice of E. (A lattice in a real vector space V is a subgroup consisting of all $\mathbb{Z}$-linear combinations of a vector space basis of V.) One has $E(\mathbb{C}) \approx \mathbb{C}/\Lambda$. One can make this map more explicit. Let P be any point on $E(\mathbb{C})$ and γ a path from O to P. Map P to the integral along γ of dx/y. The resulting map is not well defined, but it is well defined modulo Λ. This yields the preceding isomorphism.

These considerations lead to a painless definition of the notion of complex multiplication in the case of an elliptic curve defined over a subfield of the complex numbers. To any such curve one associates a lattice Λ by the process we have just described. One then considers the set $\mathbb{O} = \{z \in \mathbb{C} | z\Lambda \subseteq \Lambda\}$. $\mathbb{O}$ is a ring, as is easily seen. It always contains the integers $\mathbb{Z}$, and it usually consists precisely of $\mathbb{Z}$. When $\mathbb{O}$ is bigger than $\mathbb{Z}$, we say that E has complex multiplication. This makes some sense since anything in $\mathbb{O}$ that is not in $\mathbb{Z}$ must be complex. To see this, let λ_1 and λ_2 be a $\mathbb{Z}$ basis of Λ, i.e., $\Lambda = \mathbb{Z}\lambda_1 + \mathbb{Z}\lambda_2$. If $\omega \in \mathbb{O}$, then $\omega\lambda_i = \Sigma a_{ij}\lambda_j$ for $i = 1, 2$, and the $a_{ij} \in \mathbb{Z}$. Let $\tau = \lambda_2/\lambda_1$. Since λ_1 and λ_2 generate $\mathbb{C}$ over $\mathbb{R}$, we must have that τ is not real. Since $\omega = a_{11} + a_{12}\tau$ and $\omega\tau = a_{21} + a_{22}\tau$, we see that τ satisfies a quadratic equation with coefficients in $\mathbb{Z}$. Thus, $\mathbb{Q}(\tau)$ is an imaginary quadratic number field. Moreover, $\omega^2 - (a_{11} + a_{22})\omega + (a_{11}a_{22} - a_{12}a_{21}) = 0$, so ω is an algebraic integer in $\mathbb{Q}(\tau)$. We have shown that either $\mathbb{O} = \mathbb{Z}$ or $\mathbb{O}$ is an order in an imaginary quadratic number field, i.e., a subring of the ring of algebraic integers in $\mathbb{Q}(\tau)$ that generates $\mathbb{Q}(\tau)$ over $\mathbb{Q}$.

The curves we dealt with in Chapter 18 have complex multiplication. If $y^2 = x^3 - Dx$, it can be shown that there is a real number λ such that λ and $i\lambda$ generate the period lattice (here $i = \sqrt{-1}$). Thus, $\Lambda = \mathbb{Z}[i]\lambda$ and $\mathbb{O} = \mathbb{Z}[i]$, the ring of Gaussian integers. In the case of an elliptic curve defined by $y^2 = x^3 + D$, it can be shown that there is a real number λ such that λ and $\omega\lambda$ generate the period lattice (here ω is a primitive cube root of 1). Thus, $\Lambda = \mathbb{Z}[\omega]\lambda$ and $\mathbb{O} = \mathbb{Z}[\omega]$, the ring of Eisenstein integers.

The notion of complex multiplication can be given a completely algebraic definition. One has to define the notion of an algebraic endomorphism of an algebraic group. Then, if E is an elliptic curve, one can define the ring $\text{End}(E)$ of all algebraic endomorphisms of E. For example, if (x, y) is a point on $y^2 = x^3 - Dx$, we can define $i(x, y)$ to be $(-x, iy)$ and verify that this action gives an endomorphism on $E(\overline{K})$. Similarly, $\omega(x, y) = (\omega x, y)$ yields an algebraic endomorphism of $y^2 = x^3 + D$. In general there are three possibilities for the structure of $\text{End}(E)$; it is isomorphic to $\mathbb{Z}$, or to an order in an imaginary quadratic number field, or to an order in a quaternion algebra (the last can occur only in characteristic $p \neq 0$). If $\text{End}(E) \neq \mathbb{Z}$, we say E has complex multiplication. We will not pursue these ideas further here.

§3 Modular Curves

It is impossible to fully appreciate some of the new developments in the theory of elliptic curves without some background in the theory of modular curves. We will give a very brief introduction to these curves and their properties.

One caveat before we begin. Up to now we have not been dealing with the most general notion of an algebraic curve. We have defined a curve as the solution set of a homogeneous polynomial in the projective plane. Curves also occur as one-dimensional subvarieties of higher dimensional projective spaces, and not every such curve "fits" into the plane. In §1 we wrote down a formula for the genus of a plane curve which showed that a nonsingular plane curve must have a genus of the form $(n - 1)(n - 2)/2$. It follows that, for example, a plane curve of genus 2 must have singularities. But there are nonsingular genus 2 curves in $\mathbb{P}^3$. In what follows we use the word *curve* somewhat loosely but hope nevertheless to convey a good idea of what is going on.

Modular curves parameterize families of elliptic curves with certain extra structure. We begin by considering pairs (E, P), where E is an elliptic curve and P is a point on E of order N. We say two pairs (E, P) and (E', P'), are isomorphic if there is an algebraic isomorphism ϕ from E to E' such that $\phi(P) = P'$. There is a curve $Y_1(N)$ defined over $\mathbb{Q}$ whose points are in 1-to-1 correspondence with isomorphism classes of pairs (E, P) of the type just described. Moreover, if (E, P) corresponds to a point in $Y_1(N)(K)$, where K is an extension of $\mathbb{Q}$, then (E, P) is equivalent to a pair (E', P'), where E' is defined over K and $P' \in E(K)$. (See [La5] and [Shim].)

The curve $Y_1(N)$ is not complete in a sense we will not make precise. To make it complete it is necessary to add a finite number of points called cusps. The resulting complete curve is called $X_1(N)$. It is possible to compute the genus of $X_1(N)$, and it turns out that the genus is 0 if and only if $1 \leq N \leq 10$ or $N = 12$. This fact is essential in the proof of Mazur's theorem on the structure of $E_{\text{tors}}(\mathbb{Q})$, where E is any elliptic curve defined over $\mathbb{Q}$ (see Theorem 18.1.2). One big step in the proof is to show $Y_1(N)(\mathbb{Q})$ is empty if N is outside of the above range. For such N it follows that an elliptic curve over $\mathbb{Q}$ cannot have a rational point of order N.

A second family of modular curves parameterize isomorphism classes of pairs of the form (E, C), where E is an elliptic curve and C is a cyclic subgroup of E of order N. As before, (E, C) and (E', C') are said to be isomorphic if there is an algebraic isomorphism ϕ from E to E' such that $\phi(C) = C'$. There is an algebraic curve $Y_0(N)$ whose points are in 1-to-1 correspondence with isomorphism classes of the pairs (E, C). If (E, C) corresponds to a point on $Y_0(N)(K)$, then (E, C) is equivalent to a pair

(E', C'), where E' is defined over K and C' is also defined over K (we say a subset of $E(\overline{K})$ is defined over K is $\sigma(S) = S$ for all automorphisms σ of $\overline{K}/K$; see [Shim]).

$Y_0(N)$ is not complete and requires the addition of finitely many points (cusps) to make it into a complete curve $X_0(N)$. The genus can be computed, and one finds that the genus is 0 for $1 \leq N \leq 10$ and $N = 12, 13, 16, 18, 25$. The genus is 1 for $N = 11, 14, 15, 17, 19, 20, 21, 24, 27, 32, 36, 49$. Curves in this latter set are themselves elliptic curves. As we will see, the curves $X_0(N)$ form the key ingredient in the very important conjecture of Taniyama–Weil.

It is interesting to see what these curves look like over the complex numbers. Let $\mathcal{H} = \{z \in \mathbb{C} \mid z = x + iy, y > 0\}$, the classical upper half-plane. The group $SL(2, \mathbb{R})$ of 2×2 matrices with coefficients in $\mathbb{R}$ and with determinant 1 acts on $\mathcal{H}$ by fractional linear transformations. If $A = \begin{pmatrix} a & b \\ c & d \end{pmatrix}$, we define $A(z)$ to be $(az + b)/(cz + d)$. The discrete subgroup $\Gamma = SL(2, \mathbb{Z})$ acts on $\mathcal{H}$ in a properly discontinuous manner (definition omitted) and the quotient space $\mathcal{H}/\Gamma$ has the structure of a one-dimensional complex manifold. In fact, $\mathcal{H}/\Gamma \approx \mathbb{C}$ and so $\mathcal{H}/\Gamma$ can be compactified by adding one point to yield the Riemann sphere, which is isomorphic to $\mathbb{P}^1(\mathbb{C})$. If Γ' is any subgroup of Γ of finite index, we can also form $\mathcal{H}/\Gamma'$ and by adding finitely many points in an appropriate manner compactify it to a compact Riemann surface $C(\Gamma')$. The natural map $\mathcal{H}/\Gamma' \to \mathcal{H}/\Gamma$ extends to an analytic map from $C(\Gamma')$ to $\mathbb{P}^1(\mathbb{C})$, which realizes $C(\Gamma')$ as a branched covering of the Riemann sphere.

Define two families of subgroups of finite index in Γ:

$$\Gamma_0(N) = \left\{ \begin{pmatrix} a & b \\ c & d \end{pmatrix} \middle| c \equiv 0 (\mathrm{mod}\ N) \right\}$$

$$\Gamma_1(N) = \left\{ \begin{pmatrix} a & b \\ c & d \end{pmatrix} \middle| \begin{pmatrix} a & b \\ c & d \end{pmatrix} \equiv \begin{pmatrix} 1 & * \\ 0 & 1 \end{pmatrix} (\mathrm{mod}\ N) \right\}.$$

It is now not too hard to prove the following result.

Proposition. $Y_0(N)(\mathbb{C}) \approx \mathcal{H}/\Gamma_0(N)$ and $Y_1(N)(\mathbb{C}) \approx \mathcal{H}/\Gamma_1(N)$. Moreover, $X_0(N)(\mathbb{C}) \approx C(\Gamma_0(N))$ and $X_1(N)(\mathbb{C}) \approx C(\Gamma_1(N))$.

In this proposition "$\approx$" means "is analytically isomorphic to." The group Γ is sometimes called the modular group. The proposition shows the connection between certain subgroups of the modular group and the modular curves we discussed earlier.

A very readable introduction to the modular group and its properties is given by Serre [Se]. Subgroups of the modular group are discussed in [Ko], [La5], [Ogg], and [Shim].

We are now in a position to state one of the most important conjectures in the whole subject.

The Taniyama–Weil Conjecture. *Let E be an elliptic curve defined over $\mathbb{Q}$. Then there is an integer N and a nonconstant rational map $\phi: X_0(N) \to E$ with ϕ defined over $\mathbb{Q}$.*

If $E/\mathbb{Q}$ is the algebraic image of some $X_0(N)$, we say that E is modular. The conjecture may be paraphrased as saying that every elliptic curve over $\mathbb{Q}$ is modular.

This conjecture was first put forward by Taniyama at a conference on algebraic number theory held in Japan in 1955. In 1967 Weil [We] refined the conjecture by specifying that the integer N can be taken to be the conductor of E, a notion to be discussed later, and also proved an important theorem that made the conjecture very plausible. In 1971 G. Shimura proved, using Weil's theorem, that every elliptic curve over $\mathbb{Q}$ that has complex multiplication is modular. There is a finite algorithm that allows one to check in any given case if an elliptic curve over $\mathbb{Q}$ is modular. This has been done in hundreds, perhaps by now thousands, of cases. The evidence in its favor seems overwhelming.

This conjecture seems to have nothing at all to do with Fermat's famous conjecture, his so-called last theorem. Nevertheless, the mathematician G. Frey discovered a connection. If $a^p + b^p = c^p$ is a solution in positive integers a, b, and c, where p is a prime different from 2, Frey associates to such a solution the elliptic curve $E: y^2 = x(x - a^p)(x + b^p)$. He then shows this curve has such remarkable properties that it shouldn't exist. J.-P. Serre had previously formulated a conjecture about modular functions that would prove this nonexistence if E were modular. K. Ribet in 1986 proved a special case of Serre's conjecture that was powerful enough to yield the following theorem.

Theorem (Frey, Serre, Ribet). *The Taniyama–Weil conjecture implies Fermat's Last Theorem.*

This, together with the results of Faltings discussed in §1, represents truly amazing and unexpected progress toward a resolution of Fermat's last theorem. Oesterlé discusses this theorem and gives a sketch of the proof in [Oes1]. (For new developments the reader is referred to Notes for a New Printing on page vi.)

§4 Heights and the Height Regulator

The theory of heights plays a very important role in the subject of diophantine equations. As we saw in Chapter 19, it is a key ingredient in the proof of the Mordell–Weil theorem. In this section we briefly introduce the more general theory. One of our principal motivations is to give a definition and discussion of the height regulator, which is an important quantity associated with an elliptic curve defined over a number field. The height regulator also plays a role in the more refined version of the conjecture of Birch and Swinnerton-Dyer.

Let us recall the definition of the height of a rational number. If $r \in \mathbb{Q}$, write $r = a/b$, where a and b are relatively prime integers. Define $H(r) = \max(|a|, |b|)$. This has the following two properties; $H(r) \geq 1$ for all $r \in \mathbb{Q}$, and for every C the set $\{r \in \mathbb{Q} | H(r) \leq C\}$ is finite. We would like to extend H to a function on all of $\overline{\mathbb{Q}}$, the algebraic closure of $\mathbb{Q}$, in such a way that both these properties continue to hold. This turns out to be almost possible. The construction of such an extension is not too hard, but it would take us too far afield to give all the details here. We will show how to extend H to a function on all algebraic integers and refer the interested reader to some of the references given at the end of this chapter for the method in the general case of algebraic numbers (see [Sil], [Hu], [La3], or [La4]).

Suppose $\alpha \in K$ is an algebraic integer in some algebraic number field $K \subset \overline{\mathbb{Q}}$. Let $\sigma_1, \sigma_2, \ldots, \sigma_n$ be the imbeddings of K into the complex numbers, arranged in such a way that the first s of them are real imbeddings, the next t of them are distinct nonconjugate complex imbeddings, and σ_{s+i} is the complex conjugate of σ_{s+t+i} for $1 \leq i \leq t$. We then define the normalized absolute values as follows:

$$\|\alpha\|_i = |\sigma_i \alpha| \quad \text{if } 1 \leq i \leq s$$
$$\|\alpha\|_i = |\sigma_i \alpha|^2 \quad \text{if } s + 1 \leq i \leq s + t.$$

Definition. Let α be an algebraic integer in an algebraic number field K of degree n over $\mathbb{Q}$. The height of α is defined by

$$H(\alpha)^n = \Pi_i \max(1, \|\alpha\|_i).$$

It is not hard to check that $H(\alpha)$ is well defined and that if $\alpha \in \mathbb{Z}$, $H(\alpha)$ reduces to $\max(1, |\alpha|)$ as it should.

Proposition 20.4.1. *For all $\alpha \in \overline{\mathbb{Q}}$, $H(\alpha) \geq 1$. Moreover, if C and n are given, the set $\{\alpha \in \overline{\mathbb{Q}} \mid H(\alpha) \leq C$ and $\deg(\alpha) \leq n\}$ is finite.*

PROOF. We cannot give the proof of the full result since we have defined $H(\alpha)$ only in the case α is an algebraic integer. We give the proof in this special case and remark that the proof of the general result is quite similar.

The first assertion is clear from the definition. Now assume α is an algebraic integer and that $d = \deg(\alpha) = [\mathbb{Q}(\alpha):\mathbb{Q}] \leq n$. Then α satisfies a monic polynomial equation of degree d with coefficients in $\mathbb{Z}$: $x^d + a_1 x^{d-1} + \cdots + a_d = (x - \alpha_1)(x - \alpha_2) \cdots (x - \alpha_d)$. From the definition of height, we find that $|\alpha_i| \leq C$ for all i, and it follows that $|a_i| \leq \binom{d}{i} C^i$ for $1 \leq i \leq d$. Since d is bounded and the coefficients of the polynomial are bounded, there are only finitely many possible polynomials involved, and thus α must be one of only finitely many algebraic integers.   □

Let E/K be an elliptic curve defined over an algebraic number field, and suppose it is given in affine form by a Weierstrass equation $y^2 = x^3 + ax + b$, with $a, b \in K$. If $P \in E(\overline{K})$, write $P = (x(P), y(P))$. As usual, denote by O the point at infinity on E.

Definition. The height on $E(\overline{K})$ is a function $h: E(\overline{K}) \to \mathbb{R}$ given by $h(O) = 0$, and $h(P) = \log H(x(P))$ for $P \neq O$.

The "log" on the definition denotes the natural logarithm. As will be seen, passing to the logarithm of H has a number of advantages. Note that for all P, $h(P) \geq 0$. Also, since $-(x, y) = (x, -y)$, it follows that $h(P) = h(-P)$. The following simple consequence of Proposition 20.4.1 will be important.

Proposition 20.4.2. *Let E/K be an elliptic curve defined over an algebraic number field K. For all C, the set $\{P \in E(K) \mid h(P) \leq C\}$ is finite.*

PROOF. By Proposition 20.4.1, the set $(\alpha \in K \mid H(\alpha) \leq e^C\}$ is finite. Since for each $\alpha \in K$ there are at most two values of β such that $(\alpha, \beta) \in E(K)$, the result follows.   □

Before going further, we introduce some useful notation. If f and g are functions from some set X to $\mathbb{R}$, we define $f(x) = g(x) + O(1)$ to mean that $|f(x) - g(x)|$ is bounded above by a constant that may depend on f and g. Similarly, $f(x) \leq g(x) + O(1)$ means that there is a constant C such that $f(x) \leq g(x) + C$ for all $x \in X$.

In Chapter 19, §4, two important properties of height on elliptic curves defined over $\mathbb{Q}$ were proved. In the present context, these may be reformulated as follows: For $P \in E(\mathbb{Q})$, $4h(P) \leq h(2P) + O(1)$, and if $Q \in E(\mathbb{Q})$ is fixed, then $h(P + Q) \leq 2h(P) + O(1)$. The first follows from equation (12), and the second is derived from equation (8) (it is not hard to see that $2P$ can be replaced with P in equation (8), and one can then replace P with $P + Q$). We now present an important generalization.

Proposition 20.4.3. *Let E/K be an elliptic curve defined over a number field K. For all $P, Q \in E(\overline{K})$ we have*

$$h(P + Q) + h(P - Q) = 2h(P) + 2h(Q) + O(1).$$

PROOF. We sketch the proof, referring to Silverman [Sil] for details. Assuming that $K = \mathbb{Q}$, one can use the methods of Chapter 19 to establish that $h(P + Q) + h(P - Q) \le 2h(P) + 2h(Q) + O(1)$. The problem is to show the reverse inequality. Set $P = R + S$ and $Q = R - S$. One finds $h(2R) + h(2S) \le 2h(R + S) + 2h(R - S) + O(1)$. By previous results, we know that $4h(R) + 4h(S) \le h(2R) + h(2S) + O(1)$. Combining these inequalities and dividing by 2 yields the result. □

If we set $P = Q$, the relation $h(2P) = 4h(P) + O(1)$ falls right out. It is in fact not hard to show that $h(mP) = m^2h(P) + O(1)$ for all integers m. In Proposition 20.4.3 substitute mP for P and P for Q. Assuming the result for integers k such that $1 \le k \le m$, we find that

$$h((m + 1)P) + (m - 1)^2h(P) = 2m^2h(P) + 2h(P) + O(1).$$

Thus, $h((m + 1)P) = (m + 1)^2h(P) + O(1)$, and so we are done by induction.

All of this makes it plain that the height function on an elliptic curve behaves very much like a quadratic form, aside from the $O(1)$ terms. Both A. Neron and J. Tate found ways to modify the definition so as to get a quadratic form on $E(\overline{K})$, which behaves like the height function. Both methods have advantages, but we will present Tate's because it is more elementary.

Definition. Let E/K be an elliptic curve defined over an algebraic number field K. For $P \in E(\overline{K})$ define $\hat{h}(P)$, the canonical height of P, by the formula $\hat{h}(P) = \lim_{n \to \infty} 4^{-n}h(2^nP)$.

To show the limit in this definition exists, it is sufficient to show that the terms define a Cauchy sequence. Suppose $n > m \ge 0$. Then

$$|4^{-m}h(2^mP) - 4^{-n}h(2^nP)| \le \Sigma |4^{-m-i}h(2^{m+i}P) - 4^{-m-i-1}h(2^{m+i+1}P)|, \quad (1)$$

where the sum is from $i = 0$ to $n - m - 1$. There is a constant C such that $|4^{-1}h(2Q) - h(Q)| \le C$ for all $Q \in E(\overline{K})$. The ith term in the sum is $\le 4^{-m-i}C$. Thus, the sum is dominated by $(4^{-m} + 4^{-m-1} + \cdots + 4^{-n+1})C < 4^{-m+1}C$. This shows the terms form a Cauchy sequence.

The important properties of the canonical height are summarized in the following theorem.

Theorem 20.4.4. *The canonical height $\hat{h}(P)$ satisfies*

(i) $\hat{h}(P) = h(P) + O(1)$.
(ii) $\langle P, Q \rangle = 1/2 (\hat{h}(P + Q) - \hat{h}(P) - \hat{h}(Q))$ *is bi-additive.*
(iii) $\hat{h}(mP) = m^2\hat{h}(P)$ *for all* $m \in \mathbb{Z}$.
(iv) $\hat{h}(P) \ge 0$, *with equality holding if and only if P is a torsion point.*
(v) *If $g(P)$ is any function satisfying (i) and (iii), then $g = \hat{h}$.*

PROOF. We sketch the proof, referring the reader to the references for the details (we note parenthetically that the canonical height in [Sil] is half the one defined here).

In equation (1) set $m = 0$ and take the limit of both sides as n tends to infinity. Since we have shown that the right-hand side is dominated by $4C$, it follows that $|h(P) - \hat{h}(P)| \leq 4C$, which proves (i).

In Proposition 20.4.3, replace P and Q by $2^n P$ and $2^n Q$, respectively, divide through by 4^n, and pass to the limit as n tends to infinity. The result is that the canonical height satisfies the parallelogram law: $\hat{h}(P + Q) + \hat{h}(P - Q) = 2\hat{h}(P) + 2\hat{h}(Q)$. Property (ii) follows from this identity by an exercise in pure algebra. We omit the details.

Property (iii) can be derived in two ways. One can start with the fact that the height function h satisfies the property up to $O(1)$ terms and then get the result by replacing P by $2^n P$, dividing by 4^n, and passing to the limit. Alternatively, it follows by a formal induction using property (ii).

Since $h(P) \geq 0$ for all points P, it follows that the same is true of $\hat{h}(P)$. If P is a torsion point, there is an $m \in \mathbb{Z}$, $m \neq 0$, such that $mP = O$. Thus, $0 = \hat{h}(O) = \hat{h}(mP) = m^2 \hat{h}(P)$, and so $\hat{h}(P) = 0$. Conversely, suppose $\hat{h}(P) = 0$. Then, by property (iii), $\hat{h}(mP) = 0$ for all integers m. However, using property (i) and Proposition 20.4.1, we see that $\{mP \mid m \in \mathbb{Z}\}$ is a finite set. This can happen only if P is a torsion point.

Finally, assume that $g(P)$ satisfies (i) and (iii). From (i) we see that there is a constant C such that $|\hat{h}(P) - g(P)| \leq C$ for all points P. Choose any $m \geq 1$, and replace P by $m^k P$. Then, using property (iii), we find $|\hat{h}(P) - g(P)| \leq Cm^{-2k}$. Now let k tend to infinity. The result is $\hat{h}(P) = g(P)$. (Note that one only has to assume that (iii) holds for one integer $m \geq 2$). $\qquad\square$

Definition. Let E/K be an elliptic curve defined over a number field K. Let $P_1, P_2, \ldots, P_r$ be a basis for the free part of $E(K)$, i.e., every point of $E(K)$ can be uniquely written as the sum of a torsion point and a $\mathbb{Z}$-linear combination of the P_i. Let $\mathcal{R}$ be the matrix whose ijth entry is $\langle P_i, P_j \rangle$. Then $R(E/K)$, the height regulator of E/K, is defined to be the determinant of $\mathcal{R}$.

Just as is the case with the regulator of a number field, the height regulator of an elliptic curve has a geometric interpretation. To get an idea of how this works, we have to introduce the real vector space $V(K) = \mathbb{R} \otimes E(K)$, which has dimension r over $\mathbb{R}$. For those readers who are unfamiliar with tensor products, it is possible to give a more concrete construction of $V(K)$. Its points consist of formal $\mathbb{R}$-linear combinations of the P_i, i.e., expressions of the form $\Sigma x_i P_i$ with the $x_i \in \mathbb{R}$. Addition and subtraction are performed coordinate wise, scalar multiplication by the rule $t \Sigma x_i P_i = \Sigma t x_i P_i$. The height pairing $\langle P, Q \rangle$ can be extended to $V(K)$

in the obvious manner; if $X = \Sigma x_i P_i$ and $Y = \Sigma y_i P_i$, then $\langle X, Y \rangle = \Sigma_{i,j} x_i y_j \langle P_i, P_j \rangle$. What is not as obvious as it seems is that this extended inner product is positive definite on $V(K)$. It is true. The proof involves the almost all of Theorem 20.4.4. Assuming this result, choose an orthonormal basis $e_1, e_2, \ldots, e_r$ for $V(K)$. Now put the usual measure on the Euclidean space $V(K)$ so that the unit cube $\{\Sigma t_i e_i \mid 0 \le t_i \le 1$ for $i = 1$, $2, \ldots, r\}$ has volume equal to 1.

We define a map ϕ from $E(K)$ to $V(K)$ as follows: Every $P \in E(K)$ can be uniquely written as $T + \Sigma n_i P_i$, where T is a torsion point, and the $n_i \in \mathbb{Z}$ (the sum here is not formal; it is addition on the elliptic curve E). Define $\phi(P) = \Sigma n_i P_i \in V(K)$. It is easy to see that ϕ is a homomorphism with kernel equal to $E_{\text{tors}}(K)$ and with image a lattice in $V(K)$. A fundamental domain for this lattice is given by $\{\Sigma t_i P_i \mid 0 \le t_i \le 1$ for $i = 1, 2, \ldots, r\}$. To compute the volume of this fundamental domain is a standard exercise in linear algebra. One writes $P_i = \Sigma a_{ij} e_j$ with the $a_{ij} \in \mathbb{R}$ and the volume in question is equal to $|\det[a_{ij}]|$. Let $\mathscr{A} = [a_{ij}]$. Since the e_i are orthonormal, one sees that $\mathscr{R} = \mathscr{A}^t \mathscr{A}$ ($^t\mathscr{A}$ is the transpose of $\mathscr{A}$) and it follows that $R(E/K) = (\det \mathscr{A})^2$. We have proved Proposition 20.4.5.

Proposition 20.4.5. *The height regulator, $R(E/K)$, is the square of the volume of a fundamental domain for the lattice $\phi(E(K))$ in the vector space $V(K)$.*

Using this geometric interpretation and some standard arguments from the geometry of numbers, we can deduce a very interesting result about the distribution of rational points on an elliptic curve.

Theorem 20.4.6. *Let E/K be an elliptic curve defined over a number field K. Suppose that $E(K)$ has rank r. Let $N(R)$ be the number of elements $P \in E(K)$ such that $h(P) \le R$. Then there is a constant C such that $N(R) \sim CR^{r/2}$ (here the "$\sim$" means that the ratio of the two sides tends to 1 as R tends to infinity). More precisely, the constant C is equal to $\gamma_r |E(K)_t|/\sqrt{R(E/K)}$, where γ_r is the volume of the unit sphere in Euclidean r-space ($\gamma_r = \pi^{r/2}/\Gamma(1 + r/2)$).*

PROOF. Let L be a lattice in Euclidean n-space, $\mathbb{R}^n$. Then the number of elements in the set $\{\lambda \in L \mid \|\lambda\| \le R\}$ is asymptotic to the volume of the sphere of radius R divided by the volume of a fundamental domain for the lattice L. This is a standard result that is intuitively clear and not too hard to establish. We now apply it to the lattice $\phi(E(K))$ in $V(K)$.

First, notice that $\|\phi(P)\|^2 = \langle \phi(P), \phi(P) \rangle = 1/2 \, (\hat{h}(2P) - 2\hat{h}(P)) = \hat{h}(P)$, and by Theorem 20.4.4, part (i), $\hat{h}(P)$ differs from $h(P)$ by a bounded amount. Thus, $N(R)$ differs from the product of $|E(K)_{\text{tors}}|$ and the number of points in the set $\{P \in E(K) \mid \|\phi(P)\| \le R^{1/2}\}$ by a bounded amount ($|E(K)_{\text{tors}}|$ enters into this because it is equal to the order of the kernel of ϕ). The number of elements in the latter set is asymptotic to

$\gamma_r R^{r/2}$ divided by the volume of the lattice $\phi(E(K))$, which is, by Proposition 20.4.5, the square root of $R(E/K)$. The proposition follows. $\qquad\square$

There are many interesting open problems concerning the canonical height. For example, here is a conjecture due to Serge Lang (see [La3], p. 92). Let E be an elliptic curve defined by $y^2 = x^3 + ax + b$ with a and b in $\mathbb{Z}$. Assume $E(\mathbb{Q})$ has positive rank. Since $\phi(E(\mathbb{Q}))$ is a lattice in $V(\mathbb{Q})$, there is a nontorsion point $P_1 \in E(\mathbb{Q})$ such that $\hat{h}(P_1)$ is least. Lang conjectures that there is a constant C independent of E such that $\hat{h}(P_1) > C \log |\tilde{\Delta}_E|$, where $\tilde{\Delta}_E$ is the minimal discriminant of E. This number divides the discriminant of E. (For a precise definition, see page 224 of [Sil].) This conjecture has been proved by J. Silverman [Sil2] in special cases. For example, let j_E be defined by $1728(4a)^3/\Delta_E$. Silverman has shown that Lang's conjecture is true if one considers only elliptic curves E such that j_E is an integer (this holds automatically if $E/\mathbb{Q}$ is an elliptic curve with complex multiplication). As a result, he is able to prove [Sil3] that there is a constant C such that for all curves E with $j_E \in \mathbb{Z}$, $|E(\mathbb{Z})| < C^{r_E}$, where r_E is the rank of $E(\mathbb{Q})$. Notice that this shows that if you could find elliptic curves with $j_E \in \mathbb{Z}$ and many integral points, you would force the rank to be large. It is conjectured (also by Lang; [La3], page 140) that inequalities of this type hold without restriction on j_E, and also appropriately formulated, over any number field. (For more conjectures about the canonical heights of elements of a basis for $E(K)$, see [La6].)

We end this section by noting that Noam Elkies used some of the ideas discussed in this section to provide examples of lattices in Euclidean space with extraordinarily good sphere-packing properties. Instead of number fields, he works over rational function fields $\mathbb{F}(T)$, where $\mathbb{F}$ is a finite field. By choosing the elliptic curve E and the finite field $\mathbb{F}$ very carefully, he is able to produce lattices that equal or better the best known examples, at least in all dimensions less than or equal to 1024. Once again, this illustrates the fact that the arithmetic theory of elliptic curves has deep and surprising applications in other areas of mathematics.

§5 New Results on the Birch–Swinnerton-Dyer Conjecture

In this section we begin by reviewing the definitions that go into the Birch–Swinnerton-Dyer conjecture. The discussion is similar to that of §2 of Chapter 18. Here we work over a general number field K and also make the conjecture more precise.

Let E be defined by an equation $y^2 = x^3 + ax + b$ with coefficients in O_K, the ring of integers in K. Let $\mathcal{P}$ be a prime ideal in O_K and let $N_{\mathcal{P}} - 1$

be the number of solutions to the congruence $y^2 \equiv x^3 + ax + b$ (mod $\mathcal{P}$). Let $\mathbb{N}\mathcal{P} = |O_K/\mathcal{P}|$ and define $C_{\mathcal{P}} = \mathbb{N}\mathcal{P} + 1 - N_{\mathcal{P}}$.

Definition. $L^*(E/K, s) = \Pi(1 - C_{\mathcal{P}}\mathbb{N}\mathcal{P}^{-s} + \mathbb{N}\mathcal{P}^{1-2s})^{-1}$, where the product is over all nonzero prime ideals of O_K not dividing Δ_E. By multiplying by suitable factors at the primes dividing Δ_E, one arrives at $L(E/K, s)$, the L-function of E over K.

The idea of considering $L(E/K, s)$ is that, since it contains information about all of the foregoing congruences, it should contain a lot of information about the arithmetic of E.

First a word about the convergence of $L(E/K,s)$. If $\mathcal{P}$ does not divide Δ_E, then it can be proved that $|C_{\mathcal{P}}| \le 2(\mathbb{N}\mathcal{P})^{1/2}$. This was proved by Hasse in the 1930s. Examples of this phenomenon go back to Gauss. Here we are concerned with elliptic curves over finite fields. As we noted earlier in this book, Weil conjectured similar results for nonsingular algebraic varieties over finite fields and proved his conjectures in several important cases. The general Weil conjecture, the congruence Riemann hypothesis, was proved by Deligne in 1973.

To get back to our story, the inequality $|C_{\mathcal{P}}| \le 2(\mathbb{N}\mathcal{P})^{1/2}$ easily implies that $L(E/K,s)$ converges for Re$(s) > 3/2$. Another conjecture of Weil, closely related to the Taniyama–Weil conjecture, is the following:

Conjecture (Weil). $L(E/K,s)$ can be analytically continued to the entire complex plane and satisfies a functional equation.

For simplicity we state the conjectured functional equation in the special case when E is defined over $\mathbb{Q}$. There is an integer N_E, called the conductor of E. N_E divides Δ_E and is divisible only by primes where E has "bad" reduction. We omit the precise definition. Let

$$\Lambda_E(s) = N_E^{s/2}(2\pi)^{-s}\Gamma(s)L(E/\mathbb{Q},s).$$

Then (conjecturally) $\Lambda_E(s)$ can be analytically continued to an entire function on all of $\mathbb{C}$, and $\Lambda_E(s) = \varepsilon\Lambda_E(2 - s)$, where $\varepsilon = 1$ or -1 is called the sign of the functional equation.

Weil proved this in special cases. In 1954 Deuring proved it when E has complex multiplication. Eichler (1954) and Shimura (1958) proved it when $E/\mathbb{Q}$ is a modular elliptic curve. Thus, if the Taniyama–Weil conjecture is correct, the preceding conjecture would follow over $\mathbb{Q}$. In any case, if E has complex multiplication, or more generally if E is modular, one can consider $L(E/K,s)$ as an analytic function around the point $s = 1$.

The Conjecture of Birch and Swinnerton-Dyer. Assuming the analytic continuation, $L(E/K,s)$ has a zero of order r, the $\mathbb{Z}$-rank of $E(K)$, at $s = 1$.

Moreover, $(s - 1)^{-r}L(E/K,s) \to M_E$ as $s \to 1$, where M_E is a constant with the structure

$$M_E = 2^t |D_K|^{-1/2} |\text{Ш}(E/K)| R(E/K) |E(K)_{\text{tors}}|^{-2} \Pi d_{\mathscr{P}}.$$

Here t is half the number of complex embeddings of K over $\mathbb{Q}$, D_K is the discriminant of K, $\text{Ш}(E/K)$ is the Tate–Shaferevich group of E over K, $R(E/K)$ is the canonical height regulator of $E(K)$ (described in the last section), and the $d_{\mathscr{P}}$ are numbers that are 1 unless $\mathscr{P}$ is a prime of bad reduction or an archimedean prime. If $\mathscr{P}$ is nonarchimedean, $d_{\mathscr{P}}$ is a positive integer; if $\mathscr{P}$ is archimedean, then $d_{\mathscr{P}}$ is given by a period integral.

$\text{Ш}(E/K)$ is a very important group associated to E. It arises in connection with the problem of computing the rank of a given elliptic curve. The definition is somewhat technical, and we shall not give it here (see, e.g., Chapter 10, §4 of [Sil]). $\text{Ш}(E/K)$ is conjectured to be finite, but until recently this was not known to be true for any single case. If one could find an effective upper bound for $|\text{Ш}|$ a consequence would be the existence of a finite algorithm for determining the rank of $E(K)$ in any given case. Around 1972 Tate made the following comment on the Birch–Swinnerton-Dyer conjecture: "This remarkable conjecture relates the behavior of a function L where it is not known to be defined, to the order of a group Ш not known to be finite."

There has been dramatic progress on this conjecture in recent years. Until further notice we will assume that E is defined over $\mathbb{Q}$. If E has complex multiplication, we will say that E has CM.

Coates–Wiles *(1977)*. If E has CM, then $L(E/\mathbb{Q},1) \neq 0$ implies $E(\mathbb{Q})$ is finite [114].

Gross–Zagier *(1986)*. If E is modular and $L(E/\mathbb{Q},s)$ has a simple zero at $s = 1$, then $E(\mathbb{Q})$ is infinite [Gr-Za].

Rubin *(1987)*.

(a) If E has CM and $L(E/\mathbb{Q},1) \neq 0$, then $\text{Ш}(E/\mathbb{Q})$ is finite.
(b) If E has CM and $r_E \geq 2$, then $L(E/\mathbb{Q},s)$ has a zero at $s = 1$ of order 2 or greater. [Ru1].

Rubin's result (a) gave the first known examples of Ш being finite. For example, for $y^2 = x^3 - x$, Ш is trivial; for the curve $y^2 = x^3 + 17x$, $\text{Ш} \cong \mathbb{Z}/2\mathbb{Z} \oplus \mathbb{Z}/2\mathbb{Z}$.

Combining the preceding results leads to Theorem 20.5.1.

Theorem 20.5.1. *If E has CM and $\text{ord}_{s=1}L(E/\mathbb{Q},s) = \rho_E \leq 1$, then $\rho_E = r_E$.*

This rather spectacular result was pushed much further by V.A. Kolyvagin in 1988. It turns out that the theorem remains true when the hypothesis that E has CM is replaced by the much weaker hypothesis that E is modular. According to the Taniyama–Weil conjecture, this covers all elliptic curves defined over $\mathbb{Q}$. To explain this work, it is necessary to develop in more detail what it is that Gross and Zagier were able to prove, and to do this one must define the notion of a Heegner point on the modular curve $X_0(N)$.

Recall that the points on the modular curve $X_0(N)$ correspond to isomorphism classes of pairs (E, C), where C is a cyclic subgroup of E of order N. Let K be an imaginary quadratic number field with discriminant $D < 0$, and assume $(N,D) = 1$. We further assume that every rational prime p dividing N splits in O_K, i.e., $pO_K = \mathcal{P}_1\mathcal{P}_2$. From this assumption it is not too hard to show that there exist ideals $\mathcal{N} \subset O_K$ such that $O_K/\mathcal{N} \cong \mathbb{Z}/N\mathbb{Z}$. Consider the pair $(\mathbb{C}/O_K, \mathcal{N}^{-1}/O_K)$, where $\mathcal{N}^{-1}$ is the fractional ideal inverse to $\mathcal{N}$ in K. $\mathbb{C}/O_K$ defines an elliptic curve over $\mathbb{C}$, and $\mathcal{N}^{-1}/O_K \cong O_K/\mathcal{N} \cong \mathbb{Z}/N\mathbb{Z}$ is a cyclic subgroup of order N. Thus, we have defined a point x_K on $X_0(N)(\mathbb{C})$. It is a fact that this point has coordinates in H, the Hilbert class field of K. Recall that H is the maximal unramified extension of K whose Galois group, $\mathrm{Gal}(H/K)$, is abelian. The point x_K is called a Heegner point in honor of Kurt Heegner, who first defined such points and investigated their properties.

Now suppose that $\varphi\colon X_0(N) \to E$ is a modular parameterization of an elliptic curve E defined over $\mathbb{Q}$. If x_K is a Heegner point, define $y_K = \Sigma\, \varphi(x_K)^\sigma$, where the sum is over all automorphisms in $\mathrm{Gal}(H/K)$, the sum denotes group addition on E. Clearly, $y_K \in E(K)$. The first part of the following result was conjectured earlier, around 1983, by Birch and Stephens.

Kolyvagin *(1988).* Assume y_K has infinite order in $E(K)$. Then

(a) The group $E(K)$ has rank 1.
(b) The group $\mathrm{III}(E/K)$ is finite.

Of course, we are ultimately interested in $E(\mathbb{Q})$ and $\mathrm{III}(E/\mathbb{Q})$. By combining Kolyvagin's theorem with the work of Gross–Zagier and some analytic results (to be discussed later) we can deduce the following theorem.

Theorem 20.5.2. *Suppose $E/\mathbb{Q}$ is a modular elliptic curve. Then*

(*a*) *If $L(E/\mathbb{Q},s)$ has a simple zero at $s = 1$, then $E(\mathbb{Q})$ has rank 1, and $\mathrm{III}(E/\mathbb{Q})$ is finite.*
(*b*) *If $L(E/\mathbb{Q},1) \neq 0$, then $E(\mathbb{Q})$ is finite, and $\mathrm{III}(E/\mathbb{Q})$ is finite.*

The deduction of this theorem from the theorem of Kolyvagin is quite difficult. We will just sketch some of the ideas involved.

The first step is to connect Heegner points with the theory of L-functions. That such a connection should exist was also conjectured by B.J. Birch and N.M. Stephens.

Theorem 20.5.3 (Gross and Zagier). *Let $E/\mathbb{Q}$ be a modular elliptic curve, and $\varphi\colon X_0(N) \to E$ be a modular parameterization. Let $D < 0$ be the discriminant of an imaginary quadratic number field K. Assume $(N,D) = 1$ and that every rational prime p dividing N splits in K. Then $L'(E/K,1) = C\,\hat{h}(y_K)$, where C is a nonzero constant (which can be explicitly given) and $\hat{h}$ is the canonical height on $E(K)$.*

It follows from this theorem and the properties of the canonical height discussed in §4 that y_K has infinite order if and only if $L'(E/K,1) \neq 0$, which brings in the L-function. Now we want to relate $L(E/K,s)$ to $L(E/\mathbb{Q},s)$. To do this it is necessary to define the quadratic twist of an elliptic curve (see [Sil], Chap. 10, §5).

Suppose E is defined by $y^2 = x^3 + ax + b$. For $D \in \mathbb{Z}$, $D \neq 0$, we define E_D, the quadratic twist of E by D, by the equation $Dy^2 = x^3 + ax + b$. E_D is again an elliptic curve over $\mathbb{Q}$, and it is not too hard to prove the following proposition.

Proposition 20.5.4. *Let K be a quadratic number field with discriminant D, and E an elliptic curve over $\mathbb{Q}$. Then*

(a) rank $E(K) = $ rank $E(\mathbb{Q}) + $ rank $E_D(\mathbb{Q})$ and
(b) $L(E/K,s) = L(E/\mathbb{Q},s)\, L(E_D/\mathbb{Q},s)$.

We are now in a position to sketch the proof of Theorem 20.5.2. Let's consider part (a). The assumption is that $L(E/\mathbb{Q}, s)$ has a simple zero at $s = 1$, i.e., $L(E/\mathbb{Q}, 1) = 0$ and $L'(E/\mathbb{Q}, 1) \neq 0$. From Proposition 20.5.4, part (b), we find that $L'(E/K, 1) = L'(E/\mathbb{Q}, 1)\, L(E_D/\mathbb{Q}, 1)$. By a theorem of J.L. Waldspurger, there exist infinitely many fundamental discriminants $D < 0$ that satisfy the hypotheses of Theorem 20.5.3 and such that $L(E_D/\mathbb{Q}, 1) \neq 0$. For such a D we must have $L'(E/K, 1) \neq 0$, and so by Theorems 20.5.3 and 20.4.4, the point $y_K \in E(K)$ has infinite order. By Kolyvagin's theorem this implies rank $E(K) = 1$, and $\text{Ш}(E/K)$ is finite. By Proposition 20.5.4, part (a), either $E(\mathbb{Q})$ has rank 1, or $E_D(\mathbb{Q})$ has rank 1. Let bar denote complex conjugation. With the assumptions we have made it can be shown that $\bar{y}_K = y_K$. Thus, $2y_K = y_K + \bar{y}_K \in E(\mathbb{Q})$, and we conclude that $E(\mathbb{Q})$ has rank 1, as claimed. The fact that $\text{Ш}(E/\mathbb{Q})$ is finite follows easily from the fact that $\text{Ш}(E/K)$ is finite (provided that one knows the definition of either, of course).

To prove Theorem 20.5.2, part (b), we can use similar reasoning. The main difficulty remaining is to show the existence of discriminants D of the type we need which satisfy $L'(E_D, 1) \neq 0$. The existence of infinitely many such discriminants was shown by D. Bump, S. Friedberg, and J. Hoffstein in 1989 [Bu-Fr-Hof]. Independently, and at about the same time, this result was also obtained by M.R. Murty and V.K. Murty [Mur-Mur].

Surprisingly, these beautiful new results in the arithmetic theory of elliptic curves have led to the resolution of an old problem of Gauss on the class numbers of imaginary quadratic number fields. This is the topic of the next, and final, section of this chapter.

§6 Applications to Gauss's Class Number Conjecture

A large percentage of Gauss's number theoretic masterpiece, *Disquisitiones Arithmeticae* [136], is taken up with the theory of binary quadratic forms. In Article 303 of that work he describes the results of extensive calculations of class numbers of definite quadratic forms. These calculations can be reinterpreted as calculations of class numbers of imaginary quadratic number fields. If $D < 0$ is the discriminant of such a field, let $h(D)$ denote its class number. Gauss observed that apparently $h(D) \to \infty$ as $|D| \to \infty$. In fact, the last D for which $h(D) = 1$ seemed to be -163, the last for which $h(D) = 2$ seemed to be -427, and the last for which $h(D) = 3$ seemed to be -907 (Gauss uses a somewhat different normalization for class numbers and so his values are different from these). These observations led to two problems. First, prove the assertion that $h(D) \to \infty$ as $|D| \to \infty$. Second, prove an effective version of the same result, namely, for every integer n, produce an integer $C(n)$ such that if $|D| \geq C(n)$, $h(D) \geq n$. One would hope that the constants $C(n)$ would be small enough to show that Gauss succeeded in finding all imaginary quadratic number fields of class number 1, 2, and 3.

The first problem was solved affirmatively in the 1930s by the combined efforts of several mathematicians. The story is amusing and is connected with the Riemann hypothesis, so we pause to recall what that is about.

Let $\zeta(s) = \Sigma \, n^{-s}$ denote the zeta function of Riemann. $\zeta(s)$, as was proved in Chapter 16, can be analytically continued to the whole complex plane and is holomorphic everywhere except for a simple pole at $s = 1$. Riemann conjectured that the only zeros of $\zeta(s)$ in the strip $0 \leq \mathrm{Re}(s) \leq 1$ are on the line $\mathrm{Re}(s) = 1/2$. This assertion is known as the Riemann hypothesis and is one of the most famous unsolved conjectures in all of mathematics. There is a generalization of this assertion known as the generalized Riemann hypothesis. Dedekind associated a zeta function to an arbitrary number field K by setting $\zeta_K(s) = \Sigma \, \mathbb{N}A^{-s}$ where the sum is over all integral ideals $A \subseteq O_K$ and $\mathbb{N}A = [O_K : A]$. E. Hecke showed that this function could be analytically continued to all of $\mathbb{C}$ with only one pole, a simple pole at $s = 1$, that it satisfied a functional equation, etc. The generalized Riemann hypothesis asserts that the only zeros of $\zeta_K(s)$ in the strip $0 < \text{Re}(s) \le 1$ are on the line $\text{Re}(s) = 1/2$. In what follows, we use this assertion only as it applies to imaginary quadratic number fields.

The first major step forward toward a resolution of Gauss's conjecture was made by Hecke.

Hecke *(1918).* Let $D < 0$ be the discriminant of an imaginary quadratic number field K. Assume the generalized Riemann hypothesis. Then, there is an absolute constant C such that

$$h(D) > C \sqrt{|D|}/\log |D|.$$

This certainly shows that $h(D) \to \infty$ as $|D| \to \infty$, but it assumes a result that is far from proven even today. The next developments were really unexpected.

Deuring *(1933).* If the Riemann hypothesis is false, then $h(D) > 1$ if $|D|$ is sufficiently large.

Shortly thereafter, Mordell strengthened this result as follows:

Mordell *(1934).* If the Riemann hypothesis is false, then $h(D) \to \infty$ as $|D| \to \infty$.

Finally, H. Heilbronn completed this circle of ideas:

Heilbronn *(1934).* If the generalized Riemann hypothesis is false, then $h(D) \to \infty$ as $|D| \to \infty$.

Putting it all together gives a proof of the qualitative version of Gauss's conjecture.

Theorem 20.6.1 (Hecke, Deuring, Mordell, Heilbronn).

$$h(D) \to \infty \text{ as } |D| \to \infty.$$

The method of proof here is truly amazing. If the generalized Riemann hypothesis is true, then the theorem is true. If the generalized Riemann hypothesis is false, then the theorem is true. Thus, the theorem is true!!

C.L. Siegel took this approach one step further and proved the definitive theorem along these lines. His proof makes no use of the Riemann hypothesis one way or another.

Siegel *(1935).* Given $\varepsilon > 0$, there is a constant $C(\varepsilon) > 0$ such that

$$h(D) > C(\varepsilon) |D|^{1/2-\varepsilon}.$$

This is certainly a wonderful result, but it does not solve the problem of finding an effective version of Gauss's conjecture, because there is no way to compute the constant $C(\varepsilon)$ whose existence is asserted.

The next important step was taken almost 20 years later by Kurt Heegner, who, in 1952, published a paper entitled "Diophantische Analy-

sis und Modulfunktionen" (Diophantine Analysis and Modular Functions). In this paper Heegner claims to have solved Gauss's class number
1 conjecture by introducing new methods from the theory of modular
functions. That is, he claims to have shown that the only negative discriminants D with $h(D) = 1$ are -3, -4, -7, -8, -11, -19, -43, -67, and
-163. Although his paper was published in a reputable journal, the
Mathematische Zeitschrift, his claim was generally discounted. The paper was quite obscure in places, and it did contain some mistakes. As it
turned out, this neglect was completely unwarranted. His claim was later
vindicated. Unfortunately, he died before his accomplishment was generally recognized.

The first accepted proof of the class number 1 conjecture was given by
H. Stark in 1967. Soon thereafter A. Baker found another proof based on
the theory of transcendental numbers. The matter now being firmly established, people went back to look at Heegner's work and discovered that
the "gap" in his proof was not too hard to fill. Papers by Deuring, Siegel,
and Stark, among others, appeared showing how this could be done.

In 1971, Baker and Stark independently resolved the class number 2
problem. The largest (in absolute value) negative discriminant with class
number 2 was -427, as predicted by Gauss. However, there seemed to be
little hope that their methods could be extended to cover the case $h = 3$,
not to speak of larger class numbers.

This subject is full of surprises, and in 1976 D. Goldfeld proved a result
which connected the conjecture of Birch and Swinnerton-Dyer with the
conjecture of Gauss, although on the face of it, these conjectures are
completely unrelated.

Theorem (Goldfeld [1976]). *Suppose there exists an elliptic curve $E/\mathbb{Q}$ whose
L-function, $L(E/\mathbb{Q}, s)$, can be analytically continued to all of $\mathbb{C}$ and which
satisfies a functional equation of the predicted type (see §5 of this chapter) and
has a zero of order 3 or greater at $s = 1$. Then, given $\varepsilon > 0$, there is an effec-
tively computable constant $C(\varepsilon)$ such that $h(D) > C(\varepsilon)(\log |D|)^{1-\varepsilon}$ [Go2],
[Go3].*

If the sign of the functional equation for $L(E/\mathbb{Q}, s)$ is -1, it follows that
$L(E/\mathbb{Q}, s)$ has a zero of odd order at $s = 1$. Thus to ensure a zero of order 3
or greater in such a case, it is only necessary to prove that $L'(E/\mathbb{Q}, 1) = 0$.
If E is a modular elliptic curve, then its L-function has the required analytic continuation and functional equation. Moreover, the work of Gross
and Zagier discussed in §5 related the derivative at $s = 1$ to the height of a
Heegner point. Exploiting these connections, Gross and Zagier were able
to prove the following theorem.

Theorem (Gross–Zagier [1986]). *The curve $-139y^2 = x^3 + 10x^2 - 20x + 8$
satisfies all the hypotheses of Goldfeld's theorem. In particular, it has a
zero of order exactly 3 at $s = 1$ [Gr-Za].*

Taken together, these results of Goldfeld and Gross–Zagier finally give a positive resolution of the effective version of Gauss's class number conjecture some two hundred years after it was made.

The curve in the preceding theorem has conductor 714,877. The constant $C(\varepsilon)$ in Goldfeld's theorem is dependent on the size of the conductor, and this conductor is too large to resolve the case of class number 3. Brumer and Kramer had shown that the curve $y^2 + y = x^3 - 7x + 6$ of conductor 5077 has rank 3. If one could prove that it was modular, its L-function would have the required analytic properties and Birch-Swinnerton–Dyer would predict a zero of order 3 at $s = 1$. Assuming it to be modular, Buhler, Gross, and Zagier [Bu-Gr-Zag] proved its L-function had a zero of order 3 at $s = 1$. Then, Mestre and Serre verified that it was a modular elliptic curve. Working with this curve J. Oesterlé [Oes2] was able to prove that $h(D) > 1/55 \log(|D|)$ if D is prime. Together with earlier work of Montgomery and Weinberger, this was enough to show that -907 was the largest (in absolute value) negative discriminant of class number 3. Once again, Gauss was right!

It is perhaps fitting to end with an open problem. Throughout this section we have been discussing imaginary quadratic number fields. If one considers real quadratic number fields, the situation is much more mysterious. Gauss had already noticed that many real quadratic number fields have class number 1. Considering such fields which have prime discriminant, computations show that about 80% of them have class number 1. It is an open problem to prove that infinitely many real quadratic number fields have class number 1. In fact, it is not even known if there are infinitely many number fields with class number 1. In spite of all the successes recorded in this chapter, much remains to be done.

NOTES

In this section, numbered references refer to items in the general bibliography at the end of the book. New references relevant to the subject matter of this chapter are cited here by acronyms.

A major portion of this chapter consists of an expanded version of the expository article by M. Rosen [Ro]. For an elementary introduction to the algebraic theory of curves, the book by W. Fulton [135] is recommended. At present, the standard introduction to algebraic geometry is R. Hartshorne's book [144]. A somewhat less demanding, and very readable text, is the book by I. R. Shafarevich [Shaf].

B. Mazur has provided an excellent introduction to the ideas surrounding Faltings's resolution of the Mordell conjecture [Maz]. A very good expository article on the proof itself appears in S. Bloch [B]. Two recent volumes are devoted to providing the (extensive) mathematical background necessary to understanding the proof; [Co-Sil] and [Fa-Wu]. The first, [Co-Sil], contains an English translation of Faltings's original paper as well as a short historical article by Faltings on how he was led to the proof.

For conjectures of Mordell type in higher dimensions, the reader should consult the article by P. Vojta, "A Higher Dimensional Mordell Conjecture," in [Co-Sil]. In a somewhat different direction, an approach requiring an extensive amount of differential geometry can be found in an article by S. Lang [La1].

For an influential survey article on the arithmetic of elliptic curves, discussed in §2, see J. Tate [Ta]. There are now several texts devoted to this topic. Probably the best general introduction is by J. Silverman [Sil]. Other books, which overlap with the material in [Sil] but contain valuable discussions of other topics, are by D. Husemöller [Hu], N. Koblitz [Ko], and S. Lang [La2], [La3].

For an elegant introduction to the subject of modular forms, the reader should consult the last chapter of J.-P. Serre [Se]. More extensive introductions are given by T. Apostol [Ap], S. Lang [La5], and G. Shimura [Shim]. These are listed in increasing order of sophistication. Shimura's book contains a careful construction of the curves $X_0(N)$ and $X_1(N)$. The book by Lang has an introduction to the connection between modular forms and Galois representations, a theory used by Serre and Ribet in the proof of the theorem connecting the Taniyama–Weil conjecture and Fermat's last theorem. The book by Koblitz [Ko] is also recommended. In addition to containing an introduction to modular forms, this book presents the proof of a beautiful theorem of J. Tunnell, which virtually solves an old problem about congruent numbers (integers equal to the area of a right triangle with rational sides) by relating the problem to the conjecture of Birch and Swinnerton-Dyer.

A. Ogg [Ogg] provides an introduction to the theory of modular forms which includes an exposition of the famous 1967 paper of Weil [We]. J. Oesterlé discusses the theorem linking the Taniyama–Weil conjecture with Fermat's last theorem, and much else besides [Oes1].

For introductions to the theory of heights on elliptic curves, the reader can consult the books by Silverman [Sil] and Husemöller [Hu]. For an introduction to the theory in a more general context, see Silverman's article "The Theory of Height Functions," which appears as Chapter VI in [Co-Sil]. For the theory of heights, as well as many other things of interest in arithmetic geometry, the reader should consult S. Lang [La4]. This book appeared just before Faltings's proof of the Mordell conjecture and represented the state of the art in the subject "before the revolution."

For the theorem on lower bounds for the canonical height, and the subsequent application to bounding the number of integral points, see J. Silverman [Sil2], [Sil3].

For a more detailed series of conjectures about the canonical heights of the elements of a basis for $E(K)$, see Lang [La6].

The writing of §5 and the next section was heavily influenced by the survey article by D. Zagier [Zag]. It is amazing how much information this article condenses into just four pages.

The Coates–Wiles theorem appears in [114]. The basic new results which we discuss in this section appear in [Gr-Za], [Kol], and [Ru1].

A survey of the work of Gross–Zagier is given by J. Coates [Co].

For a somewhat simplified exposition of a portion of the theorem of Kolyvagin we have discussed, see K. Rubin [Ru2].

It is difficult to locate a reference to the analytic result of Waldspurger mentioned in the text. D. Bump, S. Friedberg, and J. Hoffstein, [Bru-Fr-Hof] discuss both his work and their new results on derivatives of L-functions. The paper containing the proof of their main theorem has not yet appeared. The same is true of the proof of M.R. Murty and V.K. Murty [Mur-Mur].

Section 6 follows rather closely the exposition given by D. Goldfeld [Go1]. We refer the reader there for an extensive bibliography of articles on this subject. The theorem of Goldfeld which we discussed is contained in two papers [Go2], [Go3].

A simplification of Goldfeld's proof, an exposition of the class number problem, and a detailed discussion of the application of the theorem of Gross–Zagier to the problem are provided by J. Oesterlé [Oes2]. The reader should also consult the introduction to [Gr-Za] as well as the expository paper of Zagier [Zag] mentioned previously. The proof that the L-function of $y^2 + y = x^3 - 7x + 6$ has a zero of order 3 at $s = 1$, subject to the assumption that it is modular, is given by J. Buhler, B. Gross, and D. Zagier [Bu-Gr-Zag].

The paper of Montgomery and Weinberger which was mentioned in connection with the class number 3 problem is "Notes on Small Class Numbers" [Mo-We]. A very interesting paper by Buell resulted from the calculation of all class numbers of imaginary quadratic number fields with discriminant of absolute value less than 4 million [Bue]. Up to 4 million $h(D) = 1$ for 9 values of $|D|$, the smallest being 3 and the largest 163. In the same range there are 18 values of $|D|$ such that $h(D) = 2$, the smallest being 15 and the largest being 427. There are 16 values of $|D|$ such that $h(D) = 3$, the smallest being 23 and the largest 907. We now know that these lists contain all discriminants with class numbers 1, 2, or 3. Buell presents similar statistics for many other values of $h(D)$. As this book goes to press, there is a rumor that the class number 4 problem has been solved.

Bibliography

Ap. T. Apostol. *Modular Functions and Dirichlet Series in Number Theory*. New York: Springer-Verlag, 1976.

B. S. Bloch. The proof of the Mordell conjecture. *Math. Intelligencer*, **6**(2), (1984), 41–47.

Bue. D.A. Buell. Small class numbers and extreme values of L-functions. *Math. Comp.*, **31** (1977), 786–796.

Bu-Fr-Hof. D. Bump, S. Friedberg, and J. Hoffstein. A non-vanishing theorem for derivatives of automorphic L-functions with applications to elliptic curves. *Bull. Am. Math. Soc.*, **21**(1), (1989), 89–93.

Bu-Gr-Zag. J. Buhler, B. Gross, and D. Zagier. On the conjecture of Birch and Swinnerton-Dyer for an elliptic curve of rank 3. *Math. Comp.*, **44**(170), (1985), 471–481.

Co. J. Coates. The Work of Gross–Zagier on Heegner Points and the Derivatives of L-series, Sem. Bourbaki, No. 635, 1984–85. In *Astérisque*, Vol. 133–34, 1986.

Co-Sil. G. Cornell and J. Silverman. *Arithmetic Geometry*. New York: Springer-Verlag, 1986.

Fa. G. Faltings. Diophantine approximation on abelian varieties, to appear.

Fa-Wu. G. Faltings, G. Wüstholtz, et al. Rational Points, Vieweg, *Aspects of Mathematics*, Vol. E6, 1984.

Go1. D. Goldfeld. Gauss's class number problem for imaginary quadratic fields. *Bull. Am. Math. Soc.*, **13**(1), (1985), 23–37.

Go2. D. Goldfeld. The class number of quadratic fields and the conjectures of Birch and Swinnerton-Dyer. *Ann. Scuola Norm. Sup., Pisa* **3**(4), (1976), 623–663.

Go3. D. Goldfeld. The conjectures of Birch and Swinnerton-Dyer and the class number of quadratic fields. *Arith. de Caen, Astérisque*, **41–42** (1977), 219–227.

Gr-Za. B. Gross and D. Zagier. Heegner points and derivatives of L-series. *Invent. Math.*, **84** (1986), 225–320.

Hu. D. Husemöller. *Elliptic Curves*. New York: Springer-Verlag, 1987. Graduate Texts in Mathematics, Vol. 111.

Ko. N. Koblitz, *Introduction to Elliptic Curves and Modular Forms*. New York: Springer-Verlag, 1984. Graduate Texts in Mathematics, Vol. 97.

Kol. V.A. Kolyvagin. Finiteness of $E(\mathbb{Q})$ and $Ш(E, \mathbb{Q})$ for a class of Weil curves. *Math. Nauk. SSSR Ser. Mat.*, **52** (1988), 522–540 (Russian); *Math. of the USSR Izvestiya*, **32** (1989), 523–542 (English).

La1. S. Lang. Hyperbolic and Diophantine analysis. *Bull. Am. Math. Soc.*, **14**(2), (1986), 159–205.

La2. S. Lang. *Elliptic Functions*. Reading, Mass.: Addison-Wesley, 1973.

La3. S. Lang. *Elliptic Curves: Diophantine Analysis*. New York: Springer-Verlag, 1978.

La4. S. Lang. *Fundamentals of Diophantine Geometry*. New York: Springer-Verlag, 1983.

La5. S. Lang. *Introduction to Modular Forms*. New York: Springer-Verlag, 1976.

La6. S. Lang. Conjectured Diophantine estimates on elliptic curves. *Progress in Mathematics*, Vol. 35. Cambridge, Mass.: Birkhäuser, 1983.

Maz. B. Mazur. Arithmetic on curves. *Bull. Am. Math. Soc.* **14**(2), (1986), 207–259.

Mo-We. H.L. Montgomery and P.J. Weinberger. Notes on small class numbers. *Acta Arith.*, **24** (1973), 529–542.

Mur-Mur. M.R. Murty and V.K. Murty. Mean values of derivatives of modular L-series. To appear in *Ann. of Math.*

Oes1. J. Oesterlé. Nouvelles Approches du "Théorème" de Fermat, Sem. Bourbaki, No. 694, 1987. In *Astérisque*, Vol. 161–62, 1988.

Oes2. J. Oesterlé. Nombres de classes de corps quadratiques imaginaire, Sem. Bourbaki, No. 631, 1983–84. In *Astérisque*, Vol. 121–22, 1985.

Ogg. A. Ogg. *Modular Forms and Dirichlet Series*. Menlo Park, Calif.: W. A. Benjamin, 1969.

Ro. M. Rosen. New results on the arithmetic of elliptic curves. *Les Gazette des Sciences Math. du Quebec*, forthcoming.

Ru1. K. Rubin. Tate–Shaferevich groups and L-functions of elliptic curves with complex multiplication. *Invent. Math.*, **89** (1987), 527–560.

Ru2. K. Rubin. The work of Kolyvagin on the arithmetic of elliptic curves. In: The Arithmetic of Complex Manifolds. *Lecture Notes in Mathematics,* Vol. 1399. New York: Springer-Verlag, 1989.

Se. J.-P. Serre. *A Course in Arithmetic.* New York: Springer-Verlag, 1973.

Shaf. I.R. Shaferevich. *Basic Algebraic Geometry.* New York: Springer-Verlag, 1977.

Shim. G. Shimura. *Arithmetic Theory of Automorphic Forms.* Tokyo: Iwanami Shoten and Princeton, N.J.: Princeton University Press, 1971.

Si1. J. Silverman. *The Arithmetic of Elliptic Curves.* New York: Springer-Verlag, 1986. Graduate Texts in Mathematics, Vol. 106.

Si2. J. Silverman. Lower bounds for the canonical height on elliptic curves. *Duke Math. J.,* **48** (1981), 633–648.

Si3. J. Silverman. A quantitative version of Siegel's theorem. *J. Reine und Angew. Math.,* **378** (1987), 60–100.

Ta. J. Tate. The arithmetic of elliptic curves. *Invent. Math.,* **23** (1974), 179–206.

We. A. Weil. Über die Bestimmung Dirichletscher Reihen durch Funktionalgleichungen. *Math. Ann.,* **168** (1967), 149–156.

Zag. D. Zagier. L-Series of elliptic curves, the Birch-Swinnerton–Dyer conjecture, and the class number problem of Gauss. *Notices Am. Math. Soc.,* **31**(7), (1984), 739–743.

Selected Hints for the Exercises

Chapter 1

6. Use Exercise 4.
8. Do it for the case $d = 1$ and then use Exercise 7 to do it in general.
9. Use Exercise 4.
15. Here is a generalization; a is an nth power iff $n \mid \text{ord}_p a$ for all primes p.
16. Use Exercise 15.
17. Use Exercise 15 to show that $a^2 = 2b^2$ implies that 2 is the square of an integer.
23. Begin by writing $4(a/2)^2 = (c - b)(c + b)$.
28. Show that $n^5 - n$ is divisible by 2, 3, and 5. Then use Exercise 9.
30. Let s be the largest integer such that $2^s \leq n$, and consider $\sum_{k=1}^{n} 2^{s-1}/k$. Show that this sum can be written in the form $a/b + \frac{1}{2}$ with b odd. Then use Exercise 29.
31. $2 = (1 + i)(1 - i) = -i(1 + i)^2$.
34. Since $\omega^2 = -1 - \omega$ we have $(1 - \omega)^2 = 1 - 2\omega + \omega^2 = -3\omega$, so $3 = -\omega^2(1 - \omega)^2$.

Chapter 2

1. Imitate the classical proof of Euclid.
2. Use $\text{ord}_p(a + b) \geq \min(\text{ord}_p a, \text{ord}_p b)$.
3. If $p_1, p_2, \ldots, p_t$ were all the primes, then $\phi(p_1 p_2 \cdots p_t) = 1$. Now use the formula for ϕ and derive a contradiction.
5. Consider $2^2 + 1, 2^4 + 1, 2^8 + 1, \ldots$. No prime that divides one of these numbers can divide any other, by the previous exercise.
6. Count! Consider the set of pairs (s, t) with $p^s t \leq n$.
12. In each case the summand is multiplicative. Hence evaluate first at prime powers and then use multiplicavity.

17. Use the formula for $\sigma(n)$.
20. If $d\,|\,n$, then n/d also divides n.
22. If $(t, n) = 1$, then $(n - t, n) = 1$, so you can pair those numbers relatively prime to n in such a way that the sum of each pair is n.

Chapter 3

1. Suppose that $p_1, p_2, \ldots, p_t$ are all congruent to -1 modulo 6. Consider $N = 6p_1p_2\cdots p_t - 1$.
3. 10^k is congruent to 1 modulo 3 and 9 and congruent to $(-1)^k$ modulo 11.
5. If a solution exists, then $x^3 \equiv 2\,(7)$ has a solution. Show that it does not.
10. If n is not a prime power, write $n = ab$ with $(a, b) = 1$. If $n = p^s$ with $s > 1$, then $(n - 1)!$ is divisible by $p \cdot p^{s-1} = p^s = n$. If $n = p^2$ and $p \neq 2$; then $(n - 1)!$ is divisible by $p \cdot 2p = 2n$.
13. Show that $n^p \equiv n\,(p)$ for all n by induction. If $(n, p) = 1$, then one can cancel n and get Fermat's formula.
17. Let x_i be a solution to $f(x) \equiv 0\,(p_i^{a_i})$ and solve the system $x \equiv x_i\,(p_i^{a_i})$.
23. Since $i \equiv -1\,(1 + i)$, we have $a + bi \equiv a - b\,(1 + i)$. Write $a - b = 2c + d$, where $d = 0$ or 1. Then $a + ib \equiv d\,(1 + i)$.
25. Write $\alpha = 1 + \beta\lambda$, cube both sides and take congruence modulo λ^4 to get $\alpha^3 \equiv 1 + (\beta^3 - \omega^2\beta)\lambda^3\,(\lambda^4)$. Then show that the term in parentheses is divisible by λ.

Chapter 4

4. If $(-a)^n \equiv 1$, and n is even, then $p - 1\,|\,n$. If n is odd, then $p - 1\,|\,2n$, which implies that $2\,|\,n$ is a contradiction.
6. This is a bit tricky. If 3 is not a primitive element, show that 3 is congruent to a square. Use Exercise 4 to show there is an integer a such that $-3 \equiv a^2\,(p)$. Now solve $2u \equiv -1 + a(p)$ and show that u has order 3. This would imply that $p = 1\,(3)$, which cannot be true.
7. Use the fact that 2 is not a square modulo p.
9. See Exercise 22 of Chapter 2 and use the fact that $g^{(p-1)/2} \equiv -1\,(p)$ for a primitive root g.
11. Express the numbers between 1 and $p - 1$ as the powers of a primitive root and use the formula for the sum of a geometric progression.
14. If $(ab)^s = e$, then $a^{ns} = 1$, implying that $m\,|\,ns$. Thus $m\,|\,s$. Similarly, $n\,|\,s$. Thus $mn\,|\,s$.
18. Choose a primitive element (e.g., 2) and construct the elements of order 7.
22. Show first that $1 + a + a^2 \equiv 0\,(p)$.
23. Use Proposition 4.2.1.

Chapter 5

3. Use the identity $4(ax^2 + bx + c) = (2ax + b)^2 - (b^2 - 4ac)$.
9. Using $k \equiv -(p - k)\,(p)$, show first that $2 \cdot 4 \cdot \ldots \cdot (p - 1) \equiv (-1)^{(p-1)/2}1 \cdot 3 \cdot 5 \cdot \ldots \cdot p - 2\,(p)$.
10. Use Exercise 9.

13. If $x^4 - x^2 + 1 \equiv 0$ (p), then $(2x^2 - 1)^2 \equiv -3$ (p) and $(x^2 - 1)^2 \equiv -x^2$ (p). Conclude that $p \equiv 1$ (3) and $p \equiv 1$ (4) by using quadratic reciprocity.

18. Let $D = p_1 p_2 \cdots p_m$ and suppose that n is a nonresidue modulo p_1. Find a number b such that $b \equiv 1$ (p_i) and $b \equiv n$ (p_1) for $1 < i \leq m$. Then use the definition of the Jacobi symbol to show that $(b/D) = -1$.

23. Since $s^2 + 1 = (s + i)(s - i)$, if p is prime in $\mathbb{Z}[i]$, then either $p|s + i$ $p|s - i$, but neither alternative is true.

26. To prove (b) notice that $a + b$ is odd, so from $2p = (a + b)^2 + (a - b)^2$ we see that $(2p/a + b) = 1$. Now use the properties of the Jacobi symbol.

29. It is useful to consider the cases $p \equiv 1$ (4) and $p \equiv 3$ (4) separately.

30. To evaluate the sum notice that $(n(n + 1)/p) = ((2n + 1)^2 - 1/p)$.

Chapter 6

1. Find an equation of degree 4.

2. If $a_0 \alpha^s + a_1 \alpha^{s-1} + \cdots + a_s = 0$, with $a_i \in \mathbb{Z}$, multiply both sides with a_0^{s-1} and conclude that $a_0 \alpha$ is an algebraic integer.

3. Suppose that α and β satisfy monic equations with integer coefficients of degree m and n, respectively. Let γ be a root of $x^2 + \alpha x + \beta$ and show that the $\mathbb{Z}$ module generated by $\alpha^i \beta^j \gamma^k$, where $0 \leq i < m, 0 \leq j < n$, and $k = 0$ or 1, is mapped into itself by γ.

10. Use $g_a = (a/p)g$ and the fact that $\sum_a (a/p) = 0$.

11. Remember that $1 + (t/p)$ is the number of solutions to $x^2 \equiv t$ (p) and that $\sum_t \zeta^t = 0$.

13. Use Exercise 12.

16. Show that otherwise $f'(\alpha) = 0$ and apply Proposition 6.1.7.

23. Use Exercise 4 to show that it is enough to show that $f(x)$ is irreducible in $\mathbb{Z}[x]$. Then write $f(x) = g(x)h(x)$, reduce modulo p, and use the fact that $F_p[x]$ is a unique factorization domain.

Chapter 7

3. Since $q \equiv 1$ (n), there are n solutions to $x^n = 1$. If $\beta^n = \alpha$, then the other solutions to $x^n = \alpha$ are given by $\gamma\beta$, where γ runs through the solutions of $x^n = 1$.

5. $q^n - 1 = (q - 1)(q^{n-1} + \cdots + q + 1)$. Since $q \equiv 1$ (n), we have $q^{n-1} + \cdots + q + 1 \equiv n \equiv 0$ (n). Thus $n(q - 1)$ divides $q^n - 1$.

7. Let $m = [K : F]$. α is a square in K iff $\alpha^{(q^m - 1)/2} = 1$. If α is not a square in F, then $\alpha^{(q-1)/2} = -1$. Show that $\alpha^{(q^m - 1)/2} = (-1)^m$. This formula yields the result.

9. Use the method of Exercise 7.

14. One can prove this by exactly the same method as for F_p. Alternatively, suppose that $q = p^m$. Let $f(x) \in F_p[x]$ be an irreducible of degree mn and let $g(x)$ be an irreducible factor of $f(x)$ in $F_q[x]$. Let α be a root of $g(x)$ and show that $F_q \subset F_p(\alpha)$. Conclude that $F_q(\alpha) = F_p(\alpha)$ and that $[F_q(\alpha) : F_q] = n$. It follows that $g(x)$ has degree n.

15. If $x^n - 1$ splits into linear factors in E, where $[E:F] = f$, then E has q^f elements and $n | q^f - 1$ since the roots of $x^n - 1$ form a subgroup of E^* of order n.

23. If β is a root of $x^p - x - \alpha$, then so are $\beta + 1, \beta + 2, \ldots, \beta + (p-1)$. Using this, one can show the statement about irreducibility. To prove the final assertion, notice that $\beta^p = \beta + \alpha$ implies that $\beta^{p^2} = \beta^p + \alpha^p = \beta + \alpha + \alpha^p$, etc. Thus $\beta^{p^n} = \beta + \text{tr}(\alpha)$ and so $\beta \in F$ iff $\text{tr}(\alpha) = 0$.

Chapter 8

1. Use the Corollary to Proposition 8.1.3 and Proposition 8.1.4.

4. Make the substitution $t = (k/2)(u + 1)$ and use Exercise 3.

6. It follows from Exercise 5 together with part (d) of Theorem 1, or directly from Exercise 4 by substituting $k = 1$.

8. Use Proposition 8.1.5 and imitate the proof of Exercise 3.

14. Use Proposition 8.3.3.

19. First show that the number of solutions is given by $p^{r-1} + J_0(\chi, \chi, \ldots, \chi)$, where χ is a character of order 2 and there are r components in J_0. Then use Proposition 8.5.1 and Theorem 3. Notice in particular that if r is odd, the answer is simply p^{r-1}.

28. For (a): Write

$$\sum_{x=1}^{p-1} x\chi(x) = \sum_{x=1}^{(p-1)/2} x\chi(x) + \sum_{x=1}^{(p-1)/2} (p-x)\chi(p-x).$$

For (b): Write

$$\sum_{x=1}^{p-1} x\chi(x) = \sum_{x=1}^{(p-1)/2} 2x\chi(2x) + \sum_{x=1}^{(p-1)/2} (p-2x)\chi(p-2x).$$

For (c) and (d): Equate (a) and (b).

Chapter 9

3. Use the fact that $N\gamma = a^2 - ab + b^2 \equiv 3(m+n) + 1$ (9).

4. Rewrite γ as $3(m+n) - 1 - 3n\lambda$. Thus $\gamma \equiv 3(m+n) - 1(3\lambda)$.

5. Remember that $3 = -\omega^2\lambda^2$.

7. $2 + 3\omega, -7 - 3\omega$, and $-4 - 3\omega$.

10. $D/5D$ has 25 elements. Thus $x^{24} - 1$ factors completely into linear factors in D.

13. Use Exercise 9 to show that the elements listed represent all the cubes in $D/5D$.

15. Remember that every element in $D/\pi D$ is represented by a rational integer.

19. Use Exercise 18, the law of cubic reciprocity, and induction on the number of primary primes dividing γ.

23. Let $p = \pi\bar{\pi}$, where π is primary. By Exercise 15 $x^3 \equiv 3$ (p) is solvable iff $\chi_\pi(3) = 1$. By Exercise 5 $\chi_\pi(3) = \omega^{2n}$, where $\pi = a + b\omega$ and $b = 3n$. It follows that $x^3 \equiv 3$ (p) is solvable iff $9 | b$.

24. (c) Use cubic reciprocity with $\pi \equiv b\omega$ (a).
    (d) Write $(a + b) = (a + b)\omega \cdot \omega^{-1}$ and note that $a + b\omega \equiv a(1 - \omega)(\pi)$.
25. (a) Use Exercise 18 and the corollary to Proposition 9.3.4 to show that $\chi_{a+b}(b) = 1$. Note that $\pi \equiv -b(1 - \omega)(a + b)$.
    (b) $\chi_{a+b}(1 - \omega) = (\chi_{a+b}(1 - \omega)^2)^2$
    $\qquad\qquad = (\chi_{a+b}(-3\omega))^2$ etc.
39. Combine Exercises 6 and 27 of Chapter 8 with Proposition 9.6.1.
40. See the hint to the previous exercise.
43. Use Exercise 23, Chapter 6.

Chapter 10

2. Map $[x_0, x_1, \ldots, x_{n-1}]$ to $[0, x_0, x_1, \ldots, x_{n-1}]$.
3. Since the number of points in $A^n(F)$ is q^n, the decomposition of $P^n(F)$ shows that the number of points in $P^n(F)$ is q^n plus the number of points in $P^{n-1}(F)$. One now proceeds by induction.
4. It is no loss of generality to assume that $a_0 \neq 0$. If $[x_0, x_1, \ldots, x_n]$ is a solution, map it to the point $[x_1, x_2, \ldots, x_n]$ of $P^{n-1}(F)$. Show this map is well defined, one to one, and onto.
5. Substitute, "dehomogenize," and use the fact that a polynomial of degree n has at most n roots.
9. The kth partial derivative is $ma_k x_k^{m-1}$. Since each $a_k \neq 0$ and m is prime to the characteristic, the only common zero of all the partial derivatives has all its components zero. This, however, does not correspond to a point of projective space.
12. The "homogenized" equation is $t^2x^2 + t^2y^2 + x^2y^2 = 0$. Setting $t = 0$ we see that the points at infinity are $(0, 0, 1)$ and $(0, 1, 0)$. Calculating partial derivatives and substituting shows that both these points are singular.
14. Consider the associated homogeneous equation and calculate the three partial derivatives. Assuming that a common solution exists, show that $4a^3 + 27b^2 = 0$.
19. The trace is identically zero on F_p iff $p \mid n$.
20. Consider the mapping $h(x) = x^p - x$ from F_q to F_q. Prove that it is a homomorphism and that its image has q/p elements. Prove also that the image of h is contained in the kernel of the trace mapping. Show that the latter map has less than or equal to q/p elements in its kernel. The result follows.
21. Count the number of such maps.
23. Substitute and calculate.

Chapter 11

4. In F_q there are $2q + 1$ points at infinity and q^2 finite points. Thus $N_s = p^{2s} + 2p^s + 1$.

7. The number of lines in $P^n(F)$ is equal to the number of planes $A^{n+1}(F)$ which pass through the origin. The answer is $(q^{n+1} - 1)(q^{n+1} - q)(q^2 - 1)^{-1}(q^2 - q)^{-1}$.

9. There is one point at infinity. For $x = 0$ there is only one point $(0, 0)$ on the curve. If $x \neq 0$, let $t = y/x$ and consider $t^2 = x + 1$. This has $p - 2$ solutions with $x \neq 0$. Altogether there are p solutions in F_p. Similarly, there are q solutions in F_q. Thus the answer is $(1 - pu)^{-1}$.

12. To begin with, calculate the number of solutions to $u^2 - v^4 = 4D$.

16. The important facts are that $N_{F_s/F}$ is a homomorphism which is onto, and that the group of multiplicative characters of a finite field is cyclic.

18. Use the relation between Gauss sums and Jacobi sums and the Hasse–Davenport relation.

19. After expanding the terms of the product into geometric series, the result reduces to the fact that every monic polynomial is the product of monic irreducible polynomials in a unique way.

20. Use the identity $1 - T^s = \prod_{k=0}^{s-1}(1 - \zeta^k T)$, where $\zeta = e^{2\pi i/s}$.

Chapter 12

7. $21 = (1 + 2\sqrt{-5})(1 + 2\sqrt{-5})$.

8. Write $\det(\omega_i^{(j)})$ as $P - N$, where P is the sum of terms corresponding to the even permutations and N is the corresponding sum for odd permutations. Then notice that $(P - N)^2 = (P + N)^2 - 4PN$. A standard argument shows that $P + N$ and PN are integers.

9. Use Proposition 12.1.4 and elementary symmetric functions.

14. Consider $\zeta + \zeta^{-1}$ where ζ is a primitive seventh root of unity.

21–23. Let $\{\beta_j\}$ be a basis for F over $\mathbb{Q}(\alpha)$. Use the basis $\{\alpha^i \beta_j\}$ for F over $\mathbb{Q}$.

26. Choose a primitive g for the residue field. Lift it to D and consider the corresponding minimal polynomial over the fixed field of the decomposition group (see [207], p. 223).

Chapter 13

1. Show that $\phi(n)$ is even if $n > 2$.

2. Use Proposition 13.1.3.

3. $\mathbb{Q}(\sqrt{p}) \subset \mathbb{Q}(\zeta_p)$.

24. The discriminant of a quadratic field is 0 or 1 modulo 4.

27. The order of σ_p cannot be 4. See Theorem 2.

Chapter 14

1. (a) Use the definition of $J(\chi, \psi)$, the binomial theorem and Exercise 11, Chapter 4. See also Lemma 1, Chapter 9, p. 115.

12. See Exercise 17(e).

14. Let P be a prime ideal dividing p. Show $(\alpha/P)(\alpha/\bar{P}) = 1$. See [166], Satz 1034.

17. (b) Examine the ramification of l in the diagram

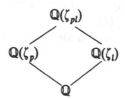

(c) Note that $\zeta_l^{\sigma_l} = \zeta_l^t = (1 - (1 - \zeta_l))^t$.

(e) Use Theorem 1, Chapter 8 and the fact that $g(\chi_P^t) = g(\chi_P)^{\sigma_t}$.

Chapter 15

2. Use Theorem 3.

3. Use Theorem 3 and Proposition 15.2.4.

9. As a function of a complex variable $(e^t - 1)^{-1}$ is analytic for $|t| < 2\pi$.

13. Use Exercise 12.

21. Set $F = 2$ in Exercise 19.

Chapter 16

4. For another evaluation note that $\int_0^1 t^{3k}(1 - t)\, dt = 1/[(3k + 1)(3k + 2)]$.

7. Show that if $p \nmid m$ and $p \mid \Phi_m(N)$ for an integer N then $p \equiv 1 \ (m)$.

11. For an integer m choose a prime $p \equiv 1 \ (m)$ and consider subfields of $\mathbb{Q}(\zeta_p)$.

12. If $p \equiv t \ (m)$ then $p \mid f(\zeta^p) = f(\zeta^t)$ where ζ is a primitive mth root of unity and $f(x) \in \mathbb{Z}[x], f(\zeta) = 0$.

14. Use Theorem 1, Chapter 6.

Chapter 17

2. $y^2 + 4 = x^3 - 27$.

3. Imitate the proof of Proposition 17.8.1 ([60], Theorem 121).

8. $(y + 2i)(y - 2i) = x^3$.

12. Consider $(x_1 + y_1\sqrt{d})^2$ for a solution (x_1, y_1) of $x^2 - dy^2 = -1$.

13. $1^3 + 2^3 + \cdots + n^3 = (n(n + 1)/2)^2$.

16. Consider the map

$$(x_1, x_2, x_3, x_4) \rightarrow \left(\frac{x_1 + x_2}{2}, \frac{x_1 - x_2}{2}, \frac{x_3 + x_4}{2}, \frac{x_3 - x_4}{2} \right).$$

18. $\binom{4}{2} = 6$.

19. Consider the hint for Problem 16.

Chapter 18

4. If t is the order of the torsion subgroup of E then for $p \equiv 2 \ (3)$, $p \equiv -1 \ (t)$. The density of the set of primes $\equiv -1 \ (t)$ is $1/\phi(t)$ while the density of primes $p \equiv 2 \ (3)$ is $\frac{1}{2}$.

8. (a) Prove first for $\mathfrak{A} = P$ using $(N(P) - 2)(N(P)) = (N(P) - 1)^2 - 1$.
   (b) See Exercise 4, Chapter 14. For $|u(a, b)| = 1$, apply σ_{-1} (cf. Lemma 4, Section 5, Chapter 14).
   (c) Show that $\hat{u}$ is invariant under the action of the appropriate Galois group.
12. (a) See Chapter 11.
   (b) See Exercise 4.
   (c) See Exercise 17.

Bibliography

First Bibliography

1. A. Albert. *Fundamental Concepts of Higher Algebra*. Chicago: University of Chicago Press, 1956.
2. E. Artin. *The Collected Papers of Emil Artin*. Reading, Mass.: Addison-Wesley, 1965.
3. J. Ax. Zeros of polynomials over finite fields. *Am. J. Math.*, **86** (1964), 255-261.
4. P. Bachman. *Niedere Zahlentheorie*, Vol. 1. Leipzig, 1902, p. 83.
5. P. Bachman. *Die Lehre von der Kreisteilung*. Leipzig, 1872.
6. P. Bachman. Über Gauss' Zahltheoretische Arbeiten. *Gott. Nach.* (1911), 455-508.
7. A. Beck, M. N. Bleicher, and D. W. Crowe. *Excursions into Mathematics*. New York: Worth, 1969.
8. H. Bilharz. Primdivisor mit vorgegebener Primitivwurzel. *Math. Ann.*, **114** (1937), 476-492.
9. Z. I. Borevich and I. R. Shafarevich. *Number Theory*. Transl. by N. Greenleaf. New York: Academic Press, 1966.
10. L. Carlitz. The arithmetic of polynomials in a Galois field. *Am. J. Math.*, **54** (1932), 39-50.
11. L. Carlitz. Some applications of a theorem of Chevalley. *Duke Math. J.*, **18** (1951), 811-819.
12. L. Carlitz. Some problems involving primitive roots in a finite field. *Proc. Nat. Acad. Sci. U.S.A.*, **38** (1952), 314-318.
13. L. Carlitz. Kloosterman sums and finite field extensions. *Acta Arithmetica*, **16** (1969), 179-193.
14. P. Cartier. Sur une généralisation des symboles de Legendre-Jacobi. *L'Enseignement Math.*, **15** (1970), 31-48.
15. J. W. S. Cassels. On Kummer sums. *Proc. London Math. Soc.*, **21**, no. 3 (1970), 19-27.
16. C. Chevalley. Démonstration d'une hypothèse de M. Artin. *Abhand. Math. Sem. Hamburg*, **11** (1936), 73-75.
17. S. Chowla. The last entry in Gauss' diary. *Proc. Nat. Acad. Sci. U.S.A.*, **35** (1949), 244-246.

18. S. Chowla. *The Riemann Hypothesis and Hilbert's Tenth Problem.* New York: Gordon & Breach, 1965.
19. S. Chowla. A note on the construction of finite Galois fields $GF(p^n)$. *J. Math. Anal. Appl.*, **15** (1966), 53–54.
20. S. Chowla. An algebraic proof of the law of quadratic reciprocity. *Norske Vid. Selsk. Forh. (Trondheim)*, **39** (1966), 59.
21. H. Davenport. On the distribution of quadratic residues mod p. *London Math. Soc. J.*, **5–6** (1930–1931), 49–54.
22. H. Davenport. *The Higher Arithmetic.* London: Hutchinson, 1968.
23. H. Davenport and H. Hasse. Die Nullstellen der Kongruenz Zetafunktion in gewissen zyklischen Fällen. *J. Reine und Angew. Math.*, **172** (1935), 151–182.
24. M. Deuring. The zeta functions of algebraic curves and varieties. *Indian J. Math.* (1955), 89–101.
25. L. Dickson. *Linear Algebraic Groups and an Exposition of the Galois Field Theory. 1900.* New York: Dover, 1958.
26. B. Dwork. On the rationality of the zeta function. *Am. J. Math.*, **82** (1959), 631–648.
27. G. Eisenstein. Beiträge zur Kreisteilung. *J. Reine und Angew. Math.* (1844), 269–278.
28. G. Eisenstein. Beweis des Reciprocitätssatzes für die kubischen Reste.... *J. Reine und Angew. Math.* (1844), 289–310.
29. G. Eisenstein. Nachtrag zum kubischen Reciprocitätssatze. *J. Reine und Angew. Math.*, **28** (1844), 28–35.
30. G. Eisenstein. Beiträge zur Theorie der elliptischen Funktionen. *J. Reine und Angew. Math.*, **35** (1847), 135–274.
31. P. Erdös. Some recent advances and current problems in number theory. *Lectures in Modern Mathematics*, Vol. 3. New York: Wiley, 1965.
32. A. Frankel. Integers and the theory of numbers. *Scripta Math. Studies*, **5**, 1955.
33. E. Galois. *Oeuvres Mathematiques.* Paris: Gauthier-Villars, 1897.
34. C. F. Gauss. *Arithmetische Untersuchungen.* New York: Chelsea, 1965.
35. L. Goldstein. Density questions in algebraic number theory. *Am. Math. Monthly*, **78** (April 1971), 342–351.
36. R. Graham. On quadruples of consecutive kth power residues. *Proc. Am. Math. Soc.* (1964), 196–197.
37. M. Greenberg. *Forms in Many Variables.* Menlo Park, Calif.: W. A. Benjamin, 1969.
38. G. H. Hardy. *Prime Numbers.* Manchester: British Association, 1915, pp. 350–354 and *Collected Papers*, Vol. 2.
39. G. H. Hardy. An introduction to the theory of numbers. *Bull. Am. Math. Soc.*, **35** (1929), 778–818.
40. G. H. Hardy and E. M. Wright. *An Introduction to the Theory of Numbers.* 4th ed., New York: Oxford University Press, 1960.
41. H. Hasse. *Vorlesungen über Zahlentheorie.* Berlin: Springer-Verlag, 1964.
42. H. Hasse. *The Riemann Hypothesis in Function Fields.* Philadelphia: Pennsylvania State University Press, 1969.
43. A. Hausner. On the law of quadratic reciprocity. *Archiv der Math.*, **12** (1961), 182–183.
44. E. Hecke. *Algebraische Zahlentheorie.* Leipzig: 1929. Reprinted by Chelsea Publishing Company, Inc., New York.
45. L. Holzer. *Zahlentheorie.* Leipzig: Teubner Verlagsgesellschaft, 1958.
46. C. Hooley. On Artin's conjecture. *J. Reine und Angew. Math.*, **225** (1967), 209–220.
47. C. Jacobi. Über die Kreisteilung .... *J. Reine und Angew. Math.* (1846), 254–274.
48. E. Jacobsthal. Über die Darstellung der Primzahlen der Form $4n + 1$ als Summe zweier Quadrate. *J. Reine und Angew. Math.*, **132** (1907), 238–245.

49. C. Jordan. *Traite des substitutions*. Paris: 1870.
50. H. Kornblum. Uber die Primfunktionen in einer Arithmetischen Progression. *Math. Z.*, **5** (1919), 100–111.
51. E. Kummer. Über die allgemeinen Reciprocitatsgesetz . . . . *Math. Abh. Akad. Wiss. zu Berlin* (1859), 19–160.
52. E. Landau, *Elementary Number Theory*. 2nd ed. New York: Chelsea, 1966.
53. S. Lang. Some theorems and conjectures on diophantine equations. *Bull. Am. Math. Soc.*, **66** (1960), 240–249.
54. D. H. Lehmer. A note on primitives. *Scripta Mathematica*, **26** (1963), 117–119.
55. E. Lehmer. On the quintic character of 2 and 3, *Duke Math. J.*, **18** (1951), 11–18.
56. E. Lehmer. Criteria for cubic and quartic residuacity. *Mathematika*, **6** (1958), 20–29.
57. P. Leonard. On constructing quartic extensions of *GF(p)*. *Norske Vid. Selsk. Forh. (Trondheim)*, **40** (1967), 41–52.
58. H. B. Mann. *Introduction to Number Theory*. Columbus, Ohio: Ohio State University Press, 1955.
59. W. H. Mills. Bounded consecutive residues and related problems. *Proc. Symp. Pure Math.*, **8** (1965).
60. T. Nagell. *Introduction to Number Theory*. New York: Wiley, 1951. Reprinted by Chelsea Publishing Company, Inc., New York.
61. I. Niven and H. S. Zuckerman. *An Introduction to the Theory of Numbers*. 2nd ed. New York: Wiley, 1966.
62. C. Pisot. Introduction à la theorie des nombres algébriques. *L'Enseignement Math.*, **8**, no. 2 (1962), 238–251.
63. H. Pollard and H. Diamond. *The Theory of Algebraic Numbers*. New York: Wiley, 1950. 2nd ed., 1975.
64. H. Rademacher. *Lectures on Elementary Number Theory*. Lexington, Mass.: Xerox College Publishing, 1964.
65. H. Rademacher and O. Toeplitz. *The Enjoyment of Mathematics*. Princeton, N.J.: Princeton University Press, 1951.
66. G. Rieger. *Die Zahlentheorie bei C. F. Gauss*. From *Gauss Gedenkband*. Berlin: Haude and Sperner, 1960.
67. P. Samuel. Unique factorization. *Am. Math. Monthly*, **75** (1968), 945–952.
68. P. Samuel. *Théorie Algébrique des Nombres*. Paris: Hermann & Cie, 1967.
69. J. P. Serre. *Compléments d'Arithmétiques, Rédigés par J. P. Ramis et G. Ruget*. Paris: Ecoles Normales Supérieures, 1964. English version, Springer-Verlag, 1973.
70. D. Shanks. *Solved and Unsolved Problems in Number Theory*. New York: Spartan Books, 1962.
71. W. Sierpinski. *A Selection of Problems in the Theory of Numbers*. Oxford: Pergamon Press, 1964.
72. H. J. S. Smith. *Report on the Theory of Numbers*, 1894. Reprinted by Chelsea Publishing Company, Inc., New York, 1965.
73. H. Stark, *An Introduction to Number Theory*. Cambridge, Mass.: M.I.T. Press, 1979.
74. T. Storer. *Cyclotomy and Difference Sets*. Chicago: Markham, 1967.
75. R. Swan. Factorization of polynomials over finite fields. *Pacific J. Math.*, **12** (1962), 1099–1106.
76. E. Vegh. Primitive roots modulo a prime as consecutive terms of an arithmetic progression. *J. Reine und Angew. Math.*, **235** (1969), 185–188.
77. I. M. Vinogradov. *Elements of Number Theory*. Transl. by S. Kravetz. New York: Dover, 1954.
78. E. Warning. Bemerkung zur vorstehenden Arbeit von Herrn Chevalley. *Agh. Math. Sem. Hamburg*, **11** (1936), 76–83.

79. W. Waterhouse. The sign of the Gauss sum. *J. Number Theory*, **2**, no. 3 (1970), 363.
80. A. Weil. Number of solutions of equations in a finite field. *Bull. Am. Math. Soc.*, **55** (1949), 497–508.
81. A. Weil. Jacobi sums as "Grössencharaktere." *Trans. Am. Math. Soc.*, **73** (1952), 487–495.
82. K. Yamamoto. On a conjecture of Hasse concerning multiplicative relations of Gauss sums. *J. Combin. Theory*, **1** (1966), 476–489.
83. A. Yokoyama. On the Gaussian sum and the jacobi sum with its applications. *Tohoku Maths. J.* (2), **16** (1964), 142–153.

Second Bibliography

84. W. W. Adams and L. J. Goldstein. *Introduction to Number Theory*. Englewood Cliffs, N.J.: Prentice-Hall, 1976.
85. L. Ahlfors. *Complex Analysis*. 2nd ed. New York: McGraw-Hill, 1966.
86. N. C. Ankeny, E. Artin, and S. Chowla. The class numbers of real quadratic fields. *Ann. Math.* (2), **56** (1952), 479–493.
87. N. Arthaud, On Birch and Swinnerton-Dyer's conjecture for elliptic curves with complex multiplication. *Comp. Math.*, **37**, Fasc. 2 (1978), 209–232.
88. R. Ayoub. Euler and the zeta function. *Am. Math. Monthly*, **81** (1974), 1067–1086.
89. A. Baker. *Transcendental Number Theory*. Cambridge: Cambridge University Press, 1975.
90. A. Baker. On the class number of imaginary quadratic fields. *Bull. Amer. Math. Soc.*, **77** (1971), 678–684.
91. G. Bergmann. Über Eulers Beweis des grossen Fermatschen Satzes für den Exponenten 3. *Math. Ann.*, **164** (1966), 159–175.
92. B. C. Berndt. Sums of Gauss, Jacobi and Jacobsthal. *J. Number Theory*, **11** (1979), 349–398.
93. B. C. Berndt and R. J. Evans. The determination of Gauss Sums. *Bull. Am. Math. Soc.*, **5** (2) (1981), 107–129.
94. B. C. Berndt. Classical theorems on quadratic residues. *L'Enseignement Math.* **22**, fasc. 3–4 (1976).
95. B. C. Berndt and R. C. Evans. Sums of Gauss, Eisenstein, Jacobi, Jacobsthal, and Brewer. *Ill. Math.*, **23**, no. 3 (1979), 374–437.
96. B. J. Birch and H. P. F. Swinnerton-Dyer. Notes on elliptic curves, I. *J. Reine und Angew. Math.*, **212** (1963), 7–25; II, **218** (1965), 79–108.
97. B. J. Birch. Conjectures on elliptic curves. In: *Theory of Numbers*, Amer. Math. Soc., Proc. of Symposia in Pure Math., Vol. 8. Pasadena, 1963.
98. E. Bombieri. Counting points on Curves over Finite Fields (d'après S. A. Stepanov) Sem. Bourbaki, Vol. 1972–73, Exposé 430. *Lecture Notes in Mathematics*, Vol. 383, pp. 234–241. New York: Springer-Verlag, 1974.
99. E. Brown. The first proof of the quadratic reciprocity law, revisited. *Am. Math. Monthly*, **88** (1981), 257–264.
100. A. Brumer and K. Kramer. The rank of elliptic curves. *Duke Math. J.*, **44** (1977), 715–742.
101. W. K. Bühler. *Gauss*. New York: Springer-Verlag, 1981.
102. K. Burde. Ein rationales biquadratisches Reciprozitätsgesetz. *J. Reine und Angew. Math.*, **235** (1969), 175–184.
103. H. S. Butts and L. Wade. Two criteria for Dedekind domains. *Am. Math. Monthly*, **73** (1966), 14–21.

104. L. Carlitz. Arithmetic properties of generalized Bernoulli numbers. *J. Reine und Angew. Math.*, **201–202** (1959), 173–182.
105. L. Carlitz. A note on irregular primes. *Proc. Am. Math. Soc.*, **5** (1954), 329–331.
106. L. Carlitz. A characterization of algebraic number fields with class number two. *Proc. Am. Math. Soc.*, **11** (1960), 391–392.
107. J. W. S. Cassels. Arithmetic on an elliptic curve. Proceedings of the International Congress of Mathematics. Stockholm, 1962. pp. 234–246.
108. J. W. S. Cassels. On Kummer sums. *Proc. London Math. Soc.* (3), **21** (1970), 19–27.
109. J. W. S. Cassels. Diophantine equations with special reference to elliptic curves. *J. London Math. Soc.*, **41** (1966), 193–291.
110. J. W. S. Cassels and A. Frölich. Algebraic number theory. Proceedings of an International Congress by the London Mathematical Society, 1967. Washington, D.C.: Thompson.
111. F. Châtelet. Les corps quadratiques. *Monographies de l'Enseignement Mathématique*, Vol. 9. Genève: 1962.
112. K. Chandrasekharan. *Introduction to Analytic Number Theory*. New York: Springer-Verlag, 1968.
113. S. Chowla. On Gaussian sums. *Proc. Nat. Acad. Sci. U.S.A.*, **48** (1962), 1127–1128.
114. J. Coates and A. Wiles. On the conjecture of Birch and Swinnerton-Dyer. *Invent. Math.*, **39** (1977), 223–251.
115. H. Cohn. *A Second Course in Number Theory*. New York: Wiley, 1962.
116. M. J. Collison. The origins of the cubic and biquadratic reciprocity laws. *Arch. Hist. Exact Sci.*, **17**, no. 1 (1977), 63–69.
117. A. Czogla. Arithmetic characterization of algebraic number fields with small class numbers. *Math. Z.* (1981), 247–253.
118. H. Davenport. The work of K. E. Roth. Proc. Int. Cong. Math., 1958, LVII–LX. Cambridge: Cambridge University Press, 1960.
119. H. Davenport. *Multiplicative Number Theory*. New York: Springer-Verlag, 1980.
120. D. Davis and O. Shisha. Simple proofs of the fundamental theorem of arithmetic. *Math. Mag.*, **54**, no. 1 (1981), 18.
121. R. Dedekind. *Mathematische Werke*, Vols. I and II. New York: Chelsea, 1969.
122. P. G. L. Dirichlet. Sur l'equation $t^2 + u^2 + v^2 + w^2 = 4m$. In: *Dirichlet's Werke*, Vol. 2, pp. 201–208. New York: Chelsea, 1969.
123. P. G. L. Dirichlet. Beweis des Satzes, dass jede unbegrenzte arithmetische Progression . . . . In: *Mathematische Werke*, pp. 313–342. New York: Chelsea, 1969.
124. P. G. L. Dirichlet. Recherches sur diverses applications de l'analyse infinitésimale a la théorie des nombres. In: *Dirichlet's Werke*, pp. 401–496. New York: Chelsea, 1969.
125. P. G. L. Dirichlet. *Werke*. 2 vols. in one. New York: Chelsea, 1969.
126. P. G. L. Dirichlet. Sur la manière de résoudre l'équation $t^2 - pu^2 = 1$ au moyen des fonctions circulaires. In: *Dirichlet's Werke*, pp. 345–350. New York: Chelsea, 1969.
127. P. G. L. Dirichlet. Dedekind, *Vorlesungen über Zahlentheorie*. New York: Chelsea, 1968.
128. H. M. Edwards. *Fermat's Last Theorem, A Genetic Introduction to Algebraic Number Theory*. New York: Springer-Verlag, 1977.
129. H. M. Edwards. The background of Kummers proof of Fermat's Last Theorem for regular exponent. *Arch. Hist. Exact Sci.*, **14** (1974), 219–326. See also postscript to the above **17** (1977), 381–394.
130. G. Eisenstein. Einfacher Beweis und Verallgemeinerung des Fundamentaltheorems für die biquadratischen Reste. In: *Mathematische Werke*, Band I, pp. 223–245. New York: Chelsea, 1975.
131. G. Eisenstein. Lois de réciprocité. In: *Mathematische Werke*, Band I, pp. 53–67. New York: Chelsea, 1975.

132. G. Eisenstein. Beweis des allgemeinsten Reciprocitätsgesetze zwischen reelen und complexen Zahlen. In: *Mathematische Werke*, Band II, pp. 189–198. New York: Chelsea, 1975.

133. P. Erdös. On a new method in elementary number theory which leads to an elementary proof of the prime number theorem. *Proc. Nat. Acad. Sci. U.S.A.*, **35** (1949), 374–384.

134. H. Flanders. Generalization of a theorem of Ankeny and Rogers. *Ann. Math.*, **57** (1953), 392–400.

135. W. Fulton. *Algebraic Curves*. New York: W. A. Benjamin, 1969.

136. C. F. Gauss. *Disquisitiones Arithmeticae*. Transl. by A. A. Clarke. New Haven, Conn.: Yale University Press, 1966.

137. C. F. Gauss. *Mathematisches Tagebuch, 1796–1814*. Edited by K.-R. Biermann. Ostwalds Klassiker 256.

138. M. Gerstenhaber. The 152nd proof of the law of quadratic reciprocity. *Am. Math. Monthly*, **70** (1963), 397–398.

139. L. J. Goldstein. A history of the prime number theorem. *Am. Math. Monthly*, **80** (1973), 599–615.

140. L. J. Goldstein. *Analytic Number Theory*. Princeton, N. J.: Prentice-Hall, 1971.

141. D. Goss. A simple approach to the analytic continuation and values at negative integers for Riemann's zeta function. *Proc. Am. Math. Soc.*, **81**, no. 4 (1981), 513–517.

142. B. H. Gross and D. E. Rohrlich. Some results on the Mordell–Weil group of the Jacobian of the Fermat curve. *Invent. Math.*, **44** (1978), 201–224.

143. T. Hall. *Carl Friedrich Gauss, a Biography*. Transl. by A. Froderberg. Cambridge, Mass.: M.I.T. Press, 1970.

144. R. Hartshorne. *Algebraic Geometry*. New York: Springer-Verlag, 1977.

145. P. G. Hartung. On the Pellian equation. *J. Number Theory*, **12** (1980), 110–112.

146. T. L. Heath. *Diophantus of Alexandria: A Study in the History of Greek Algebra*, New York: Dover, 1964.

147. D. R. Heath Brown and S. J. Patterson. The distribution of Kummer sums at prime arguments. *J. Reine und Angew. Math.*, **310** (1979), 111–136.

148. L. Heffter. Ludwig Stickelberger. *Deutsche Math. Jahr.*, **47** (1937), 79–86.

149. J. Herbrand. Sur les classes des corps circulaires. *J. Math. Pures et Appl.*, **ii** (1932), 417–441.

150. I. Herstein. *Topics in Algebra*. Lexington, Mass.: Xerox College, 1975.

151. D. Hilbert. Die Theorie der algebraischen Zahlkörper. In: *Gesammelte Abhandlungen*, Vol. 1, pp. 63–363. New York: Chelsea, 1965.

152. J. E. Hofmann. Über Zahlentheoretische Methoden Fermats und Eulers, ihre Zusammenhänge und ihre Bedeutung. *Arch. Hist. Exact Sci.* (1960–62), 122–159.

153. A. Hurwitz. Einige Eigenschaften der Dirichlet'schen Funktion $F(s) = \sum (D/n)1/n^s$, etc. . . . . In: A Hurwitz: *Mathematische Werke*, Band I, pp. 72–88, Basel and Stuttgart: Birkhäuser-Verlag, 1963.

154. A. Hurwitz. *Mathematische Werke*, Band II. Basel und Stuttgart: Birkhäuser-Verlag, 1963.

155. K. Iwasawa. Lectures on p-adic L-functions. *Ann. Math. Studies*. Princeton Press, 1974.

156. K. Iwasawa. A note on Jacobi sums. *Symp. Math.*, **15** (1975), 447–459.

157. K. Iwasawa. A note on cyclotomic fields. *Invent. Math.* **36** (1976), 115–123.

158. S. Iyanaga (Ed.). *Theory of Numbers*. Amsterdam: North-Holland, 1975.

159. W. Johnson. Irregular primes and cyclotomic invariants. *Math. Comp.*, **29** (1975), 113–120.

160. J. R. Joly. Equations et variétés algébriques sur un corps fini. *L'Enseignement Math.*, **19** (1973), 1–117.

161. N. Katz. An Overview of Deligne's proof of the Riemann hypothesis for varieties over finite fields. Proc. of Symposia in Pure Math., Vol. 28, pp. 275–305. Providence, R.I.: Am. Math. Society, 1976.

162. N. Koblitz. *P-adic Numbers, p-adic Analysis, and Zeta-Functions*, New York: Springer-Verlag, 1977.

163. E. E. Kummer. De residuis cubicis disquisitiones nonnullae analyticae. *J. Reine und Angew Math.*, **32** (1846), 341–359 (*Collected Papers*, Vol. 1, pp. 145–163. New York: Springer-Verlag, 1975.

164. E. E. Kummer. *Collected Papers*, Vol. 1. New York: Springer-Verlag, 1975.

165. E. Landau. *Einführung in die elementare und analytische Theorie der algebraischen Zahlen und der Ideale....* New York: Chelsea, 1949.

166. E. Landau, *Vorlesungen über Zahlentheorie*, Vols. 1–3. Leipzig, 1927.

167. S. Lang. *Cyclotomic Fields*. New York: Springer-Verlag, 1978.

168. S. Lang. *Algebraic Number Theory*. Reading, Mass: Addison-Wesley, 1970.

169. S. Lang. *Elliptic Functions*. Reading, Mass.: Addison-Wesley, 1973.

170. S. Lang. *Diophantine Geometry*. New York: Wiley-Interscience, 1962.

171. S. Lang. *Cyclotomic Fields*, II. New York: Springer-Verlag, 1980.

172. S. Lang. Review of L. J. Mordell's diophantine equations. *Bull. Am. Math. Soc.*, **76** (1970), 1230–1234.

173. S. Lang. Higher dimensional diophantine problems. Bull. Am. Math. Soc., **80**, no. 5 (1974), 779–787.

174. E. Lehmer. On Euler's criterion. *J. Aust. Math. Soc.* (1959/61), Part 1, 67–70.

175. E. Lehmer. Rational reciprocity laws. *Am. Math. Monthly*, **85** (1978), 467–472.

176. E. Lehmer. On the location of Gauss sums. *Math. Comp.*, **10** (1956), 194–202.

177. P. A. Leonard and S. Williams. Jacobi sums and a theorem of Brewer. *Rocky Mountain. J. Math.*, **5**, no. 2 (Spring, 1975).

178. A. W. Leopoldt. Eine Verallgemeinerung der Bernoullischen Zahlen. *Abhand. Math. Sem. Hamburg*, **22** (1958), 131–140.

179. W. J. LeVeque. *Fundamentals of Number Theory*. Reading, Mass.: Addison-Wesley, 1977.

180. W. J. LeVeque. A brief survey of diophantine equations. *M.A.A. Studies in Mathematics*, **6** (1969), 4–24.

181. H. von Lienen, Reele kubische und biquadratische Legendre Symbole. *Reine und Angew. Math.*, **305** (1979), 140–154.

182. J. H. Loxton. Some conjectures concerning Gauss sums. *J. Reine und Angew. Math.*, **297** (1978), 153–158.

183. D. A. Marcus. *Number Fields*. New York: Springer-Verlag, 1977.

184. J. M. Masley, Where are number fields with small class number?, *Lecture Notes in Mathematics*, Vol. 751, pp. 221–242. New York: Springer-Verlag, 1979.

185. J. Masley. Class groups of abelian number fields. Proc. Queen's Number Theory Conference, 1979. Edited by P. Ribenboim. Kingston, Ontario: Queen's University.

186. C. R. Matthews. Gauss sums and elliptic functions. I: The Kummer sum. *Invent. Math.*, **52** (1979), 163–185; II: The Quartic Case, **54** (1979), 23–52.

187. B. Mazur. Rational points on modular curves. In: Modular Functions of One Variable, V. *Lecture Notes in Mathematics*, Vol. 601. New York: Springer-Verlag, 1976.

188. T. Metsänkylä. Distribution of irregular prime members. *J. Reine und Angew. Math.*, **282** (1976), 126–130.

189. L. J. Mordell. *Diophantine Equations*. New York: Academic Press, 1969.

190. L. J. Mordell. Review of S. Lang's diophantine geometry. *Bull. Am. Math. Soc.*, **70** (1964), 491–498.

191. L. J. Mordell. The infinity of rational solutions of $y^2 = x^3 + k$. *J. London Math. Soc.*, **41** (1966), 523–525.

192. B. Morlaye. Demonstration élémentaire d'un théoreme de Davenport et Hasse. *L'enseignement Math.*, **18** (1973), 269–276.
193. Leo Moser. A thorem on quadratic residues. *Proc. Am. Math. Soc.*, **2** (1951), 503–504.
194. J. B. Muskat. Reciprocity and Jacobi sums. *Pacific J. Math.*, **20** (1967), 275–280.
195. T. Nagell. Sur les restes et nonrestes cubiques. *Arkiv Math.*, **1** (1952), 579–586.
196. W. Narkiewicz. *Elementary and Analytic Theory of Algebraic Numbers.* Warsaw: Polish Scientific Publications, 1974.
197. J. von Neumann and H. H. Goldstine. A numerical study of a conjecture of Kummer. *MTAC*, **7** (1953), 133–134.
198. D. J. Newman. Simple analytic proof of the prime number theorem. *Am. Math. Monthly*, **87** (1980), 693–696.
199. N. Nielson. *Traité Élémentaire des Nombres Bernoulli.* Paris: 1923.
200. L. D. Olson. The trace of Frobenius for elliptic curves with complex multiplication. *Lecture Notes in Mathematics*, Vol. 732, pp. 454–476. New York: Springer-Verlag, 1979.
201. L. D. Olson. Points of finite order on elliptic curves with complex multiplication. *Manuscripta Math.*, **14** (1974), 195–205.
202. L. D. Olson. Hasse invariants and anomolous primes for elliptic curves with complex multiplication. *J. Number Theory*, **8** (1976), 397–414.
203. H. Poincaré. Sur les propriétés des courbes algebriques planes. *J. Liouville* (v), **7** (1901), 161–233.
204. H. Rademacher. *Topics in Analytic Number Theory.* Die Grundlehren der Mathematischen, Wissenschaften. New York: Springer-Verlag, 1964.
205. H. Reichardt. Einige im Kleinen überall lösbare, im Grossen unlösbare diophantische Gleichungen. *J. Reine Angew. und Math.*, **184** (1942), 12–18.
206. P. Ribenboim. 13 Lectures on Fermat's Last Theorem. New York: Springer-Verlag, 1979.
207. P. Ribenboim. *Algebraic Numbers.* New York: Wiley, 1972.
208. K. Ribet. A modular construction of unramified p-extensions of $Q(\mu_p)$. *Invent. Math.*, **34** (1976), 151–162.
209. G. J. Rieger. Die Zahlentheorie bei C. F. Gauss. In: C. F. Gauss, *Leben und Werk.* pp. 38–77. Berlin: Haude & Spenersche Verlagsbuchhandlung, 1960.
210. A. Robert. Elliptic curves. *Lecture Notes in Mathematics*, Vol. 326. New York: Springer-Verlag, 1973.
211. M. I. Rosen and J. Kraft. Eisenstein reciprocity and nth power residues. *Am. Math. Monthly*, **88** (1981), 269–270.
212. M. I. Rosen. Abel's theorem on the lemniscate. *Am. Math. Monthly*, **88** (1981), 387–395.
213. P. Samuel. *Théorie Algébrique des Nombres.* Hermann: Paris, 1967. Transl. by A. Silberger. Boston: Houghton Mifflin, 1970.
214. P. Samuel and O. Zariski. *Commutative Algebra*, Vol. 1, New York: Springer-Verlag, 1975–1976.
215. A. Selberg. An elementary proof of the prime number theorem. *Ann. Math.*, **50** (1949), 305–319.
216. E. S. Selmer. The diophantine equation $ax^3 + by^3 + cz^3 = 0$. *Acta Math.*, **85** (1951), 203–362.
217. W. M. Schmidt. Diophantine approximation. *Lecture Notes in Mathematics*, Vol. 785. New York: Springer-Verlag, 1980.
218. W. M. Schmidt. Equations over finite fields: an elementary approach. *Lecture Notes in Mathematics*, Vol. 536. New York: Springer-Verlag, 1976.
219. I. R. Shafarevich. *Basic Algebraic Geometry.* Grundlehren der Mathematischen Wissenschaften 213. Transl. by K. A. Hirsch. New York: Springer-Verlag, 1977.
220. D. E. Smith. *Source Book in Mathematics*, Vols. 1 and 2. New York: Dover, 1959.

221. H. M. Stark. On the Riemann hypothesis in hyperelliptic function fields. *A.M.S. Proc. Symp. Pure Math.*, **24** (1973), 285–302.

222. S. A. Stepanov. Rational points on algebraic curves over finite fields (in Russian). Report of a 1972 Conference on Analytic Number Theory in Minsk, U.S.S.R., pp. 223–243.

223. N. M. Stephens. The diophantine equation $x^3 + y^3 = dz^3$ and the conjectures of Birch and Swinnerton-Dyer. *J. Reine und Angew. Math.*, 231.

224. L. Stickelberger. Über eine Verallgemeinerung der Kreistheilung. *Math. Ann.*, **37** (1890), 321–367.

225. K. B. Stolarsky. *Algebraic Numbers and Diophantine Approximation.* New York: Dekker, 1974.

226. H. P. F. Swinnerton-Dyer. The conjectures of Birch and Swinnerton-Dyer and of Tate. In: *Proceedings of a Conference on Local Fields.* Berlin–Heidelberg–New York: Springer-Verlag, 1967.

227. J. Tate. The arithmetic of elliptic curves. *Invent. Math.*, **23** (1974), 179–206.

228. A. D. Thomas. *Zeta Functions: An Introduction to Algebraic Geometry.* London, San Francisco: Pitman, 1977.

229. E. Trost. *Primzahlen.* Basel and Stuttgart: Birhäuser-Verlag, 1953.

230. J. V. Uspensky and M. A. Heaslet, New York: McGraw-Hill, 1939.

231. H. S. Vandiver. Fermat's last theorem. *Am. Math. Monthly*, **53** (1946), 555–578.

232. H. S. Vandiver. On developments in an arithmetic theory of the Bernoulli and allied numbers. *Scripta Math.*, **25** (1961), 273–303.

233. A. van der Poorten. A proof that Euler missed ... Apery's proof of the irrationality of $\zeta(3)$: An informal report. *The Mathematical Intelligencer*, **1**, no. 1 (1978) 195–203.

234. S. Wagstaff. The irregular primes to 125,000. *Math. Comp.*, **32**, no. 142 (1978), 583–591.

235. A. Weil. Two lectures on number theory: Past and present. *L'Enseignement Math.*, **XX** (1973), 81–110. Also in: A. Weil, *Oeuvres Scientifiques*, Vol. III, pp. 279–302. New York: Springer-Verlag, 1979.

236. A. Weil. Sommes de Jacobi et caractères de Hecke. *Gött. Nach.* (1974), 1–14. Also in: A. Weil, *Oeuvres Scientifiques*, Vol. III, pp. 329–342. New York: Springer-Verlag, 1979.

237. A. Weil. Sur les sommes de trois et quatre carrés. *L'Enseignement Math.*, **20** (1974), 303–310. Also in: A. Weil, *Oeuvres Scientifiques*, Vol. III, New York: Springer-Verlag, 1979.

238. A. Weil. La cyclotomie jadis et naguère. *L'Enseignement Math.*, **20** (1974), 247–263. Also in: A. Weil, *Oeuvres Scientifiques*, Vol. III, pp. 311–327. New York: Springer-Verlag, 1979.

239. A. Weil. Review of "Mathematische Werke, by Gotthold Eisenstein". In: A. Weil, *Oeuvres Scientifiques*, Vol. III, pp. 398–403. New York: Springer-Verlag, 1979.

240. A. Weil. Fermat et l'équation de Pell. In: *Oeuvres Scientifiques*, Vol. III, pp. 413–419. New York: Springer-Verlag, 1979.

241. A. Weil. *Oeuvres Scientifiques, Collected Papers,* 3 vols. Corrected second printing. New York: Springer-Verlag, 1980.

242. A. Wiles. Modular curves and the class group of $\mathbb{Q}(\zeta_p)$. *Invent. Math.*, **58** (1980), 1–35.

243. K. S. Williams. On Euler's criterion for cubic nonresidues. *Proc. Am. Math. Soc.*, **49** (1975), 277–283.

244. K. S. Williams. Note on Burde's rational biquadratic reciprocity law. *Can. Math. Bull.*, (1) **20** (1977), 145–146.

245. K. S. Williams. On Eisenstein's supplement to the law of cubic reciprocity. *Bull. Cal. Math. Soc.*, **69** (1977), 311–314.

246. B. F. Wyman. What is a reciprocity law?. *Am. Math. Monthly*, **79** (1972), 571–586.
247. H. Yokoi. On the distribution of irregular primes. *J. Number Theory*, 7 (1975), 71–76.
248. Zeta Functions. In: *Encyclopedic Dictionary of Mathematics*. Edited by S. Iyanaga and Y. Kawada. Cambridge, Mass.: M.I.T. Press, 1977. pp. i372–i393.

Index

Graduate Texts in Mathematics

(continued from page ii)

Printed in the United States
By Bookmasters